A HISTORY OF
MATHEMATICAL NOTATIONS

Volume I
NOTATIONS IN ELEMENTARY
MATHEMATICS

A HISTORY OF
MATHEMATICAL
NOTATIONS

By

FLORIAN CAJORI, Ph.D.

Professor of the History of Mathematics
University of California

Volume I

NOTATIONS IN ELEMENTARY
MATHEMATICS

The Open Court Publishing Company
LA SALLE · ILLINOIS

QA
21
.C135
1928

V. 1

31784

PREFACE

The study of the history of mathematical notations was suggested to me by Professor E. H. Moore, of the University of Chicago. To him and to Professor M.W. Haskell, of the University of California, I am indebted for encouragement in the pursuit of this research. As completed in August, 1925, the present history was intended to be brought out in one volume. To Professor H. E. Slaught, of the University of Chicago, I owe the suggestion that the work be divided into two volumes, of which the first should limit itself to the history of symbols in elementary mathematics, since such a volume would appeal to a wider constituency of readers than would be the case with the part on symbols in higher mathematics. To Professor Slaught I also owe generous and vital assistance in many other ways. He examined the entire manuscript of this work in detail, and brought it to the sympathetic attention of the Open Court Publishing Company. I desire to record my gratitude to Mrs. Mary Hegeler Carus, president of the Open Court Publishing Company, for undertaking this expensive publication from which no financial profits can be expected to accrue.

I gratefully acknowledge the assistance in the reading of the proofs of part of this history rendered by Professor Haskell, of the University of California; Professor R. C. Archibald, of Brown University; and Professor L. C. Karpinski, of the University of Michigan.

FLORIAN CAJORI

UNIVERSITY OF CALIFORNIA

v

TABLE OF CONTENTS

ILLUSTRATIONS

I

INTRODUCTION

In this history it has been an aim to give not only the first appearance of a symbol and its origin (whenever possible), but also to indicate the competition encountered and the spread of the symbol among writers in different countries. It is the latter part of our program which has given bulk to this history.

The rise of certain symbols, their day of popularity, and their eventual decline constitute in many cases an interesting story. Our endeavor has been to do justice to obsolete and obsolescent notations, as well as to those which have survived and enjoy the favor of mathematicians of the present moment.

If the object of this history of notations were simply to present an array of facts, more or less interesting to some students of mathematics—if, in other words, this undertaking had no ulterior motive—then indeed the wisdom of preparing and publishing so large a book might be questioned. But the author believes that this history constitutes a mirror of past and present conditions in mathematics which can be made to bear on the notational problems now confronting mathematics. The successes and failures of the past will contribute to a more speedy solution of the notational problems of the present time.

NUMERAL SYMBOLS AND COMBINATIONS OF SYMBOLS

BABYLONIANS

1. In the Babylonian notation of numbers a vertical wedge Υ stood for 1, while the characters $\langle$ and $\Upsilon\!\!\!\succ$ signified 10 and 100, respectively. Grotefend[1] believes the character for 10 originally to have been the picture of two hands, as held in prayer, the palms being pressed together, the fingers close to each other, but the thumbs thrust out. Ordinarily, two principles were employed in the Babylonial notation—the additive and multiplicative. We shall see that limited use was made of a third principle, that of subtraction.

2. Numbers below 200 were expressed ordinarily by symbols whose respective values were to be *added*. Thus, $\Upsilon\!\!\!\succ\langle\langle\Upsilon\Upsilon\Upsilon$ stands for 123. The principle of multiplication reveals itself in $\langle\Upsilon\!\!\!\succ$ where the smaller symbol 10, placed before the 100, is to be multiplied by 100, so that this symbolism designates 1,000.

3. These cuneiform symbols were probably invented by the early Sumerians. Their inscriptions disclose the use of a decimal scale of numbers and also of a *sexagesimal* scale.[2]

Early Sumerian clay tablets contain also numerals expressed by circles and curved signs, made with the blunt circular end of a stylus, the ordinary wedge-shaped characters being made with the pointed end. A circle ● stood for 10, a semicircular or lunar sign stood for 1. Thus, a "round-up" of cattle shows ●●DDD / DDD, or 36, cows.[3]

4. The sexagesimal scale was first discovered on a tablet by E. Hincks[4] in 1854. It records the magnitude of the illuminated portion

---

[1] His first papers appeared in *Göttingische Gelehrte Anzeigen* (1802), Stück 149 und 178; *ibid.* (1803), Stück 60 und 117.

[2] In the division of the year and of the day, the Babylonians used also the duodecimal plan.

[3] G. A. Barton, *Haverford Library Collection of Tablets*, Part I (Philadelphia, 1905), Plate 3, HCL 17, obverse; see also Plates 20, 26, 34, 35. Allotte de la Fuye, "En-e-tar-zi patési de Lagaš," *H. V. Hilprecht Anniversary Volume* (Chicago, 1909), p. 128, 133.

[4] "On the Assyrian Mythology," *Transactions of the Royal Irish Academy.* "Polite Literature," Vol. XXII, Part 6 (Dublin, 1855), p. 406, 407.

of the moon's disk for every day from new to full moon, the whole disk being assumed to consist of 240 parts. The illuminated parts during the first five days are the series 5, 10, 20, 40, 1.20, which is a geometrical progression, on the assumption that the last number is 80. From here on the series becomes arithmetical, 1.20, 1.36, 1.52, 2.8, 2.24, 2.40, 2.56, 3.12, 3.28, 3.44, 4, the common difference being 16. The last number is written in the tablet ⋎⋎⋎, and, according to Hincks's interpretation, stood for $4 \times 60 = 240$.

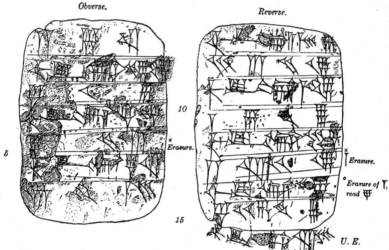

Obverse.   *Reverse.*

FIG. 1.—Babylonian tablets from Nippur, about 2400 B.C.

5. Hincks's explanation was confirmed by the decipherment of tablets found at Senkereh, near Babylon, in 1854, and called the *Tablets of Senkereh*. One tablet was found to contain a table of square numbers, from 1^2 to 60^2, a second one a table of cube numbers from 1^3 to 32^3. The tablets were probably written between 2300 and 1600 B.C. Various scholars contributed toward their interpretation. Among them were George Smith (1872), J. Oppert, Sir H. Rawlinson, Fr. Lenormant, and finally R. Lepsius.[1] The numbers 1, 4, 9, 16, 25, 36,

[1] George Smith, *North British Review* (July, 1870), p. 332 n.; J. Oppert, *Journal asiatique* (August–September, 1872; October–November, 1874); J. Oppert, *Étalon des mesures assyr. fixé par les textes cunéiformes* (Paris, 1874); Sir H. Rawlinson and G. Smith, "The Cuneiform Inscriptions of Western Asia," Vol. IV: *A Selection from the Miscellaneous Inscriptions of Assyria* (London, 1875), Plate 40; R. Lepsius, "Die Babylonisch-Assyrischen Längenmaasse nach der Tafel von Senkereh," *Abhandlungen der Königlichen Akademie der Wissenschaften zu Berlin* (aus dem Jahre 1877 [Berlin, 1878], Philosophisch-historische Klasse), p. 105–44.

and 49 are given as the squares of the first seven integers, respectively. We have next $1.4 = 8^2$, $1.21 = 9^2$, $1.40 = 10^2$, etc. This clearly indicates the use of the sexagesimal scale which makes $1.4 = 60+4$, $1.21 = 60+21$, $1.40 = 60+40$, etc. This sexagesimal system marks the earliest appearance of the all-important "principle of position" in writing numbers. In its general and systematic application, this principle requires a symbol for zero. But no such symbol has been found on early Babylonian tablets; records of about 200 B.C. give a symbol for zero, as we shall see later, but it was not used in calculation. The earliest thorough and systematic application of a symbol for zero and the principle of position was made by the Maya of Central America, about the beginning of the Christian Era.

6. An extension of our knowledge of Babylonian mathematics was made by H. V. Hilprecht who made excavations at Nuffar (the ancient Nippur). We reproduce one of his tablets[1] in Figure 1.

Hilprecht's transliteration, as given on page 28 of his text is as follows:

Line 1.	125	720		Line 9.	2,000	18
Line 2. IGI-GAL-BI		103,680		Line 10. IGI-GAL-BI		6,480
Line 3.	250	360		Line 11.	4,000	9
Line 4. IGI-GAL-BI		51,840		Line 12. IGI-GAL-BI		3,240
Line 5.	500	180		Line 13.	8,000	18
Line 6. IGI-GAL-BI		25,920		Line 14. IGI-GAL-BI		1,620
Line 7.	1,000	90		Line 15.	16,000	9
Line 8. IGI-GAL-BI		12,960		Line 16. IGI-GAL-BI		810

7. In further explanation, observe that in

Line	1. $125 = 2 \times 60 + 5$,	$720 = 12 \times 60 + 0$
Line	2. Its denominator,	$103,680 = [28 \times 60 + 48(?)] \times 60 + 0$
Line	3. $250 = 4 \times 60 + 10$,	$360 = 6 \times 60 + 0$
Line	4. Its denominator,	$51,840 = [14 \times 60 + 24] \times 60 + 0$
Line	5. $500 = 8 \times 60 + 20$,	$180 = 3 \times 60 + 0$
Line	6. Its denominator,	$25,920 = [7 \times 60 + 12] \times 60 + 0$
Line	7. $1,000 = 16 \times 60 + 40$,	$90 = 1 \times 60 + 30$
Line	8. Its denominator,	$12,960 = [3 \times 60 + 36] \times 60 + 0$

[1] *The Babylonian Expedition of the University of Pennsylvania*. Series A: "Cuneiform Texts," Vol. XX, Part 1. *Mathematical, Metrological and Chronological Tablets from the Temple Library of Nippur* (Philadelphia, 1906), Plate 15, No. 25.

Line 9. $2,000 = 33 \times 60 + 20,$ $18 = 10 + 8$
Line 10. Its denominator, $6,480 = [1 \times 60 + 48] \times 60 + 0$
Line 11. $4,000 = [1 \times 60 + 6] \times 60 + 40,$ 9
Line 12. Its denominator, $3,240 = 54 \times 60 + 0$
Line 13. $8,000 = [2 \times 60 + 13] \times 60 + 20,$ 18
Line 14. Its denominator, $1,620 = 27 \times 60 + 0$
Line 15. $16,000 = [4 \times 60 + 26] \times 60 + 40,$ 9
Line 16. Its denominator, $810 = 13 \times 60 + 30$
$IGI\text{-}GAL$ = Denominator, BI = Its, i.e., the number 12,960,000 or 60^4.

We quote from Hilprecht (*op. cit.*, pp. 28–30):

"We observe (*a*) that the first numbers of all the odd lines (1, 3, 5, 7, 9, 11, 13, 15) form an increasing, and all the numbers of the even lines (preceded by $IGI\text{-}GAL\text{-}BI$ = 'its denominator') a descending geometrical progression; (*b*) that the first number of every odd line can be expressed by a fraction which has 12,960,000 as its numerator and the closing number of the corresponding even line as its denominator, in other words,

$$125 = \frac{12,960,000}{103,680} \; ; \qquad 250 = \frac{12,960,000}{51,840} \; ; \qquad 500 = \frac{12,960,000}{25,920} \; ;$$

$$1,000 = \frac{12,960,000}{12,960} \; ; \qquad 2,000 = \frac{12,960,000}{6,480} \; ; \qquad 4,000 = \frac{12,960,000}{3,240} \; ;$$

$$8,000 = \frac{12,960,000}{1,620} \; ; \qquad 16,000 = \frac{12,960,000}{810} \; .$$

But the closing numbers of all the odd lines (720, 360, 180, 90, 18, 9, 18, 9) are still obscure to me. . . . .

"The question arises, what is the meaning of all this? What in particular is the meaning of the number 12,960,000 ($= 60^4$ or $3,600^2$) which underlies all the mathematical texts here treated . . . . ? . . . . This 'geometrical number' (12,960,000), which he [Plato in his *Republic* viii. 546*B*–*D*] calls 'the lord of better and worse births,' is the arithmetical expression of a great law controlling the Universe. According to Adam this law is 'the Law of Change, that law of inevitable degeneration to which the Universe and all its parts are subject'—an interpretation from which I am obliged to differ. On the contrary, it is the Law of Uniformity or Harmony, i.e. that fundamental law which governs the Universe and all its parts, and which cannot be ignored and violated without causing an anomaly, i.e. without resulting in a degeneration of the race." The nature of the "Platonic number" is still a debated question.

8. In the reading of numbers expressed in the Babylonian sexagesimal system, uncertainty arises from the fact that the early Babylonians had no symbol for zero. In the foregoing tablets, how do we know, for example, that the last number in the first line is 720 and not 12? Nothing in the symbolism indicates that the 12 is in the place where the local value is "sixties" and not "units." Only from the study of the entire tablet has it been inferred that the number intended is 12×60 rather than 12 itself. Sometimes a horizontal line was drawn following a number, apparently to indicate the absence of units of lower denomination. But this procedure was not regular, nor carried on in a manner that indicates the number of vacant places.

9. To avoid confusion some Babylonian documents even in early times contained symbols for 1, 60, 3,600, 216,000, also for 10, 600, 36,000.[1] Thus · was 10, ● was 3,600, ⊙ was 36,000.

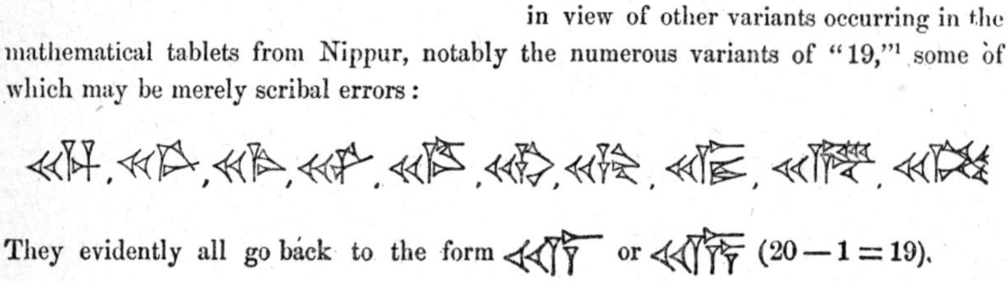

in view of other variants occurring in the mathematical tablets from Nippur, notably the numerous variants of "19,"[1] some of which may be merely scribal errors:

They evidently all go back to the form ⧼⧽ or ⧼⧽ (20 − 1 = 19).

Fig. 2.—Showing application of the principle of subtraction

10. Besides the principles of addition and multiplication, Babylonian tablets reveal also the use of the principle of subtraction, which is familiar to us in the Roman notation XIX (20−1) for the number 19. Hilprecht has collected ideograms from the Babylonian tablets which he has studied, which represent the number 19. We reproduce his symbols in Figure 2. In each of these twelve ideograms (Fig. 2), the two symbols to the left signify together 20. Of the symbols immediately to the right of the 20, one vertical wedge stands for "one" and the remaining symbols, for instance ⟩⟩, for LAL or "minus"; the entire ideogram represents in each of the twelve cases the number 20−1 or 19.

One finds the principle of subtraction used also with curved signs;[2] D ● ● ⟩⟩D meant 60+20−1, or 79.

---

[1] See François Thureau-Dangin, Recherches sur l'origine de l'écriture cunéiforme (Paris, 1898), Nos. 485–91, 509–13. See also G. A. Barton, Haverford College Library Collection of Cuneiform Tablets, Part I (Philadelphia, 1905), where the forms are somewhat different; also the Hilprecht Anniversary Volume (Chicago, 1909), p. 128 ff.

[2] G. A. Barton, op. cit., Plate 3, obverse.

11. The symbol used about the second century B.C. to designate the *absence* of a number, or a blank space, is shown in Figure 3, containing numerical data relating to the moon.[1] As previously stated, this symbol, ⪅, was not used in computation and therefore performed

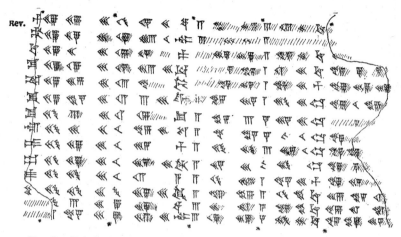

Fig. 3.—Babylonian lunar tables, reverse; full moon for one year, about the end of the second century B.C.

only a small part of the functions of our modern zero. The symbol is seen in the tablet in row 10, column 12; also in row 8, column 13. Kugler's translation of the tablet, given in his book, page 42, is shown below. Of the last column only an indistinct fragment is preserved; the rest is broken off.

REVERSE

1...	Nisannu	28°56′30″	19°16′	″ Librae	3ᶻ 6°45′	4ᴵ74ᴵᴵ10ᴵᴵᴵ	sik
2...	Airu	28 38 30	17 54 30	Scorpii	3 21 28	6 20 30	sik
3...	Simannu	28 20 30	16 15	Arcitenentis	3 31 39	3 45 30	sik
4...	Dûzu	28 18 30	14 33 30	Capri	3 34 41	1 10 30	sik
5...	Âbu	28 36 30	13 9	Aquarii	3 27 56	1 24 30	bar
6...	Ulûlu	29 54 30	13 3 30	Piscium	3 15 34	1 59 30	num
7...	Tišrîtu	29 12 30	11 16	Arietis	2 58 3	4 34 30	num
8...	Araḥ-s.	29 30 30	10 46 30	Tauri	2 40 54	6 0 10	num
9...	Kislimu	29 48 30	10 35	Geminorum	2 29 29	3 25 10	num
10...	Tebitu	29 57 30	10 32 30	Cancri	2 24 30	0 57 10	num
11...	Šabâtu	29 39 30	10 12	Leonis	2 30 53	1 44 50	bar
12...	Adâru I	29 21 30	9 33 30	Virginis	2 42 56	2 19 50	sik
13...	Adâru II	29 3 30	8 36	Librae	3 0 21	4 54 50	sik
14...	Nisannu	28 45 30	7 21 30	Scorpii	3 17 36	5 39 50	sik

[1] Franz Xaver Kugler, S. J., *Die babylonische Mondrechnung* (Freiburg im Breisgau, 1900), Plate IV, No. 99 (81–7–6), lower part.

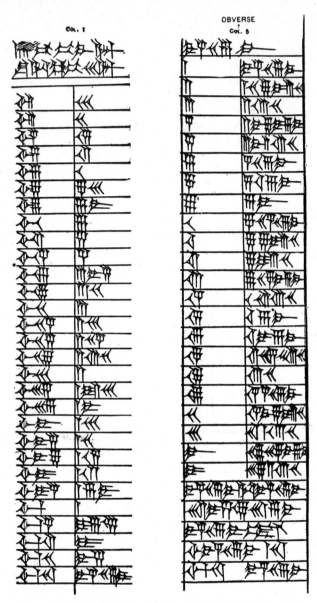

FIG. 4.—Mathematical cuneiform tablet, CBS 8536, in the Museum of the University of Pennsylvania.

12. J. Oppert pointed out the Babylonian use of a designation for the sixths, viz., $\frac{1}{6}$, $\frac{1}{3}$, $\frac{1}{2}$, $\frac{2}{3}$, $\frac{5}{6}$. These are unit fractions or fractions whose numerators are one less than the denominators.[1] He also advanced evidence pointing to the Babylonian use of sexagesimal *fractions* and the use of the sexagesimal system in weights and measures. The occurrence of sexagesimal fractions is shown in tablets recently examined. We reproduce in Figure 4 two out of twelve columns found on a tablet described by H. F. Lutz.[2] According to Lutz, the tablet "cannot be placed later than the Cassite period, but it seems more probable that it goes back even to the First Dynasty period, *ca.* 2000 B.C."

13. To mathematicians the tablet is of interest because it reveals operations with sexagesimal fractions resembling modern operations with decimal fractions. For example, 60 is divided by 81 and the quotient expressed sexagesimally. Again, a sexagesimal number with two fractional places, 44(26)(40), is multiplied by itself, yielding a product in four fractional places, namely, [32]55(18)(31)(6)(40). In this notation the [32] stands for 32×60 units, and to the (18), (31), (6), (40) must be assigned, respectively, the denominators 60, 60^2, 60^3, 60^4.

The tablet contains twelve columns of figures. The first column (Fig. 4) gives the results of dividing 60 in succession by twenty-nine different divisors from 2 to 81. The eleven other columns contain tables of multiplication; each of the numbers 50, 48, 45, 44(26)(40), 40, 36, 30, 25, 24, 22(30), 20 is multiplied by integers up to 20, then by the numbers 30, 40, 50, and finally by itself. Using our modern numerals, we interpret on page 10 the first and the fifth columns. They exhibit a larger number of fractions than do the other columns. The Babylonians had no mark separating the fractional from the integral parts of a number. Hence a number like 44(26)(40) might be interpreted in different ways; among the possible meanings are $44 \times 60^2 + 26 \times 60 + 40$, $44 \times 60 + 26 + 40 \times 60^{-1}$, and $44 + 26 \times 60^{-1} + 40 \times 60^{-2}$. Which interpretation is the correct one can be judged only by the context, if at all.

The exact meaning of the first two lines in the first column is uncertain. In this column 60 is divided by each of the integers written on the left. The respective quotients are placed on the right.

---

[1] Symbols for such fractions are reproduced also by Thureau-Dangin, *op. cit.*, Nos. 481–84, 492–508, and by G. A. Barton, *Haverford College Library Collection of Cuneiform Tablets*, Part I (Philadelphia, 1905).

[2] "A Mathematical Cuneiform Tablet," *American Journal of Semitic Languages and Literatures*, Vol. XXXVI (1920), p. 249–57.

In the fifth column the multiplicand is 44(26)(40) or 44⅓. The last two lines seem to mean "$60^2 \div 44(26)(40) = 81$, $60^2 \div 81 = 44(26)(40)$."

First Column . . . . gal (?) -bi 40 -âm šu a- na gal-bi 30 -âm			Fifth Column
			44(26)(40)
igi 2	30	1	44(26)(40)
igi 3	20	2	[1]28(53)(20)
igi 4	15	3	[2]13(20)
igi 5	12	4	[2]48(56)(40)*
igi 6	10	5	[3]42(13)(20)
igi 8	7(30)	6	[4]26(40)
igi 9	6(40)	7	[5]11(6)(40)
igi 10	6	9	[6]40
igi 12	5	10	[7]24(26)(40)
igi 15	4	11	[8]8(53)(20)
igi 16	3(45)	12	[8]53(20)
igi 18	3(20)	13	[9]27(46)(40)*
igi 20	3	14	[10]22(13)(20)
igi 24	2(30)	15	[11]6(40)
igi 25	2(24)	16	[11]51(6)(40)
igi 28*	2(13)(20)	17	[12]35(33)(20)
igi 30	2	18	[13]20
igi 35*	1(52)(30)	19	[14]4(26)(40)
igi 36	1(40)	20	[14]48(53)(20)
igi 40	1(30)	30	[22]13(20)
igi 45	1(20)	40	[29]37(46)(40)
igi 48	1(15)	50	[38]2(13)(20)*
igi 50	1(12)	44(26)(40)a-na 44(26)(40)	
igi 54	1(6)(40)	[32]55(18)(31)(6)(40)	
igi 60	1	44(26)(40) square	
igi 64	(56)(15)	igi 44(26)(40) 81	
igi 72	(50)	igi 81 44(26)(40)	
igi 80	(45)		
igi 81	(44)(26)(40)		

Numbers that are incorrect are marked by an asterisk (*).

14. The Babylonian use of sexagesimal fractions is shown also in a clay tablet described by A. Ungnad.[1] In it the diagonal of a rectangle whose sides are 40 and 10 is computed by the approximation

[1] *Orientalische Literaturzeitung* (ed. Peise, 1916), Vol. XIX, p. 363–68. See also Bruno Meissner, *Babylonien und Assyrien* (Heidelberg, 1925), Vol. II, p. 393.

$40+2\times40\times10^2\div60^2$, yielding 42(13)(20), and also by the approximation $40+10^2\div\{2\times40\}$, yielding 41(15). Translated into the decimal scale, the first answer is 42.22+, the second is 41.25, the true value being 41.23+. These computations are difficult to explain, except on the assumption that they involve sexagesimal *fractions*.

15. From what has been said it appears that the Babylonians had ideograms which, transliterated, are *Igi-Gal* for "denominator" or "division," and *Lal* for "minus." They had also ideograms which, transliterated, are *Igi-Dua* for "division," and *A-Du* and *Ara* for "times," as in $Ara-1$ 18, for "$1\times18=18$," $Ara-2$ 36 for "$2\times18=36$"; the *Ara* was used also in "squaring," as in 3 *Ara* 3 9 for "$3\times3=9$." They had the ideogram $Ba-Di-E$ for "cubing," as in 27-*E* 3 $Ba-Di-E$ for "$3^3=27$"; also $Ib-Di$ for "square," as in 9-*E* 3 *Ib-Di* for "$3^2=9$." The sign $A-An$ rendered numbers "distributive."[1]

EGYPTIANS

16. The Egyptian number system is based on the scale of 10, although traces of other systems, based on the scales of 5, 12, 20, and 60, are believed to have been discovered.[2] There are three forms of Egyptian numerals: the hieroglyphic, hieratic, and demotic. Of these the hieroglyphic has been traced back to about 3300 B.C.;[3] it is found mainly on monuments of stone, wood, or metal. Out of the hieroglyphic sprang a more cursive writing known to us as hieratic. In the beginning the hieratic was simply the hieroglyphic in the rounded forms resulting from the rapid manipulation of a reed-pen as contrasted with the angular and precise shapes arising from the use of the chisel. About the eighth century B.C. the demotic evolved as a more abbreviated form of cursive writing. It was used since that time down to the beginning of the Christian Era. The important mathematical documents of ancient Egypt were written on papyrus and made use of the hieratic numerals.[4]

[1] Hilprecht, *op. cit.*, p. 23; Arno Poebel, *Grundzüge der sumerischen Grammatik* (Rostock, 1923), p. 115; B. Meissner, *op. cit.*, p. 387–89.

[2] Kurt Sethe, *Von Zahlen und Zahlworten bei den alten Ägyptern* (Strassburg, 1916), p. 24–29.

[3] J. E. Quibell and F. W. Green, *Hierakonopolis* (London, 1900–1902), Part I, Plate 26*B*, who describe the victory monument of King *Ncr-mr;* the number of prisoners taken is given as 120,000, while 400,000 head of cattle and 1,422,000 goats were captured.

[4] The evolution of the hieratic writing from the hieroglyphic is explained in G. Möller, *Hieratische Paläographie*, Vol. I, Nos. 614 ff. The demotic writing

17. The hieroglyphic symbols were ‖ for 1, ∩ for 10, Ϲ for 100, ⁑ for 1,000, ‖ for 10,000, ☞ for 100,000, ⚐ for 1,000,000, ◯ for 10,000,000. The symbol for 1 represents a vertical staff; that for 1,000 a lotus plant; that for 10,000 a pointing finger; that for 100,000 a burbot; that for 1,000,000 a man in astonishment, or, as more recent

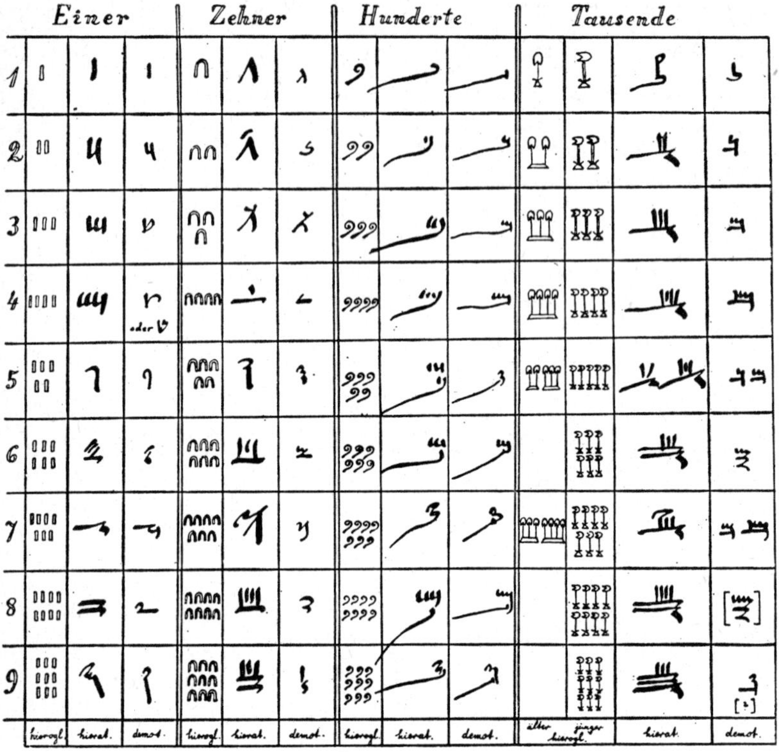

FIG. 5.—Egyptian numerals. Hieroglyphic, hieratic, and demotic numeral symbols. (This table was compiled by Kurt Sethe.)

Egyptologists claim, the picture of the cosmic deity Hh.[1] The symbols for 1 and 10 are sometimes found in a horizontal position.

18. We reproduce in Figures 5 and 6 two tables prepared by Kurt

---

is explained by F. L. Griffith, *Catalogue of the Demotic Papyri in the John Rylands Library* (Manchester, 1909), Vol. III, p. 415 ff., and by H. Brugsch, *Grammaire démotique*, §§ 131 ff.

[1] Sethe, *op. cit.*, p. 11, 12.

Sethe. They show the most common of the great variety of forms which are found in the expositions given by Möller, Griffith, and Brugsch.

Observe that the old hieratic symbol for ¼ was the cross ✕, signifying perhaps a part obtainable from two sections of a body through the center.

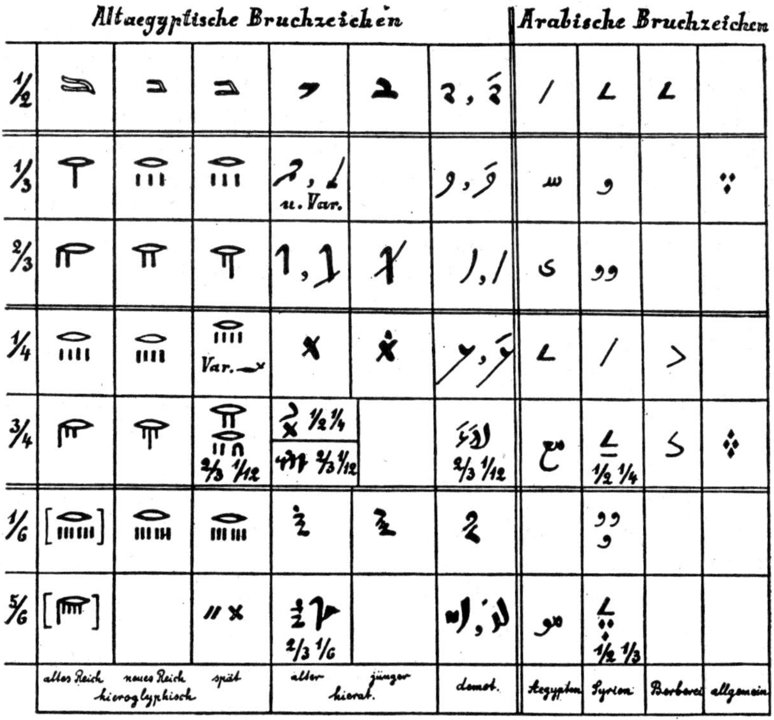

Altaegyptische Bruchzeichen						*Arabische Bruchzeichen*		

FIG. 6.—Egyptian symbolism for simple fractions. (Compiled by Kurt Sethe)

19. In writing numbers, the Egyptians used the principles of addition and multiplication. In applying the additive principle, not more than four symbols of the same kind were placed in any one group. Thus, 4 was written in hieroglyphs | | | |; 5 was not written | | | | |, but either | | | | | or $\begin{smallmatrix} | | | \\ | | \end{smallmatrix}$. There is here recognized the same need which caused the Romans to write V after IIII, L = 50 after XXXX = 40, D = 500 after CCCC = 400. In case of two unequal groups, the Egyptians always wrote the larger group before, or above the smaller group; thus, seven was written $\begin{smallmatrix} | | | | \\ | | | \end{smallmatrix}$.

20. In the *older* hieroglyphs 2,000 or 3,000 was represented by two or three lotus plants grown *in one bush*. For example, 2,000 was ℣; correspondingly, 7,000 was designated by ℣℣. The later hieroglyphs simply place two lotus plants together, to represent 2,000, without the appearance of springing from one and the same bush.

21. The multiplicative principle is not so old as the additive; it came into use about 1600–2000 B.C. In the oldest example hitherto known,[1] the symbols for 120, placed before a lotus plant, signify 120,000. A smaller number written before or below or above a symbol representing a larger unit designated multiplication of the larger by the smaller. Möller cites a case where 2,800,000 is represented by one burbot, with characters placed beneath it which stand for 28.

22. In hieroglyphic writing, unit fractions were indicated by placing the symbol ⌒ over the number representing the denominator. Exceptions to this are the modes of writing the fractions $\frac{1}{2}$ and $\frac{2}{3}$; the old hieroglyph for $\frac{1}{4}$ was ⧸, the later was ⟋; of the slightly varying hieroglyphic forms for $\frac{2}{3}$, ⋔ was quite common.[2]

23. We reproduce an algebraic example in hieratic symbols, as it occurs in the most important mathematical document of antiquity known at the present time—the *Rhind papyrus*. The scribe, Ahmes, who copied this papyrus from an older document, used black and red ink, the red in the titles of the individual problems and in writing auxiliary numbers appearing in the computations. The example which, in the Eisenlohr edition of this papyrus, is numbered 34, is hereby shown.[3] Hieratic writing was from right to left. To facilitate the study of the problem, we write our translation from right to left and in the same relative positions of its parts as in the papyrus, except that numbers are written in the order familiar to us; i.e., 37 is written in our translation 37, and not 73 as in the papyrus. Ahmes writes unit fractions by placing a dot over the denominator, except in case of

---

[1] *Ibid.*, p. 8.

[2] *Ibid.*, p. 92–97, gives detailed information on the forms representing $\frac{2}{3}$. The Egyptian procedure for decomposing a quotient into unit fractions is explained by V. V. Bobynin in *Abh. Gesch. Math.*, Vol. IX (1899), p. 3.

[3] *Ein mathematisches Handbuch der alten Ägypter* (*Papyrus Rhind des British Museum*) *übersetzt und erklärt* (Leipzig, 1877; 2d ed., 1891). The explanation of Problem 34 is given on p. 55, the translation on p. 213, the facsimile reproduction on Plate XIII of the first edition. The second edition was brought out without the plates. A more recent edition of the Ahmes papyrus is due to T. Eric Peet and appears under the title *The Rhind Mathematical Papyrus*, British Museum, Nos. 10057 and 10058, Introduction, Transcription, and Commentary (London, 1923).

$\frac{1}{2}$, $\frac{1}{3}$, $\frac{2}{3}$, $\frac{1}{4}$, each of which had its own symbol. Some of the numeral symbols in Ahmes deviate somewhat from the forms given in the two preceding tables; other symbols are not given in those tables. For the reading of the example in question we give here the following symbols:

Four	—	One-fourth	×	
Five	"l	Heap	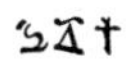 ꙑꙇ†	See Fig. 7
Seven	⟨	The whole	l3	See Fig. 7
One-half	⟍	It gives	儿	See Fig. 7

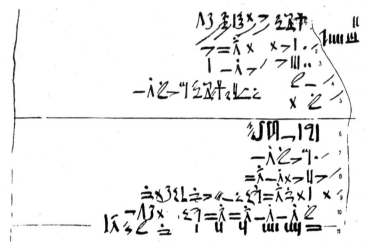

Fig. 7.—An algebraic equation and its solution in the Ahmes papyrus, 1700 B.C., or, according to recent authorities, 1550 B.C. (Problem 34, Plate XIII in Eisenlohr; p. 70 in Peet; in chancellor Chace's forthcoming edition, p. 76, as R. C. Archibald informs the writer.)

Translation (reading from right to left):
   "10 gives it, whole its, $\frac{1}{4}$ its, $\frac{1}{2}$ its, Heap   No. 34

$\frac{1}{2}$ $\frac{1}{28}\frac{1}{4}$ $\frac{1}{4}\frac{1}{2}1$

1      $\frac{1}{14}\frac{1}{2}$ $\frac{1}{2}3..$

$\frac{1}{14}\frac{1}{7}\frac{1}{2}5$ is heap the together   7   4

$\frac{1}{4}$ $\frac{1}{7}$

Proof the of Beginning

$\frac{1}{14}\frac{1}{7}\frac{1}{2}5$

$\frac{1}{28}\frac{1}{14}\frac{1}{4}\frac{1}{2}2$ $\frac{1}{2}$

$\frac{1}{8}\frac{1}{4}$ Remainder $\frac{1}{8}\frac{1}{2}9$ together $\frac{1}{56}\frac{1}{28}\frac{1}{8}\frac{1}{4}1$ $\frac{1}{4}$

14 gives $\frac{1}{4}$ $\frac{1}{56}\frac{1}{28}\frac{1}{28}\frac{1}{14}\frac{1}{14}\frac{1}{7}$

21 Together .7 gives $\frac{1}{8}$ 1 2 2 4 4 8"

24. Explanation:

The algebraic equation is $\dfrac{x}{2}+\dfrac{x}{4}+x=10$

i.e., $(1+\tfrac{1}{2}+\tfrac{1}{4})x=10$

The solution answers the question, By what must $(1\ \tfrac{1}{2}\ \tfrac{1}{4})$ be multiplied to yield the product 10? The four lines 2–5 contain on the right the following computation:

Twice $(1\ \tfrac{1}{2}\ \tfrac{1}{4})$ yields $3\tfrac{1}{2}$.

Four times $(1\ \tfrac{1}{2}\ \tfrac{1}{4})$ yields 7.

One-seventh of $(1\ \tfrac{1}{2}\ \tfrac{1}{4})$ is $\tfrac{1}{4}$.

1° UNITES.

- SIGNES		LETTRES NUMÉRALES coptes.	VALEUR des SIGNES.	NOMS DE NOMBRE en dialecte thébain.
HIEROGLYPHIQUES , creux et pleins.	HIÉRATIQUES , avec variantes.			
𝟙	ı	) ⟩ ? ?	Ⲁ̄	1	oua.
𝟙𝟙	ıı	ч ч	Ⲃ̄	2	snau.
𝟙𝟙𝟙	ııı	ꙟ ꙟ	Ⲅ̄	3	choment.
𝟙𝟙 𝟙𝟙	ıı ıı	ꙟ ꙟ 𝟒	Ⲇ̄	4	ftoou.
𝟙𝟙𝟙 𝟙𝟙	ııı ıı	ꙶ ꙷ ꙶ	Ⲉ̄	5	tiou.
𝟙𝟙𝟙 𝟙𝟙𝟙	ııı ııı	⫽ ⟝	Ⲋ̄	6	soou.
𝟙𝟙𝟙𝟙 𝟙𝟙𝟙	ıııı ııı	⌐ ꙿ ꙿ	Ⲍ̄	7	sachf.
𝟙𝟙𝟙𝟙 𝟙𝟙𝟙𝟙	ıııı ıııı	⊐ ⊐	Ⲏ̄	8	chmoun.
𝟙𝟙𝟙 𝟙𝟙𝟙 𝟙𝟙𝟙	ııı ııı ııı	⟨ ⟨	Ⲑ̄	9	psis.

[Continued on facing page]

[i.e., taking $(1\ \tfrac{1}{2}\ \tfrac{1}{4})$ once, then four times, together with $\tfrac{1}{7}$ of it, yields only 9; there is lacking 1. The remaining computation is on the four lines 2–5, on the left. Since $\tfrac{1}{7}$ of $(1\ \tfrac{1}{2}\ \tfrac{1}{4})$ yields $(\tfrac{1}{4}\ \tfrac{1}{14}\ \tfrac{1}{28})$ or $\tfrac{1}{4}$, $\tfrac{2}{7}$ or]

$(\tfrac{1}{4}\ \tfrac{1}{28})$ of $(1\ \tfrac{1}{2}\ \tfrac{1}{4})$, yields $\tfrac{1}{2}$.

And the double of this, namely, $(\tfrac{1}{2}\ \tfrac{1}{14})$ of $(1\ \tfrac{1}{2}\ \tfrac{1}{4})$ yields 1.

Adding together 1, 4, $\tfrac{1}{7}$ and $(\tfrac{1}{2}\ \tfrac{1}{14})$, we obtain Heap $=5\tfrac{1}{2}$

$\tfrac{1}{7}\ \tfrac{1}{14}$ or $5\tfrac{5}{7}$, the answer.

Proof.—5 $\frac{1}{2}$ $\frac{1}{7}$ $\frac{1}{14}$ is multiplied by (1 $\frac{1}{2}$ $\frac{1}{4}$) and the partial products are added. In the first line of the proof we have 5 $\frac{1}{2}$ $\frac{1}{7}$ $\frac{1}{14}$, in the second line half of it, in the third line one-fourth of it. Adding at first only the integers of the three partial products and the simpler fractions $\frac{1}{2}$, $\frac{1}{2}$, $\frac{1}{4}$, $\frac{1}{4}$, $\frac{1}{8}$, the partial sum is 9 $\frac{1}{2}$ $\frac{1}{8}$. This is $\frac{1}{4}$ $\frac{1}{8}$ short of 10. In the fourth line of the proof (l. 9) the scribe writes the remaining fractions and, reducing them to the common denominator 56, he writes (in

2° DIZAINES.

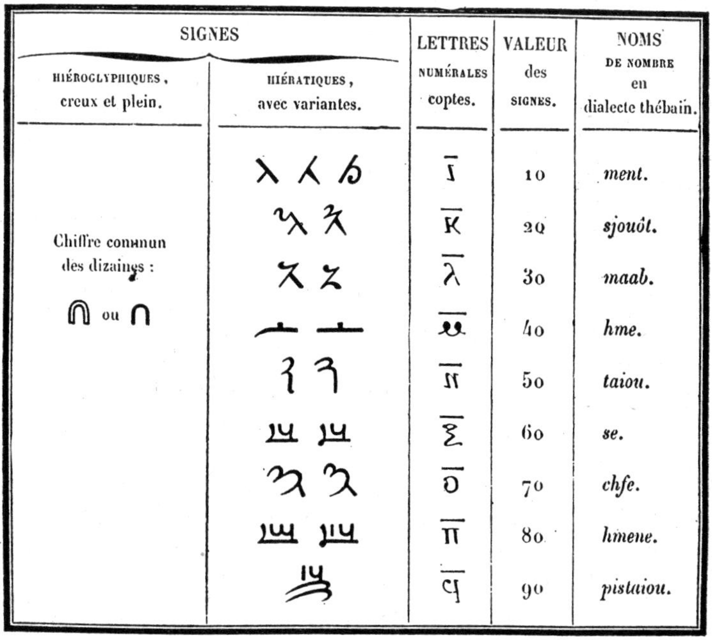

SIGNES		LETTRES NUMÉRALES coptes.	VALEUR des SIGNES.	NOMS DE NOMBRE en dialecte thébain.
HIÉROGLYPHIQUES, creux et plein.	HIÉRATIQUES, avec variantes.			
	⅄ ⋏ ℔	Ī	10	*ment.*
Chiffre commun des dizaines :	⋏ ⋏	K̄	2Q	*sjouót.*
∩ ou ∩	⋏ ⋏	λ̄	30	*maab.*
	⏕ ⏕	⁙	40	*hme.*
	⟩ ⟩	⨅	50	*taiou.*
	⎧ ⎧	ⵣ̄	60	*se.*
	⋏ ⋏	ō	70	*chfe.*
	⎧ ⎧	π̄	80	*hmene.*
	⎧	ϥ̄	90	*pistaiou.*

Fig. 8.—Hieroglyphic, hieratic, and Coptic numerals. (Taken from A. P. Pihan, *Exposé des signes de numération* [Paris, 1860], p. 26, 27.)

red color) in the last line the numerators 8, 4, 4, 2, 2, 1 of the reduced fractions. Their sum is 21. But $\frac{21}{56} = \frac{14+7}{56} = \frac{1}{4}\frac{1}{8}$, which is the exact amount needed to make the total product 10.

A pair of legs symbolizing addition and subtraction, as found in impaired form in the Ahmes papyrus, are explained in § 200.

25. The Egyptian Coptic numerals are shown in Figure 8. They are of comparatively recent date. The hieroglyphic and hieratic are

the oldest Egyptian writing; the demotic appeared later. The Coptic writing is derived from the Greek and demotic writing, and was used by Christians in Egypt after the third century. The Coptic numeral symbols were adopted by the Mohammedans in Egypt after their conquest of that country.

26. At the present time two examples of the old Egyptian solution of problems involving what we now term "quadratic equations"[1] are known. For square root the symbol ⌐ has been used in the modern hieroglyphic transcription, as the interpretation of writing in the two papyri; for quotient was used the symbol ⅜ .

PHOENICIANS AND SYRIANS

27. The Phoenicians[2] represented the numbers 1–9 by the respective number of vertical strokes. Ten was usually designated by a horizontal bar. The numbers 11–19 were expressed by the juxtaposition of a horizontal stroke and the required number of vertical ones.

Palmyrenische Zahlzeichen	I	Yᵢ	כᵢ	3ᵢ	כ', כב ', כך' ;	"yכ3כ""כך"
Varianten bei Gruter	I	⅄ᵢ	⅂ᵢ	כᵢ	⅄'; ⅄⅄', ⅄⅄⅄	"⅄⅄⅄⅄"⅄⅄"
Bedeutung	1.	5.	10.	20	100, 110. 1000	2437.

Fɪɢ. 9.—Palmyra (Syria) numerals. (From M. Cantor, *Kulturleben, etc.*, Fig. 48)

As Phoenician writing proceeded from right to left, the horizontal stroke signifying 10 was placed farthest to the right. Twenty was represented by two parallel strokes, either horizontal or inclined and sometimes connected by a cross-line as in H, or sometimes by two strokes, thus Λ. One hundred was written thus |<| or thus |כ| . Phoenician inscriptions from which these symbols are taken reach back several centuries before Christ. Symbols found in Palmyra (modern Tadmor in Syria) in the first 250 years of our era resemble somewhat the numerals below 100 just described. New in the Palmyra numer-

---

[1] See H. Schack-Schackenburg, "Der Berliner Papyrus 6619," *Zeitschrift für ägyptische Sprache und Altertumskunde*, Vol. XXXVIII (1900), p. 136, 138, and Vol. XL (1902), p. 65–66.

[2] Our account is taken from Moritz Cantor, *Vorlesungen über Geschichte der Mathematik*, Vol. I (3d ed.; Leipzig, 1907), p. 123, 124; *Mathematische Beiträge zum Kulturleben der Völker* (Halle, 1863), p. 255, 256, and Figs. 48 and 49.

als is γ for 5. Beginning with 100 the Palmyra numerals contain new forms. Placing a | to the right of the sign for 10 (see Fig. 9) signifies multiplication of 10 by 10, giving 100. Two vertical strokes || mean 10×20, or 200; three of them, 10×30, or 300.

28. Related to the Phoenician are numerals of Syria, found in manuscripts of the sixth and seventh centuries A.D. Their shapes and their mode of combination are shown in Figure 10. The Syrians employed also the twenty-two letters of their alphabet to represent the numbers 1–9, the tens 10–90, the hundreds 100–400. The following hundreds were indicated by juxtaposition: 500=400+100, 600= 400+200, ... , 900=400+400+100, or else by writing respectively 50–90 and placing a dot over the letter to express that its value is to be taken tenfold. Thousands were indicated by the letters for 1–9, with a stroke annexed as a subscript. Ten thousands were expressed

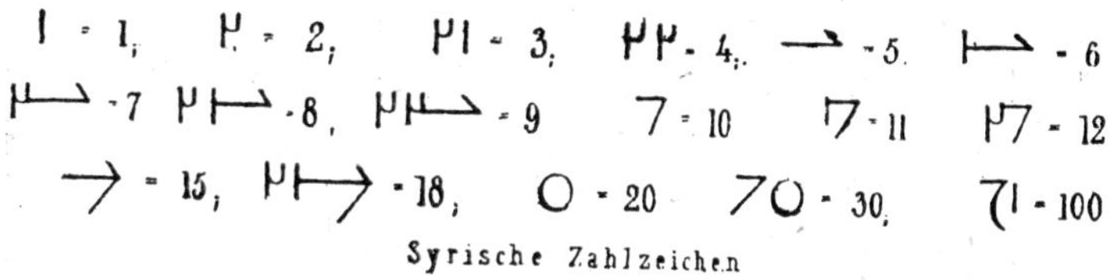

Syrische Zahlzeichen

FIG. 10.—Syrian numerals. (From M. Cantor, *Kulturleben, etc.,* Fig. 49)

by drawing a small dash below the letters for one's and ten's. Millions were marked by the letters 1–9 with two strokes annexed as subscripts (i.e., 1,000×1,000=1,000,000).

HEBREWS

29. The Hebrews used their alphabet of twenty-two letters for the designation of numbers, on the decimal plan, up to 400. Figure 11 shows three forms of characters: the Samaritan, Hebrew, and Rabbinic or cursive. The Rabbinic was used by commentators of the Sacred Writings. In the Hebrew forms, at first, the hundreds from 500 to 800 were represented by juxtaposition of the sign for 400 and a second number sign. Thus, קת stood for 500, רת for 600, שת for 700, תת for 800.

30. Later the end forms of five letters of the Hebrew alphabet came to be used to represent the hundreds 500–900. The five letters representing 20, 40, 50, 80, 90, respectively, had two forms; one of

LETTRES			NOMS ET TRANSCRIPTION DES LETTRES.		VALEURS.	NOMS DE NOMBRE.
SAMARITAINES.	HÉBRAÏQUES.	RABBINIQUES.				
Ѧ	א	ƒ	aleph,	a	1	ekhâd.
∃	ב	℥	bet,	b	2	chenaïn.
٦	ג	℥	ghimel,	gh	3	chelochâh.
ꟻ	ד ה	ד ל	dalet,	d	4	arbắâh.
∃	ה	℗	hé,	h	5	khamichâh.
Հ	ו	℩	waw,	w	6	chichâh.
ℛ	ז	℩	zaïn,	z	7	chib'âh.
Ⴓ	ח	℘	khet,	kh	8	chemonâh.
◹	ט	ꞷ	t'et',	t'	9	tich'âh.
⅏	י כ	כ	iod,	i	10	'asârâh.
ঞ	כ	כ	kaph,	k	20	'esrîm.
ⵊ	ל	ℒ	lamed,	l	3o	chelochîm.
℥	מ ס נ	℘ ℘	mem,	m	4o	arbắîm.
ℒ	נ	℘	noun,	n	5o	khamichîm.
℈	ס	℘	s'amek	s	6o	chichîm.
▽	ע	ℐ	'aïn,	'a	7o	chib'îm.
ℸ	פ	℘	phé,	ph	8o	chemonîm.
ℳ	צ	ℐ	tsadé,	ts	9o	tich'îm.
℣	ק	℘	qoph,	q	100	méâh.
℈	ר	ℛ	rech,	r	200	mâtaïm.
ⴤ	ש	℘	chin,	ch	3oo	châlôch méôt.
Ѧ	ת	℘	tau,	t	4oo	arba' méôt.

FIG. 11.—Hebrew numerals. (Taken from A. P. Pihan, *Exposé des signes de numération* [Paris, 1860], p. 172, 173.)

the forms occurred when the letter was a terminal letter of a word.
These end forms were used as follows:

ץ   ף   ז   ם   ך

900   800   700   600   500 .

To represent thousands the Hebrews went back to the beginning of
their alphabet and placed two dots over each letter. Thereby its
value was magnified a thousand fold. Accordingly, אַ represented
1,000. Thus any number less than a million could be represented by
their system.

31. As indicated above, the Hebrews wrote from right to left.
Hence, in writing numbers, the numeral of highest value appeared on
the right; אַה meant 5,001, הא meant 1,005. But 1,005 could be
written also הא, where the two dots were omitted, for when א meant
unity, it was always placed to the left of another numeral. Hence
when appearing on the right it was interpreted as meaning 1,000.
With a similar understanding for other signs, one observes here the
beginning of an imperfect application in Hebrew notation of the
principle of local value. By about the eighth century A.D., one finds
that the signs המשה signify 5,845, the number of verses in the laws
as given in the Masora. Here the sign on the extreme right means
5,000; the next to the left is an 8 and must stand for a value less than
5,000, yet greater than the third sign representing 40. Hence the
sign for 8 is taken here as 800.[1]

GREEKS

32. On the island of Crete, near Greece, there developed, under
Egyptian influence, a remarkable civilization. Hieroglyphic writing
on clay, of perhaps about 1500 B.C., discloses number symbols as
follows:) or | for 1,))))) or || ||| or $\begin{smallmatrix}|||\\||\end{smallmatrix}$ for 5, · for 10, \ or / for
100, ◇ for 1,000, V for ¼ (probably), \\\\::::))) for 483.[2] In this
combination of symbols only the additive principle is employed.
Somewhat later,[3] 10 is represented also by a horizontal dash; the

---

[1] G. H. F. Nesselmann, *Die Algebra der Griechen* (Berlin, 1842), p. 72. 494;
M. Cantor, *Vorlesungen über Geschichte der Mathematik*, Vol. I (3d ed.), p. 126. 127.

[2] Arthur J. Evans, *Scripta Minoa*, Vol. I (1909), p. 258, 256.

[3] Arthur J. Evans, *The Palace of Minos* (London, 1921). Vol. I. p 646; see
also p. 279.

sloping line indicative of 100 and the lozenge-shaped figure used for
1,000 were replaced by the forms O for 100, and ◇ for 1,000.

$$\diamondsuit\diamondsuit{}^{\circ}_{\circ}\,{}^{\circ}_{\circ} \equiv \equiv \equiv \,\Big|\Big|\Big| \quad \text{stood for 2,496 .}$$

33. The oldest strictly Greek numeral symbols were the so-called
Herodianic signs, named after Herodianus, a Byzantine grammarian
of about 200 A.D., who describes them. These signs occur frequently
in Athenian inscriptions and are, on that account, now generally
called Attic. They were the initial letters of numeral adjectives.[1]
They were used as early as the time of Solon, about 600 B.C., and continued in use for several centuries, traces of them being found as late
as the time of Cicero. From about 470 to 350 B.C. this system existed
in competition with a newer one to be described presently. The
Herodianic signs were

Ι Iota for 1   H Eta for 100
II or ᵀI or Iᵀ Pi for 5   X Chi for 1,000
Δ Delta for 10   M My for 10,000

34. Combinations of the symbols for 5 with the symbols for 10,100,
1,000 yielded symbols for 50, 500, 5,000. These signs appear on an
abacus found in 1847, represented upon a Greek marble monument on
the island of Salamis.[2] This computing table is represented in Figure 12.

The four right-hand signs I C T X, appearing on the horizontal
line below, stand for the fractions $\frac{1}{6}$, $\frac{1}{12}$, $\frac{1}{24}$, $\frac{1}{48}$, respectively. Proceeding next from right to left, we have the symbols for 1, 5, 10, 50, 100,
500, 1,000, 5,000, and finally the sign T for 6,000. The group of symbols drawn on the left margin, and that drawn above, do not contain
the two symbols for 5,000 and 6,000. The pebbles in the columns
represent the number 9,823. The four columns represented by the
five vertical lines on the right were used for the representation of the
fractional values $\frac{1}{6}$, $\frac{1}{12}$, $\frac{1}{24}$, $\frac{1}{48}$, respectively.

35. Figure 13 shows the old Herodianic numerals in an Athenian
state record of the fifth century B.C. The last two lines are: Κεφάλαιον

---

[1] See, for instance, G. Friedlein, *Die Zahlzeichen und das elementare Rechnen der
Griechen und Römer* (Erlangen, 1869), p. 8; M. Cantor, *Vorlesungen über Geschichte
der Mathematik*, Vol. I (3d ed.), p. 120; H. Hankel, *Zur Geschichte der Mathematik
im Alterthum und Mittelalter* (Leipzig, 1874), p. 37.

[2] Kubitschek, "Die Salaminische Rechentafel," *Numismatische Zeitschrift*
(Vienna, 1900), Vol. XXXI, p. 393–98; A. Nagl, *ibid.*, Vol. XXXV (1903), p. 131–
43; M. Cantor, *Kulturleben der Völker* (Halle, 1863), p. 132, 136; M. Cantor, *Vorlesungen über Geschichte der Mathematik*, Vol. I (3d ed.), p. 133.

ἀνα[λώατοσ τ] οὗ ἐπὶ τ[ης] ἀρχῆς ⊢⊢⊢⊢⊓ ⊤⊤⊤ . . . . ; i.e., "Total of expenditures during our office three hundred and fifty-three talents. . . . ."

36. The exact reason for the displacement of the Herodianic symbols by others is not known. It has been suggested that the commercial intercourse of Greeks with the Phoenicians, Syrians, and Hebrews brought about the change. The Phoenicians made one important contribution to civilization by their invention of the alphabet. The Babylonians and Egyptians had used their symbols to represent whole syllables or words. The Phoenicians borrowed hieratic

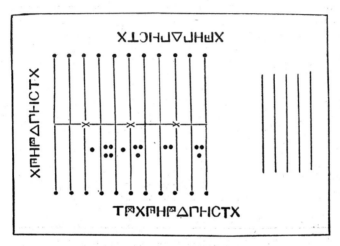

Fɪɢ. 12.—The computing table of Salamis

signs from Egypt and assigned them a more primitive function as letters. But the Phoenicians did not use their alphabet for numerical purposes. As previously seen, they represented numbers by vertical and horizontal bars. The earliest use of an entire alphabet for designating numbers has been attributed to the Hebrews. As previously noted, the Syrians had an alphabet representing numbers. The Greeks are supposed by some to have copied the idea from the Hebrews. But Moritz Cantor[1] argues that the Greek use is the older and that the invention of alphabetic numerals must be ascribed to the Greeks. They used the twenty-four letters of their alphabet, together with three strange and antique letters, ϛ (old *van*), ϙ (*koppa*), ϡ (*sampi*), and the symbol M. This change was decidedly for the worse, for the old Attic numerals were less burdensome on the memory inas-

---

[1] *Vorlesungen über Geschichte der Mathematik*, Vol. I (3d ed., 1907), p. 25.

Fig. 13.—Account of disbursements of the Athenian state, 418–415 B.C., British Museum, Greek Inscription No. 23. (Taken from R. Brown, *A History of Accounting and Accountants* [Edinburgh, 1905], p. 26.)

much as they contained fewer symbols. The following are the Greek alphabetic numerals and their respective values:

α	β	γ	δ	ϵ	ς	ζ	η	θ	ι	κ	λ	μ	ν	ξ	o	π	φ
1	2	3	4	5	6	7	8	9	10	20	30	40	50	60	70	80	90

ρ	σ	τ	υ	ϕ	χ	ψ	ω	$\mathfrak{I}$	$,a$	$,\beta$	$,\gamma,$
100	200	300	400	500	600	700	800	900	1,000	2,000	3,000

etc.

	$\overset{\beta}{\text{M}}$	$\overset{\gamma}{\text{M}},$	etc.
M	M	M,	
10,000	20,000	30,000	

37. A horizontal line drawn over a number served to distinguish it more readily from words. The coefficient for M was sometimes placed before or behind instead of over the M. Thus 43,678 was written $\overline{\delta\text{M}}{,}\gamma\chi o\eta$. The horizontal line over the Greek numerals can hardly be considered an essential part of the notation; it does not seem to have been used except in manuscripts of the Byzantine period.[1] For 10,000 or myriad one finds frequently the symbol M or Mυ, sometimes simply the dot · , as in $\beta{\cdot}o\delta$ for 20,074. Often[2] the coefficient of the myriad is found written above the symbol μ^{υ}.

38. The paradox recurs, Why did the Greeks change from the Herodianic to the alphabet number system? Such a change would not be made if the new did not seem to offer some advantages over the old. And, indeed, in the new system numbers could be written in a more compact form. The Herodianic representation of 1,739 was χ ⊓HHΔΔΔΠ||||; the alphabetic was $,a\psi\lambda\theta$. A scribe might consider the latter a great innovation. The computer derived little aid from either. Some advantage lay, however, on the side of the Herodianic, as Cantor pointed out. Consider HHHH+HH = ⊓ H, ΔΔΔΔ+ΔΔ = ⊐Δ; there is an analogy here in the addition of hundred's and of ten's. But no such analogy presents itself in the alphabetic numerals, where the corresponding steps are $\upsilon+\sigma=\chi$ and $\mu+\kappa=\xi$; adding the hundred's expressed in the newer notation affords no clew as to the sum of the corresponding ten's. But there was another still more important consideration which placed the Herodianic far above the alphabetical numerals. The former had only six symbols, yet they afforded an easy representation of numbers below 100,000; the latter demanded twenty-seven symbols for numbers below 1,000! The mental effort

---

[1] *Encyc. des scien. math.*, Tome I, Vol. I (1904), p. 12.   [2] *Ibid.*

of remembering such an array of signs was comparatively great. We are reminded of the centipede having so many legs that it could hardly advance.

39. We have here an instructive illustration of the fact that a mathematical topic may have an amount of symbolism that is a hindrance rather than a help, that becomes burdensome, that obstructs progress. We have here an early exhibition of the truth that the movements of science are not always in a forward direction. Had the Greeks not possessed an abacus and a finger symbolism, by the aid of which computations could be carried out independently of the numeral notation in vogue, their accomplishment in arithmetic and algebra might have been less than it actually was.

40. Notwithstanding the defects of the Greek system of numeral notation, its use is occasionally encountered long after far better systems were generally known. A Calabrian monk by the name of Barlaam,[1] of the early part of the fourteenth century, wrote several mathematical books in Greek, including arithmetical proofs of the second book of Euclid's *Elements*, and six books of *Logistic*, printed in 1564 at Strassburg and in several later editions. In the *Logistic* he develops the computation with integers, ordinary fractions, and sexagesimal fractions; numbers are expressed by Greek letters. The appearance of an arithmetical book using the Greek numerals at as late a period as the close of the sixteenth century in the cities of Strassburg and Paris is indeed surprising.

41. Greek writers often express fractional values in words. Thus Archimedes says that the length of a circle amounts to three diameters and a part of one, the size of which lies between one-seventh and ten-seventy-firsts.[2] Eratosthenes expresses $\frac{11}{83}$ of a unit arc of the earth's meridian by stating that the distance in question "amounts to eleven parts of which the meridian has eighty-three."[3] When expressed in symbols, fractions were often denoted by first writing the numerator marked with an accent, then the denominator marked with two accents and written twice. Thus,[4] $\iota\zeta'\ \kappa\alpha''\ \kappa\alpha''=\frac{17}{21}$. Archimedes, Eutocius, and Diophantus place the denominator in the position of the

---

[1] All our information on Barlaam is drawn from M. Cantor, *Vorlesungen über Geschichte der Mathematik*, Vol. I (3d ed.), p. 509, 510; A. G. Kästner, *Geschichte der Mathematik* (Göttingen, 1796), Vol. I, p. 45; J. C. Heilbronner, *Historia matheseos universae* (Lipsiae, 1742), p. 488, 489.

[2] *Archimedis opera omnia* (ed. Heiberg; Leipzig, 1880), Vol. I, p. 262.

[3] Ptolemäus, Μεγάλη σύνταξις (ed. Heiberg), Pars I, Lib. 1, Cap. 12, p. 68.

[4] Heron, *Stereometrica* (ed. Hultsch; Berlin, 1864), Pars I, Par. 8, p. 155.

modern exponent; thus[1] Archimedes and Eutocius use the notation $\iota\varsigma^{\overline{\kappa\alpha'}}$ or $\iota\varsigma^{\kappa\alpha}$ for $\frac{17}{21}$, and Diophantus (§§ 101–6), in expressing large numbers, writes (*Arithmetica*, Vol. IV, p. 17), $\frac{\beta_1\psi\delta^{\sim}}{\gamma\cdot\iota\varsigma\chi\kappa\alpha}$ for $\frac{36,621}{2,704}$.

Here the sign $\sim$ takes the place of the accent. Greek writers, even as late as the Middle Ages, display a preference for unit fractions, which played a dominating rôle in old Egyptian arithmetic.[2] In expressing such fractions, the Greeks omitted the α' for the numerator and wrote the denominator only once. Thus $\mu\delta'' = \frac{1}{44}$. Unit fractions in juxtaposition were added,[3] as in $\zeta''\ \kappa\eta''\ \rho\iota\beta''\ \sigma\kappa\delta'' = \frac{1}{7} + \frac{1}{28} + \frac{1}{112} + \frac{1}{224}$. One finds also a single accent,[4] as in $\delta' = \frac{1}{4}$. Frequent use of unit fractions is found in Geminus (first century B.C.), Diophantus (third century A.D.), Eutocius and Proclus (fifth century A.D.). The fraction $\frac{1}{2}$ had a mark of its own,[5] namely, L or L', but this designation was no more adopted generally among the Greeks than were the other notations of fractions. Ptolemy[6] wrote $38°50'$ (i.e., $38°\frac{1}{2}\ \frac{1}{3}$) thus, $\lambda\eta'\ L'\gamma'''$. Hultsch has found in manuscripts other symbols for $\frac{1}{2}$, namely, the semicircles C^{VI}, C, and the sign $\mathcal{S}$; the origin of the latter is uncertain. He found also a symbol for $\frac{2}{3}$, resembling somewhat the small omega (ω).[7] Whether these symbols represent late practice, but not early usage, it is difficult to determine with certainty.

42. A table for reducing certain ordinary fractions to the sum of unit fractions is found in a Greek papyrus from Egypt, described by

[1] G. H. F. Nesselmann, *Algebra der Griechen* (Berlin, 1842), p. 114.

[2] J. Baillet describes a papyrus, "Le papyrus mathématique d'Akhmîm," in *Mémoires publiés par les membres de la Mission archéologique française au Caire* (Paris, 1892), Vol. IX, p. 1–89 (8 plates). This papyrus, found at Akhmîm, in Egypt, is written in Greek, and is supposed to belong to the period between 500 and 800 A.D. It contains a table for the conversion of ordinary fractions into unit fractions.

[3] Fr. Hultsch, *Metrologicorum scriptorum reliquiae* (1864–66), p. 173–75; M. Cantor, *Vorlesungen über Geschichte der Mathematik*, Vol. I (3d ed.), p. 129.

[4] Nesselmann, *op. cit.*, p. 112.

[5] *Ibid.*; James Gow, *Short History of Greek Mathematics* (Cambridge, 1884), p. 48, 50.

[6] *Geographia* (ed. Carolus Müllerus; Paris, 1883), Vol. I, Part I, p. 151.

[7] *Metrologicorum scriptorum reliquiae* (Leipzig, 1864), Vol. I, p. 173, 174. On p. 175 and 176 Hultsch collects the numeral symbols found in three Parisian manuscripts, written in Greek, which exhibit minute variations in the symbolism. For instance, 700 is found to be ψ^π, ψ, ψ'.

L. C. Karpinski,[1] and supposed to be intermediate between the Ahmes papyrus and the Akhmim papyrus. Karpinski (p. 22) says: "In the table no distinction is made between integers and the corresponding unit fractions; thus γ' may represent either 3 or $\frac{1}{3}$, and actually $\gamma'\gamma'$ in the table represents $3\frac{1}{3}$. Commonly the letters used as numerals were distinguished in early Greek manuscripts by a bar placed above the letters but not in this manuscript nor in the Akhmim papyrus." In a third document dealing with unit fractions, a Byzantine table of fractions, described by Herbert Thompson,[2] $\frac{2}{3}$ is written ɪ̦; $\frac{1}{2}$, ϵ; $\frac{1}{3}$, ⋌ (from ʃ '); $\frac{1}{4}$, ⩘ (from Δ'); $\frac{1}{6}$, ℰ (from ϵ'); $\frac{1}{8}$, ⩗ (from H'). As late as the fourteenth century, Nicolas Rhabdas of Smyrna wrote two letters in the Greek language, on arithmetic, containing tables for unit fractions.[3] Here letters of the Greek alphabet used as integral numbers have bars placed above them.

43. About the second century before Christ the Babylonian sexagesimal numbers were in use in Greek astronomy; the letter omicron, which closely resembles in form our modern zero, was used to designate a vacant space in the writing of numbers. The Byzantines wrote it usually ō, the bar indicating a numeral significance as it has when placed over the ordinary Greek letters used as numerals.[4]

44. The division of the circle into 360 equal parts is found in Hypsicles.[5] Hipparchus employed sexagesimal fractions regularly, as did also C. Ptolemy[6] who, in his *Almagest,* took the approximate value of π to be $3+\dfrac{8}{60}+\dfrac{30}{60\times60}$. In the Heiberg edition this value is written $\bar{\gamma}\ \bar{\eta}\ \bar{\lambda}$, purely a *notation of position.* In the tables, as printed by Heiberg, the dash over the letters expressing numbers is omitted. In the edition of N. Halma[7] is given the notation $\bar{\gamma}\ \eta'\ \lambda''$, which is

[1] "The Michigan Mathematical Papyrus No. 621," *Isis,* Vol. V (1922), p. 20–25.

[2] "A Byzantine Table of Fractions," *Ancient Egypt,* Vol. I (1914), p. 52–54.

[3] The letters were edited by Paul Tannery in *Notices et extraits des manuscrits de la Bibliothèque Nationale,* Vol. XXXII, Part 1 (1886), p. 121–252.

[4] C. Ptolemy, *Almagest* (ed. N. Halma; Paris, 1813), Book I, chap. ix, p. 38 and later; J. L. Heiberg, in his edition of the *Almagest (Syntaxis mathematica)* (Leipzig, 1898; 2d ed., Leipzig, 1903), Book I, does not write the bar over the *o* but places it over all the significant Greek numerals. This procedure has the advantage of distinguishing between the *o* which stands for 70 and the *o* which stands for zero. See *Encyc. des scien. math.,* Tome I, Vol. I (1904), p. 17, n. 89.

[5] Ἀναφορικός (ed. K. Manitius), p. xxvi.

[6] *Syntaxis mathematica* (ed. Heiberg), Vol. I, Part 1, p. 513.

[7] *Composition math. de Ptolémée* (Paris, 1813), Vol. I, p. 421; see also *Encyc. des scien. math.,* Tome I, Vol. I (1904), p. 53, n. 181.

probably the older form. Sexagesimal fractions were used during the whole of the Middle Ages in India, and in Arabic and Christian countries. One encounters them again in the sixteenth and seventeenth centuries. Not only sexagesimal fractions, but also the sexagesimal notation of integers, are explained by John Wallis in his *Mathesis universalis* (Oxford, 1657), page 68, and by V. Wing in his *Astronomia Britannica* (London, 1652, 1669), Book I.

EARLY ARABS

45. At the time of Mohammed the Arabs had a script which did not differ materially from that of later centuries. The letters of the early Arabic alphabet came to be used as numerals among the Arabs

1 ‍ا	10 ى	100 ق	1000 غ	10 000 ﻲﺑ	100 000 ﻖﻗ
2 ب	20 ك	200 ر	2000 ﺞﺑ	20 000 ﻚﻛ	200 000 ﻍﺭ
3 ج	30 ل	300 ش	3000 ﺞﺟ	30 000 ﻊﻟ	300 000 ﻊﺷ
4 د	40 م	400 ت	4000 ﻍﺩ	40 000 ﻊﻣ	400 000 ﻊﺗ
5 ﻩ	50 ن	500 ث	5000 ﻮﻫ	50 000 ﻦﻧ	500 000 ﻊﺛ
6 و	60 س	600 خ	6000 ﻍﻭ	60 000 ﻊﺳ	600 000 ﻊﺧ
7 ز	70 ع	700 ذ	7000 ﻍﺯ	70 000 ﻊﻋ	700 000 ﻍﺫ
8 ح	80 ف	800 ض	8000 ﺢﺣ	80 000 ﻊﻓ	800 000 ﻊﺿ
9 ط	90 ص	900 ظ	9000 ﻊﻃ	90 000 ﻊﺻ	900 000 ﻊﻇ

FIG. 14.—Arabic alphabetic numerals used before the introduction of the Hindu-Arabic numerals.

as early as the sixth century of our era.[1] After the time of Mohammed, the conquering Moslem armies coming in contact with Greek culture acquired the Greek numerals. Administrators and military leaders used them. A tax record of the eighth century contains numbers expressed by Arabic letters and also by Greek letters.[2] Figure 14 is a table given by Ruska, exhibiting the Arabic letters and the numerical values which they represent. Taking the symbol for 1,000 twice, on the multiplicative principle, yielded 1,000,000. The Hindu-Arabic

[1] Julius Ruska, "Zur ältesten arabischen Algebra und Rechenkunst," *Sitzungsberichte d. Heidelberger Akademie der Wissensch.* (Philos.-histor. Klasse, 1917; 2. Abhandlung), p. 37.

[2] *Ibid.*, p. 40.

numerals, with the zero, began to spread among the Arabs in the ninth and tenth centuries, and they slowly displaced the Arabic and Greek numerals.[1]

ROMANS

46. We possess little definite information on the origin of the Roman notation of numbers. The Romans never used the successive letters of their alphabet for numeral purposes in the manner practiced by the Syrians, Hebrews, and Greeks, although (as we shall see) an alphabet system was at one time proposed by a late Roman writer. Before the ascendancy of Rome the Etruscans, who inhabited the country nearly corresponding to modern Tuscany and who ruled in Rome until about 500 B.c., used numeral signs which resembled letters of their alphabet and also resembled the numeral signs used by the Romans. Moritz Cantor[2] gives the Etrurian and the old Roman signs, as follows: For 5, the Etrurian Λ or V, the old Roman V; for 10 the Etrurian X or +, the old Roman X; for 50 the Etrurian ↑ or ↓, the old Roman Ⴕ or ↓ or ⊥ or ⌐ or L; for 100 the Etrurian ⊕, the old Roman ⊖; for 1,000 the Etrurian 8, the old Roman ⊕. The resemblance of the Etrurian numerals to Etrurian letters of the alphabet is seen from the following letters: V, +, ↓, O, 8. These resemblances cannot be pronounced accidental. "Accidental, on the other hand," says Cantor, "appears the relationship with the later Roman signs, I V, X, L, C, M, which from their resemblance to letters transformed themselves by popular etymology into these very letters." The origins of the Roman symbols for 100 and 1,000 are uncertain; those for 50 and 500 are generally admitted to be the result of a bisection of the two former. "There was close at hand," says G. Friedlein,[3] "the abbreviation of the word *centum* and *mille* which at an early age brought about for 100 the sign C, and for 1,000 the sign Λ and after Augustus[4] M." A view held by some Latinists[5] is that "the signs for 50, 100, 1,000 were originally the three Greek aspirate letters which the Romans did not require, viz., Ψ, ☉, ⊕, i.e., χ, θ, Φ. The Ψ was written ⊥ and abbreviated into L; ☉ from a false notion of its origin made like

---

[1] *Ibid.*, p. 47.

[2] *Vorlesungen über Geschichte der Mathematik*, Vol. I (3d ed.), p. 523, and the table at the end of the volume.

[3] *Die Zahlzeichen und das elementare Rechnen der Griechen und Römer* (Erlangen, 1869), p. 28.

[4] Theodor Mommsen, *Die unteritalischen Dialekte* (Leipzig, 1840), p. 30.

[5] Ritschl, *Rhein. Mus.*, Vol. XXIV (1869), p. 12.

the initial of centum; and ⓪ assimilated to ordinary letters CIƆ. The half of ⓪, viz., D, was taken to be ½ 1,000, i.e., 500; X probably from the ancient form of ⊖, viz., ⊗, being adopted for 10, the half of it V was taken for 5."[1]

47. Our lack of positive information on the origin and early history of the Roman numerals is not due to a failure to advance working hypotheses. In fact, the imagination of historians has been unusually active in this field.[2] The dominating feature in the Roman notation is the principle of addition, as seen in II, XII, CC, MDC, etc.

48. Conspicuous also is the frequent use of the principle of subtraction. If a letter is placed before another of greater value, its value is to be subtracted from that of the greater. One sees this in IV, IX, XL. Occasionally one encounters this principle in the Babylonian notations. Remarks on the use of it are made by Adriano Cappelli in the following passage:

"The well-known rule that a smaller number, placed to the left of a larger, shall be subtracted from the latter, as ⓪|ƆƆ = 4,000, etc., was seldom applied by the old Romans and during the entire Middle Ages one finds only a few instances of it. The cases that I have found belong to the middle of the fifteenth century and are all cases of IX, never of IV, and occurring more especially in French and Piedmontese documents. Walther, in his *Lexicon diplomaticum*, Göttingen, 1745–47, finds the notation LXL = 90 in use in the eighth century. On the other hand one finds, conversely, the numbers IIIX, VIX with the meaning of 13 and 16, in order to conserve, as Lupi remarks, the Latin terms *tertio decimo* and *sexto decimo*."[3] L. C. Karpinski points out that the subtractive principle is found on some early tombstones and on a signboard of 130 B.C., where at the crowded end of a line 83 is written XXCIII, instead of LXXXIII.

---

[1] H. J. Roby, *A Grammar of the Latin Language from Plautus to Suetonius* (4th ed.; London, 1881), Vol. I, p. 441.

[2] Consult, for example, Friedlein, *op. cit.*, p. 26–31; Nesselmann, *op. cit.*, p. 86–92; Cantor, *Mathematische Beiträge zum Kulturleben der Völker*, p. 155–67; J. C. Heilbronner, *Historia Matheseos universae* (Lipsiae, 1742), p. 732–35; Grotefend, *Lateinische Grammatik* (3d ed.; Frankfurt, 1820), Vol. II, p. 163, is quoted in the article "Zahlzeichen" in G. S. Klügel's *Mathematisches Wörterbuch*, continued by C. B. Mollweide and J. A. Grunert (Leipzig, 1831); Mommsen, *Hermes*, Vol. XXII (1887), p. 596; Vol. XXIII (1888), p. 152. A recent discussion of the history of the Roman numerals is found in an article by Ettore Bortolotti in *Bolletino della Mathesis* (Pavia, 1918), p. 60–66, which is rich in bibliographical references, as is also an article by David Eugene Smith in *Scientia* (July–August, 1926).

[3] *Lexicon Abbreviaturarum* (Leipzig, 1901), p. xlix.

49. Alexander von Humboldt[1] makes the following observations: "Summations by juxtaposition one finds everywhere among the Etruscans, Romans, Mexicans and Egyptians; subtraction or lessening forms of speech in Sanskrit among the Indians: in 19 or *unavinsati;* 99 *unusata;* among the Romans in *undeviginti* for 19 (*unus de viginti*), *undeoctoginta* for 79; *duo de quadraginta* for 38; among the Greeks *eikosi deonta henos* 19, and *pentekonta düoin deontoin* 48, i.e., 2 missing in 50. This lessening form of speech has passed over in the graphics of numbers when the group signs for 5, 10 and even their multiples, for example, 50 or 100, are placed to the left of the characters they modify (IV and IΛ, XL and XT for 4 and 40) among the Romans and Etruscans (Otfried Müller, *Etrusker*, II, 317–20), although among the latter, according to Otfried Müller's new researches, the numerals descended probably entirely from the alphabet. In rare Roman inscriptions which Marini has collected (*Iscrizioni della Villa di Albano*, p. 193; Hervas, *Aritmetica delle nazioni* [1786], p. 11, 16), one finds even 4 units placed before 10, for example, IIIIX for 6."

50. There are also sporadic occurrences in the Roman notations of the principle of multiplication, according to which VM does not stand for 1,000−5, but for 5,000. Thus, in Pliny's *Historia naturalis* (about 77 A.D.), VII, 26; XXXIII, 3; IV praef., one finds[2] LXXXIII.M, XCII.M, CX.M for 83,000, 92,000, 110,000, respectively.

51. The thousand-fold value of a number was indicated in some instances by a horizontal line placed above it. Thus, Aelius Lampridius (fourth century A.D.) says in one place, "$\overline{\text{CXX}}$, equitum Persarum fudimus: et mox $\overline{\text{X}}$ in bello interemimus," where the numbers designate 120,000 and 10,000. Strokes placed on top and also on the sides indicated hundred thousands; e.g., $|\overline{\text{X}}|$CLXXXDC stood for 1,180,600. In more recent practice the strokes sometimes occur only on the sides, as in $|$X$| \cdot$ DC.XC., the date on the title-page of Sigüenza's *Libra astronomica*, published in the city of Mexico in 1690. In antiquity, to prevent fraudulent alterations, XXXM was written for 30,000, and later still CIↃ took the place of M.[3] According to

---

[1] "Über die bei verschiedenen Völkern üblichen Systeme von Zahlzeichen, etc.," *Crelle's Journal für die reine und angewandte Mathematik* (Berlin, 1829), Vol. IV, p. 210, 211.

[2] Nesselmann, *op. cit.*, p. 90.

[3] Confer, on this point, Theodor Mommsen and J. Marquardt, *Manuel des antiquités romaines* (trans. G. Humbert), Vol. X by J. Marquardt (trans. A. Vigié; Paris, 1888), p. 47, 49.

Cappelli[1] "one finds, often in French documents of the Middle Ages, the multiplication of 20 expressed by two small x's which are placed as exponents to the numerals III, VI, VIII, etc., as in IIII^{xx} = 80, VI^{xx}XI = 131."

52. A Spanish writer[2] quotes from a manuscript for the year 1392 the following:

M   C
"IIII, IIII, LXXIII florins" for 4,473 florins.

M   XX
"III C IIII III florins" for 3,183 (?) florins.

In a Dutch arithmetic, printed in 1771, one finds[3]

c   c   m   c
i ꝛꝛiij for 123, i ꝛꝛiij iiij lѵj for 123,456.

53. For 1,000 the Romans had not only the symbol M, but also I, ∞ and CIƆ. According to Priscian, the celebrated Latin grammarian of about 500 A.D., the ∞ was the ancient Greek sign X for 1,000, but modified by connecting the sides by curved lines so as to distinguish it from the Roman X for 10. As late as 1593 the ∞ is used by C. Dasypodius[4] the designer of the famous clock in the cathedral at Strasbourg. The CIƆ was a I inclosed in parentheses (or *apostrophos*). When only the right-hand parenthesis is written, IƆ, the value represented is only half, i.e., 500. According to Priscian,[5] "quinque milia per I et duas in dextera parte apostrophos, IƆƆ. decem milia per supra dictam formam additis in sinistra parte contrariis duabus notis quam sunt apostrophi, CCIƆƆ." Accordingly, IƆƆ stood for 5,000, CCIƆƆ for 10,000; also IƆƆƆ represented 50,000; and CCCIƆƆƆ, 100,000; (∞), 1,000,000. If we may trust Priscian, the symbols that look like the letters C, or those letters facing in the opposite direction, were not really letters C, but were apostrophes or what we have called

[1] *Op. cit.*, p. xlix.

[2] Liciniano Saez, *Demostración Histórica del verdadero valor de Todas Las Monedas que corrían en Castilla durante el reynado del Señor Don Enrique III* (Madrid, 1796).

[3] *De Vernieuwde Cyfferinge* van M.ͬ Willem Bartjens. Herstelt, ... door M.ͬ Jan van Dam, ... en van alle voorgaande Fauten gezuyvert door . . . . Klaas Bosch (Amsterdam, 1771), p. 8.

[4] *Cunradi Dasypodii Institutionum Mathematicarum voluminis primi Erotemata* (1593), p. 23.

[5] "De figuris numerorum," *Henrici Keilii Grammatici Latini* (Lipsiae, 1859), Vol. III, 2, p. 407.

parentheses. Through Priscian it is established that this notation is at least as old as 500 A.D.; probably it was much older, but it was not widely used before the Middle Ages.

54. While the Hindu-Arabic numerals became generally known in Europe about 1275, the Roman numerals continued to hold a commanding place. For example, the fourteenth-century banking-house of Peruzzi in Florence—Compagnia Peruzzi—did not use Arabic numerals in their account-books. Roman numerals were used, but the larger amounts, the thousands of *lira*, were written out in words; one finds, for instance, "lb. quindicimilia CXV ∤ V ⨏ VI in fiorini" for 15,115 *lira* 5 *soldi* 6 *denari;* the specification being made that the *lira* are *lira a fiorino d'oro* at 20 *soldi* and 12 *denari*. There appears also a symbol much like ⟩, for thousand.[1]

Nagl states also: "Specially characteristic is . . . . during all the Middle Ages, the regular prolongation of the last | in the units, as VI|=VII, which had no other purpose than to prevent the subsequent addition of a further unit."

55. In a book by H. Giraua Tarragones[2] at Milan the Roman numerals appear in the running text and are usually underlined; in the title-page, the date has the horizontal line *above* the numerals. The Roman four is IIII. In the tables, columns of degrees and minutes are headed "G.M."; of hour and minutes, "H.M." In the tables, the Hindu-Arabic numerals appear; the five is printed ꝶ, without the usual upper stroke. The vitality of the Roman notation is illustrated further by a German writer, Sebastian Frank, of the sixteenth century, who uses Roman numerals in numbering the folios of his book and in his statistics: "Zimmet kumpt von Zailon .CC.VÑ LX. teütscher meil von Calicut weyter gelegen. . . . . Die Nägelin kummen von Meluza / für Calicut hinaussgelegen vij·c. vnd XL. deutscher meyl."[3] The two numbers given are 260 and 740 German miles. Peculiar is the insertion of *vnd* ("and"). Observe also the use of the principle of multiplication in vij·c. (=700). In Jakob Köbel's *Rechenbiechlin* (Augsburg, 1514), fractions appear in Roman numerals;

thus, $\dfrac{\text{II}^{\text{C}}}{\text{IIII}^{\text{C}}.\text{LX}}$ stands for $\frac{2\,0\,0}{4\,6\,0}$.

[1] Alfred Nagl, *Zeitschrift für Mathematik und Physik*, Vol. XXXIV (1889), Historisch-literarische Abtheilung, p. 164.

[2] *Dos Libros de Cosmographie*, compuestos nueuamente por Hieronymo Giraua Tarragones (Milan, M̅.D̅.L̅V̅I̅).

[3] *Weltbůch / spiegel vnd bildtnis des gantzen Erdtbodens . . . . von Sebastiano Franco Wördensi . . . .* (M.D. XXXIIII), fol. ccxx.

56. In certain sixteenth-century Portuguese manuscripts on navigation one finds the small letter b used for 5, and the capital letter R for 40. Thus, $xbiij$ stands for 18, $Riij$ for 43.[1]

(a)   (c)   (b)   (d)

FIG. 15.—Degenerate forms of Roman numerals in English archives (Common Pleas, Plea Rolls, 637, 701, and 817; also Recovery Roll 1). (Reduced.)

A curious development found in the archives of one or two English courts of the fifteenth and sixteenth centuries[2] was a special Roman

[1] J. I. de Brito Rebello, *Livro de Marinharia* (Lisboa, 1903), p. 37, 85–91, 193, 194.

[2] *Antiquaries Journal* (London, 1926), Vol. VI, p. 273, 274.

numeration for the membranes of their Rolls, the numerals assuming a degraded form which in its later stages is practically unreadable. In Figure 15 the first three forms show the number 147 as it was written in the years 1421, 1436, and 1466; the fourth form shows the number 47 as it was written in 1583.

57. At the present time the Roman notation is still widely used in marking the faces of watches and clocks, in marking the dates of books on title-pages, in numbering chapters of books, and on other occasions calling for a double numeration in which confusion might arise from the use of the same set of numerals for both. Often the Roman numerals are employed for aesthetic reasons.

58. A striking feature in Roman arithmetic is the partiality for duodecimal fractions. Why duodecimals and not decimals? We can only guess at the answer. In everyday affairs the division of units into two, three, four, and six equal parts is the commonest, and duodecimal fractions give easier expressions for these parts. Nothing definite is known regarding the time and place or the manner of the origin of these fractions. Unlike the Greeks, the Romans dealt with concrete fractions. The Roman *as*, originally a copper coin weighing one pound, was divided into 12 *unciae*. The abstract fraction $\frac{11}{12}$ was called *deuna* ($=de$ $uncia$, i.e., *as* [1] less *uncia* [$\frac{1}{12}$]). Each duodecimal subdivision had its own name and symbol. This is shown in the following table, taken from Friedlein,[1] in which S stands for *semis* or "half" of an *as*.

TABLE

as.	1	. . . .	. . . .
deunx	$\frac{11}{12}$	$S = = -$ or $S :: \cdot$	(de uncia $1-\frac{1}{12}$)
dextans $\}$ (decunx)$\int$	$\frac{5}{6}$	$S = =$ or $S ::$	$\{$(de sextans $1-\frac{1}{6}$) $\{$(decem unciae)
dodrans.	$\frac{3}{4}$	$S = -$ or $S = 1$ or $S : \cdot$	(de quadrans $1-\frac{1}{4}$)
bes.	$\frac{2}{3}$	$S =$ or $ - S -$ or $S:$	(duae assis *sc.* partes)
septunx.	$\frac{7}{12}$	$S -$ or $S \cdot$	(septem unciae)
semis.	$\frac{1}{2}$	S	. . . .
quincunx.	$\frac{5}{12}$	$= = -$ or $= - =$ or $:: \cdot$	(quinque unciae)
triens.	$\frac{1}{3}$	$= =$ or $\sum \sum$ or $::$	. . . .
quadrans.	$\frac{1}{4}$	$= -$ or $= 1$ or $: \cdot$	. . . .
sextans	$\frac{1}{6}$	$=$ or Z or $:$	. . . .
sescuncia $1\frac{1}{2}$.	$\frac{1}{12}=\frac{1}{8}$	$- \text{Ŀ} \text{Ƚ} \text{Ɫ} \text{Ʃ} \text{Ƚ}$	. . . .
uncia.	$\frac{1}{12}$	$-$ or $\cdot$ or on bronze abacus $\ominus$	

In place of straight lines $-$ occur also curved ones $\sim$.

----

[1] *Op. cit.*, Plate 2, No. 13; see also p. 35.

59. Not all of these names and signs were used to the same extent. Since $\frac{1}{2}+\frac{1}{3}=\frac{5}{6}$, there was used in ordinary life $\frac{1}{2}$ and $\frac{1}{3}$ (*semis et triens*) in place of $\frac{5}{6}$ or $\frac{10}{12}$ (*decunx*). Nor did the Romans confine themselves to the duodecimal fractions or their simplified equivalents $\frac{1}{2}$, $\frac{1}{3}$, $\frac{1}{4}$, $\frac{1}{6}$, etc., but used, for instance, $\frac{1}{10}$ in measuring silver, a *libella* being $\frac{1}{10}$ *denarius*. The *uncia* was divided in 4 *sicilici*, and in 24 *scripuli* etc.[1] In the *Geometry* of Boethius the Roman symbols are omitted and letters of the alphabet are used to represent fractions. Very probably this part of the book is not due to Boethius, but is an interpolation by a writer of later date.

60. There are indeed indications that the Romans on rare occasions used letters for the expression of integral numbers.[2] Theodor Mommsen and others discovered in manuscripts found in Bern, Einsiedeln, and Vienna instances of numbers denoted by letters. Tartaglia gives in his *General trattato di nvmeri*, Part I (1556), folios 4, 5, the following:

A	500	I	1	R	80
B	300	K	51	S	70
C	100	L	50	T	160
D	500	M	1,000	V	5
E	250	N	90	X	10
F	40	O	11	Y	150
G	400	P	400	Z	2,000
H	200	Q	500		

61. Gerbert (Pope Sylvestre II) and his pupils explained the Roman fractions. As reproduced by Olleris,[3] Gerbert's symbol for $\frac{1}{2}$ does not resemble the capital letter S, but rather the small letter ς.

[1] For additional details and some other symbols used by the Romans, consult Friedlein, p. 33–46 and Plate 3; also H. Hankel, *op. cit.*, p. 57–61, where computations with fractions are explained. Consult also Fr. Hultsch, *Metrologic. scriptores Romani* (Leipzig, 1866).

[2] Friedlein, *op. cit.*, p. 20, 21, who gives references. In the *Standard Dictionary of the English Language* (New York, 1896), under S, it is stated that $\overline{S}$ stood for 7 or 70.

[3] *Œuvres de Gerbert* (Paris, 1867), p. 343–48, 393–96, 583, 584.

PERUVIAN AND NORTH AMERICAN KNOT RECORDS[1]
ANCIENT *QUIPU*

62. "The use of knots in cords for the purpose of reckoning, and recording numbers" was practiced by the Chinese and some other ancient people; it had a most remarkable development among the Inca of Peru, in South America, who inhabited a territory as large as the United States east of the Rocky Mountains, and were a people of superior mentality. The period of Inca supremacy extended from about the eleventh century A.D. to the time of the Spanish conquest in the sixteenth century. The *quipu* was a twisted woolen cord, upon which other smaller cords of different colors were tied. The color, length, and number of knots on them and the distance of one from another all had their significance. Specimens of these ancient *quipu* have been dug from graves.

63. We reproduce from a work by L. Leland Locke a photograph of one of the most highly developed *quipu*, along with a line diagram of the two right-hand groups of strands. In each group the top strand usually gives the sum of the numbers on the four pendent strands. Thus in the last group, the four hanging strands indicate the numbers 89, 258, 273, 38, respectively. Their sum is 658; it is recorded by the top string. The repetition of units is usually expressed by a long knot formed by tying the overhand knot and passing the cord through the loop of the knot as many times as there are units to be denoted. The numbers were expressed on the decimal plan, but the *quipu* were not adopted for calculation; pebbles and grains of maize were used in computing.

64. Nordenskiöld shows that, in Peru, 7 was a magic number; for in some *quipu*, the sums of numbers on cords of the same color, or the numbers emerging from certain other combinations, are multiples of 7 or yield groups of figures, such as 2777, 777, etc. The *quipu* disclose also astronomical knowledge of the Peruvian Indians.[2]

65. Dr. Leslie Spier, of the University of Washington, sends me the following facts relating to Indians in North America: "The data that I have on the *quipu*-like string records of North-American Indians indicate that there are two types. One is a long cord with knots and

---

[1] The data on Peru knot records given here are drawn from a most interesting work, *The Ancient Quipu or Peruvian Knot Record*, by L. Leland Locke (American Museum of Natural History, 1923). Our photographs are from the frontispiece and from the diagram facing p. 16. See Figs. 16 and 17.

[2] Erland Nordenskiöld, *Comparative Ethnographical Studies*, No. 6, Part 1 (1925), p. 36.

bearing beads, etc., to indicate the days. It is simply a string record. This is known from the Yakima of eastern Washington and some Interior Salish group of Nicola Valley,[1] B.C.

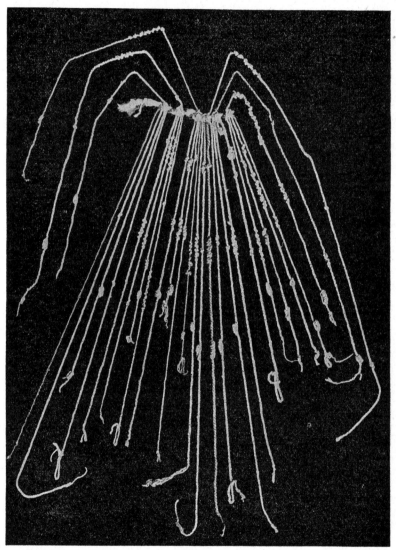

Fig. 16.—A *quipu*, from ancient Chancay in Peru, now kept in the American Museum of Natural History (Museum No. B8713) in New York City.

[1] J. D. Leechman and M. R. Harrington, *String Records of the Northwest, Indian Notes and Monographs* (1921).

"The other type I have seen in use among the Havasupai and Walapai of Arizona. This is a cord bearing a number of knots to indicate the days until a ceremony, etc. This is sent with the messenger who carries the invitation. A knot is cut off or untied for each day that elapses; the last one indicating the night of the dance. This is also used by the Northern and Southern Maidu and the Miwok of California.[1] There is a mythical reference to these among the Zuñi of New Mexico.[2] There is a note on its appearance in San Juan Pueblo in the same state in the seventeenth century, which would indicate that its use was widely known among the Pueblo Indians. 'They directed him (the leader of the Pueblo rebellion of 1680) to make a rope of the palm leaf and tie in it a number of knots to represent the number of days before the rebellion was to take place; that he must send the rope to all the Pueblos in the Kingdom, when each should signify its approval of, and union with, the conspiracy by untying one of the knots.'[3] The Huichol of Central Mexico also have knotted strings to keep count of days, untieing them as the days elapse. They also keep records of their lovers in the same way.[4] The Zuñi also keep records of days worked in this fashion.[5]

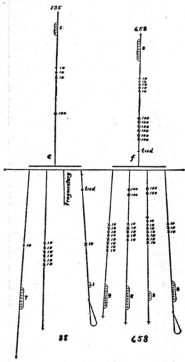

Fig. 17.—Diagram of the two right-hand groups of strands in Fig. 16.

[1] R. B. Dixon, "The Northern Maidu," *Bulletin of the American Museum of Natural History*, Vol. XVII (1905), p. 228, 271; P.-L. Faye, "Notes on the Southern Maidu," *University of California Publications of American Archaeology and Ethnology*, Vol. XX (1923), p. 44; Stephen Powers, "Tribes of California," *Contributions to North American Ethnology*, Vol. III (1877), p. 352.

[2] F. H. Cushing, "Zuñi Breadstuff," *Indian Notes and Monographs*, Vol. VIII (1920), p. 77.

[3] Quoted in J. G. Bourke, "Medicine-Men of the Apache," *Ninth Annual Report, Bureau of American Ethnology* (1892), p. 555.

[4] K. Lumholtz, *Unknown Mexico*, Vol. II, p. 218–30.

[5] Leechman and Harrington, *op. cit.*

"Bourke[1] refers to medicine cords with olivella shells attached among the Tonto and Chiricahua Apache of Arizona and the Zuñi. This may be a related form.

"I think that there can be no question the instances of the second type are historically related. Whether the Yakima and Nicola Valley usage is connected with these is not established."

AZTECS

66. "For figures, one of the numerical signs was the dot (·), which marked the units, and which was repeated either up to 20 or up to the figure 10, represented by a lozenge. The number 20 was represented by a flag, which, repeated five times, gave the number 100, which was

Fig. 18.—Aztec numerals

marked by drawing quarter of the barbs of a feather. Half the barbs was equivalent to 200, three-fourths to 300, the entire feather to 400. Four hundred multiplied by the figure 20 gave 8,000, which had a purse for its symbol."[2] The symbols were as shown in the first line of Figure 18.

The symbols for 20, 400, and 8,000 disclose the number 20 as the base of Aztec numeration; in the juxtaposition of symbols the additive principle is employed. This is seen in the second line[3] of Figure 18, which represents

$$2\times8,000+400+3\times20+3\times5+3=16,478 .$$

67. The number systems of the Indian tribes of North America, while disclosing no use of a symbol for zero nor of the principle of

---

[1] *Op. cit.*, p. 550 ff.

[2] Lucien Biart, *The Aztecs* (trans. J. L. Garner; Chicago, 1905), p. 319.

[3] Consult A. F. Pott, *Die quinäre und vigesimale Zählmethode bei Völkern aller Welttheile* (Halle, 1847).

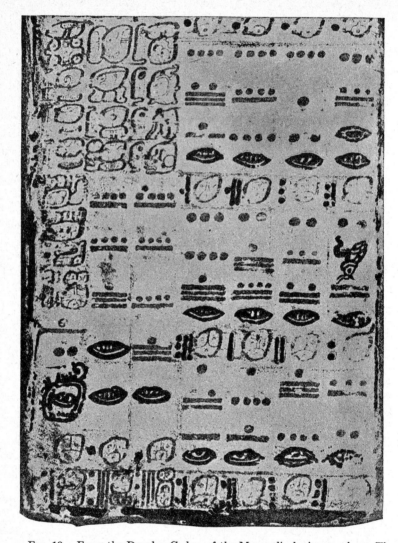

Fig. 19.—From the Dresden Codex, of the Maya, displaying numbers. The second column on the left, from above down, displays the numbers 9, 9, 16, 0, 0, which stand for $9 \times 144,000 + 9 \times 7,200 + 16 \times 360 + 0 + 0 = 1,366,560$. In the third column are the numerals 9, 9, 9, 16, 0, representing 1,364,360. The original appears in black and red colors. (Taken from Morley, *An Introduction to the Study of the Maya Hieroglyphs*, p. 266.)

local value, are of interest as exhibiting not only quinary, decimal, and vigesimal systems, but also ternary, quaternary, and octonary systems.[1]

MAYA

68. The Maya of Central America and Southern Mexico developed hieroglyphic writing, as found on inscriptions and codices, dating apparently from about the beginning of the Christian Era, which discloses the use of a remarkable number system and chronology.[2] The number system discloses the application of the principle of local value, and the use of a symbol for zero centuries before the Hindus began to use their symbol for zero. The Maya system was vigesimal, except in one step. That is, 20 units (*kins*, or "days") make 1 unit of the next higher order (*uinals*, or 20 days), 18 *uinals* make 1 unit of the third order (*tun*, or 360 days), 20 *tuns* make 1 unit of the fourth order (*Katun*, or 7,200 days), 20 *Katuns* make 1 unit of the fifth order (*cycle*, or 144,000 days), and finally 20 *cycles* make 1 *great cycle* of 2,880,000 days. In the Maya codices we find symbols for 1–19, expressed by bars and dots. Each bar stands for 5 units, each dot for 1 unit. For instance,

$$\overset{.}{\underset{1}{}} \quad \overset{..}{\underset{2}{}} \quad \overset{::}{\underset{4}{}} \quad \overset{}{\underset{5}{\rule{1em}{0.4pt}}} \quad \overset{..}{\underset{7}{\rule{1em}{0.4pt}}} \quad \overset{.}{\underset{11}{\overline{\overline{\rule{1em}{0pt}}}}} \quad \overset{....}{\underset{19}{\equiv}}.$$

The zero is represented by a symbol that looks roughly like a half-closed eye. In writing 20 the principle of local value enters. It is expressed by a dot placed over the symbol for zero. The numbers are written vertically, the lowest order being assigned the lowest position (see Fig. 19). The largest number found in the codices is 12,489,781.

CHINA AND JAPAN

69. According to tradition, the oldest Chinese representation of number was by the aid of *knots in strings*, such as are found later among the early inhabitants of Peru. There are extant two Chinese tablets[3] exhibiting knots representing numbers, odd numbers being designated by white knots (standing for the complete, as day, warmth,

[1] W. C. Eells, "Number-Systems of North-American Indians," *American Mathematical Monthly*, Vol. XX (1913), p. 263–72, 293–99; also *Bibliotheca mathematica* (3d series, 1913), Vol. XIII, p. 218–22.

[2] Our information is drawn from S. G. Morley, *An Introduction to the Study of the Maya Hieroglyphs* (Washington, 1915).

[3] Paul Perny, *Grammaire de la langue chinoise orale et écrite* (Paris, 1876), Vol. II, p. 5–7; Cantor, *Vorlesungen über Geschichte der Mathematik*, Vol. 1 (3d ed.), p. 674.

the sun) while even numbers are designated by black knots (standing for the incomplete, as night, cold, water, earth). The left-hand tablet shown in Figure 20 represents the numbers 1–10. The right-hand tablet pictures the magic square of nine cells in which the sum of each row, column, and diagonal is 15.

70. The Chinese are known to have used three other systems of writing numbers, the Old Chinese numerals, the mercantile numerals, and what have been designated as scientific numerals. The time of the introduction of each of these systems is uncertain.

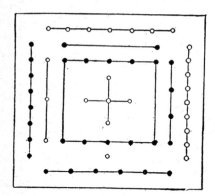 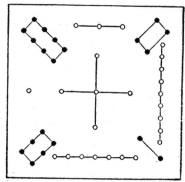

Fig. 20.—Early Chinese knots in strings, representing numerals

71. The *Old Chinese numerals* were written vertically, from above down. Figure 21 shows the Old Chinese numerals and mercantile numerals, also the Japanese cursive numerals.[1]

72. The *Chinese scientific numerals* are made up of vertical and horizontal rods according to the following plan: The numbers 1–9 are represented by the rods |, ||, |||, ||||, |||||, T, TT, TTT, TTTT; the numbers 10–90 are written thus _ = ≡ ≣ ⊥ ⊥ ⊥ ⊥. According to the Chinese author Sun-Tsu, units are represented, as just shown, by vertical rods, ten's by horizontal rods, hundred's again by vertical rods, and so on. For example, the number 6,728 was designated by ⊥ TT = TTT .

73. The Japanese make use of the Old Chinese numerals, but have two series of names for the numeral symbols, one indigenous, the other derived from the Chinese language, as seen in Figure 21.

[1] See also Ed. Biot, *Journal asiatique* (December, 1839), p. 497–502; Cantor, *Vorlesungen über Geschichte der Mathematik*, Vol. I, p. 673; Biernatzki, *Crelle's Journal*, Vol. LII (1856), p. 59–94.

HINDU-ARABIC NUMERALS

74. *Introduction.*—It is impossible to reproduce here all the forms of our numerals which have been collected from sources antedating 1500 or 1510 A.D. G. F. Hill, of the British Museum, has devoted a

CHIFFRES			VALEURS.	NOMS DE NOMBRE	
CHINOIS kiäi-chôu.	JAPONAIS cursifs.	DU COMMERCE.		EN JAPONAIS PUR.	EN SINICO-JAPONAIS.
一	一	1	1	*fitots.*	*itsi.*
二	二	11	2	*foutats.*	*ni.*
三	三	川	3	*mits.*	*san.*
四	四	メ	4	*yots.*	*si.*
五	五	ぶ	5	*itsouts.*	*go.*
六	六	上	6	*mouts.*	*rok.*
七	七	亠	7	*nanats.*	*sitsi.*
八	八	三	8	*yats.*	*fats.*
九	九	久	9	*kokonots.*	*kou.*
十	十	十	10	*towo.*	*zyou.*
百	百	日	100	*momo.*	*fak ou fyak.*
千	千	千	1,000	*tsidzi.*	*sen.*
萬	弟	万	10,000	*yorodz.*	*man.*

FIG. 21.—Chinese and Japanese numerals. (Taken from A. P. Pihan, *Exposé des signes de numération* [Paris, 1860], p. 15.)

whole book[1] of 125 pages to the early numerals in Europe alone. Yet even Hill feels constrained to remark: "What is now offered, in the shape of just 1,000 classified examples, is nothing more than a *vinde-*

[1] *The Development of Arabic Numerals in Europe* (exhibited in 64 tables; Oxford, 1915).

miatio prima." Add to the Hill collection the numeral forms, or supposedly numeral forms, gathered from other than European sources,
and the material would fill a volume very much larger than that of
Hill. We are compelled, therefore, to confine ourselves to a few of the
more important and interesting forms of our numerals.[1]

75. One feels the more inclined to insert here only a few tables of
numeral forms because the detailed and minute study of these forms
has thus far been somewhat barren of positive results. With all the
painstaking study which has been given to the history of our numerals
we are at the present time obliged to admit that we have not even
settled the time and place of their origin. At the beginning of the
present century the Hindu origin of our numerals was supposed to
have been established beyond reasonable doubt. But at the present
time several earnest students of this perplexing question have expressed grave doubts on this point. Three investigators—G. R. Kaye
in India, Carra de Vaux in France, and Nicol. Bubnov in Russia—
working independently of one another, have denied the Hindu origin.[2]
However, their arguments are far from conclusive, and the hypothesis
of the Hindu origin of our numerals seems to the present writer to
explain the known facts more satisfactorily than any of the substitute
hypotheses thus far advanced.[3]

[1] The reader who desires fuller information will consult Hill's book which is
very rich in bibliographical references, or David Eugene Smith and Louis Charles
Karpinski's *The Hindu-Arabic Numerals* (Boston and London, 1911). See also an
article on numerals in English archives by H. Jenkinson in *Antiquaries Journal*,
Vol. VI (1926), p. 263–75. The valuable original researches due to F. Woepcke
should be consulted, particularly his great "Mémoire sur la propagation des
chiffres indiens" published in the *Journal asiatique* (6th series; Paris, 1863), p. 27–
79, 234–90, 442–529. Reference should be made also to a few other publications of
older date, such as G. Friedlein's *Zahlzeichen und das elementare Rechnen der
Griechen und Römer* (Erlangen, 1869), which touches questions relating to our
numerals. The reader will consult with profit the well-known histories of mathematics by H. Hankel and by Moritz Cantor.

[2] G. R. Kaye, "Notes on Indian Mathematics," *Journal and Proceedings of the
Asiatic Society of Bengal* (N.S., 1907), Vol. III, p. 475–508; "The Use of the Abacus
in Ancient India," *ibid.*, Vol. IV (1908), p. 293–97; "References to Indian Mathematics in Certain Mediaeval Works," *ibid.*, Vol. VII (1911), p. 801–13; "A Brief
Bibliography of Hindu Mathematics," *ibid.*, p. 679–86; *Scientia*, Vol. XXIV
(1918), p. 54; "Influence grecque dans le développement des mathématiques
hindoues," *ibid.*, Vol. XXV (1919), p. 1–14; Carra de Vaux, "Sur l'origine des
chiffres," *ibid.*, Vol. XXI (1917), p. 273–82; Nicol. Bubnov, *Arithmetische Selbst-
ständigkeit der europäischen Kultur* (Berlin, 1914) (trans. from Russian ed.; Kiev,
1908).

[3] F. Cajori, "The Controversy on the Origin of Our Numerals," *Scientific
Monthly*, Vol. IX (1919), p. 458–64. See also B. Datta in *Amer. Math. Monthly*,
Vol. XXXIII, p. 449; *Proceed. Benares Math. Soc.*, Vol. VII.

76. Early Hindu mathematicians, Aryabhata (b. 476 A.D.) and Brahmagupta (b. 598 A.D.), do not give the expected information about the Hindu-Arabic numerals.

Āryabhaṭa's work, called *Aryabhatiya*, is composed of three parts, in only the first of which use is made of a special notation of numbers. It is an alphabetical system[1] in which the twenty-five consonants represent 1–25, respectively; other letters stand for 30, 40, ... , 100, etc.[2] The other mathematical parts of Āryabhaṭa consists of rules without examples. Another alphabetical system prevailed in Southern India, the numbers 1–19 being designated by consonants, etc.[3]

In Brahmagupta's *Pulverizer*, as translated into English by H. T. Colebrooke,[4] numbers are written in our notation with a zero and the principle of local value. But the manuscript of Brahmagupta used by Colebrooke belongs to a late century. The earliest commentary on Brahmagupta belongs to the tenth century; Colebrooke's text is later.[5] Hence this manuscript cannot be accepted as evidence that Brahmagupta himself used the zero and the principle of local value.

77. Nor do inscriptions, coins, and other manuscripts throw light on the origin of our numerals. Of the old notations the most important is the Brahmi notation which did not observe place value and in which 1, 2, and 3 are represented by _ , $=$, $\equiv$. The forms of the Brahmi numbers do not resemble the forms in early place-value notations[6] of the Hindu-Arabic numerals.

Still earlier is the Kharoshthi script,[7] used about the beginning of the Christian Era in Northwest India and Central Asia. In it the first three numbers are | || |||, then X=4, |X=5, ||X=6, XX=8, $\daleth$=10, $\daleth$=20, $\daleth\daleth$=40, $\daleth\daleth\daleth$=50, X|=100. The writing proceeds from right to left.

78. *Principle of local value.*—Until recently the preponderance of authority favored the hypothesis that our numeral system, with its concept of local value and our symbol for zero, was wholly of Hindu origin. But it is now conclusively established that the principle of

---

[1] M. Cantor, *Vorlesungen über Geschichte der Mathematik*, Vol. I (3d ed.), p. 606.

[2] G. R. Kaye, *Indian Mathematics* (Calcutta and Simla, 1915), p. 30, gives full explanation of Aryabhata's notation.

[3] M. Cantor, *Math. Beiträge z. Kulturleben der Völker* (1863), p. 68, 69.

[4] *Algebra with Arithmetic and Mensuration from the Sanscrit* (London, 1817), p. 326 ff.

[5] *Ibid.*, p. v, xxxii.

[6] See forms given by G. R. Kaye, *op. cit.*, p. 29. [7] *Ibid.*

local value was used by the Babylonians much earlier than by the Hindus, and that the Maya of Central America used this principle and symbols for zero in a well-developed numeral system of their own and at a period antedating the Hindu use of the zero (§ 68).

79. The earliest-known reference to Hindu numerals outside of India is the one due to Bishop Severus Sebokht of Nisibis, who, living in the convent of Kenneshre on the Euphrates, refers to them in a fragment of a manuscript (MS Syriac [Paris], No. 346) of the year 662 A.D. Whether the numerals referred to are the ancestors of the modern numerals, and whether his Hindu numerals embodied the principle of local value, cannot at present be determined. Apparently hurt by the arrogance of certain Greek scholars who disparaged the Syrians, Sebokht, in the course of his remarks on astronomy and mathematics, refers to the Hindus, "their valuable methods of calculation; and their computing that surpasses description. I wish only to say that this computation is done by means of nine signs."[1]

80. Some interest attaches to the earliest dates indicating the use of the perfected Hindu numerals. That some kind of numerals with a zero was used in India earlier than the ninth century is indicated by Brahmagupta (b. 598 A.D.), who gives rules for computing with a zero.[2] G. Bühler[3] believes he has found definite mention of the decimal system and zero in the year 620 A.D. These statements do not necessarily imply the use of a decimal system based on the principle of local value. G. R. Kaye[4] points out that the task of the antiquarian is complicated by the existence of forgeries. In the eleventh century in India "there occurred a specially great opportunity to regain confiscated endowments and to acquire fresh ones." Of seventeen citations of inscriptions before the tenth century displaying the use of place value in writing numbers, all but two are eliminated as forgeries; these two are for the years 813 and 867 A.D.; Kaye is not sure of the reliability even of these. According to D. E. Smith and L. C. Karpinski,[5] the earliest authentic document unmistakably containing the numerals with the zero in India belongs to the year 876 A.D. The earli-

---

[1] See M. F. Nau, *Journal asiatique* (10th ser., 1910), Vol. XVI, p. 255; L. C. Karpinski, *Science*, Vol. XXXV (1912), p. 969–70; J. Ginsburg, *Bulletin of the American Mathematical Society*, Vol. XXIII (1917), p. 368.

[2] Colebrooke, *op. cit.*, p. 339, 340.

[3] "Indische Paläographie," *Grundriss d. indogerman. Philologie u. Altertumskunde*, Band I, Heft 11 (Strassburg, 1896), p. 78.

[4] *Journal of the Asiatic Society of Bengal* (N.S., 1907), Vol. III, p. 482–87.

[5] *The Hindu-Arabic Numerals* (New York, 1911), p. 52.

est Arabic manuscripts containing the numerals are of 874[1] and 888
A.D. They appear again in a work written at Shiraz in Persia[2] in 970 A.D.
A church pillar[3] not far from the Jeremias Monastery in Egypt has

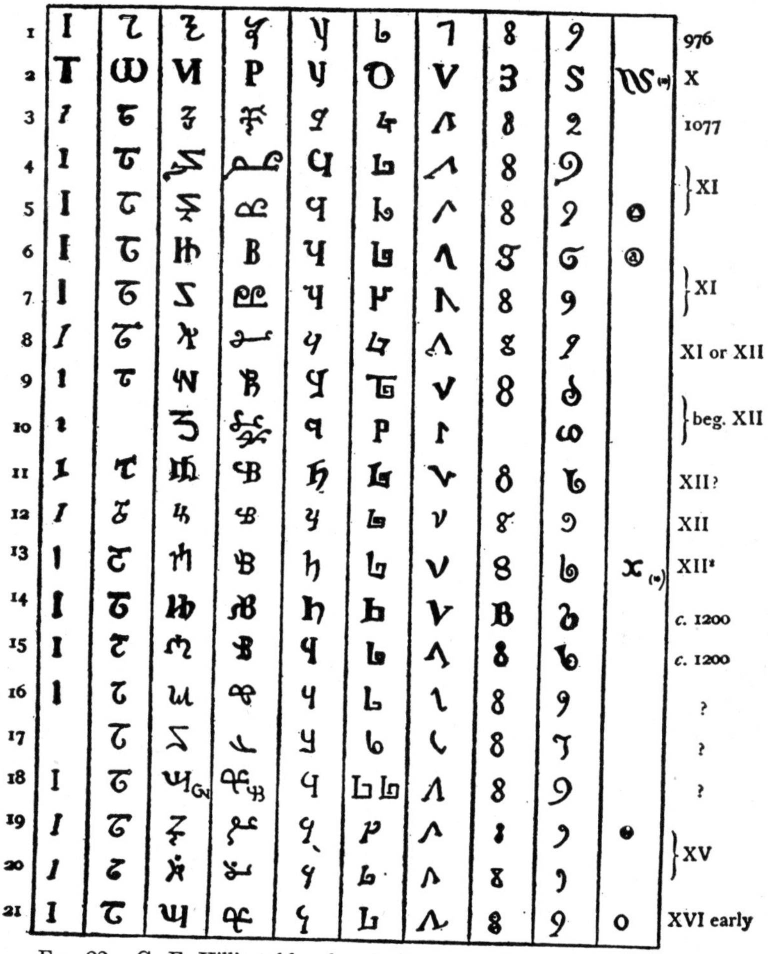

FIG. 22.—G. F. Hill's table of early European forms and Boethian apices.
(From G. F. Hill, *The Development of Arabic Numerals in Europe* [Oxford, 1915],
p. 28. Mr. Hill gives the MSS from which the various sets of numerals in this table
are derived: [1] Codex Vigilanus; [2] St. Gall MS now in Zürich; [3] Vatican MS
3101, etc. The Roman figures in the last column indicate centuries.)

[1] Karabacek, *Wiener Zeitschrift für die Kunde des Morgenlandes*, Vol. II (1897),
p. 56.

[2] L. C. Karpinski, *Bibliotheca mathematica* (3d ser., 1910–11), p. 122.

[3] Smith and Karpinski, *op. cit.*, p. 138–43.

the date 349 A.H. (=961 A.D.). The oldest definitely dated European manuscript known to contain the Hindu-Arabic numerals is the Codex Vigilanus (see Fig. 22, No. 1), written in the Albelda Cloister in Spain in 976 A.D. The nine characters without the zero are given, as an addition, in a Spanish copy of the *Origines* by Isidorus of Seville, 992 A.D. A tenth-century manuscript with forms differing materially from those in the Codex Vigilanus was found in the St. Gall manuscript (see Fig. 22, No. 2), now in the University Library at Zürich. The numerals are contained in a Vatican manuscript of 1077 (see Fig. 22, No. 3), on a Sicilian coin of 1138, in a Regensburg (Bavaria)

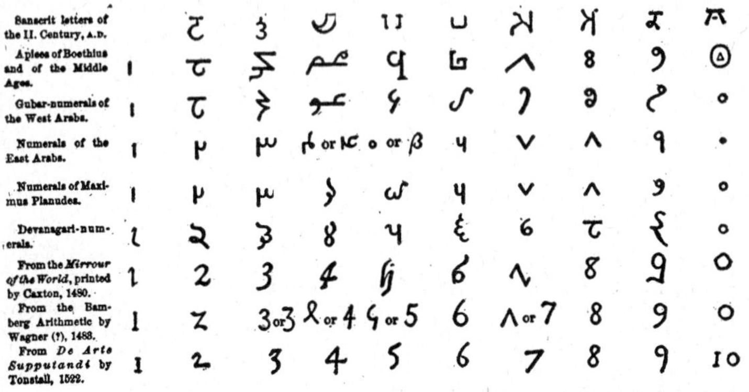

FIG. 23.—Table of important numeral forms. (The first six lines in this table are copied from a table at the end of Cantor's *Vorlesungen über Geschichte der Mathematik*, Vol. I. The numerals in the Bamberg arithmetic are taken from Friedrich Unger, *Die Methodik der praktischen Arithmetik in historischer Entwickelung* [Leipzig, 1888], p. 39.)

chronicle of 1197. The earliest manuscript in French giving the numerals dates about 1275. In the British Museum one English manuscript is of about 1230–50; another is of 1246. The earliest undoubted Hindu-Arabic numerals on a gravestone are at Pforzheim in Baden of 1371 and one at Ulm of 1388. The earliest coins outside of Italy that are dated in the Arabic numerals are as follows: Swiss 1424, Austrian 1484, French 1485, German 1489, Scotch 1539, English 1551.

81. *Forms of numerals.*—The Sanskrit letters of the second century A.D. head the list of symbols in the table shown in Figure 23. The implication is that the numerals have evolved from these letters. If such a connection could be really established, the Hindu origin of our numeral forms would be proved. However, a comparison of the forms appearing in that table will convince most observers that an origin

from Sanskrit letters cannot be successfully demonstrated in that way; the resemblance is no closer than it is to many other alphabets.

The forms of the numerals varied considerably. The 5 was the most freakish. An upright 7 was rare in the earlier centuries. The symbol for zero first used by the Hindus was a dot.[1] The symbol for zero (0) of the twelfth and thirteenth centuries is sometimes crossed by a horizontal line, or a line slanting upward.[2] The Boethian apices, as found in some manuscripts, contain a triangle inscribed in the circular zero. In Athelard of Bath's translation of Al-Madjrītī's revision of Al-Khowarizmi's astronomical tables there are in different manuscripts three signs for zero,[3] namely, the $\ominus$ (=theta?) referred to above, then Υ (=$teca$),[4] and $\bar{0}$. In one of the manuscripts 38 is written several times XXXO, and 28 is written XXO; the O being intended most likely as the abbreviation for $octo$ ("eight").

82. The symbol Υ for zero is found also in a twelfth-century manuscript[5] of N. Ocreatus, addressed to his master Athelard. In that century it appears especially in astronomical tables as an abbreviation for $teca$, which, as already noted, was one of several names for zero;[6] it is found in those tables by itself, without connection with other numerals. The symbol occurs in the $Algorismus\ vulgaris$ ascribed to Sacrobosco.[7] C. A. Nallino found $\bar{o}$ for zero in a manuscript of Escurial, used in the preparation of an edition of Al-Battani. The symbol $\ominus$ for zero occurs also in printed mathematical books.

The one author who in numerous writings habitually used θ for zero was the French mathematician Michael Rolle (1652–1719). One finds it in his $Traité\ d'algèbre$ (1690) and in numerous articles in the publications of the French Academy and in the $Journal\ des\ sçavans$.

[1] Smith and Karpinski, $op.\ cit.$, p. 52, 53.

[2] Hill, $op.\ cit.$, p. 30–60.

[3] H. Suter, $Die\ astronomischen\ Tafeln\ des\ Muḥammed\ ibn\ Mūsā\ Al-Khwārizmī$ $in\ der\ Bearbeitung\ des\ Maslama\ ibn\ Aḥmed\ Al-Madjrītī\ und\ der\ lateinischen\ Ueber$-$setzung\ des\ Athelhard\ von\ Bath$ (Københąvn, 1914), p. xxiii.

[4] See also M. Curtze, $Petri\ Philomeni\ de\ Dacia\ in\ Algorismum\ vulgarem$ $Johannis\ de\ Sacrobosco\ Commentarius$ (Hauniae, 1897), p. 2, 26.

[5] "Prologus N. Ocreati in Helceph ad Adelardum Batensem Magistrum suum. Fragment sur la multiplication et la division publié pour la première fois par Charles Henry," $Abhandlungen\ zur\ Geschichte\ der\ Mathematik$, Vol. III (1880), p. 135–38.

[6] M. Curtze, $Urkunden\ zur\ Geschichte\ der\ Mathematik\ im\ Mittelalter\ und\ der$ $Renaissance$ (Leipzig, 1902), p. 182.

[7] M. Curtze, $Abhandlungen\ zur\ Geschichte\ der\ Mathematik$, Vol. VIII (Leipzig, 1898), p. 3–27.

Manuscripts of the fifteenth century, on arithmetic, kept in the
Ashmolean Museum[1] at Oxford, represent the zero by a circle, crossed
by a vertical stroke and resembling the Greek letter ϕ. Such forms
for zero are reproduced by G. F. Hill[2] in many of his tables of numerals.

83. In the fifty-six philosophical treatises of the brothers Iḫwān
aṣ-ṣafā (about 1000 A.D.) are shown Hindu-Arabic numerals and the
corresponding Old Arabic numerals.

The forms of the Hindu-Arabic numerals, as given in Figure 24,
have maintained themselves in Syria to the present time. They appear with almost identical form in an Arabic school primer, printed

Fɪɢ. 24.—In the first line are the Old Arabic numerals for 10, 9, 8, 7, 6, 5, 4, 3,
2, 1. In the second line are the Arabic names of the numerals. In the third line
are the Hindu-Arabic numerals as given by the brothers Iḫwān aṣ-ṣafā. (Reproduced from J. Ruska, *op. cit.*, p. 87.)

at Beirut (Syria) in 1920. The only variation is in the 4, which in 1920
assumes more the form of a small Greek epsilon. Observe that 0 is
represented by a dot, and 5 by a small circle. The forms used in modern Arabic schoolbooks cannot be recognized by one familiar only with
the forms used in Europe.

84. In fifteenth-century Byzantine manuscripts, now kept in the
Vienna Library,[3] the numerals used are the Greek letters, but the
principle of local value is adopted. Zero is γ or in some places $\cdot$; aa
means 11, $\beta\gamma$ means 20, $a\gamma\gamma$ means 1,000. "This symbol γ for zero
means elsewhere 5," says Heiberg, "conversely, o stands for 5 (as now
among the Turks) in Byzantine scholia to Euclid. . . . . In Constantinople the new method was for a time practiced with the retention of

[1] Robert Steele, *The Earliest Arithmetics in English* (Oxford, 1922), p. 5.

[2] *Op. cit.*, Tables III, IV, V, VI, VIII, IX, XI, XV, XVII, XX, XXI, XXII.
See also E. Wappler, *Zur Geschichte der deutschen Algebra im XV. Jahrhundert*
(Zwickauer Gymnasialprogramm von 1887), p. 11–30.

[3] J. L. Heiberg, "Byzantinische Analekten," *Abhandlungen zur Geschichte der
Mathematik*, Vol. IX (Leipzig, 1899), p. 163, 166, 172. This manuscript in the
Vienna Library is marked "Codex Phil. Gr. 65."

the old letter-numerals, mainly, no doubt, in daily intercourse." At the close of one of the Byzantine manuscripts there is a table of numerals containing an imitation of the Old Attic numerals. The table gives also the Hindu-Arabic numerals, but apparently without recognition of the principle of local value; in writing 80, the 0 is placed over the 8. This procedure is probably due to the ignorance of the scribe.

85. A manuscript[1] of the twelfth century, in Latin, contains the symbol Ⱶ for 3 which Curtze and Nagl[2] declare to have been found only in the twelfth century. According to Curtze, the foregoing strange symbol for 3 is simply the symbol for *tertia* used in the notation for sexagesimal fractions which receive much attention in this manuscript.

86. Recently the variations in form of our numerals have been summarized as follows: "The form[3] of the numerals 1, 6, 8 and 9 has not varied *much* among the [medieval] Arabs nor among the Christians of the Occident; the numerals of the Arabs of the Occident for 2, 3 and 5 have forms offering some analogy to ours (the 3 and 5 are originally reversed, as well among the Christians as among the Arabs of the Occident); but the form of 4 and that of 7 have greatly modified themselves. The numerals 5, 6, 7, 8 of the Arabs of the Orient differ distinctly from those of the Arabs of the Occident (Gobar numerals). For *five* one still writes 5 and ꝫ." The use of *i* for 1 occurs in the first printed arithmetic (Treviso, 1478), presumably because in this early stage of printing there was no type for 1. Thus, 9,341 was printed 934*i*.

87. Many points of historical interest are contained in the following quotations from the writings of Alexander von Humboldt. Although over a century old, they still are valuable.

"In the Gobar[4] the group signs are dots, that is zeroes, for in India, Tibet and Persia the zeroes and dots are identical. The Gobar symbols, which since the year 1818 have commanded my whole attention, were discovered by my friend and teacher, Mr. Silvestre de Sacy, in a manuscript from the Library of the old Abbey St. Germain du Près. This great orientalist says: 'Le Gobar a un grand rapport

[1] Algorithmus-MSS Clm 13021, fols. 27–29, of the Munich Staatsbibliothek. Printed and explained by Maximilian Curtze, *Abhandlungen zur Geschichte der Mathematik*, Vol. VIII (Leipzig, 1898), p. 3–27.

[2] *Zeitschrift für Mathematik und Physik* (Hist. Litt. Abth.), Vol. XXXIV (Leipzig, 1889), p. 134.

[3] *Encyc. des Scien. math.*, Tome I, Vol. I (1904), p. 20, n. 105, 106.

[4] Alexander von Humboldt, *Crelle's Journal*, Vol. IV (1829), p. 223, 224.

avec le chiffre indien, mais il n'a pas de zéro (S. *Gramm. arabe*, p. 76, and the note added to Pl. 8).' I am of the opinion that the zero-symbol is present, but, as in the Scholia of Neophytos on the units, it stands over the units, not by their side. Indeed it is these very zero-symbols or dots, which give these characters the singular name *Gobar* or dust-writing. At first sight one is uncertain whether one should recognize therein a transition between numerals and letters of the alphabet. One distinguishes with difficulty the Indian 3, 4, 5 and 9. *Dal* and *ha* are perhaps ill-formed Indian numerals 6 and 2. The notation by dots is as follows:

$$3^{\cdot}\ \text{for}\ 30\ ,$$
$$4^{\cdot\cdot}\ \text{for}\ 400\ ,$$
$$6^{\cdot\cdot\cdot}\ \text{for}\ 6,000\ .$$

These dots remind one of an old-Greek but rare notation (Ducange, *Palaeogr.*, p. xii), which begins with the myriad: $a^{\cdot\cdot}$ for 10,000, $\beta^{\cdot\cdot}$ for 200 millions. In this system of geometric progressions a single dot, which however is not written down, stands for 100. In Diophantus and Pappus a dot is placed between letter-numerals, instead of the initial Mv (myriad). A dot multiplies what lies to its left by 10,000. . . . . A real zero symbol, standing for the absence of some unit, is applied by Ptolemy in the descending sexagesimal scale for missing degrees, minutes or seconds. Delambre claims to have found our symbol for zero also in manuscripts of Theon, in the Commentary to the Syntaxis of Ptolemy.[1] It is therefore much older in the Occident than the invasion of the Arabs and the work of Planudes on *arithmoi indikoi*." L. C. Karpinski[2] has called attention to a passage in the Arabic biographical work, the *Fihrist* (987 A.D.), which describes a Hindu notation using dots placed *below* the numerals; one dot indicates tens, two dots hundreds, and three dots thousands.

88. There are indications that the magic power of the principle of local value was not recognized in India from the beginning, and that our perfected Hindu-Arabic notation resulted from gradual evolution. Says Humboldt: "In favor of the successive perfecting of the designation of numbers in India testify the Tamul numerals which, by means

---

[1] J. B. J. Delambre, *Histoire de l'astron. ancienne*, Vol. I, p. 547; Vol. II, p. 10. The alleged passage in the manuscripts of Theon is not found in his printed works. Delambre is inclined to ascribe the Greek sign for zero either as an abbreviation of *ouden* or as due to the special relation of the numeral omicron to the sexagesimal fractions (*op. cit.*, Vol. II, p. 14, and *Journal des sçavans* [1817], p. 539).

[2] *Bibliotheca mathematica*, Vol. XI (1910–11), p. 121–24.

of the nine signs for the units and by signs of the groups 10, 100, or
1,000, express all values through the aid of multipliers placed on the

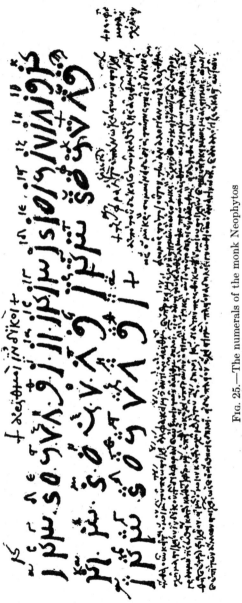

FIG. 25.—The numerals of the monk Neophytos

left. This view is supported also by the singular *arithmoi indikoi* in
the scholium of the monk *Neophytos,* which is found in the Parisian

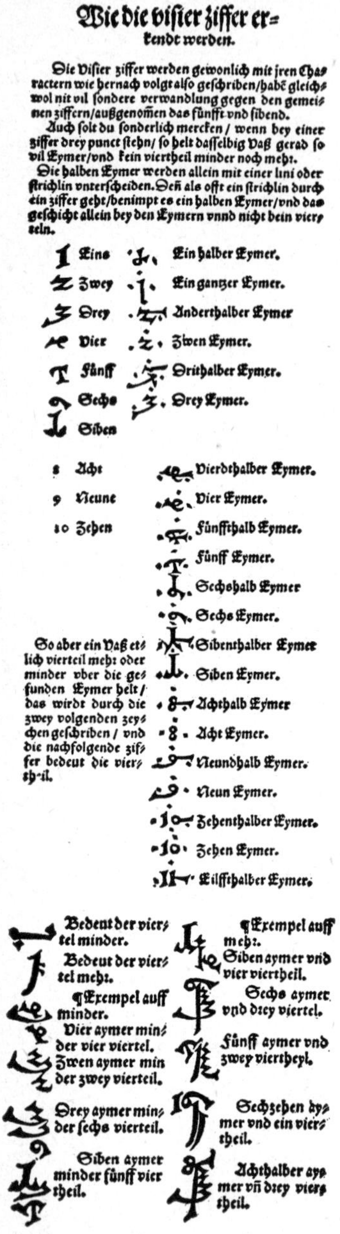

Wie die visier ziffer er=
kendt werden.

Die Visier ziffer werden gewonlich mit jren Cha=
ractern wie hernach volgt also geschriben/habe gleich=
wol nit vil sondere verwandlung gegen den gemei=
nen ziffern/außgenomen das fünfft vnd sibend.
Auch solt du sonderlich mercken / wenn bey einer
ziffer drey punct stehn/ so helt dasselbig Daß gerad so
vil Eymer/vnd kein viertheil minder noch mehr.
Die halben Eymer werden allein mit einer lini oder
strichlin vnterscheiden. Deñ als offt ein strichlin durch
ein ziffer geht/benimpt es ein halben Eymer/vnd das
geschicht allein bey den Eymern vnnd nicht bein vier=
teln.

1 Eins · **J.** , Ein halber Eymer.
2 Zwey · **j.** , Ein gantzer Eymer.
3 Drey · **Z.** Anderthalber Eymer
4 Vier · **z.** , Zwen Eymer.
5 Fünff · **Z.** Drithalber Eymer.
6 Sechs · **Z.** Drey Eymer.
7 Siben

8 Acht   **Z.** Vierdthalber Eymer.
9 Neune   **Z.** Vier Eymer.
10 Zehen   **Z.** Fünffthalb Eymer.
           T. Fünff Eymer.
           Z. Sechshalb Eymer
           Z. Sechs Eymer.

So aber ein Daß et=   **Z** Sibenthalber Eymer
lich viertel mehr oder **Z.** Siben Eymer.
minder vber die ge=
funden Eymer helt /   **Z** Achthalb Eymer
das wirdt durch die   **8** · Acht Eymer.
zwey volgenden zey=
chen geschriben / vnd   **Z** Neundhalb Eymer.
die nachfolgende zif=
fer bedeut die vier=   **Z** · Neun Eymer.
theil.
           Jo· Zehenthalber Eymer.
           Jo· Zehen Eymer.
           JL· Eilffthalber Eymer.

Bedeut der vier=   Exempel auff
tel minder.   mehr.
Bedeut der vier=   Siben aymer vnd
tel mehr.   vier viertheil.
Exempel auff   Sechs aymer
minder.   vnd drey viertel.
Vier aymer min=
der vier viertel.   Fünff aymer vnd
Zwen aymer min   zwey vierthepl.
der zwey vierteil.
Drey aymer min=   Sechzehen hy=
der sechs vierteil. mer vnd ein viers
                    theil.
Siben aymer
minder fünff vier   Achthalber ay=
theil.   mer vñ drey viers
                    theil.

Fig. 26.—From Christoff Rudolff's *Künstliche Rechnung mit der
Ziffer* (Augsburg, 1574[?]).

Library (Cod. Reg., fol. 15), for an account of which I am indebted to Prof. Brandis. The nine digits of Neophytos wholly resemble the Persian, except the 4. The digits 1, 2, 3 and 9 are found even in Egyptian number inscriptions (Kosegarten, *de Hierogl. Aegypt.*, p. 54). The nine units are enhanced tenfold, 100 fold, 1,000 fold by writing above them one, two or three zeros, as in:

$$\overset{\circ}{2}=20,\quad \overset{\circ}{24}=24,\quad \overset{\circ\circ}{5}=500,\quad \overset{\circ\circ\circ}{6}=6,000.$$

If we imagine dots in place of the zero symbols, then we have the arabic Gobar numerals."[1] Humboldt copies the scholium of Neophytos. J. L. Heiberg also has called attention to the scholium of Neophytos and to the numbering of scholia to Euclid in a Greek manuscript of the twelfth century (Codex Vindobonensis, Gr. 103), in which numerals resembling the Gobar numerals occur.[2] The numerals of the monk Neophytos (Fig. 25), of which Humboldt speaks, have received the special attention of P. Tannery.[3]

89. *Freak forms.*—We reproduce herewith from the Augsburg edition of Christoff Rudolff's *Künstliche Rechnung* a set of our numerals, and of symbols to represent such fractions

[1] *Op. cit.*, p. 227.

[2] See J. L. Heiberg's edition of *Euclid* (Leipzig, 1888), Vol. V; P. Tannery, *Revue archéol.* (3d ser., 1885), Vol. V, p. 99, also (3d ser., 1886), Vol. VII, p. 355; *Encycl des scien. math.*, Tome I, Vol. I (1904), p. 20, n. 102.

[3] *Mémoires scientifiques*, Vol. IV (Toulouse and Paris, 1920), p. 22.

and mixed numbers as were used in Vienna in the measurement of wine. We have not seen the first edition (1526) of Rudolff's book, but Alfred Nagl[1] reproduces part of these numerals from the first edition. "In the Viennese wine-cellars," says Hill, "the casks were marked according to their contents with figures of the forms given."[2] The symbols for fractions are very curious.

90. *Negative numerals.*—J. Colson[3] in 1726 claimed that, by the use of negative numerals, operations may be performed with "more ease and expedition." If 8605729398715 is to be multiplied by 389175836438, reduce these to small numbers $\overline{1}4\overline{1}433\overline{1}40\overline{1}315$ and $4\overline{1}\overline{1}2\overline{2}4\overline{1}24444\overline{2}$. Then write the multiplier on a slip of paper and place it in an inverted position, so that its first figure is just over the left-hand figure of the multiplicand. Multiply $4\times1=4$ and write down 4. Move the multiplier a place to the right and collect the two products, $4\times\overline{1}+\overline{1}\times1=\overline{5}$; write down $\overline{5}$. Move the multiplier another place to the right, then $4\times\overline{4}+\overline{1}\times\overline{1}+\overline{1}\times1=\overline{16}$; write the $\overline{1}$ in the second line. Similarly, the next product is 11, and so on. Similar processes and notations were proposed by A. Cauchy,[4] E. Selling,[5] and W. B. Ford,[6] while J. P. Ballantine[7] suggests 1 inverted, thus ı, as a sign for negative 1, so that ı$\times7=$ı3 and the logarithm $9.69897-10$ may be written ı9.69897 or ı.69897. Negative logarithmic characteristics are often marked with a negative sign placed over the numeral (Vol. II, § 476).

91. *Grouping digits in numeration.*—In the writing of numbers containing many digits it is desirable to have some symbol separating the numbers into groups of, say, three digits. Dots, vertical bars, commas, arcs, and colons occur most frequently as signs of separation.

In a manuscript, Liber algorizmi,[8] of about 1200 A.D., there appear

---

[1] *Monatsblatt der numismatischen Gesellschaft in Wien*, Vol. VII (December, 1906), p. 132.

[2] G. F. Hill, *op. cit.*, p. 53.

[3] *Philosophical Transactions*, Vol. XXXIV (1726), p. 161–74; *Abridged Transactions*, Vol. VI (1734), p. 2–4. See also G. Peano, *Formulaire mathématique*, Vol. IV (1903), p. 49.

[4] *Comptes rendus*, Vol. XI (1840), p. 796; *Œuvres* (1st ser.), Vol. V, p. 434–55.

[5] *Eine neue Rechenmaschine* (Berlin, 1887), p. 16; see also *Encyklopädie d. Math. Wiss.*, Vol. I, Part 1 (Leipzig, 1898–1904), p. 944.

[6] *American Mathematical Monthly*, Vol. XXXII (1925), p. 302.

[7] *Op. cit.*, p. 302.

[8] M. Cantor, *Zeitschrift für Mathematik*, Vol. X (1865), p. 3; G. Eneström, *Bibliotheca mathematica* (3d ser., 1912–13), Vol. XIII, p. 265.

dots to mark periods of three. Leonardo of Pisa, in his *Liber Abbaci* (1202), directs that the hundreds, hundred thousands, hundred millions, etc., be marked with an accent above; that the thousands, millions, thousands of millions, etc., be marked with an accent below.

In the 1228 edition,[1] Leonardo writes 678 935 784 105 296. Johannes de Sacrobosco (d. 1256), in his *Tractatus de arte numerandi*, suggests that every third digit be marked with a dot.[2] His commentator, Petrus de Dacia, in the first half of the fourteenth century, does the same.[3] Directions of the same sort are given by Paolo Dagomari[4] of Florence, in his *Regoluzze di Maestro Paolo dall Abbaco* and Paolo of Pisa,[5] both writers of the fourteenth century. Luca Pacioli, in his *Summa* (1494), folio 19*b*, writes 8 659 421 635 894 676; Georg Peurbach (1505),[6] 3790528614. Adam Riese[7] writes 86789325178. M. Stifel (1544)[8] writes 2329089562800. Gemma Frisius[9] in 1540 wrote 24 456 345 678. Adam Riese (1535)[10] writes 86·7·89·3·25·178. The Dutch writer, Martinus Carolus Creszfeldt,[11] in 1557 gives in his *Arithmetica* the following marking of a number:

"Exempel. || 5 8 7 4 9 3 6 2 5 3 4 || ."

---

[1] *El liber abbaci di Leonardo Pisano . . . .* da B. Boncompagni (Roma, 1857), p. 4.

[2] J. O. Halliwell, *Rara mathematica* (London, 1839), p. 5; M. Cantor, *Vorlesungen*, Vol. II (2d ed., 1913), p. 89.

[3] *Petri Philomeni de Dacia in Algorismum vulgarem Iohannis de Sacrobosco commentarius* (ed. M. Curtze; Kopenhagen, 1897), p. 3, 29; J. Tropfke, *Geschichte der Elementarmathematik* (2d ed., 1921), Vol. I, p. 8.

[4] Libri, *Histoire des sciences mathématiques en Italie*, Vol. III, p. 296–301 (Rule 1).

[5] *Ibid.*, Vol. II, p. 206, n. 5, and p. 526; Vol. III, p. 295; see also Cantor, *op. cit.*, Vol. II (2d ed., 1913), p. 164.

[6] *Opus algorithmi* (Herbipoli, 1505). See Wildermuth, "Rechnen," *Encyklopaedie des gesammten Erziehungs- und Unterrichtswesens* (Dr. K. A. Schmid, 1885).

[7] *Rechnung auff der Linien vnnd Federn* (1544); Wildermuth, "Rechnen," *Encyklopaedie* (Schmid, 1885), p. 739.

[8] Wildermuth, *op. cit.*, p. 739.

[9] *Arithmeticae practicae methodus facilis* (1540); F. Unger, *Die Methodik der praktischen Arithmetik in historischer Entwickelung* (Leipzig, 1888), p. 25, 71.

[10] *Rechnung auff d. Linien u. Federn* (1535). Taken from H. Hankel, *op. cit.* (Leipzig, 1874), p. 15.

[11] *Arithmetica* (1557). Taken from Bierens de Haan, *Bouwstoffen voor de Geschiedenis der Wis-en Natuurkundige Wetenschappen*, Vol. II (1887), p. 3.

Thomas Blundeville (1636)[1] writes 5|936|649. Tonstall[2] writes
. . . .⠀⠀⠀⠀⠀⠀⠀4 3 2 1 0
3210987654321. Clavius[3] writes 42329089562800. Chr. Rudolff[4] writes
23405639567. Johann Caramuel[5] separates the digits, as in "34:252,-
Integri.⠀⠀Partes.
341;154,329"; W. Oughtred,[6] 9|876|543|21012|345|678|9; K. Schott[7],

‖‖‖⠀⠀‖⠀⠀⠀|
769743232908956
2436; N. Barreme,[8] 254.567.804.652; W. J. G.

⠀⠀⠀III⠀⠀⠀⠀II⠀⠀I⠀⠀⠀0
Karsten,[9] 872 094,826 152,870 364,008; I. A. de Segner,[10] 5|329″|870|
325′|743|297°, 174; Thomas Dilworth,[11] 789 789 789; Nicolas Pike,[12]
⠀⠀3⠀⠀⠀⠀2⠀⠀⠀1
356;809,379;120,406;129,763; Charles Hutton,[13] 281,427,307; E.
Bézout,[14] 23, 456, 789, 234, 565, 456.

In M. Lemos' *Portuguese encyclopedia*[15] the population of New

[1] *Mr. Blundevil, His Exercises contayning eight Treatises* (7th ed., Ro. Hartwell; London, 1636), p. 106.

[2] *De Arte Svppvtandi, libri qvatvor Cvtheberti Tonstalli* (Argentorati), Colophon 1544, p. 5.

[3] *Christophori Clavii epitome arithmeticae practicae* (Romae, 1583), p. 7.

[4] *Künstliche Rechnung mit der Ziffer* (Augsburg, 1574[?]), Aiij B.

[5] *Joannis Caramvelis mathesis biceps, vetus et nova* (Companiae [southeast of Naples], 1670), p. 7. The passage is as follows: "Punctum finale (.) est, quod ponitur post unitatem: ut cùm scribimus 23. viginti tria. Comma (,) post millenarium scribitur . . . . ut cùm scribimus, 23,424. . . . . Millenarium à centenario distinguere alios populos docent Hispani, qui utuntur hoc charactere Ɩƒ, ... Hypocolon (;) millionem à millenario separat, ut cùm scribimus 2;041,311. Duo puncta ponuntur post billionem, seu millionem millionum, videlicet 34:252,341;154,329." Caramuel was born in Madrid. For biographical sketch see *Revista matemática Hispano-American*, Vol. I (1919), p. 121, 178, 203.

[6] *Clavis mathematicae* (London, 1652), p. 1 (1st ed., 1631).

[7] *Cursus mathematicus* (Herbipoli, 1661), p. 23.

[8] *Arithmétique* (new ed.; Paris, 1732), p. 6.

[9] *Mathesis theoretica elementaris atqve svblimior* (Rostochii, 1760), p. 195.

[10] *Elementa arithmeticae geometriae et calcvli geometrici* (2d ed.; Halle, 1767), p. 13.

[11] *Schoolmaster's Assistant* (22d ed.; London, 1784), p. 3.

[12] *New and Complete System of Arithmetic* (Newburyport, 1788), p. 18.

[13] "Numeration," *Mathematical and Philosophical Dictionary* (London, 1795).

[14] *Cours de mathématiques* (Paris, 1797), Vol. I, p. 6.

[15] "Portugal," *Encyclopedia Portugueza Illustrada* . . . . de Maximiano Lemos (Porto).

York City is given as "3.437:202"; in a recent Spanish encyclopedia,[1] the population of America is put down as "150·979,995."

In the process of extracting square root, two early commentators[2] on Bhāskara's *Lilavati*, namely Rama-Crishna Deva and Gangad'hara (*ca.* 1420 A.D.), divide numbers into periods of two digits in this manner, 8 8̄ 2̇ 0̄ 9̇. In finding cube roots Rama-Crishna Deva writes 1̇ 9 5̄ 3̇ 1̄ 2̄ 5̇.

92. *The Spanish "calderón."*—In Old Spanish and Portuguese numeral notations there are some strange and curious symbols. In a contract written in Mexico City in 1649 the symbols "7U291*e*" and "VIIUCCXCI*ps*" each represent 7,291 *pesos*. The U, which here resembles an O that is open at the top, stands for "thousands."[3] I. B. Richman has seen Spanish manuscripts ranging from 1587 to about 1700, and Mexican manuscripts from 1768 to 1855, all containing symbols for "thousands" resembling U or D, often crossed by one or two horizontal or vertical bars. The writer has observed that after 1600 this U is used freely both with Hindu-Arabic and with Roman numerals; *before 1600* the U occurs more commonly with Roman numerals. Karpinski has pointed out that it is used with the Hindu-Arabic numerals as early as 1519, in the accounts of the Magellan voyages. As the Roman notation does not involve the principle of local value, U played in it a somewhat larger rôle than merely to afford greater facility in the reading of numbers. Thus VIUCXV equals 6×1,000+115. This use is shown in manuscripts from Peru of 1549 and 1543,[4] in manuscripts from Spain of 1480[5] and 1429.[6]

We have seen the corresponding *type symbol* for 1,000 in Juan Perez de Moya,[7] in accounts of the coining in the Real Casa de Moneda de

[1] "América," *Enciclopedia illustrada segui Diccionario universal* (Barcelona).

[2] Colebrooke, *op. cit.*, p. 9, 12, xxv, xxvii.

[3] F. Cajori, "On the Spanish Symbol U for 'thousands,' " *Bibliotheca mathematica*, Vol. XII (1912), p. 133.

[4] *Cartas de Indias publícalas por primera vez el Ministerio de Fomento* (Madrid, 1877), p. 502, 543, *facsimiles X* and *Y*.

[5] José Gonzalo de las Casas, *Anales de la Paleografia Española* (Madrid, 1857), Plates 87, 92, 109, 110, 113, 137.

[6] Liciniano Saez, *Demostración Histórica del verdadero valor de todas las monedas que corrian en Castilla durante el Reynado del Señor Don Enrique III* (Madrid, 1796), p. 447. See also Colomera y Rodríguez, Venancio, *Paleografia castellana* (1862).

[7] *Aritmetica practica* (14th ed.; Madrid, 1784), p. 13 (1st ed., 1562).

Mexico (1787), in eighteenth-century books printed in Madrid,[1] in the *Gazetas de Mexico* of 1784 (p. 1), and in modern reprints of seventeenth-century documents.[2] In these publications the printed symbol resembles the Greek sampi ϡ for 900, but it has no known connection with it. In books printed in Madrid[3] in 1760, 1655, and 1646, the symbol is a closer imitation of the written U, and is curiously made up of the two small printed letters, *l*, *f*, each turned halfway around. The two inverted letters touch each other below, thus ↄʃ. Printed symbols representing a distorted U have been found also in some Spanish arithmetics of the sixteenth century, particularly in that of Gaspard de Texeda[4] who writes the number 103,075,102,300 in the Castellanean form c.iijU.75q̂s c.ijU300 and also in the algoristic form 103U075q̂s 102U300. The Spaniards call this symbol and also the sampi-like symbol a *calderón*.[5] A non-Spanish author who explains the *calderón* is Johann Caramuel,[6] in 1670.

93. The present writer has been able to follow the trail of this curious symbol U from Spain to Northwestern Italy. In Adriano Cappelli's *Lexicon* is found the following: "In the liguric documents of the second half of the fifteenth century we found in frequent use, to indicate the multiplication by 1,000, in place of M, an O crossed by a horizontal line."[7] This closely resembles some forms of our Spanish symbol U. Cappelli gives two facsimile reproductions[8] in

[1] Liciniano Saez, *op. cit.*

[2] Manuel Danvila, *Boletin de la Real Academia de la Historia* (Madrid, 1888), Vol. XII, p. 53.

[3] *Cuentas para todas, compendio arithmético, e Histórico* . . . . su autor D. Manuel Recio, Oficial de la contaduría general de postos del Reyno (Madrid, 1760); *Teatro Eclesiástico de la primitiva Iglesia de las Indias Occidentales* . . . . el M. Gil Gonzalez Davila, su Coronista Mayor de las Indias, y de los Reynos de las dos Castillas (Madrid, 1655), Vol. II; *Memorial, y Noticias Sacras, y reales del Imperio de las Indias Occidentales* . . . . Escriuiale por el año de 1646, Juan Diez de la Calle, Oficial Segundo de la Misma Secretaria.

[4] *Suma de Arithmetica pratica* (Valladolid, 1546), fol. iiijr.; taken from D. E. Smith, *History of Mathematics*, Vol. II (1925), p. 88. The q̂s means *quentos* (*cuentos*, "millions").

[5] In Joseph Aladern, *Diccionari popular de la Llengua Catalana* (Barcelona, 1905), we read under "Caldero": "Among ancient copyists a sign (ↄʃ) denoted a thousand."

[6] *Joannis Caramvelis Mathesis biceps vetus et nova* (Companiae, 1670), p. 7.

[7] *Lexicon Abbreviaturarum* (Leipzig, 1901), p. l.

[8] *Ibid.*, p. 436, col. 1, Nos. 5 and 6.

which the sign in question is small and is placed in the position of an exponent to the letters XL, to represent the number 40,000. This corresponds to the use of a small *c* which has been found written to the right of and above the letters XI, to signify 1,100. It follows, therefore, that the modified U was in use during the fifteenth century in Italy, as well as in Spain, though it is not known which country had the priority.

What is the origin of this *calderón?* Our studies along this line make it almost certain that it is a modification of one of the Roman

FIG. 27.—From a contract (Mexico City, 1649). The right part shows the sum of 7,291 *pesos,* 4 *tomines,* 6 *granos,* expressed in Roman numerals and the *calderón.* The left part, from the same contract, shows the same sum in Hindu-Arabic numerals and the *calderón.*

symbols for 1,000. Besides M, the Romans used for 1,000 the symbols CIↃ, T, ∞, and ⊤. These symbols are found also in Spanish manuscripts. It is easy to see how in the hands of successive generations of amanuenses, some of these might assume the forms of the *calderón.* If the lower parts of the parentheses in the forms CIↃ or CIIↃ are united, we have a close imitation of the U, crossed by one or by two bars.

94. *The Portuguese "cifrão."*—Allied to the distorted Spanish U is the Portuguese symbol for 1,000, called the *cifrão*.[1] It looks somewhat like our modern dollar mark, $. But its function in writing numbers was identical with that of the *calderón*. Moreover, we have seen forms of this Spanish "thousand" which need only to be turned through a right angle to appear like the Portuguese symbol for 1,000. Changes of that sort are not unknown. For instance, the Arabic numeral 5 appears upside down in some Spanish books and manuscripts as late as the eighteenth and nineteenth centuries.

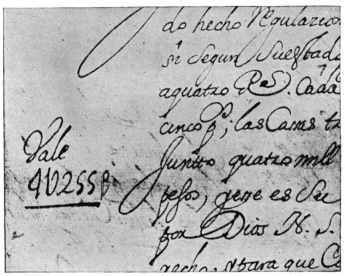

Fig. 28.—Real estate sale in Mexico City, 1718. The sum written here is 4,255 *pesos*.

95. *Relative size of numerals in tables.*—André says on this point: "In certain numerical tables, as those of Schrön, all numerals are of the same height. In certain other tables, as those of Lalande, of Callet, of Houël, of Dupuis, they have unequal heights: the 7 and 9 are prolonged downward; 3, 4, 5, 6 and 8 extend upward; while 1 and 2 do not reach above nor below the central body of the writing. . . . . The unequal numerals, by their very inequality, render the long train of numerals easier to read; numerals of uniform height are less legible."[2]

[1] See the word *cifrão* in Antonio de Moraes Silva, *Dicc. de Lingua Portuguesa* (1877); in Vieira, *Grande Dicc. Portuguez* (1873); in *Dicc. Comtemp. da Lingua Portuguesa* (1881).

[2] D. André, *Des notations mathématiques* (Paris, 1909), p. 9.

96. *Fanciful hypotheses on the origin of the numeral forms.*—A problem as fascinating as the puzzle of the origin of language relates to the evolution of the forms of our numerals. Proceeding on the tacit assumption that each of our numerals contains within itself, as a skeleton so to speak, as many dots, strokes, or angles as it represents units, imaginative writers of different countries and ages have advanced hypotheses as to their origin. Nor did these writers feel that they were indulging simply in pleasing pastime or merely contributing to mathematical recreations. With perhaps only one exception, they were as convinced of the correctness of their explanations as are circle-squarers of the soundness of their quadratures.

The oldest theory relating to the forms of the numerals is due to the Arabic astrologer Aben Ragel[1] of the tenth or eleventh century. He held that a circle and two of its diameters contained the required forms as it were in a nutshell. A diameter represents 1; a diameter and the two terminal arcs on opposite sides furnished the 2. A glance at Part I of Figure 29 reveals how each of the ten forms may be evolved from the fundamental figure.

On the European Continent, a hypothesis of the origin from dots is the earliest. In the seventeenth century an Italian Jesuit writer, Mario Bettini,[2] advanced such an explanation which was eagerly accepted in 1651 by Georg Philipp Harsdörffer[3] in Germany, who said: "Some believe that the numerals arose from points or dots," as in Part II. The same idea was advanced much later by Georges Dumesnil[4] in the manner shown in the first line of Part III. In cursive writing the points supposedly came to be written as dashes, yielding forms resembling those of the second line of Part III. The two horizontal dashes for 2 became connected by a slanting line yielding the modern form. In the same way the three horizontal dashes for 3 were joined by two slanting lines. The 4, as first drawn, resembled the 0; but confusion was avoided by moving the upper horizontal stroke into a

---

[1] J. F. Weidler, *De characteribus numerorum vulgaribus dissertatio mathematica-critica* (Wittembergae, 1737), p. 13; quoted from M. Cantor, *Kulturleben der Völker* (Halle, 1863), p. 60, 373.

[2] *Apiaria universae philosophiae, mathematicae*, Vol. II (1642), Apiarium XI, p. 5. See Smith and Karpinski, *op. cit.*, p. 36.

[3] *Delitae mathematicae et physicae* (Nürnberg, 1651). Reference from M. Sterner, *Geschichte der Rechenkunst* (München and Leipzig [1891]), p. 138, 524.

[4] "Note sur la forme des chiffres usuels," *Revue archéologique* (3d ser.; Paris, 1890), Vol. XVI, p. 342–48. See also a critical article, "Prétendues notations Pythagoriennes sur l'origine de nos chiffres," by Paul Tannery, in his *Mémoires scientifiques*, Vol. V (1922), p. 8.

vertical position and placing it below on the right. To avoid confounding the 5 and 6, the lower left-hand stroke of the first 5 was

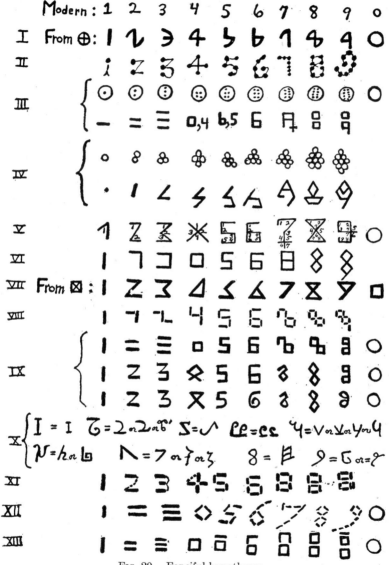

Fig. 29.—Fanciful hypotheses

changed from a vertical to a horizontal position and placed at the top of the numeral. That all these changes were accepted as historical, without an atom of manuscript evidence to support the different steps in the supposed evolution, is an indication that Baconian inductive methods of research had not gripped the mind of Dumesnil. The origin from dots appealed to him the more strongly because points played a rôle in Pythagorean philosophy and he assumed that our numeral system originated with the Pythagoreans.

Carlos le-Maur,[1] of Madrid, in 1778 suggested that lines joining the centers of circles (or pebbles), placed as shown in the first line of Part IV, constituted the fundamental numeral forms. The explanation is especially weak in accounting for the forms of the first three numerals.

A French writer, P. Voizot,[2] entertained the theory that originally a numeral contained as many angles as it represents units, as seen in Part V. He did not claim credit for this explanation, but ascribed it to a writer in the Genova Catholico Militarite. But Voizot did originate a theory of his own, based on the number of strokes, as shown in Part VI.

Édouard Lucas[3] entertains readers with a legend that Solomon's ring contained a square and its diagonals, as shown in Part VII, from which the numeral figures were obtained. Lucas may have taken this explanation from Jacob Leupold[4] who in 1727 gave it as widely current in his day.

The historian Moritz Cantor[5] tells of an attempt by Anton Müller[6] to explain the shapes of the digits by the number of strokes necessary to construct the forms as seen in Part VIII. An eighteenth-century writer, Georg Wachter,[7] placed the strokes differently, somewhat as in Part IX. Cantor tells also of another writer, Piccard,[8] who at one time had entertained the idea that the shapes were originally deter-

---

[1] *Elementos de Matématica pura* (Madrid, 1778), Vol. I, chap. i.

[2] "Les chiffres arabes et leur origine," *La nature* (2d semestre, 1899), Vol. XXVII, p. 222.

[3] *L'Arithmétique amusante* (Paris, 1895), p. 4. Also M. Cantor, *Kulturleben der Völker* (Halle, 1863), p. 60, 374, n. 116; P. Treutlein, *Geschichte unserer Zahlzeichen* (Karlsruhe, 1875), p. 16.

[4] *Theatrvm Arithmetico-Geometricvm* (Leipzig, 1727), p. 2 and Table III.

[5] *Kulturleben der Völker*, p. 59, 60.

[6] *Arithmetik und Algebra* (Heidelberg, 1833). See also a reference to this in P. Treutlein, *op. cit.* (1875), p. 15.

[7] *Naturae et Scripturae Concordia* (Lipsiae et Hafniae, 1752), chap. iv.

[8] Mémoire sur la forme et de la provenance des chiffres, *Société Vaudoise des sciences naturelles* (séances du 20 Avril et du 4 Mai, 1859), p. 176, 184. M. Cantor reproduces the forms due to Piccard; see Cantor, *Kulturleben, etc.*, Fig. 44.

mined by the number of strokes, straight or curved, necessary to express the units to be denoted. The detailed execution of this idea, as shown in Part IX, is somewhat different from that of Müller and some others. But after critical examination of his hypothesis, Piccard candidly arrives at the conclusion that the resemblances he pointed out are only accidental, especially in the case of 5, 7, and 9, and that his hypothesis is not valid.

This same Piccard offered a special explanation of the forms of the numerals as found in the geometry of Boethius and known as the "Apices of Boethius." He tried to connect these forms with letters in the Phoenician and Greek alphabets (see Part X). Another writer whose explanation is not known to us was J. B. Reveillaud.[1]

The historian W. W. R. Ball[2] in 1888 repeated with apparent approval the suggestion that the nine numerals were originally formed by drawing as many strokes as there are units represented by the respective numerals, with dotted lines added to indicate how the writing became cursive, as in Part XI. Later Ball abandoned this explanation. A slightly different attempt to build up numerals on the consideration of the number of strokes is cited by W. Lietzmann.[3] A still different combination of dashes, as seen in Part XII, was made by the German, David Arnold Crusius, in 1746.[4] Finally, C. P. Sherman[5] explains the origin by numbers of short straight lines, as shown in Part XIII. "As time went on," he says, "writers tended more and more to substitute the easy curve for the difficult straight line and not to lift the pen from the paper between detached lines, but to join the two—which we will call cursive writing."

These hypotheses of the origin of the forms of our numerals have been barren of results. The value of any scientific hypothesis lies in co-ordinating known facts and in suggesting new inquiries likely to advance our knowledge of the subject under investigation. The hypotheses here described have done neither. They do not explain the very great variety of forms which our numerals took at different times

---

[1] *Essai sur les chiffres arabes* (Paris, 1883). Reference from Smith and Karpinski, *op. cit.*, p. 36.

[2] *A Short Account of the History of Mathematics* (London, 1888), p. 147.

[3] *Lustiges und Merkwürdiges von Zahlen und Formen* (Breslau, 1922), p. 73, 74. He found the derivation in Raether, *Theorie und Praxis des Rechenunterrichts* (1. Teil, 6. Aufl.; Breslau, 1920), p. 1, who refers to H. von Jacobs, *Das Volk der Siebener-Zähler* (Berlin, 1896).

[4] *Anweisung zur Rechen-Kunst* (Halle, 1746), p. 3.

[5] *Mathematics Teacher*, Vol. XVI (1923), p. 398–401.

and in different countries. They simply endeavor to explain the numerals as they are printed in our modern European books. Nor have they suggested any fruitful new inquiry. They serve merely as entertaining illustrations of the operation of a pseudo-scientific imagination, uncontrolled by all the known facts.

97. *A sporadic artificial system.*—A most singular system of numeral symbols was described by Agrippa von Nettesheim in his *De occulta philosophia* (1531) and more fully by Jan Bronkhorst of Nimwegen in Holland who is named after his birthplace Noviomagus.[1] In 1539 he published at Cologne a tract, *De numeris*, in which he describes numerals composed of straight lines or strokes which, he claims, were used by *Chaldaei et Astrologi*. Who these Chaldeans are whom he mentions it is difficult to ascertain; Cantor conjectures that they were late Roman or medieval astrologers. The symbols are given again in a document published by M. Hostus in 1582 at Antwerp. An examination of the symbols indicates that they enable one to write numbers up into the millions in a very concise form. But this conciseness is attained at a great sacrifice of simplicity; the burden on the memory is great. It does not appear as if these numerals grew by successive steps of time; it is more likely that they are the product of some inventor who hoped, perhaps, to see his symbols supersede the older (to him) crude and clumsy contrivances.

An examination, in Figure 30, of the symbols for 1, 10, 100, and 1,000 indicates how the numerals are made up of straight lines. The same is seen in 4, 40, 400, and 4,000 or in 5, 50, 500, and 5,000.

98. *General remarks.*—Evidently one of the earliest ways of recording the small numbers, from 1 to 5, was by writing the corresponding number of strokes or bars. To shorten the record in expressing larger numbers new devices were employed, such as placing the bars representing higher values in a different position from the others, or the introduction of an altogether new symbol, to be associated with the primitive strokes on the additive, or multiplicative principle, or in some cases also on the subtractive principle.

After the introduction of alphabets, and the observing of a fixed sequence in listing the letters of the alphabets, the use of these letters

[1] See M. Cantor, *Vorlesungen über Geschichte der Mathematik*, Vol. II (2d ed.; Leipzig, 1913), p. 410; M. Cantor, *Mathemat. Beiträge zum Kulturleben der Völker* (Halle, 1863), p. 166, 167; G. Friedlein, *Die Zahlzeichen und das elementare Rechnen der Griechen und Römer* (Erlangen, 1869), p. 12; T. H. Martin, *Annali di matematica* (B. Tortolini; Rome, 1863), Vol. V, p. 298; J. C. Heilbronner, *Historia Matheseos universae* (Lipsiae, 1742), p. 735–37; J. Ruska, *Archiv für die Geschichte der Naturwissenschaften und Technik*, Vol. IX (1922), p. 112–26.

for the designation of numbers was introduced among the Syrians, Greeks, Hebrews, and the early Arabs. The alphabetic numeral systems called for only very primitive powers of invention; they made

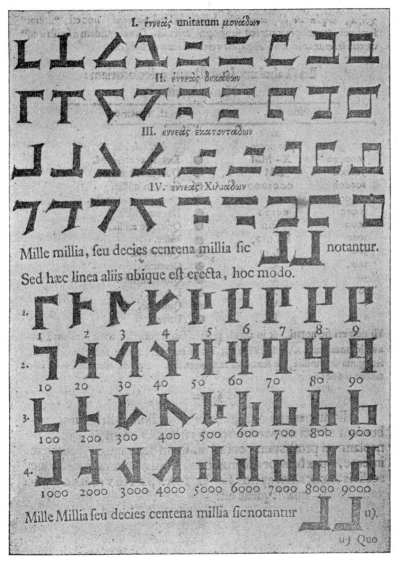

Fɪɢ. 30.—The numerals described by Noviomagus in 1539. (Taken from J. C. Heilbronner, *Historia matheseos* [1742], p. 736.)

unnecessarily heavy demands on the memory and embodied no attempt to aid in the processes of computation.

The highest powers of invention were displayed in the systems employing the principle of local value. Instead of introducing new symbols for units of higher order, this principle cleverly utilized the position of one symbol relative to others, as the means of designating different orders. Three important systems utilized this principle: the Babylonian, the Maya, and the Hindu-Arabic systems. These three were based upon different scales, namely, 60, 20 (except in one step), and 10, respectively. The principle of local value applied to a scale with a small base affords magnificent adaptation to processes of computation. Comparing the processes of multiplication and division which we carry out in the Hindu-Arabic scale with what the alphabetical systems or the Roman system afforded places the superiority of the Hindu-Arabic scale in full view. The Greeks resorted to abacal computation, which is simply a primitive way of observing local value in computation. In what way the Maya or the Babylonians used their notations in computation is not evident from records that have come down to us. The scales of 20 or 60 would call for large multiplication tables.

The origin and development of the Hindu-Arabic notation has received intensive study. Nevertheless, little is known. An outstanding fact is that during the past one thousand years no uniformity in the shapes of the numerals has been reached. An American is sometimes puzzled by the shape of the number 5 written in France. A European traveler in Turkey would find that what in Europe is a 0 is in Turkey a 5.

99. *Opinion of Laplace.*—Laplace[1] expresses his admiration for the invention of the Hindu-Arabic numerals and notation in this wise: "It is from the Indians that there has come to us the ingenious method of expressing all numbers, in ten characters, by giving them, at the same time, an absolute and a place value; an idea fine and important, which appears indeed so simple, that for this very reason we do not sufficiently recognize its merit. But this very simplicity, and the extreme facility which this method imparts to all calculation, place our system of arithmetic in the first rank of the useful inventions. How difficult it was to invent such a method one can infer from the fact that it escaped the genius of Archimedes and of Apollonius of Perga, two of the greatest men of antiquity."

[1] *Exposition du système du monde* (6th ed.; Paris, 1835), p. 376.

III

SYMBOLS IN ARITHMETIC AND ALGEBRA
(ELEMENTARY PART)

100. In ancient Babylonian and Egyptian documents occur certain ideograms and symbols which are not attributable to particular individuals and are omitted here for that reason. Among these signs is ⌐ for square root, occurring in a papyrus found at Kahun and now at University College, London,[1] and a pair of walking legs for squaring in the Moscow papyrus.[2] These symbols and ideograms will be referred to in our "Topical Survey" of notations.

A. GROUPS OF SYMBOLS USED BY INDIVIDUAL WRITERS

GREEK: DIOPHANTUS, THIRD CENTURY A.D.

101. The unknown number in algebra, defined by Diophantus as containing an undefined number of units, is represented by the Greek letter s with an accent, thus s', or in the form s°. In plural cases the symbol was doubled by the Byzantines and later writers, with the addition of case endings. Paul Tannery holds that the evidence is against supposing that Diophantus himself duplicated the sign.[3] G. H. F. Nesselmann[4] takes this symbol to be final sigma and remarks that probably its selection was prompted by the fact that it was the only letter in the Greek alphabet which was not used in writing numbers. Heath favors "the assumption that the sign was a mere tachygraphic abbreviation and not an algebraical symbol like our x, though discharging much the same function."[5] Tannery suggests that the sign is the ancient letter koppa, perhaps slightly modified. Other views on this topic are recorded by Heath.

[1] Moritz Cantor, *Vorlesungen über Geschichte der Mathematik*, Vol. I, 3d ed., Leipzig, p. 94.

[2] B. Touraeff, *Ancient Egypt* (1917), p. 102.

[3] *Diophanti Alexandrini opera omnia cum Graecis commentariis* (Lipsiae, 1895), Vol. II, p. xxxiv–xlii; Sir Thomas L. Heath, *Diophantus of Alexandria* (2d ed.; Cambridge, 1910), p. 32, 33.

[4] *Die Algebra der Griechen* (Berlin, 1842), p. 290, 291.

[5] *Op. cit.*, p. 34–36.

A square, x^2, is in Diophantus' *Arithmetica* Δ^Y

A cube, x^3, is in Diophantus' *Arithmetica* K^Y

A square-square, x^4, is in Diophantus' *Arithmetica* $\Delta^Y\Delta$

A square-cube, x^5, is in Diophantus' *Arithmetica* ΔK^Y

A cube-cube, x^6, is in Diophantus' *Arithmetica* $K^Y K$

In place of the capital letters kappa and delta, small letters are sometimes used.[1] Heath[2] comments on these symbols as follows: "There is
no obvious connection between the symbol Δ^Y and the symbol s
of which it is the square, as there is between x^2 and x, and in this lies
the great inconvenience of the notation. But upon this notation no
advance was made even by late editors, such as Xylander, or by
Bachet and Fermat. They wrote N (which was short for *Numerus*) for
the s of Diophantus, Q (*Quadratus*) for Δ^Y, C (*Cubus*) for K^Y, so that we
find, for example, $1Q + 5N = 24$, corresponding to $x^2 + 5x = 24$.[3] Other
symbols were however used even before the publication of Xylander's
Diophantus, e.g., in Bombelli's *Algebra*."

102. Diophantus has no symbol for multiplication; he writes down
the numerical results of multiplication without any preliminary step
which would necessitate the use of a symbol. Addition is expressed

---

[1] From Fermat's edition of Bachet's *Diophantus* (Toulouse, 1670), p. 2,
Definition II, we quote: "Appellatvr igitur Quadratus, Dynamis, & est illius nota
δ` superscriptum habens $\bar{v}$ sic $\delta^{\bar{v}}$. Qui autem sit ex quadrato in suum latus cubus
est, cuius nota est $\dot{\kappa}$, superscriptum habens $\bar{v}$ hoc pacto $\kappa^{\bar{v}}$. Qui autem sit ex quadrato in seipsum multiplicato, quadrato-quadratus est, cuius nota est geminum δ`
habens superscriptum $\dot{v}$, hac ratione $\delta\delta^{\bar{v}}$. Qui sit quadrato in cubum qui ab eodem
latere profectus est, ducto, quadrato-cubus nominatur, nota eius $\delta\bar{\kappa}$ superscriptum
habens $\bar{v}$ sic $\delta\kappa^{\bar{v}}$. Qui ex cubo in se ducto nascitur, cubocubus vocatur, & est eius
nota geminum $\bar{\kappa}$ superscriptum habens $\bar{v}$, hoc pacto $\kappa\kappa^{\bar{v}}$. Cui vero nulla harum
proprietatum obtigit, sed constat multitudine vnitatem rationis experte, numerus
vocatur, nota eius `ς. Est et aliud signum immutabile definitorum, vnitas, cuius
nota $\bar{\mu}$ superscriptum habens $\bar{o}$ sic $\mu^{\bar{o}}$." The passage in Bachet's edition of 1621 is
the same as this.

[2] *Op. cit.*, p. 38.

[3] In Fermat's edition of Bachet's *Diophantus* (Toulouse, 1670), p. 3, Definition
II, we read: "Haec ad verbum exprimenda esse arbitratus sum potiùs quàm cum
Xilandro nescio quid aliud comminisci. Quamuis enim in reliqua versione nostra
notis ab eodem Xilandro excogitatis libenter vsus sim, quas tradam infrà. Hîc
tamen ab ipso Diophanto longiùs recedere nolui, quòd hac definitione notas ex-
plicet quibus passim libris istis vtitur ad species omnes compendio designandas, &
qui has ignoret ne quidem Graeca Diophanti legere possit. Porrò quadratum Dy-
namin vocat, quae vox potestatem sonat, quia videlicet quadratus est veluti
potestas cuius libet lineae, & passim ab Euclide, per id quod potest linea, quadratus
illius designatur. Itali, Hispanique eadem ferè de causa Censum vocant, quasi

by mere juxtaposition. Thus the polynomial X^3+13x^2+5x+2 would be in Diophantine symbols $K^Y \bar{a} \Delta^Y \overline{\iota \gamma} s \bar{\epsilon} \overset{o}{M} \bar{\beta}$, where $\overset{o}{M}$ is used to represent units and shows that $\bar{\beta}$ or 2 is the absolute term and not a part of the coefficient of s or x. It is to be noted that in Diophantus' "square-cube" symbol for x^5, and "cube-cube" symbol for x^6, the *additive* principle for exponents is employed, rather than the *multiplicative* principle (found later widely prevalent among the Arabs and Italians), according to which the "square-cube" power would mean x^6 and the "cube-cube" would mean x^9.

103. Diophantus' symbol for subtraction is "an inverted Ψ with the top shortened, $\wedge$." Heath pertinently remarks: "As Diophantus used no distinct sign for $+$, it is clearly necessary, in order to avoid confusion, that all the negative terms in an expression, should be placed together after all the positive terms. And so in fact he does place them."[1] As regards the origin of this sign $\wedge$, Heath believes that the explanation which is quoted above from the Diophantine text as we have it is not due to Diophantus himself, but is "an explanation made by a scribe of a symbol which he did not understand." Heath[2] advances the hypothesis that the symbol originated by placing a | within the uncial form $\wedge$, thus yielding $\wedge$. Paul Tannery,[3] on the other hand, in 1895 thought that the sign in question was adapted from the old letter sampi $\wp$, but in 1904 he[4] concluded that it was rather a conventional abbreviation associated with the root of a certain Greek verb. His considerations involve questions of Greek grammar and were prompted by the appearance of the Diophantine sign

---

dicas redditum, prouentúmque, quòd à latere seu radice, tanquam à feraci solo quadratus oriatur. Inde factum vt Gallorum nonnulli & Germanorum corrupto vocabulo zenzum appellarint. Numerum autem indeterminatum & ignotum, qui & aliarum omnium potestatum latus esse intelligitur, Numerum simpliciter Diophantus appellat. Alij passim Radicem, vel latus, vel rem dixerunt, Itali patrio vocabulo Cosam. Caeterùm nos in versione nostra his notis *N. Q. C. QQ. QC. CC.* designabimus Numerum, Quadratum, Cubum, Quadratoquadratum, Quadratocubum, Cubocubum. Nam quod ad vnitates certas & determinatas spectat, eis notam aliquam adscribere superuacaneum duxi, quòd hae seipsis absque vlla ambiguitate sese satis indicent. Ecquis enim cùm audit numerum 6. non statim cogitat sex vnitates? Quid ergo necesse est sex vnitates dicere, cùm sufficiat dicere, sex? . . . . " This passage is the same as in Bachet's edition of 1621.

[1] Heath, *op. cit.*, p. 42.

[2] *Ibid.*, p. 42, 43.

[3] Tannery, *op. cit.*, Vol. II, p. xli.

[4] *Bibliotheca mathematica* (3d ser.), Vol. V, p. 5–8.

of subtraction in the critical notes to Schöne's edition[1] of the *Metrica* of Heron.

For equality the sign in the archetypal manuscripts seems to have been ι^σ; "but copyists introduced a sign which was sometimes confused with the sign ЧΙ" (Heath).

104. The notation for division comes under the same head as the notation for fractions (see § 41). In the case of unit fractions, a double accent is used with the denominator: thus $\gamma'' = \frac{1}{3}$. Sometimes a simple accent is used; sometimes it appears in a somewhat modified form as $\wedge$, or (as Tannery interprets it) as χ : thus $y^\chi = \frac{1}{3}$. For $\frac{1}{2}$ appear the symbols $\angle'$ and $\smile$, the latter sometimes without the dot. Of fractions that are not unit fractions, $\frac{2}{3}$ has a peculiar sign ω of its own, as was the case in Egyptian notations. "Curiously enough," says Heath, "it occurs only four times in Diophantus." In some old manuscripts the denominator is written above the numerator, in some rare cases. Once we find $\iota\epsilon^\delta = \frac{15}{14}$, the denominator taking the position where we place exponents. Another alternative is to write the numerator first and the denominator after it in the same line, marking the denominator with a submultiple sign in some form: thus, $\overline{\gamma}\delta' = \frac{3}{4}$.[2] The following are examples of fractions from Diophantus:

From v. 10: $\dfrac{\iota\beta}{\iota\zeta} = \dfrac{17}{12}$ From v. 8, Lemma: $\bar{\beta}\angle's' = 2\,\tfrac{1}{2}\,\tfrac{1}{6}$

From iv. 3: $s\,\chi\bar{\eta} = \dfrac{8}{x}$ From iv. 15: $\Delta^Y\chi\,\overline{\sigma\nu} = \dfrac{250}{x^2}$

From vi. 12: $\Delta^Y\bar{\xi}\overset{\circ}{M},\overline{\beta\phi\kappa}\ \dot{\epsilon}\nu\ \gamma o\rho\dot{\iota}\omega\ \Delta^Y\Delta\bar{a}\overset{\circ}{M}\,)\,\wedge\Delta^Y\bar{\xi}$

$$= (60x^2 + 2{,}520)/(x^4 + 900 - 60x^2)\ .$$

105. The fact that Diophantus had only one symbol for unknown quantity affected considerably his mode of exposition. Says Heath: "This limitation has made his procedure often very different from our modern work." As we have seen, Diophantus used but few symbols. Sometimes he ignored even these by describing an operation in words, when the symbol would have answered as well or better. Considering the amount of symbolism used, Diophantus' algebra may be designated as "syncopated."

---

[1] *Heronis Alexandrini opera*, Vol. III (Leipzig, 1903), p. 156, l. 8, 10. The manuscript reading is $\mu o\nu\acute{a}\delta\omega\nu\ o\delta\tau\iota\delta'$, the meaning of which is $74 - \frac{1}{14}$.

[2] Heath, *op. cit.*, p. 45, 47.

106. We begin with a quotation from H. T. Colebrooke on Hindu algebraic notation:[1] "The Hindu algebraists use abbreviations and initials for symbols: they distinguish negative quantities by a dot, but have not any mark, besides the absence of the negative sign, to discriminate a positive quantity. No marks or symbols (other than abbreviations of words) indicating operations of addition or multiplication, etc., are employed by them: nor any announcing equality[2] or relative magnitude (greater or less). . . . . A fraction is indicated by placing the divisor under the dividend, but without a line of separation. The two sides of an equation are ordered in the same manner, one under the other. . . . . The symbols of unknown quantity are not confined to a single one: but extend to ever so great a variety of denominations: and the characters used are the initial syllables of the names of colours, excepting the first, which is the initial of *yávat-távat*, as much as."

107. In Brahmagupta,[3] and later Hindu writers, abbreviations occur which, when transliterated into our alphabet, are as follows:

ru for *rupa*, the absolute number
ya for *yávat-távat*, the (first) unknown
ca for *calaca* (black), a second unknown
ni for *nílaca* (blue), a third unknown
pi for *pítaca* (yellow), a fourth unknown
pa for *pandu* (white), a fifth unknown
lo for *lohita* (red), a sixth unknown
c for *carani*, surd, or square root
ya v for x^2, the *v* being the contraction for
varga, square number

108. In Brahmagupta,[4] the division of *ru* 3 *c* 450 *c* 75 *c* 54 by *c* 18 *c* 3 (i.e., $3 + \sqrt{450} + \sqrt{75} + \sqrt{54}$ by $\sqrt{18} + \sqrt{3}$) is carried out as follows: "Put *c* 18 *c* 3. The dividend and divisor, multiplied by this, make *ru* 75 *c* 625. The dividend being then divided by the single surd
ru 15
constituting the divisor, the quotient is *ru* 5 *c* 3."

---

[1] H. T. Colebrooke, *Algebra, with Arithmetic and Mensuration from the Sanscrit of Bramegupta and Bháscara* (London, 1817), p. x, xi.

[2] The Bakhshālī MS (§ 109) was found after the time of Colebrooke and has an equality sign.

[3] *Ibid.*, p. 339 ff.

[4] *Brahme-sphuṭa-sidd'hánta*, chap. xii. Translated by H. T. Colebrooke in *op. cit.* (1817), p. 277–378; we quote from p. 342.

In modern symbols, the statement is, substantially: Multiply dividend and divisor by $\sqrt{18}-\sqrt{3}$; the products are $75+\sqrt{675}$ and 15; divide the former by the latter, $5+\sqrt{3}$.

"Question 16.[1] When does the residue of revolutions of the sun, less one, fall, on a Wednesday, equal to the square root of two less than the residue of revolutions, less one, multiplied by ten and augmented by two?

"The value of residue of revolutions is to be here put square of *yávat-távat* with two added: *ya v* 1 *ru* 2 is the residue of revolutions.

Sanskrit characters or letters, by which the Hindus denote the unknown quantities in their notation, are the following: पा, का, नी, पी, लो.

Fig. 31.—Sanskrit symbols for unknowns. (From Charles Hutton, *Mathematical Tracts*, II, 167.) The first symbol, *pa*, is the contraction for "white"; the second, *ca*, the initial for "black"; the third, *ni*, the initial for "blue"; the fourth, *pi*, the initial for "yellow"; the fifth, *lo*, for "red."

This less two is *ya v* 1; the square root of which is *ya* 1. Less one, it is *ya* 1 *ru* $\overset{\bullet}{1}$; which multiplied by ten is *ya* 10 *ru* $\overset{\bullet}{10}$; and augmented by two, *ya* 10 *ru* $\overset{\bullet}{8}$. It is equal to the residue of revolutions *ya v* 1 *ru* 2 less one; viz. *ya v* 1 *ru* 1. Statement of both sides $\dfrac{ya\ v\ 0\ ya\ 10\ ru\ \overset{\bullet}{8}}{ya\ v\ 1\ ya\ \ 0\ ru\ 1}$. Equal subtraction being made conformably to rule 1 there arises *ya v* 1 $\dfrac{ru\ \overset{\bullet}{9}}{ya\ \overset{\bullet}{10}}$.

Now, from the absolute number ($\overset{\bullet}{9}$), multiplied by four times the [coefficient of the] square ($3\overset{\bullet}{6}$), and added to (100) the square of the [coefficient of the] middle term (making consequently 64), the square root being extracted (8), and lessened by the [coefficient of the] middle term ($1\overset{\bullet}{0}$), the remainder is 18 divided by twice the [coefficient of the] square (2), yields the value of the middle term 9. Substituting with this in the expression put for the residue of revolutions, the answer comes out, residue of revolutions of the sun 83. Elapsed period of days deduced from this, 393, must have the denominator in least terms added so often until it fall on Wednesday."

[1] Colebrooke, *op. cit.*, p. 346. The abbreviations *ru, c, ya, ya v, ca, ni*, etc., are transliterations of the corresponding letters in the Sanskrit alphabet.

Notice that $\dfrac{ya\ v\ 0\ ya\ 10\ ru\ \overset{\cdot}{8}}{ya\ v\ 1\ ya\ \ 0\ ru\ 1}$ signifies $0x^2+10x-8=x^2+0x+1$.

Brahmagupta gives[1] the following equation in three unknown quantities and the expression of one unknown in terms of the other two:

"ya 197 ca 1$\overset{\cdot}{6}$44 ni $\overset{\cdot}{1}$ ru 0

ya 0 ca 0 ni 0 ru 6302.

Equal subtraction being made, the value of $y\acute{a}vat$-$t\acute{a}vat$ is

ca 1644 ni 1 ru 6302 ."

(ya) 197

In modern notation:

$$197x-1644y-z+0=0x+0y+0z+6302\ ,$$

whence,

$$x=\frac{1644y+z+6302}{197}\ .$$

HINDU: THE BAKHSHĀLĪ MS

109. The so-called Bakhshālī MS, found in 1881 buried in the earth near the village of Bakhshālī in the northwestern frontier of India, is an arithmetic written on leaves of birch-bark, but has come down in mutilated condition. It is an incomplete copy of an older manuscript, the copy having been prepared, probably about the eighth, ninth, or tenth century. "The system of notation," says A. F. Rudolph Hoernle,[2] "is much the same as that employed in the arithmetical works of Brahmagupta and Bhāskara. There is, however, a very important exception. The *sign for the negative* quantity is a cross ($+$). It looks exactly like our modern sign for the positive quantity, but it is placed after the number which it qualifies. Thus $\dfrac{12\ 7+}{1\ \ 1}$ means $12-7$ (i.e. 5). This is a sign which I have not met with in any other Indian arithmetic. . . . . The sign now used is a dot placed over the number to which it refers. Here, therefore, there appears to be a mark of great antiquity. As to its origin I am not able to suggest any satisfactory explanation. . . . . A *whole* number, when it occurs in an arithmetical operation, as may be seen from the above given example, is indicated by placing the number 1 under it. This, however, is

[1] Colebrooke, *op. cit.*, p. 352.

[2] "The Bakhshālī Manuscript," *Indian Antiquary*, Vol. XVII (Bombay, 1888), p. 33–48, 275–79; see p. 34.

a practice which is still occasionally observed in India. . . . . The following statement from the first example of the twenty-fifth *sùtra* affords a good example of the system of notation employed in the Bakhshālī arithmetic:

$$\begin{array}{|ccc l|} \dot{1} & 1 & 1 & 1 & \textit{bhâ } 32 \\ & 1 & 1 & 1 & \\ & 3+ & 3+ & 3+ & \end{array} \quad \textit{phalaṁ } 108$$

Here the initial dot is used much in the same way as we use the letter x to denote the unknown quantity, the value of which is sought. The number 1 under the dot is the sign of the whole (in this case, unknown) number. A fraction is denoted by placing one number under the other without any line of separation; thus $\frac{1}{3}$ is $\frac{1}{3}$, i.e. one-third. A mixed number is shown by placing the three numbers under one another; thus $\begin{smallmatrix}1\\1\\3\end{smallmatrix}$ is $1+\frac{1}{3}$ or $1\frac{1}{3}$, i.e. one and one-third. Hence $\begin{smallmatrix}1\\1\\3+\end{smallmatrix}$ means $1-\frac{1}{3}$ $\left(\text{i.e. } \frac{2}{3}\right)$. Multiplication is usually indicated by placing the numbers side by side; thus

$$\begin{array}{|cc|} 5 & 32 \\ 8 & 1 \end{array} \quad \textit{phalaṁ } 20$$

means $\frac{5}{8}\times 32 = 20$. Similarly $\begin{smallmatrix}1 & 1 & 1\\1 & 1 & 1\\3+ & 3+ & 3+\end{smallmatrix}$ means $\frac{2}{3}\times\frac{2}{3}\times\frac{2}{3}$ or $\left(\frac{2}{3}\right)^3$, i.e. $\frac{8}{27}$. *Bhâ* is an abbreviation of *bhâga*, 'part,' and means that the number preceding it is to be treated as a denominator. Hence $\begin{smallmatrix}1 & 1 & 1\\1 & 1 & 1\\3+ & 3+ & 3+\end{smallmatrix}$ *bhâ* means $1\div\frac{8}{27}$ or $\frac{27}{8}$. The whole statement, therefore,

$$\begin{array}{|cccc l|} \dot{1} & 1 & 1 & 1 & \\ & 1 & 1 & 1 & \textit{bhâ } 32 \quad \textit{phalaṁ } 108 \ , \\ & 3+ & 3+ & 3+ & \end{array}$$

means $\frac{27}{8}\times 32 = 108$, and may be thus explained,—'a certain number is found by dividing with $\frac{8}{27}$ and multiplying with 32; that number is 108.' The dot is also used for another purpose, namely as one of the ten fundamental figures of the decimal system of notation, or the zero (0123456789). It is still so used in India for both purposes, to indicate the unknown quantity as well as the naught. . . . . The Indian dot, unlike our modern zero, is not properly a numerical figure at all. It is simply a sign to indicate an empty place or a hiatus. This is clearly shown by its name *sûnya*, 'empty.' . . . . Thus the two figures 3 and 7, placed in juxtaposition (37), mean 'thirty-seven,' but with an 'empty space' interposed between them (3 7), they mean 'three hundred and seven.' To prevent misunderstanding the presence of the 'empty space' was indicated by a dot (3.7); or by what is now the zero (307). On the other hand, occurring in the statement of a problem, the 'empty place' could be filled up, and here the dot which marked its presence signified a 'something' which was to be discovered and to be put in the empty place. . . . . In its double signification, which still survives in India, we can still discern an indication of that country as its birthplace. . . . . The operation of multiplication alone is not indicated by any special sign. Addition is indicated by *yu* (for *yuta*), subtraction by + (*ka* for *kanita?*) and division by *bhâ* (for *bhâga*). The whole operation is commonly enclosed between lines (or sometimes double lines), and the result is set down outside, introduced by *pha* (for *phala*)." Thus, *pha* served as a sign of equality.

FIG. 32.—From Bakhshālī arithmetic (G. R. Kaye, *Indian Mathematics* [1915], p. 26; R. Hoernle, *op. cit.*, p. 277).

The problem solved in Figure 32 appears from the extant parts to have been: Of a certain quantity of goods, a merchant has to pay, as duty, $\frac{1}{3}$, $\frac{1}{4}$, and $\frac{1}{5}$ on three successive occasions. The total duty is 24. What was the original quantity of his goods? The solution appears in the manuscript as follows: "Having subtracted the series from one," we get $\frac{2}{3}$, $\frac{3}{4}$, $\frac{4}{5}$; these multiplied together give $\frac{2}{5}$; that again, subtracted from 1 gives $\frac{3}{5}$; with this, after having divided (i.e., inverted, $\frac{5}{3}$), the total duty (24) is multiplied, giving 40; that is the original amount. Proof: $\frac{2}{5}$ multiplied by 40 gives 16 as the remainder. Hence the original amount is 40. Another proof: 40 multiplied by $1-\frac{1}{3}$ and $1-\frac{1}{4}$ and $1-\frac{1}{5}$ gives the result 16; the deduction is 24; hence the total is 40.

110. Bhāskara speaks in his *Lilavati*[1] of squares and cubes of numbers and makes an allusion to the raising of numbers to higher powers than the cube. Ganesa, a sixteenth-century Indian commentator of Bhāskara, specifies some of them. Taking the words *varga* for square of a number, and *g'hana* for cube of a number (found in Bhāskara and earlier writers), Ganesa explains[2] that the product of four like numbers is the square of a square, *varga-varga;* the product of six like numbers is the cube of a square, or square of a cube, *varga-g'hana* or *g'hana-varga;* the product of eight numbers gives *varga-varga-varga;* of nine, gives the cube of a cube, *g'hana-g'hana.* The fifth power was called *varga-g'hana-gháta;* the seventh, *varga-varga-g'hana-ghâta.*

111. It is of importance to note that the higher powers of the unknown number are built up on the principle of involution, except the powers whose index is a prime number. According to this principle, indices are multiplied. Thus *g'hana-varga* does not mean $n^3 \cdot n^2 = n^5$, but $(n^3)^2 = n^6$. Similarly, *g'hana-g'hana* does not mean $n^3 \cdot n^3 = n^6$, but $(n^3)^3 = n^9$. In the case of indices that are prime, as in the fifth and seventh powers, the multiplicative principle became inoperative and the additive principle was resorted to. This is indicated by the word *ghâta* ("product"). Thus, *varga-g'hana-ghâta* means $n^2 \cdot n^3 = n^5$.

In the application, whenever possible, of the multiplicative principle in building up a symbolism for the higher powers of a number, we see a departure from Diophantus. With Diophantus the symbol for x^2, followed by the symbol for x^3, meant x^5; with the Hindus it meant x^6. We shall see that among the Arabs and the Europeans of the thirteenth to the seventeenth centuries, the practice was divided, some following the Hindu plan, others the plan of Diophantus.

112. In Bhāskara, when unlike colors (dissimilar unknown quantities, like x and y) are multiplied together, the result is called *bhavita* ("product"), and is abbreviated *bha*. Says Colebrooke: "The product of two unknown quantities is denoted by three letters or syllables, as *ya.ca bha, ca.ni bha,* etc. Or, if one of the quantities be a higher power, more syllables or letters are requisite; for the square, cube, etc., are likewise denoted by the initial syllables, *va, gha, va-va, va-gha, gha-gha,*[3] etc. Thus *ya va · ca gha bha* will signify the square of the

---

[1] Colebrooke, *op. cit.*, p. 9, 10.

[2] *Ibid.*, p. 10, n. 3; p. 11.

[3] *Gha-gha* for the sixth, instead of the ninth, power, indicates the use here of the additive principle.

first unknown quantity multiplied by the cube of the second. A dot is, in some copies of the text and its commentaries, interposed between the factors, without any special direction, however, for this notation."[1] Instead of *ya va* one finds in Brahmagupta and Bhāskara also the severer contraction *ya v;* similarly, one finds *cav* for the square of the second unknown.[2]

It should be noted also that "equations are not ordered so as to put all the quantities positive; nor to give precedence to a positive term in a compound quantity: for the negative terms are retained, and even preferably put in the first place."[3]

According to N. Ramanujacharia and G. R. Kaye,[4] the content of the part of the manuscript shown in Figure 33 is as follows: The

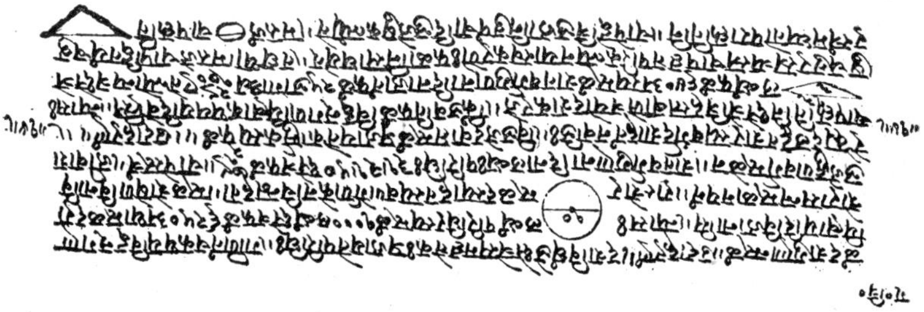

Fig. 33.—Śridhara's *Triśātikā*. Śridhara was born 991 A.D. He is cited by Bhāskara; he explains the "Hindu method of completing the square" in solving quadratic equations.

circumference of a circle is equal to the square root of ten times the square of its diameter. The area is the square root of the product of ten with the square of half the diameter. Multiply the quantity whose square root cannot be found by any large number, take the square root of the product, leaving out of account the remainder. Divide it by the square root of the factor. To find the segment of a circle, take the sum of the chord and arrow, multiply it by the arrow, and square the product. Again multiply it by ten-ninths and extract its square root. Plane figures other than these areas should be calculated by considering them to be composed of quadrilaterals, segments of circles, etc.

[1] *Op. cit.*, p. 140, n. 2; p. 141. In this quotation we omitted, for simplicity, some of the accents found in Colebrooke's transliteration from the Sanskrit.

[2] *Ibid.*, p. 63, 140, 346.

[3] *Ibid.*, p. xii.

[4] *Bibliotheca mathematica* (3d ser.), Vol. XIII (1912–13), p. 206, 213, 214.

113. *Bhāskara Áchábrya, "Lilavati,"*[1] 1150 *A.D.*—"Example: Tell me the fractions reduced to a common denominator which answer to three and a fifth, and one-third, proposed for addition; and those which correspond to a sixty-third and a fourteenth offered for subtraction. Statement:

$$\begin{array}{ccc} 3 & 1 & 1 \\ 1 & 5 & 3 \end{array}$$

Answer: Reduced to a common denominator

$$\frac{45}{15} \quad \frac{3}{15} \quad \frac{5}{15}\cdot \quad \text{Sum } \frac{53}{15}\cdot$$

Statement of the second example:

$$\frac{1}{63} \quad \frac{1}{14}\cdot$$

Answer: The denominator being abridged, or reduced to least terms, by the common measure seven, the fractions become

$$\frac{1}{9} \quad \frac{1}{2}\cdot$$

Numerator and denominator, multiplied by the abridged denominators, give respectively $\frac{2}{126}$ and $\frac{9}{126}$. Subtraction being made, the difference is $\frac{7}{126}$."

114. *Bhāskara Áchábrya, "Vija-Ganita."*[2]—"Example: Tell quickly the result of the numbers three and four, negative or affirmative, taken together: .... The characters, denoting the quantities known and unknown, should be first written to indicate them generally; and those, which become negative, should be then marked with a dot over them. Statement:[3] 3·4. Adding them, the sum is found 7. Statement: 3̇·4̇. Adding them, the sum is 7̇. Statement: 3·4̇. Taking the difference, the result of addition comes out 1̇.

" 'So much as' and the colours 'black, blue, yellow and red,'[4] and others besides these, have been selected by venerable teachers for names of values of unknown quantities, for the purpose of reckoning therewith.

---

[1] Colebrooke, *op. cit.*, p. 13, 14.   [2] *Ibid.*, p. 131.

[3] In modern notation, 3+4=7, (−3)+(−4) = −7, 3+(−4) = −1.

[4] Colebrooke, *op. cit.*, p. 139.

"Example:[1] Say quickly, friend, what will affirmative one un-known with one absolute, and affirmative pair unknown less eight absolute, make, if addition of the two sets take place? . . . . State-ment:[2]

$$ya\ 1 \quad ru\ 1$$
$$ya\ 2 \quad ru\ \overset{\cdot}{8}$$

Answer: The sum is $ya\ 3 \quad ru\ \overset{\cdot}{7}$.

"When absolute number and colour (or letter) are multiplied one by the other, the product will be colour (or letter). When two, three or more homogeneous quantities are multiplied together, the product will be the square, cube or other [power] of the quantity. But, if unlike quantities be multiplied, the result is their (*bhávita*) 'to be' product or factum.

"23. Example:[3] Tell directly, learned sir, the product of the multiplication of the unknown (*yávat-távat*) five, less the absolute num-ber one, by the unknown (*yávat-távat*) three joined with the absolute two: . . . . Statement:[4]

$$ya\ 5 \quad ru\ \overset{\cdot}{1}$$
$$ya\ 3 \quad ru\ 2$$
Product: $ya\ v\ 15 \quad ya\ 7 \quad ru\ \overset{\cdot}{2}$.

"Example:[5] 'So much as' three, 'black' five, 'blue' seven, all affirmative: how many do they make with negative two, three, and one of the same respectively, added to or subtracted from them? Statement:[6]

$$ya\ 3 \quad ca\ 5 \quad ni\ 7 \quad \text{Answer: Sum } ya\ 1 \quad ca\ 2 \quad ni\ 6.$$
$$ya\ \overset{\cdot}{2} \quad ca\ \overset{\cdot}{3} \quad ni\ \overset{\cdot}{1} \quad \text{Difference } ya\ 5 \quad ca\ 8 \quad ni\ 8.$$

"Example:[7] Say, friend, [find] the sum and difference of two ir-rational numbers eight and two: . . . . after full consideration, if thou be acquainted with the sixfold rule of surds. Statement:[8] $c\ 2\ c\ 8$.

---

[1] *Ibid.*　　　[2] In modern notation, $x+1$ and $2x-8$ have the sum $3x-7$

[3] Colebrooke, *op. cit.*; p. 141, 142.

[4] In modern notation $(5x-1)(3x+2)=15x^2+7x-2$.

[5] Colebrooke, *op. cit.*, p. 144.

[6] In modern symbols, $3x+5y+7z$ and $-2x-3y-z$ have the sum $x+2y+6z$, and the difference $5x+8y+8z$.

[7] Colebrooke, *op. cit.*, p. 146.

[8] In modern symbols, the example is $\sqrt{8}+\sqrt{2}=\sqrt{18}$, $\sqrt{8}-\sqrt{2}=\sqrt{2}$. The same example is given earlier by Brahmagupta in his *Brahme-sputa-sidd'hánta*, chap. xviii, in Colebrooke, *op. cit.*, p. 341.

Answer: Addition being made, the sum is c 18. Subtraction taking place, the difference is c 2."

ARABIC: aL-KHOWÂRIZMÎ, NINTH CENTURY A.D.

115. In 772 Indian astronomy became known to Arabic scholars. As regards algebra, the early Arabs failed to adopt either the Diophantine or the Hindu notations. The famous *Algebra* of al-Khowârizmî of Bagdad was published in the original Arabic, together with an English translation, by Frederic Rosen,[1] in 1831. He used a manuscript preserved in the Bodleian Collection at Oxford. An examination of this text shows that the exposition was altogether rhetorical, i.e., devoid of all symbolism. "Numerals are in the text of the work always expressed by words: [Hindu-Arabic] figures are only used in some of the diagrams, and in a few marginal notes."[2] As a specimen of al-Khowârizmî's exposition we quote the following from his *Algebra*, as translated by Rosen:

"What must be the amount of a square, which, when twenty-one dirhems are added to it, becomes equal to the equivalent of ten roots of that square? Solution: Halve the number of the roots; the moiety is five. Multiply this by itself; the product is twenty-five. Subtract from this the twenty-one which are connected with the square; the remainder is four. Extract its root; it is two. Subtract this from the moiety of the roots, which is five; the remainder is three. This is the root of the square which you required, and the square is nine. Or you may add the root to the moiety of the roots; the sum is seven; this is the root of the square which you sought for, and the square itself is forty-nine."[3]

By way of explanation, Rosen indicates the steps in this solution, expressed in modern symbols, as follows: Example:

$$x^2+21=10x \; ; \quad x=\tfrac{10}{2}\pm\sqrt{[(\tfrac{10}{2})^2-21]}=5\pm\sqrt{25-21}=5\pm\sqrt{4}=5\pm2 \, .$$

ARABIC: aL-KARKHÎ, EARLY ELEVENTH CENTURY A.D.

116. It is worthy of note that while Arabic algebraists usually build up the higher powers of the unknown quantity on the multiplicative principle of the Hindus, there is at least one Arabic writer, al-Karkhî of Bagdad, who followed the Diophantine additive principle.[4]

[1] *The Algebra of Mohammed Ben Musa* (ed. and trans. Frederic Rosen; London, 1831). See also L. C. Karpinski, *Robert of Chester's Latin Translation of the Algebra of Al-Khowarizmi* (1915).

[2] Rosen, *op. cit..*, p. xv.   [3] *Ibid.*, p. 11.

[4] See Cantor, *op. cit.*, Vol. I (3d ed.), p. 767, 768; Heath, *op. cit.*, p. 41.

In al-Kharkî's work, the *Fakhrī*, the word *mal* means x^2, *ka^{c}b* means x^3; the higher powers are *māl māl* for x^4, *māl ka^{c}b* for x^5 (not for x^6), *ka^{c}b ka^{c}b* for x^6 (not for x^9), *māl māl ka^{c}b* for x^7 (not for x^{12}), and so on.

Cantor[1] points out that there are cases among Arabic writers where *māl* is made to stand for x, instead of x^2, and that this ambiguity is reflected in the early Latin translations from the Arabic, where the word *census* sometimes means x, and not x^2.[2]

BYZANTINE: MICHAEL PSELLUS, ELEVENTH CENTURY A.D.

117. Michael Psellus, a Byzantine writer of the eleventh century who among his contemporaries enjoyed the reputation of being the first of philosophers, wrote a letter[3] about Diophantus, in which he gives the names of the successive powers of the unknown, used in Egypt, which are of historical interest in connection with the names used some centuries later by Nicolas Chuquet and Luca Pacioli. In Psellus the successive powers are designated as the first number, the second number (square), etc. This nomenclature appears to have been borrowed, through the medium of the commentary by Hypatia, from Anatolius, a contemporary of Diophantus.[4] The association of the successive powers of the unknown with the series of natural numbers is perhaps a partial recognition of exponential values, for which there existed then, and for several centuries that followed Psellus, no adequate notation. The next power after the fourth, namely, x^5, the Egyptians called "the first undescribed," because it is neither a square nor a cube; the sixth power they called the "cube-cube"; but the seventh was "the second undescribed," as being the product of the square and the "first undescribed." These expressions for x^5 and x^7 are closely related to Luca Pacioli's *primo relato* and *secondo relato*, found in his *Summa* of 1494.[5] Was Pacioli directly or indirectly influenced by Michael Psellus?

ARABIC: IBN ALBANNA, THIRTEENTH CENTURY A.D.

118. While the early Arabic algebras of the Orient are characterized by almost complete absence of signs, certain later Arabic works on

[1] *Op. cit.*, p. 768. See also Karpinski, *op. cit.*, p. 107, n. 1.

[2] Such translations are printed by G. Libri, in his *Histoire des sciences mathématiques*, Vol. I (Paris, 1838), p. 276, 277, 305.

[3] Reproduced by Paul Tannery, *op. cit.*, Vol. II (1895), p. 37–42.

[4] See Heath, *op. cit.*, p. 2, 18.

[5] See *ibid.*, p. 41; Cantor, *op. cit.*, Vol. II (2d ed.), p. 317.

algebra, produced in the Occident, particularly that of al-Qalasâdî of Granada, exhibit considerable symbolism. In fact, as early as the thirteenth century symbolism began to appear; for example, a notation for continued fractions in al-Ḥaṣṣâr (§ 391). Ibn Khaldûn[1] states that Ibn Albanna at the close of the thirteenth century wrote a book when under the influence of the works of two predecessors, Ibn Almunᶜim and Alaḥdab. "He [Ibn Albanna] gave a summary of the demonstrations of these two works and of other things as well, concerning the technical employment of symbols[2] in the proofs, which serve at the same time in the abstract reasoning and the representation to the eye, wherein lies the secret and essence of the explication of theorems of calculation with the aid of signs." This statement of Ibn Khaldûn, from which it would seem that symbols were used by Arabic mathematicians before the thirteenth century, finds apparent confirmation in the translation of an Arabic text into Latin, effected by Gerard of Cremona (1114–87). This translation contains symbols for x and x^2 which we shall notice more fully later. It is, of course, quite possible that these notations were introduced into the text by the translator and did not occur in the original Arabic. As regards Ibn Albanna, many of his writings have been lost and none of his extant works contain algebraic symbolism.

CHINESE: CHU SHIH-CHIEH

(1303 A.D.)

119. Chu Shih-Chieh bears the distinction of having been "instrumental in the advancement of the Chinese abacus algebra to the highest mark it has ever attained."[3] The Chinese notation is interesting as being decidedly unique. Chu Shih-Chieh published in 1303 a treatise, entitled *Szu-yuen Yü-chien*, or "The Precious Mirror of the Four Elements," from which our examples are taken. An expression like $a+b+c+d$, and its square, $a^2+b^2+c^2+d^2+2ab+2ac+2ad+$

---

[1] Consult F. Woepcke, "Recherches sur l'histoire des sciences mathématiques chez les orientaux," *Journal asiatique* (5th ser.), Vol. IV (Paris, 1854), p. 369–72; Woepcke quotes the original Arabic and gives a translation in French. See also Cantor, *op. cit.*, Vol. I (3d ed.), p. 805.

[2] Or, perhaps, letters of the alphabet.

[3] Yoshio Mikami, *The Development of Mathematics in China and Japan* (Leipzig, 1912), p. 89. All our information relating to Chinese algebra is drawn from this book, p. 89–98.

$2bc+2bd+2cd$, were represented as shown in the following two illustrations:

```
                                           1
                                        2  0  2
                                             2
         1                          1  0  ✳  0  1
                                        2
     1   ✳   1                          2  0  2
                                           1
         1
```

Where we have used the asterisk in the middle, the original has the character *t'ai* ("great extreme"). We may interpret this symbolism by considering *a* located one space to the right of the asterisk (✳), *b* above, *c* to the left, and *d* below. In the symbolism for the square of $a+b+c+d$, the 0's indicate that the terms *a*, *b*, *c*, *d* do not occur in the expression. The squares of these letters are designated by the 1's two spaces from ✳. The four 2's farthest from ✳ stand for $2ab$, $2ac$, $2bc$, $2bd$, respectively, while the two 2's nearest to ✳ stand for $2ac$ and $2bd$. One is impressed both by the beautiful symmetry and by the extreme limitations of this notation.

120. Previous to Chu Shih-Chieh's time algebraic equations of only one unknown number were considered; Chu extended the process to as many as four unknowns. These unknowns or elements were called the "elements of heaven, earth, man, and thing." Mikami states that, of these, the heaven element was arranged below the known quantity (which was called "the great extreme"), the earth element to the left, the man element to the right, and the thing element above. Letting ✳ stand for the great extreme, and x, y, z, u, for heaven, earth, man, thing, respectively, the idea is made plain by the following representations:

$$\boxed{\begin{matrix} ✳ \\ 1 \end{matrix}} = x , \qquad \boxed{1 \mid ✳} = y , \qquad \boxed{✳ \mid 1} = z , \qquad \boxed{\begin{matrix} 1 \\ ✳ \end{matrix}} = u .$$

Mikami gives additional illustrations:

$$\boxed{\begin{matrix} 0_2 \mid ✳ \\ 0 \mid 0 \end{matrix}} \qquad \boxed{\begin{matrix} ✳ \mid 0 \mid 1 \\ 0 \mid 1 \end{matrix}} \qquad \boxed{\begin{matrix} 0_{-2} \mid ✳ \mid 0 \mid 1 \\ 0 \mid 1 \end{matrix}} \qquad \boxed{\begin{matrix} -2 \mid ✳ \mid 1 \\ 1 \end{matrix}} .$$

$$+2yz \qquad\qquad xz+z^2 \qquad -2yz+xz+z^2=0 \qquad -2y+x+z=0$$

Using the Hindu-Arabic numerals in place of the Chinese calculating pieces or rods, Mikami represents three equations, used by Chu, in the following manner:

	4	✳	4
−1	0	2	1
		−1	

a

1	0	✳	0	−1
		0		
		1		

b

	−1	
2	✳	0
	2	

c

In our notation, the four equations are, respectively,

$$a)\ 2x\ -\ x^2 + 4y - xy^2 + 4z + xy = 0\ ,$$
$$b)\ x^2 + \ y^2 - \ z^2 = 0\ ,$$
$$c)\ 2x + 2y\ -\ u = 0\ .$$

No sign of equality is used here. All terms appear on one side of the equation. Notwithstanding the two-dimensional character of the notation, which permits symbols to be placed above and below the starting-point, as well as to left and right, it made insufficient provision for the representation of complicated expressions and for easy methods of computation. The scheme does not lend itself easily to varying algebraic forms. It is difficult to see how, in such a system, the science of algebra could experience a rapid and extended growth. The fact that Chinese algebra reached a standstill after the thirteenth century may be largely due to its inelastic and faulty notation.

BYZANTINE: MAXIMUS PLANUDES, FOURTEENTH CENTURY A.D.

121. Maximus Planudes, a monk of the first half of the fourteenth century residing in Constantinople, brought out among his various compilations in Greek an arithmetic,[1] and also scholia to the first two books of Diophantus' *Arithmetica*.[2] These scholia are of interest to us, for, while Diophantus evidently wrote his equations in the running text and did not assign each equation a separate line, we find in Planudes the algebraic work broken up so that each step or each equation is assigned a separate line, in a manner closely resembling modern practice. To illustrate this, take the problem in Diophantus (i. 29),

[1] *Das Rechenbuch des Maximus Planudes* (Halle: herausgegeben von C. I. Gerhardt, 1865).

[2] First printed in Xylander's Latin translation of Diophantus' *Arithmetica* (Basel, 1575). These scholia in Diophantus are again reprinted in P. Tannery, *Diophanti Alexandrini opera omnia* (Lipsiae, 1895), Vol. II, p. 123–255; the example which we quote is from p. 201.

"to find two numbers such that their sum and the difference of their squares are given numbers." We give the exposition of Planudes and its translation.

Planudes		Translation	
$\bar{\kappa}$	$\bar{\pi}$	[Given the numbers],	20, 80
ἔκϑ · $\ s\bar{a}\mu^{\circ}\bar{\iota}$	$\mu^{\circ}\bar{\iota} \wedge s\bar{a}$......	Putting for the numbers,	$x+10$,
			$10-x$
τετρ · $\Delta^{Y}\bar{a}ss\bar{\kappa}\mu^{\circ}\bar{\rho}$	$\Delta^{Y}\bar{a}\mu^{\circ}\rho \wedge ss\bar{\kappa}$...	Squaring,	$x^2+20x+100$,
			$x^2+100-20x$
ὑπεροχ· $\ ss\bar{\mu}$	$\bar{\iota}^{\sigma}$· $\ \mu^{\circ}\bar{\pi}$	Taking the difference,	$40x=80$
μερ · $\ s\bar{a}$	$\bar{\iota}^{\sigma}$· $\ \mu^{\circ}\bar{\beta}$	Dividing,	$x=2$
ὑπ · $\ \mu^{\circ}\bar{\iota\beta}$	$\mu^{\circ}\bar{\eta}$·........	Result,	12, 8

ITALIAN: LEONARDO OF PISA
(1202 A.D.)

122. Leonardo of Pisa's mathematical writings are almost wholly rhetorical in mode of exposition. In his *Liber abbaci* (1202) he used the Hindu-Arabic numerals. To a modern reader it looks odd to see expressions like $\frac{1}{13} \frac{2}{11} \frac{3}{5} 42$, the fractions written before the integer in the case of a mixed number. Yet that mode of writing is his invariable practice. Similarly, the coefficient of x is written after the name for x, as, for example,[1] —"radices $\frac{1}{2}12$" for $12\frac{1}{2}x$. A computation is indicated, or partly carried out, on the margin of the page, and is inclosed in a rectangle, or some irregular polygon whose angles are right angles. The reason for the inverted order of writing coefficients or of mixed numbers is due, doubtless, to the habit formed from the study of Arabic works; the Arabic script proceeds from right to left. Influenced again by Arabic authors, Leonardo uses frequent geometric figures, consisting of lines, triangles, and rectangles to illustrate his arithmetic or algebraic operations. He showed a partiality for unit fractions; he separated the numerator of a fraction from its denominator by a fractional line, but was probably not the first to do this (§ 235). The product of a and b is indicated by *factus ex.a.b.* It has been stated that he denoted multiplication by juxtaposition,[2] but G. Eneström shows by numerous quotations from the *Liber abbaci* that such is not the case.[3] Cantor's quotation from the *Liber abbaci*,

[1] *Il liber abbaci di Leonardo Pisano* (ed. B. Boncompagni), Vol. I (Rome, 1867), p. 407.

[2] Cantor, *op. cit.*, Vol. II (2d ed.), p. 62.

[3] *Bibliotheca mathematica* (3d ser.), Vol. XII (1910–11), p. 335, 336.

"sit numerus .*a.e.c.* quaedam coniunctio quae uocetur prima, numeri vero .*d.b.f.* sit coniunctio secunda,"[1] is interpreted by him as a product, the word *coniunctio* being taken to mean "product." On the other hand, Eneström conjectures that *numerus* should be *numeri*, and translates the passage as meaning, "Let the numbers *a, e, c* be the first, the numbers *d, b, f* the second combination." If Eneström's interpretation is correct, then *a.e.c* and *d.b.f* are not products. Leonardo used in his *Liber abbaci* the word *res* for *x*, as well as the word *radix*. Thus, he speaks, "et intellige pro re summam aliquam ignotam, quam inuenire uis."[2] The following passage from the *Liber abbaci* contains the words *numerus* (for a given number), *radix* for *x*, and *census* for x^2: "Primus enim modus est, quando census et radices equantur numero. . . . . Verbi gratia: duo census, et decem radices equantur denariis 30,"[3] i.e., $2x^2 + 10x = 30$. The use of *res* for *x* is found also in a Latin translation of al-Khowârizmî's algebra,[4] due perhaps to Gerard of Cremona, where we find, "res in rem fit census," i.e., $x.x = x^2$. The word *radix* for *x* as well as *res*, and *substantia* for x^2, are found in Robert of Chester's Latin translation of al-Khowârizmî's algebra.[5] Leonardo of Pisa calls x^3 *cubus*, x^4 *census census*, x^6 *cubus cubus*, or else *census census census;* he says, " . . . . est multiplicare per cubum cubi, sicut multiplicare per censum census census."[6] He goes even farther and lets x^8 be *census census census census*. Observe that this phraseology is based on the additive principle $x^2 \cdot x^2 \cdot x^2 \cdot x^2 = x^8$. Leonardo speaks also of *radix census census.*[7]

The first appearance of the abbreviation R or R for *radix* is in his *Practica geometriae* (1220),[8] where one finds the R meaning "square root" in an expression "et minus R. 78125 dragme, et diminuta radice 28125 dragme." A few years later, in Leonardo's *Flos*,[9] one finds marginal notes which are abbreviations of passages in the text relating to square root, as follows:

[1] *Op. cit.*, Vol. I (3d ed.), p. 132.

[2] *Ibid.*, Vol. I, p. 191.

[3] *Ibid.*, Vol. I, p. 407.

[4] Libri, *Histoire des sciences mathématiques en Italie*, Vol. I (Paris, 1838), p. 268.

[5] L. C. Karpinski, *op. cit.*, p. 68, 82.

[6] *Il liber abbaci*, Vol. I, p. 447.

[7] *Ibid.*, Vol. I, p. 448.

[8] *Scritti di Leonardo Pisano* (ed. B. Boncompagni), Vol. II (Rome, 1862), p. 209.

[9] *Op. cit.*, Vol. II, p. 231. For further particulars of the notations of Leonardo of Pisa, see our §§ 219, 220, 235, 271–73, 290, 292, 318, Vol. II, § 389.

$.R.^x p^i$. $Bino.^{ij}$ for *primi* [*quidem*] *binomij radix*
$2.^i B. R.^x$ for *radix* [*quippe*] *secundi binomij*
$.Bi. 3^i. R.^x$ for *Tertij* [*autem*] *binomij radix*
$.Bi. 4^i. R.^x$ for *Quarti* [*quoque*] *binomij radix*

FRENCH: NICOLE ORESME, FOURTEENTH CENTURY A.D.

123. Nicole Oresme (*ca.* 1323–82), a bishop in Normandy, prepared a manuscript entitled *Algorismus proportionum*, of which several copies are extant.[1] He was the first to conceive the notion of fractional powers which was afterward rediscovered by Stevin. More than this, he suggested a notation for fractional powers. He considers powers of ratios (called by him *proportiones*). Representing, as does Oresme himself, the ratio 2:1 by 2, Oresme expresses $2^{\frac{1}{2}}$ by the symbolism $\boxed{\begin{array}{c} 1.p \\ \hline 2.2 \end{array}}$ and reads this *medietas* [*proportionis*] *duplae;* he expresses $(2\frac{1}{2})^{\frac{1}{4}}$ by the symbolism $\boxed{\begin{array}{c} 1.p.1 \\ \hline 4.2.2 \end{array}}$ and reads it *quarta pars* [*proportionis*] *duplae sesquialterae.* The fractional exponents $\frac{1}{2}$ and $\frac{1}{4}$ are placed to the left of the ratios affected.

H. Wieleitner adds that Oresme did not use these symbols in computation. Thus, Oresme expresses in words, ". . . . proponatur proportio, que sit due tertie quadruple; et quia duo est numerator, ipsa erit vna tertia quadruple duplicate, sev sedecuple,"[2] i.e., $4^{\frac{2}{3}} = (4^2)^{\frac{1}{3}} = 16^{\frac{1}{3}}$. Oresme writes[3] also: "Sequitur quod *.a.* moueatur velocius *.b.* in proportione, que est medietas proportionis .50. ad .49.," which means, "the velocity of a:velocity of $b = \sqrt{50}:\sqrt{49}$," the word *medietas* meaning "square root."[4]

The transcription of the passage shown in Figure 34 is as follows:

"Una media debet sic scribi $\boxed{\begin{array}{c} 1 \\ \hline 2 \end{array}}$, una tertia sic $\boxed{\begin{array}{c} 1 \\ \hline 3. \end{array}}$ et due tertie sic $\boxed{\begin{array}{c} 2 \\ \hline 3. \end{array}}$; et sic de alijs. et numerus, qui supra uirgulam, dicitur

---

[1] Maximilian Curtze brought out an edition after the MS R. 4° 2 of the Gymnasiat-Bibliothek at Thorn, under the title *Der Algorithmus Proportionum des Nicolaus Oresme* (Berlin, 1868). Our photographic illustration is taken from that publication.

[2] Curtze, *op. cit.*, p. 15.   [3] *Ibid.*, p. 24.

[4] See Eneström, *op. cit.*, Vol. XII (1911–12), p. 181. For further details see also Curtze, *Zeitschrift für Mathematik und Physik*, Vol. XIII (Suppl. 1868), p. 65 ff.

numerator, iste uero, qui est sub uirgula, dicitur denominator. 2. Proportio dupla scribitur isto modo 2.^{la}, et tripla isto modo 3.^{la}; et sic
de alijs. Proportio sesquialtera sic scribitur $\boxed{\frac{p\ 1.}{1\ 2.}}$, et sesquitert'e

$\boxed{\frac{p\ 1}{1\ 3}}$. Proportio superpartiens duas tertias scribitur sic $\boxed{\frac{p\ 2}{1\ 3.}}$.

Proportio dupla superpartiens duas quartas scribitur sic $\boxed{\frac{p\ 2}{2\ 4}}$; et

sic de alijs. 3. Medietas duple scribitur sic $\boxed{\frac{1\ p}{2\cdot 2}}$, quarta pars

duple sesquialtere scribitur sic $\boxed{\frac{1\cdot p\ 1}{4\cdot 2\cdot 2}}$; et sic de alijs."

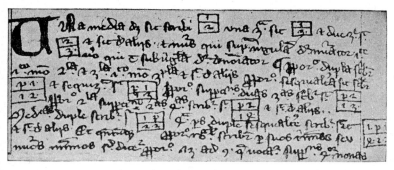

FIG. 34.—From the first page of Oresme's *Algorismus proportionum* (fourteenth century).

A free translation is as follows:

"Let a half be written $\boxed{\frac{1}{2}}$, a third $\boxed{\frac{1}{3}}$, and two-thirds $\boxed{\frac{2}{3}}$,
and so on. And the number above the line is called the 'numerator,'
the one below the line is called the 'denominator.' 2. A double ratio
is written in this manner 2.^{la}, a triple in this manner 3.^{la}, and thus in

other cases. The ratio one and one-half is written $\boxed{\frac{p\ 1.}{1\ 2.}}$, and one and

one-third is written $\boxed{\frac{p\ 1}{1\ 3}}$. The ratio one and two-thirds is written

$\boxed{\frac{p\ 2}{1\ 3.}}$. A double ratio and two-fourths are written $\boxed{\frac{p\ 2}{2\ 4}}$, and thus in other cases. 3. The square root of two is written thus $\boxed{\dfrac{1.\ p}{2\ \ 2}}$, the

fourth root of two and one-half is written thus $\boxed{\dfrac{1.\ p\ 1}{4.\ 2.\ 2}}$, and thus in other cases."

ARABIC: AL-QALASÂDÎ, FIFTEENTH CENTURY A.D.

124. Al-Qalasâdî's *Raising of the Veil of the Science of Gubar* appeared too late to influence the progress of mathematics on the European Continent. Al-Qalasâdî used ـ, the initial letter in the Arabic word *jidr*, "square root"; the symbol was written above the number whose square root was required and was usually separated from it by a horizontal line. The same symbol, probably considered this time as the first letter in *jahala* ("unknown"), was used to represent the unknown term in a proportion, the terms being separated by the sign ∴ . But in the part of al-Qalasâdî's book dealing more particularly with algebra, the unknown quantity x is represented by the letter ش, x^2 by the letter ص, x^3 by the letter ك ; these are written above their respective coefficients. Addition is indicated by juxtaposition. Subtraction is ; the equality sign, , is seen to resemble the Diophantine ι, if we bear in mind that the Arabs wrote from right to left, so that the curved stroke faced in both cases the second member of the equation. We reproduce from Woepcke's article a few samples of al-Qalasâdî's notation. Observe the peculiar shapes of the Hindu-Arabic numerals (Fig. 35).

Woepcke[1] reproduces also symbols from an anonymous Arabic manuscript of unknown date which uses symbols for the powers of x and for the powers of the reciprocal of x, built up on the additive principle of Diophantus. The total absence of data relating to this manuscript diminishes its historic value.

GERMAN: REGIOMONTANUS

(ca. 1473)

125. Regiomontanus died, in the prime of life, in 1476. After having studied in Rome, he prepared an edition of Ptolemy[2] which was issued in 1543 as a posthumous publication. It is almost purely rhetorical, as appears from the following quotation on pages 21 and 22.

[1] *Op. cit.*, p. 375–80.

[2] *Ioannis de Monte Regio et Georgii Pvrbachii epitome, in Cl. Ptolemaei magnam compositionem* (Basel, 1543). The copy examined belongs to Mr. F. E. Brasch.

By the aid of a quadrant is determined the angular elevation ACE, "que erit altitudo tropici hiemalis," and the angular elevation ACF, "que erit altitudo tropici aestivalis," it being required to find the arc EF between the two. "Arcus itaque EF, fiet distantia duorum tropi-

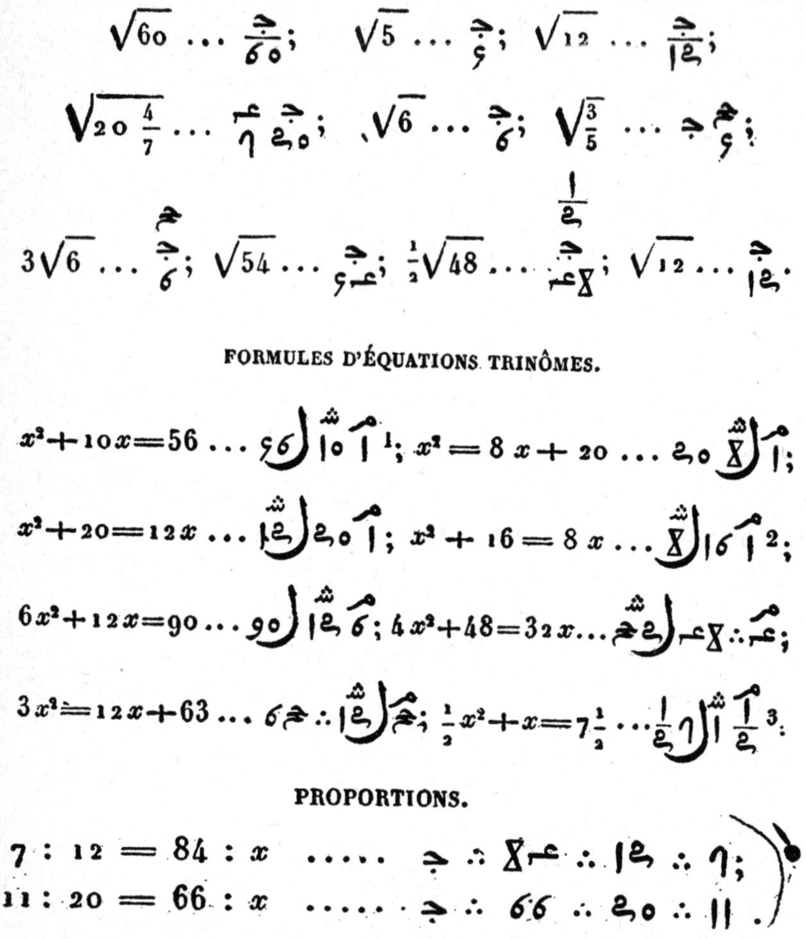

FORMULES D'ÉQUATIONS TRINÔMES.

$x^2 + 10x = 56$ ...

$x^2 = 8x + 20$ ...

$x^2 + 20 = 12x$ ...

$x^2 + 16 = 8x$ ...

$6x^2 + 12x = 90$ ...

$4x^2 + 48 = 32x$ ...

$3x^2 = 12x + 63$ ...

$\frac{1}{2}x^2 + x = 7\frac{1}{2}$ ...

PROPORTIONS.

$7 : 12 = 84 : x$ .....

$11 : 20 = 66 : x$ .....

Fig. 35.—Al-Qalasâdî's algebraic symbols. (Compiled by F. Woepcke, *Journal asiatique* [Oct. and Nov., 1854], p. 363, 364, 366.)

corum quęsita. Hâc Ptolemaeus reperit 47. graduum 42. minutorum 40. secundorum. Inuenit enim proportionem eius ad totum circulū sicut 11. ad 83, postea uerò minorem inuenerunt. Nos autem inuenimus arcum AF 65. graduum 6. minutorum, & arcum AE 18. graduum 10.

minutorum. Ideoq. nunc distantia tropicorum est 46. graduum 56. minutorum, ergo declinatio solis maxima nostro tempore est 23. graduum 28. minutorum."

126. We know, however, that in some of his letters and manuscripts symbols appear. They are found in letters and sheets containing computations, written by Regiomontanus to Giovanni Bianchini, Jacob von Speier, and Christian Roder, in the period 1463–71. These documents are kept in the Stadtbibliothek of the city of Nürnberg.[1] Regiomontanus and Bianchini designate angles thus: $\overset{u}{gr}$ 35 $\overset{u}{m}$ 17; Regiomontanus writes also: $\overline{44}$. 42′. 4″ (see also § 127).

In one place[2] Regiomontanus solves the problem: Divide 100 by a certain number, then divide 100 by that number increased by 8; the sum of the quotients is 40. Find the first divisor. Regiomontanus writes the solution thus:

In Modern Symbols

"$$\frac{100}{1\,\mathcal{C}} \qquad \frac{100}{1\,\mathcal{C}\ \text{et}\ 8} \qquad\qquad \frac{100}{x} \qquad \frac{100}{x+8}$$

$$100\,\mathcal{C}\ \text{et}\ 800 \qquad\qquad\qquad 100x+800$$
$$100\,\mathcal{C} \qquad\qquad\qquad\qquad\quad 100x$$

$$\frac{200\,\mathcal{C}\ \text{et}\ 800}{1\,\mathcal{c}\!\ell\ \text{et}\ 8\,\mathcal{C}} - 40 \qquad\qquad \frac{200x+800}{x^2+8x} = 40$$

$$40\,\mathcal{c}\!\ell\ \text{et}\ 320\,\mathcal{C} - 200\,\mathcal{C}\ \text{et}\ 800 \qquad 40x^2+320x = 200x+800.$$
$$40\,\mathcal{c}\!\ell\ \text{et}\ 120\,\mathcal{C} - 800 \qquad\qquad\ 40x^2+120x = 800$$
$$1\,\mathcal{c}\!\ell\ \text{et}\quad 3\,\mathcal{C} - \quad 20 \qquad\qquad\quad x^2+\quad 3x = \quad 20$$

$\tfrac{3}{2} \cdot \tfrac{9}{4}$ addo numerum $20\tfrac{9}{4}$—$\tfrac{8\,9}{4}$ $\qquad$ $\tfrac{3}{2} \mid \tfrac{9}{4}$ add the no. $20\tfrac{9}{4} = \tfrac{8\,9}{4}$

Radix quadrata de $\tfrac{8\,9}{4}$ minus $\tfrac{3}{2}$—$1\,\mathcal{C}$ $\qquad$ $\sqrt{\tfrac{8\,9}{4}} - \tfrac{3}{2} = x$

Primus ergo divisor fuit ℞ de $22\tfrac{1}{4}$ $\qquad$ Hence the first divisor was
$\overline{i9}\ 1\tfrac{1}{2}$." $\qquad\qquad\qquad\qquad\qquad$ $\sqrt{22\tfrac{1}{4}} - 1\tfrac{1}{2}$.

Note that "plus" is indicated here by *et*; "minus" by $\overline{i9}$, which is probably a ligature or abbreviation of "minus." The unknown quantity is represented by $\mathcal{C}$ and its square by $\mathcal{c}\!\ell$. Besides, he had a sign for equality, namely, a horizontal dash, such as was used later in Italy by Luca Pacioli, Ghaligai, and others. See also Fig. 36.

[1] Curtze, *Urkunden zur Geschichte der Mathematik im Mittelalter und der Renaissance* (Leipzig, 1902), p. 185–336 = *Abhandlungen zur Geschichte der Mathematik*, Vol. XII. See also L. C. Karpinski, *Robert of Chester's Translation of the Algebra of Al-Khowarizmi* (1915), p. 36, 37.

[2] Curtze, *op. cit.*, p. 278.

127. Figure 37[1] illustrates part of the first page of a calendar issued by Regiomontanus. It has the heading *Janer* ("January"). Farther to the right are the words *Sunne—Monde—Stainpock* ("Sun—Moon—Capricorn"). The first line is *1 A. Kl. New Jar* (i.e., "first day, A. calendar, New Year"). The second line is *2. b. 4. nō. der achtet S. Stephans.* The seven letters A, b, c, d, e, F, g, in the second column on the left, are the dominical letters of the calendars. Then come the days of the Roman calendar. After the column of saints' days comes a double column for the place of the sun. Then follow two double columns for the moon's longitude; one for the mean, the other for the

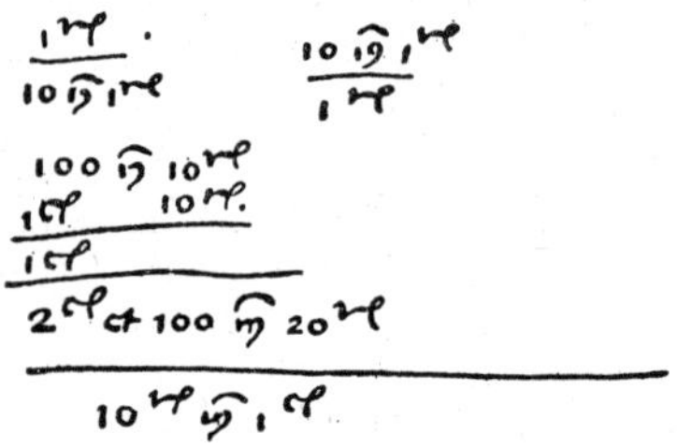

$$\frac{x}{10-x} \qquad \frac{10-x}{x}$$

$$\frac{100-10x}{x^2-10x}$$

$$x^2$$

$$\frac{2x^2+100-20x}{10x-x^2}$$

Fig. 36.—Computations of Regiomontanus, in letters of about 1460. (From manuscript, Nürnberg, fol. 23. (Taken from J. Tropfke, *Geschichte der Elementar-Mathematik* (2d ed.), Vol. II [1921], p. 14.)

true. The *S* signifies *signum* (i.e., 30°); the *G* signifies *gradus*, or "degree." The numerals, says De Morgan, are those facsimiles of the numerals used in manuscripts which are totally abandoned before the end of the fifteenth century, except perhaps in reprints. Note the shapes of the 5 and 7. This almanac of Regiomontanus and the *Compotus* of Anianus are the earliest almanacs that appeared in print.

ITALIAN: THE EARLIEST PRINTED ARITHMETIC

(1478)

128. The earliest arithmetic was printed anonymously at Treviso, a town in Northeastern Italy. Figure 38 displays the method of solving proportions. The problem solved is as follows: A courier travels from Rome to Venice in 7 days; another courier starts at the same time and travels from Venice to Rome in 9 days. The distance between Rome and Venice is 250 miles. In how many days will the

[1] Reproduced from Karl Falkenstein, *Geschichte der Buchdruckerkunst* (Leipzig, 1840), Plate XXIV, between p. 54 and 55. A description of the almanac of Regiomontanus is given by A. de Morgan in the *Companion to the British Almanac*, for 1846, in the article, "On the Earliest Printed Almanacs," p. 18–25.

couriers meet, and how many miles will each travel before meeting?
Near the top of Figure 38 is given the addition of 7 and 9, and the

Calender des Magister Johann von Kunsperk.

(Johannes Regiomontanus.)

FIG. 37.—"Calendar des Magister Johann von Kunsperk (Johannes Regiomontanus) Nürnberg um 1473."

division of 63 by 16, by the scratch method.[1] The number of days is
$3\frac{15}{16}$. The distance traveled by the first courier is found by the pro-

[1] Our photograph is taken from the *Atti dell'Accademia Pontificia de' nuovi
Lincei*, Vol. XVI (Roma, 1863), p. 570.

portion $7:250 = \frac{63}{16}:x$. The mode of solution is interesting. The 7 and
250 are written in the form of fractions. The two lines which cross

Fig. 38.—From the earliest printed arithmetic, 1478

and the two horizontal lines on the right, connecting the two numerators and the two denominators, respectively, indicate what numbers

are to be multiplied together: $7 \times 1 \times 16 = 112$; $1 \times 250 \times 63 = 15,750$. The multiplication of 250 and 63 is given; also the division of 15,750

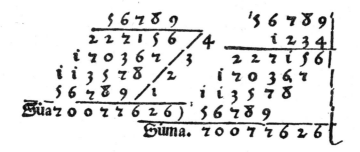

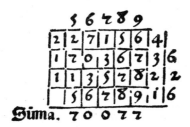

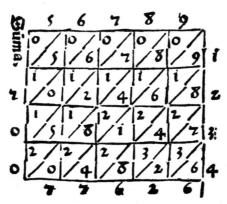

Fig. 39.—Multiplications in the Treviso arithmetic; four multiplications of 56,789 by 1,234 as given on one page of the arithmetic.

by 112, according to thè scratch method. Similarly is solved the proportion $9:250 = \frac{69}{18}:x$. Notice that the figure 1 is dotted in the same way as the Roman I is frequently dotted. Figure 39 represents other examples of multiplication.[1]

FRENCH: NICOLAS CHUQUET

(1484)

129. Over a century after Oresme, another manuscript of even greater originality in matters of algebraic notation was prepared in France, namely, *Le triparty en la science des nombres* (1484), by Nicolas Chuquet, a physician in Lyons.[2] There are no indications that he had seen Oresme's manuscripts. Unlike Oresme, he does not use fractional exponents, but he has a notation involving integral, zero, and negative exponents. The only possible suggestion for such exponential notation known to us might have come to Chuquet from the Gobar numerals, the Fihrist, and from the scholia of Neophytos (§§ 87, 88) which are preserved in manuscript in the National Library at Paris. Whether such connection actually existed we are not able to state. In any case, Chuquet elaborates the exponential notation to a completeness apparently never before dreamed of. On this subject Chuquet was about one hundred and fifty years ahead of his time; had his work been printed at the time when it was written, it would, no doubt, have greatly accelerated the progress of algebra. As it was, his name was known to few mathematicians of his time.

Under the head of "Numeration," the *Triparty* gives the Hindu-Arabic numerals in the inverted order usual with the Arabs: ".0.9.8.7.6.5.4.3.2.1." and included within dots, as was customary in late manuscripts and in early printed books. Chuquet proves addition by "casting out the 9's," arranging the figures as follows:

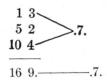

$$
\begin{array}{l}
1\ 3 \\
5\ 2 \\
10\ 4
\end{array} \Big\rangle .7.
$$

$$16\ 9.\underline{\quad\quad\quad}.7.$$

---

[1] *Ibid.*, p. 550.

[2] *Op. cit.* (publié d'après le manuscrit fonds Français N. 1346 de la Bibliothèque nationale de Paris et précédé d'une notice, par M. Aristide Marre), *Bullettino di Bibliog. e di Storia delle scienze mat. et fisiche*, Vol. XIII (1880), p. 555–659, 693–814; Vol. XIV, p. 413–60.

The addition of $\frac{2}{3}$ and $\frac{3}{4}$ is explained in the text, and the following arrangement of the work is set down by itself:[1]

130. In treating of roots he introduces the symbol R, the first letter in the French word *racine* and in the Latin *radix*. A number, say 12, he calls *racine premiere*, because 12, taken as a factor once, gives 12; 4 is a *racine seconde* of 16, because 4, taken twice as a factor, gives 16. He uses the notations $R^1.12.$ equal $.12.$, $R^2.16.$ equal $.4.$, $R^4.16.$ equal $.2.$, $R^5.32.$ equal $.2.$ To quote: "Il conuiendroit dire que racine p̅miere est entendue pour tous nombres simples Cōme qui diroit la racine premiere de $.12.$ que lon peult ainsi noter en mettant $.1.$ dessus R. en ceste maniere $R^1.12.$ cest $.12.$ Et $R^1.9.$ est $.9.$ et ainsi de tous aultres nobres. Racine seconde est celle qui posee en deux places lune soubz laultre et puys multipliee lune par laultre pduyt le nombre duquel elle est racine seconde Comme 4. et .4. qui multipliez lung par laultre sont $.16.$ ainsi la racine seconde de $.16.$ si est $.4.$ ... on le peult ainsi mettre $R^216.$ ... Et $R^5.32.$ si est $.2.$ Racine six^e se doit ainsi mettre $R^6.$ et racine septiesme ainsi $R^7.$ ... Aultres manieres de racines sont que les simples devant dictes que lon peult appeller racines composees Cōme de 14. plus $R^2180.$ dont sa racine seconde si est $.3.$ $\overline{p}.$ $R^25.$ [i.e., $\sqrt{14+\sqrt{180}}=3+\sqrt{5}$] ... cōe la racine seconde de $.14.$ $\overline{p}$ $R^2.180.$ se peult ainsi mettre $R^2.14.\overline{p}.R^2.180.$"[2]

Not only have we here a well-developed notation for roots of integers, but we have also the horizontal line, drawn underneath the binomial $14+\sqrt{180}$, to indicate aggregation, i.e., to show that the square root of the entire binomial is intended.

Chuquet took a position in advance of his time when he computed with zero as if it were an actual quantity. He obtains,[3] according to his rule, $x=2\pm\sqrt{4-4}$ as the roots of $3x^2+12=12x$. He adds: "... reste $.0.$ Donc $R^2.0.$ adioustee ou soustraicte avec $.2.$ ou de $.2.$ monte $.2.$ qui est le nōb^e que lon demande."

131. Chuquet uses $\overline{p}$ and $\overline{m}$ to designate the words *plus* and *moins*. These abbreviations we shall encounter among Italian writers. Proceeding to the development of his exponential theory and notation,

[1] Boncompagni, *Bullettino*, Vol. XIII, p. 636.

[2] *Ibid.*, p. 655.

[3] *Ibid.*, p. 805; Eneström, *Bibliotheca mathematica*, Vol. VIII (1907–8), p. 203.

he states first that a number may be considered from different points of view.[1] One is to take it without any denomination (*sans aulcune denomîacion*), or as having the denomination 0, and mark it, say, .12⁰ and .13⁰ Next a number may be considered the primary number of a continuous quantity, called "linear number" (*nombre linear*), designated .12^1 .13^1 .20^1, etc. Third, it may be a secondary or superficial number, such as 12^2. 13^2. 19^2., etc. Fourth, it may be a cubical number, such as .12^3. 15^3. 1^3., etc. "On les peult aussi entendre estre nombres quartz ou quarrez de quarrez qui seront ainsi signez . 12^4. 18^4. 30^4., etc." This nomenclature resembles that of the Byzantine monk Psellus of the eleventh century (§ 117).

Chuquet states that the ancients called his primary numbers "things" (*choses*) and marked them .ρ.; the secondary numbers they called "hundreds" and marked them .ᶜ.; the cubical numbers they indicated by □; the fourth they called "hundreds of hundreds" (*champs de champ*), for which the character was t ᶜ. This ancient nomenclature and notation he finds insufficient. He introduces a symbolism "que lon peult noter en ceste maniere $R^2.12^1$. $R^2.12^2$. $R^2.12^3$. $R^2.12^4$. etc. $R^3.12^1$. $R^3.12^2$. $R^3.12^3$. $R^3.12^4$. etc. $R^4.13^5$. $R^6.12^5$. etc." Here "$R^4.13^5$." means $\sqrt[4]{13x^5}$. He proceeds further and points out "que lon peult ainsi noter .12$^{1\,·\tilde{m}·}$ ou moins 12.," thereby introducing the notion of an exponent "minus one." As an alternative notation for this last he gives ".$\tilde{m}$.12$^1_·$," which, however, is not used again in this sense, but is given another interpretation in what follows.

From what has been given thus far, the modern reader will probably be in doubt as to what the symbolism given above really means. Chuquet's reference to the ancient names for the unknown and the square of the unknown may have suggested the significance that he gave to his symbols. His 12^2 does not mean 12×12, but our $12x^2$; *the exponent is written without its base.* Accordingly, his ".12$^{1\,·\tilde{m}·}$" means $12x^{-1}$. This appears the more clearly when he comes to "adiouster 8^1 avec $\tilde{m}$.5^1 monte tout .3.^{1} Ou .10.^{1} avec .$\tilde{m}$.16.^{1} mõte tout $\tilde{m}$.6.$^1_·$," i.e., $8x-5x=3x$, $10x-16x=-6x$. Again, ".8.^{2} avec .12.^{2} montent .20.^{2}" means $8x^2+12x^2=20x^2$; subtracting ".$\tilde{m}$.16⁰" from ".12.^{2}" leaves "12.^{2} $\tilde{m}$. $\tilde{m}$. 16⁰ qui valent autant cõme .12.^{2} $\bar{p}$. 16⁰"[2] The meaning of Chuquet's ".12⁰" appears from his "Example. qui multiplie .12⁰ par .12⁰ montent .144. puis qui adiouste .0. avec .0. monte 0. ainsi monte ceste multiplicacion .144⁰,"[3] i.e., $12x^0\times12x^0=$

[1] Boncompagni, *op. cit.*, Vol. XIII, p. 737.

[2] *Ibid.*, p. 739.   [3] *Ibid.*, p. 740.

$144x^0$. Evidently, $x^0 = 1$; he has the correct interpretation of the exponent zero. He multiplies $.12^2$ by $.10^2$ and obtains 120^2; also $.5^1$ times $.8^1$ yields $.40^2$; $.12^3$ times $.10^5$ gives $.120^8$; $.8^1$ times $.7^{1 \cdot \tilde{m}}$ gives $.56^2$ or $.56.$; $.8^3$ times $.7^{1 \cdot \tilde{m}}$ gives $.56^2$ Evidently algebraic multiplication, involving the product of the coefficients and the sum of the exponents, is a familiar process with Chuquet. Nevertheless, he does not, in his notation, apply exponents to given numbers, i.e., with him "3^2" never means 9, it always means $3x^2$. He indicates (p. 745) the division of $30 - x$ by $x^2 + x$ in the following manner:

$$\frac{30. \ \tilde{m}. \ 1^1}{1^2 \ \bar{p}. \ 1^1} \ .$$

As a further illustration, we give $R^2 1.\frac{1}{2}.\bar{p}R^2 24.\bar{p}.R^2 1.\frac{1}{2}.$ multiplied by $R^2 1.\frac{1}{2} \ \bar{p} \ R^2 \ 24.$ $\tilde{m} R^2 1.\frac{1}{2}.$ gives $R^2 24.$ This is really more compact and easier to print than our $\sqrt{1\frac{1}{2}} + \sqrt{24} + \sqrt{1\frac{1}{2}}$ times $\sqrt{1\frac{1}{2}} + \sqrt{24} - \sqrt{1\frac{1}{2}}$ equals $\sqrt{24}$.

FRENCH: ÉSTIENNE DE LA ROCHE
(1520)

132. Éstienne de la Roche, Villefranche, published *Larismethique*, at Lyon in 1520, which appeared again in a second edition at Lyon in 1538, under the revision of Gilles Huguetan. De la Roche mentions Chuquet in two passages, but really appropriates a great deal from his distinguished predecessor, without, however, fully entering into his spirit and adequately comprehending the work. It is to be regretted that Chuquet did not have in De la Roche an interpreter acting with sympathy and full understanding. De la Roche mentions the Italian Luca Pacioli.

De la Roche attracted little attention from writers antedating the nineteenth century; he is mentioned by the sixteenth-century French writers Buteo and Gosselin, and through Buteo by John Wallis. He employs the notation of Chuquet, intermixed in some cases, by other notations. He uses Chuquet's $\bar{p}$ and $\tilde{m}$ for *plus* and *moins*, also Chuquet's radical notation R^2, R^3, R^4, ... , but gives an alternative notation: $R \ \square$ for R^3, HR for R^4, $HR \ \square$ for R^6. His strange uses of the geometric square are shown further by his writing $\square$ to indicate the cube of the unknown, an old procedure mentioned by Chuquet.

The following quotation is from the 1538 edition of De la Roche, where, as does Chuquet, he calls the unknown and its successive powers by the names of primary numbers, secondary numbers, etc.:

"... vng chascun nombre est considere comme quantité continue que aultrement on dit nombre linear qui peult etre appelle chose ou premier: et telz nombres seront notez apposition de une unite au dessus deulx en ceste maniere 12^1 ou 13^1, etc., ou telz nombres seront signes dung tel characte apres eux comme 12.ρ. ou 13.ρ. ... cubes que lon peut ainsi marquer $12.^3$ ou $13.^3$ et ainsi 12 □ ou 13 □."[1]

The translation is as follows:

"And a number may be considered as a continuous quantity, in other words, a linear number, which may be designated a thing or as primary, and such numbers are marked by the apposition of unity above them in this manner 12^1 or 13^1, etc., or such numbers are indicated also by a character after them, like 12.ρ. or 13.ρ. ... Cubes one

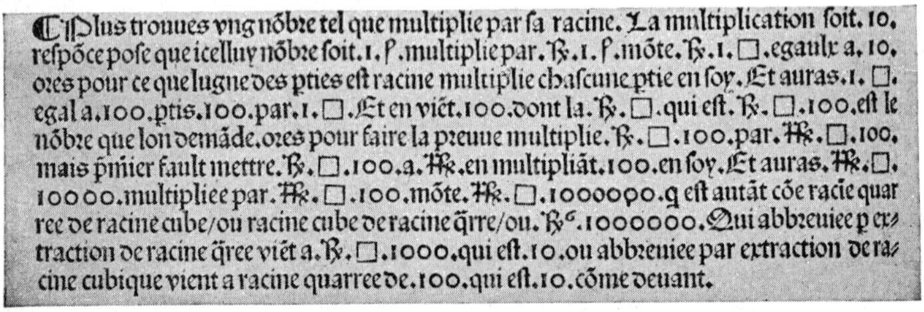
¶ Plus trouues vng nõbre tel que multiplie par sa racine. La multiplication soit. 10.
respõce pose que icelluy nõbre soit. 1. ƿ. multiplie par. ℞. 1. ƿ. mõte. ℞. 1. □. egaulx a. 10.
ores pour ce que lugne des pties est racine multiplie chascune ptie en soy. Et auras. 1. □.
egal a. 100. ptis. 100. par. 1. □. Et en viẽt. 100. dont la. ℞. □. qui est. ℞. □. 100. est le
nõbre que lon demãde. ores pour faire la preuue multiplie. ℞. □. 100. par. ℞℞. □. 100.
mais pmier fault mettre. ℞. □. 100. a. ℞℞. en multipliãt. 100. en soy. Et auras. ℞℞. □.
10000. multipliee par. ℞℞. □. 100. mõte. ℞℞. □. 1000000. q est autãt cõe racie quar
ree de racine cube/ou racine cube de racine qrre/ou. ℞⁶. 1000000. Qui abbreuiee p extraction de racine qree viẽt a. ℞. □. 1000. qui est. 10. ou abbreuiee par extraction de racine cubique vient a racine quarree de. 100. qui est. 10. cõme deuant.

FIG. 40.—Part of fol. 60*B* of De la Roche's *Larismethique* of 1520

may mark $12.^3$ or $13.^3$ and also 12 □. or 13□." (We have here $12^1 = 12x$, $12.^3 = 12x^3$, etc.)

A free translation of the text shown in Figure 40 is as follows:

"Next find a number such that, multiplied by its root, the product is 10. Solution: Let the number be x. This multiplied by $\sqrt{x}$ gives $\sqrt{x^3} = 10$. Now, as one of the sides is a radical, multiply each side by itself. You obtain $x^3 = 100$. Solve. There results the cube root of 100, i.e., $\sqrt[3]{100}$ is the required number. Now, to prove this, multiply $\sqrt[3]{100}$ by $\sqrt[6]{100}$. But first express $\sqrt[3]{100}$ as $\sqrt[6]{}$, by multiplying 100 by itself, and you have $\sqrt[6]{10{,}000}$. This multiplied by $\sqrt[6]{100}$ gives $\sqrt[6]{1{,}000{,}000}$, which is the square root of the cube root, or the cube root of the square root, or $\sqrt[6]{1{,}000{,}000}$. Extracting the square root gives $\sqrt[3]{1{,}000}$ which is 10, or reducing by the extraction of the cube root gives the square root of 100, which is 10, as before."

[1] See an article by Terquem in the *Nouvelles annales de mathématiques* (Terquem et Gerono), Vol. VI (1847), p. 41, from which this quotation is taken. For extracts from the 1520 edition, see Boncompagni, *op. cit.*, Vol. XIV (1881), p. 423.

The end of the solution of the problem shown in Figure 41 is in modern symbols as follows:

$$\text{first } \sqrt{34}+7\,.$$

$$x \text{——} x+1 \qquad\qquad x+4$$

$$2\tfrac{1}{2}x \qquad\qquad 1\tfrac{1}{2}x+4\tfrac{1}{2}$$

$$2\tfrac{1}{2}x^2+2\tfrac{1}{2}x \text{——} 1\tfrac{1}{2}x^2+10\tfrac{1}{2}x+18$$

$$x^2 \text{————————} 8x+18$$

$$4 \qquad 16$$

$$16 \quad \sqrt{34}+4 \text{ second.}$$

[i.e., $x=\sqrt{34}+4$]

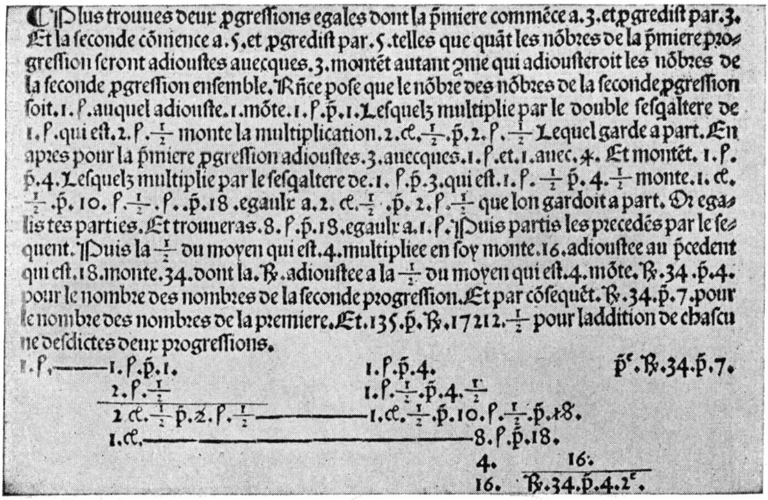

¶ Plus trouues deux ꝑgressions egales dont la p̃miere commẽce a.3.et ꝑgredist par.3. Et la seconde cõmence a.5.et ꝑgredist par.5.telles que quãt les nõbres de la p̃miere progression seront adioustes auecques.3.montẽt autant cõme qui adiousteroit les nõbres de la seconde ꝑgression ensemble. Rñce pose que le nõbre des nõbres de la seconde ꝑgression soit.1.ρ.auquel adiouste.1.mõte.1.ρ.p̃.1.Lesquelz multiplie par le double sesqaltere de 1.ρ.qui est.2.ρ.½ monte la multiplication.2.ce.½.p̃.2.ρ.½ Lequel garde a part. En apres pour la p̃miere ꝑgression adioustes.3.auecques.1.ρ.et.1.auec.4. Et montẽt. 1.ρ. p̃.4.Lesquelz multiplie par le sesqaltere de.1.ρ.p̃.3.qui est.1.ρ.½ p̃.4.½ monte.1.ce. ½.p̃.10.ρ.½.ρ..p̃.18.egaulx a.2.ce.½.p̃.2.ρ.½ que lon gardoit a part. Or egalis tes parties. Et trouueras.8.ρ.p̃.18.egaulx a.1.ρ. Puis partis les precedẽs par le sequent. Puis la ½ du moyen qui est.4.multipliee en soy monte.16.adioustee au p̃cedent qui est.18.monte.34.dont la.℞.adioustee a la ½ du moyen qui est.4.mõte.℞.34.p̃.4. pour le nombre des nombres de la seconde progression. Et par cõsequẽt.℞.34.p̃.7.pour le nombre des nombres de la premiere. Et.135.p̃.℞.17212.½ pour laddition de chascune desdictes deux progressions.

1.ρ.——1.ρ.p̃.1. 1.ρ.p̃.4. p̃ᵉ.℞.34.p̃.7.
2.ρ.½ 1.ρ.½.p̃.4.½
2.ce.½ p̃.2.ρ.½——————1.ce.½.p̃.10.ρ.½.p̃.18.
1.ce.————————————8.ρ.p̃.18.
4. 16.
16. ℞.34.p̃.4.2ᵉ.

FIG. 41.—Part of fol. 66 of De la Roche's *Larismethique* of 1520

ITALIAN: PIETRO BORGI (OR BORGHI) (1484, 1488)

133. Pietro Borgi's *Arithmetica* was first printed in Venice in 1484; we use the edition of 1488. The book contains no algebra. It displays the scratch method of division and the use of dashes in operating with fractions (§§ 223, 278). We find in this early printed *Arithmetica* the use of curved lines in the solution of problems in alligation. Such graphic aids became frequent in the solution of the indeterminate problems of alligation, as presented in arithmetics. Pietro Borgi, on the unnumbered folio 79*B*, solves the following problem: Five sorts of spirits, worth per *ster*, respectively, 44, 48, 52, 60, 66 *soldi*,

are to be mixed so as to obtain 50 *ster*, worth each 56 *soldi*. He solves this by taking the qualities of wine in pairs, always one quality dearer and the other cheaper than the mixture, as indicated by the curves in the example.

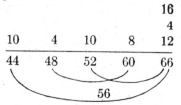

Then $56-44=12$; $66-56=10$; write 12 above 66 and 10 above 44. Proceed similarly with the pairs 48 and 60, 52 and 66. This done, add 10, 4, 10, 8, 16. Their sum is 48, but should be 50. Hence multiply each by $\frac{50}{48}$ and you obtain $10\frac{5}{12}$ as the number of *ster* of wine worth 44 *soldi* to be put into the mixture, etc.

ITALIAN: LUCA PACIOLI

(1494, 1523)

134. *Introduction.*—Luca Pacioli's *Summa de arithmetica geometria proportioni et proportionalita* (Venice, 1494)[1] is historically important because in the first half of the sixteenth century it served in Italy as the common introduction to mathematics and its influence extended to other European countries as well. The second edition (1523) is a posthumous publication and differs from the first edition

---

[1] Cosmo Gordon ("Books on Accountancy, 1494–1600," *Transactions of the Bibliographical Society* [London], Vol. XIII, p. 148) makes the following remarks on the edition of 1494: "The *Summa de arithmetica* occurs in two states. In the first the body of the text is printed in Proctor's type 8, a medium-sized gothic. On sig. *a* 1, on which the text begins, there is the broad wood-cut border and portrait-initial *L* already described. In the second state of the *Summa*, of which the copy in the British Museum is an example, not only do the wood-cut border and initial disappear from *a* 1, but sigs. *a–c* with the two outside leaves of sigs. *d* and *e*, and the outside leaf of sig. *a*, are printed in Proctor's type 10**, a type not observed by him in any other book from Paganino's press. There are no changes in the text of the reprinted pages, but that they are reprinted is clear from the fact that incorrect head-lines are usually corrected, and that the type of the remaining pages in copies which contain the reprints shows signs of longer use than in copies where the text type does not vary. It may be supposed that a certain number of the sheets of the signatures in question were accidentally destroyed, and that type 8 was already in use. The sheets had, therefore, to be supplied in the nearest available type." The copy of the 1494 edition in the Library of the University of California exhibits the type 10.

only in the spelling of some of the words. References to the number of the folio apply to both editions.

In the *Summa* the words "plus" and "minus," in Italian *più* and *meno*, are indicated by $\tilde{p}$ and $\tilde{m}$. The unknown quantity was called "thing," in the Italian *cosa*, and from this word were derived in Germany and England the words *Coss* and "cossic art," which in the sixteenth and seventeenth centuries were synonymous with "algebra." As pointed out more fully later, *co.* (*cosa*) meant our x; *ce.* (*censo*) meant our x^2; *cu.* (*cubo*) meant our x^3. Pacioli used the letter $\mathcal{R}$ for *radix*. *Censo* is from the Latin *census* used by Leonardo of Pisa and Regiomontanus. Leonardo of Pisa used also the word *res* ("thing").

135. *Different uses of the symbol* $\mathcal{R}$.—The most common use of $\mathcal{R}$, the abbreviation for the word *radix* or *radici*, was to indicate roots. Pacioli employs for the same purpose the small letter $\curlywedge\gamma$, sometimes in the running text,[1] but more frequently when he is pressed for space in exhibiting algebraic processes on the margin.[2] He writes in Part I of his *Summa:*

(Fol. 70*B*)	$\mathcal{R}$ .200. for $\sqrt{200}$
(Fol. 119*B*)	$\mathcal{R}$.*cuba. de* .64. for $\sqrt[3]{64}$
(Fol. 182*A*)[3]	$\mathcal{R}$.*relato.* for fifth root
(Fol. 182*A*)	$\mathcal{R}\ \mathcal{R}\ \mathcal{R}$. *cuba.* for seventh root
(Fol. 86*A*)	$\mathcal{R}$ .6.$\tilde{m}$.$\mathcal{R}$.2. for $\sqrt{6}-\sqrt{2}$
(Fol. 131*A*)	$\mathcal{R}\ \mathcal{R}$.120. for $\sqrt[4]{120}$
(Fol. 182*A*)	$\mathcal{R}$. *cuba. de* $\mathcal{R}$. *cuba.* for sixth root
(Fol. 182*A*)	$\mathcal{R}\ \mathcal{R}$. *cuba. de* $\mathcal{R}$. *cuba.* for eighth root.

The use of the $\mathcal{R}v$. for the designation of the roots of expressions containing two or more terms is shown in the following example:

(Fol. 149*A*) $\mathcal{R}v.\ \mathcal{R}.20\tfrac{1}{4}.\tilde{m}.\tfrac{1}{2}.$ for $\sqrt{\sqrt{20\tfrac{1}{4}}-\tfrac{1}{2}}$.

The following are probably errors in the use of $\mathcal{R}v.$:

(Fol. 93*A*) $\mathcal{R}v.$ 50000.$\tilde{m}$.200. for $\sqrt{50,000-200}$,

(Fol. 93*A*) $\mathcal{R}\ \mathcal{R}v.$ 50000.$\tilde{m}$.200. for $\sqrt{\sqrt{50,000-200}}$.

In combining symbols to express the higher roots, Pacioli uses the additive principle of Diophantus, while in expressing the higher powers

[1] Part I (1523), fol. 86 *A*.

[2] *Ibid.*, fol. 124 *A*.

[3] On the early uses of *radix relata* and *primo relato* see Eneström, *Bibliotheca mathematica*, Vol. XI (1910–11), p. 353.

he uses the multiplication principle of the Hindus. Thus Pacioli indicates the seventh root by R R R. *cuba.* (2+2+3), but the eighth power by *ce.ce.ce.* (2×2×2). For the fifth, seventh, and eleventh powers, which are indicated by prime numbers, the multiplication principle became inapplicable. In that case he followed the notation of wide prevalence at that time and later: $p^o.r^o$ (*primo relato*) for the fifth power, $2^o.r^o$ (*secundo relato*) for the seventh power, $3^o.r^o$ (*terzo relato*) for the eleventh power.[1] Whenever the additive principle was used in marking powers or roots, these special symbols became superfluous. Curiously, Pacioli applies the additive principle in his notation for roots, yet does not write $R.R$ *cuba* (2+3) for the fifth root, but R. *relata.* However, the seventh root he writes R R R. *cuba* (2+2+3) and not $R2^o.r^o$.[2]

136. In other parts of Pacioli's *Summa* the sign R is assigned altogether different meanings. Apparently, his aim was to describe the various notations of his day, in order that readers might select the symbols which they happened to prefer. Referring to the prevailing diversity, he says, "tante terre: tante vsanze."[3] Some historians have noted only part of Pacioli's uses of R, while others have given a fuller account but have fallen into the fatal error of interpreting certain powers as being roots. Thus far no one has explained all the uses of the sign R in Pacioli's *Summa.* It was Julius Rey Pastor and Gustav Eneström who briefly pointed out an inaccuracy in Moritz Cantor, when he states that Pacioli indicated by R 30 the thirtieth *root,* when Pacioli really designated by R .30^a the twenty-ninth *power.* This point is correctly explained by J. Tropfke.[4]

We premise that Pacioli describes two notations for representing powers of an unknown, x^2, x^3, ... , and three notations for x. The one most commonly used by him and by several later Italian writers of the sixteenth century employs for x, x^2, x^3, x^4, x^5, x^6, x^7, ... , the abbreviations *co.* (*cosa*), *ce.* (*censo*), *cu.* (*cubo*), *ce.ce.* (*censo de censo*), $p^o.r^o$ (*primo relato*), *ce.cu.* (*censo de cubo*), $2^o.r^o$ (*secundo relato*), ...[5]

Pacioli's second notation for powers involves the use of R, as already indicated. He gives: $R.p^a$ (*radix prima*) for x^0, $R.2^a$ (*radix secunda*) for x, $R.3^a$ (*radix terza*) for x^2, ... , $R.30^a$ (*nono relato*) for x^{29}.[6] When Eneström asserts that folio 67B deals, not with roots,

[1] Part I, fol. 67B.

[2] *Ibid.*, fol. 182A.

[3] *Ibid.*, fol. 67B.   [4] *Op. cit.* (2d ed.), Vol. II (1921), p. 109.

[5] *Op. cit.*, Part I, fol. 67B.   [6] *Ibid.*

but exclusively with the powers x^0, x, x^2, ... , x^{29}, he is not quite accurate, for besides the foregoing symbols placed on the margin of the page, he gives on the margin also the following: "Rx. Radici; R R. Radici de Radici; Rv. Radici vniuersale. Ouer radici legata. O voi dire radici vnita; R. cu. Radici cuba; $\tilde{\wp}\beta^a$ quantita." These expressions are used by Pacioli in dealing with roots as well as with powers, except that Rv. is employed with roots only; as we have seen, it signifies the root of a binomial or polynomial. In the foregoing two uses of R, how did Pacioli distinguish between roots and powers? The ordinal number, *prima, secunda, terza*, etc., placed after the R, always signifies a "power," or a *dignita*. If a root was intended, the number affected was written after the R; for example, $R.200.$ for $\sqrt{200}$. In folio 143AB Pacioli dwells more fully on the use of R in the designation of powers and explains the multiplication of such expressions as $R.$ 5^a *via*. $R.$ 11^a *fa* $R.$ 15^a, i.e., $x^4 \times x^{10} = x^{14}$. In this notation one looks in vain for indications of the exponential concepts and recognition of the simple formula $a^m \cdot a^n = a^{m+n}$. Pacioli's results are in accordance with the formula $a^m \cdot a^n = a^{m+n-1}$. The ordinal numbers in R 11^a, etc., exceed by unity the power they represent. This clumsy designation made it seem necessary to Pacioli to prepare a table of products, occupying one and one-half pages, and containing over two hundred and sixty entries; the tables give the various combinations of factors whose products do not exceed x^{29}. While Eneström and Rey Pastor have pointed out that expressions like $R.28^a$ mark powers and not roots, they have failed to observe that Pacioli makes no use whatever of this curious notation in the working of problems. Apparently his aim in inserting it was encyclopedial.

137. In working examples in the second part of the *Summa*, Pacioli exhibits a third use of the sign R not previously noted by historians. There R is used to indicate powers of numbers, but in a manner different from the notation just explained. We quote from the *Summa* a passage[1] in which R refers to powers as well as to roots. Which is meant appears from the mode of phrasing: "... $R.108.$ e questo mēa con laxis ch' $R.16.$ fa. $R.1728$ piglia el .⅓. cioe recca .3. a. $R.$ fa .9. parti .1728 in. 9. neuien. 192. e. $R.192.$ . . . ." ($\therefore$ $\sqrt{108}$ and multiplying this with the axis which is $\sqrt{16}$ gives $\sqrt{1,728}$. Take ⅓, i.e., raising 3 to the second power gives 9; dividing 1,728 by 9 gives 192, and the $\sqrt{192}$. . . . .) Here "recca. 3. a. $R.$ fa. 9." identifies R with a power. In Part I, folio 186A, one reads, "quando fia recata prima. 1.

---

[1] *Ibid.*, Part II, fol. 72 *B*.

co. a. R. fa. 1. ce" ("raising the x to the second power gives x^2"). Such phrases are frequent as, Part II, folio 72B, "reca. 2. a. R. cu. fa. 8" ("raise 2 to the third power; it gives 8"). Observe that R. *cu.* means the "third" power, while[1] R. 3^a and R. *terza.* refer to the "second" power. The expression of powers by the Diophantine additive plan (2+3) is exhibited in "reca. 3. a. R R. cuba fa. 729" ("raise 3 to the fifth power; it gives 729").[2]

A fourth use of R is to mark the unknown x. We have previously noted Pacioli's designation of x by *co.* (*cosa*) and by R. 2^a. In Part II, folio 15B, he gives another way: "la mita dun censo e .12. dramme: sonno equali a .5.R. E questo cõme a dire .10. radici sonno equali a vn censo e. 24. dramme" ("Half of x^2 and the number 12 are equal to 5x; and this amounts to saying 10x are equal to x^2 and the number 24").

In Part I, folio 60B, the sign R appears on the margin twice in a fifth rôle, namely, as the abbreviation for *rotto* ("fraction"), but this use is isolated. From what we have stated it is evident that Pacioli employed R in five different ways; the reader was obliged to watch his step, not to get into entanglements.

138. *Sign of equality.*—Another point not previously noted by historians is that Pacioli used the dash (—) as a symbol for equality. In Part I, folio 91A, he gives on the margin algebraic expressions relating to a problem that is fully explained in the body of the page. We copy the marginal notes and give the modern equivalents:

Summa (Part I, fol. 91A)		Modern Equivalents	
$\bar{p}^a$ 1. *co.* $\tilde{m}$. 1. $\wp\beta^a$		1st	$x - y$
3^a 1. *co.* $\bar{p}$. 1. $\wp\beta^a$		3d	$x + y$
1. *co.* $\tilde{m}$. 1. *ce. de.* $\wp\beta^a$ 36			$x^2 - y^2 = 36$
Rv. 1. *ce.* $\tilde{m}$ 36 1. *ce. de* $\wp\beta^a$		$\sqrt{x^2 - 36} = y$,	
Valor quantitatis.		the value of y .	
p^a 1. *co.* $\tilde{m}$ Rv. 1. *ce.* $\tilde{m}$ 36		1st $x - \sqrt{x^2 - 36}$	
2^a 6		2d 6	
3^a 1. *co.* $\bar{p}$ Rv. 1. *ce.* $\tilde{m}$ 36 .		3d $x + \sqrt{x^2 - 36}$	
2. *co.* $\bar{p}$. 6.——216		$2x + 6$ $= 216$	
2. *co.* 210		$2x$ $= 210$	
Valor rei. 105		Value of x 105	

[1] Part I, fol. 67B.   [2] Part II, fol. 72B.

Notice that the *co.* in the third expression should be *ce.*, and that the
.1. *ce. de* $\overline{\gamma\beta}{}^{a}$ in the fourth expression should be .1. *co. de* $\gamma\beta{}^{a}$. Here, the
short lines or dashes express equality. Against the validity of this
interpretation it may be argued that Pacioli uses the dash for several
different purposes. The long lines above are drawn to separate the
sum or product from the parts which are added or multiplied. The
short line or dash occurs merely as a separator in expressions like

$$\text{\textit{Simplices} \quad \textit{Quadrata}}$$
$$3 \text{\underline{\hspace{2cm}}} 9$$

in Part I, folio 39A. The dash is used in Part I, folio 54 B, to indicate
multiplication, as in

$$
\begin{array}{cc}
14 & 15 \\
\frac{2}{5} & \hspace{-0.3cm}\times\hspace{-0.3cm} & \hspace{-0.3cm}\frac{3}{7}
\end{array}
$$

where the dash between 5 and 7 expresses 5×7, one slanting line
means 2×7, the other slanting line 5×3. In Part II, folio 37A, the
dash represents some line in a geometrical figure; thus $d\underline{\hspace{0.2cm}3\hspace{0.2cm}}k$ means
that the line dk in a complicated figure is 3 units long. The fact that
Pacioli uses the dash for several distinct purposes does not invalidate
the statement that one of those purposes was to express equality. This
interpretation establishes continuity of notation between writers preceding and following Pacioli. Regiomontanus,[1] in his correspondence
with Giovanni Bianchini and others, sometimes used a dash for equality. After Pacioli, Francesco Ghaligai, in his *Pratica d'arithmetica*, used
the dash for the same purpose. Professor E. Bortolotti informs me that
a manuscript in the Library of the University of Bologna, probaby
written between 1550 and 1568, contains two parallel dashes (=) as a
symbol of equality. The use of two dashes was prompted, no doubt,
by the desire to remove ambiguity arising from the different interpretations of the single dash.

Notice in Figure 42 the word *cosa* for the unknown number, and
its abbreviation, *co.; censo* for the square of the unknown, and its contraction, *ce.; cubo* for the cube of the unknown; also .$\bar{p}$. for "plus"
and .$\tilde{m}$. for "minus." The explanation given here of the use of *cosa,
censo, cubo*, is not without interest.

[1] See Maximilian Curtze, *Urkunden zur Geschichte der Mathematik im Mittelalter und der Renaissance* (Leipzig, 1902), p. 278.

The first part of the extract shown in Figure 43 gives $\sqrt{\sqrt{40}+6}+\sqrt{\sqrt{40}-6}$ and the squaring of it. The second part gives $\sqrt{\sqrt{20}+2}+\sqrt{\sqrt{20}-2}$ and the squaring of it; the simplified result is given as $\sqrt{80}+4$, but it should be $\sqrt{80}+8$. Remarkable in this second example is the omission of the *v* to express *vniversale*. From the computation as well as from the explanation of the text it appears that the first R̸ was intended to express universal root, i.e., $\sqrt{\sqrt{20}+2}$ and not $\sqrt[4]{20}+2$.

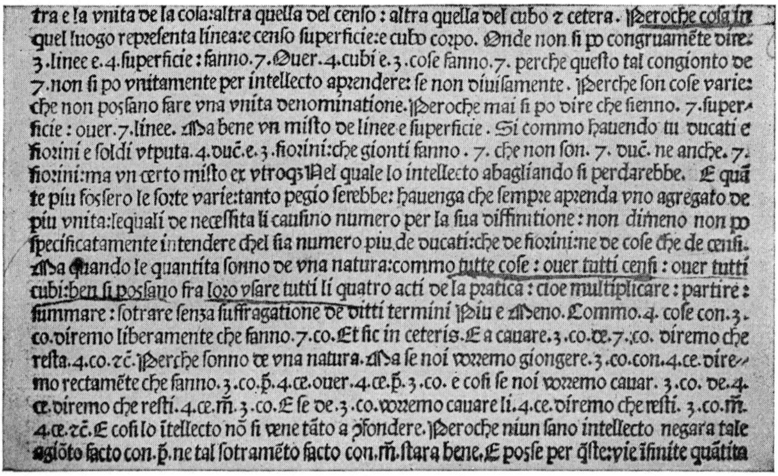
tra e la vnita de la cosa: altra quella del censo: altra quella del cubo z cetera. Peroche cosa in quel luogo representa linea: e censo superficie: e cubo corpo. Onde non si po congruamēte dire: 3. linee e. 4. superficie: fanno. 7. Ouer. 4. cubi e. 3. cose fanno. 7. perche questo tal congionto de 7. non si po vnitamente per intellecto aprendere: se non diuisamente. Perche son cose varie: che non possano fare vna vnita denominatione. Peroche mai si po dire che sienno. 7. superficie: ouer. 7. linee. Ma bene vn misto de linee e superficie. Si commo hauendo tu ducati e fiorini e soldi vtputa. 4. duc. e. 3. fiorini: che gionti fanno. 7. che non son. 7. duc. ne anche. 7. fiorini: ma vn certo misto ex vtroq; Nel quale lo intellecto abagliando si perdarebbe. E quā te piu fossero le sorte varie: tanto pegio serebbe: hauenga che sempre aprenda vno agregato de piu vnita: lequali de necessita li causino numero per la sua diffinitione: non dimeno non po specificatamente intendere chel sia numero piu de ducati: che de fiorini: ne de cose che de censi. Ma quando le quantita sonno de vna natura: commo tutte cose: ouer tutti censi: ouer tutti cubi: ben si possano fra loro vsare tutti li quatro acti de la pratica: cioe multiplicare: partire: summare: sotrare senza suffragatione de ditti termini Piu e Meno. Commo. 4. cose con. 3. co. diremo liberamente che fanno. 7. co. Et sic in ceteris. E a cauare. 3. co. de. 7. co. diremo che resta. 4. co. zc. Perche sonno de vna natura. Ma se noi vorremo giongere. 3. co. con. 4. ce. diremo rectamēte che fanno. 3. co. p̃. 4. ce. ouer. 4. ce. p̃. 3. co. e cosi se noi vorremo cauar. 3. co. de. 4. ce. diremo che resti. 4. ce. m̃. 3. co. E se de. 3. co. vorremo cauare li. 4. ce. diremo che resti. 3. co. m̃. 4. ce. zc. E cosi lo ītellecto nō si vene tāto a confondere. Peroche niun sano intellecto negara tale agiōto facto con. p̃. ne tal sotramēto facto con. m̃. stara bene. E posse per qste vie īfinite quātita

FIG. 42.—Part of a page in Luca Pacioli's *Summa*, Part I (1523), fol. 112*A*

ITALIAN: F. GHALIGAI

(1521, 1548, 1552)

139. Ghaligai's *Pratica d'arithmetica*[1] appeared in earlier editions, which we have not seen, in 1521 and 1548. The three editions do not differ from one another according to Riccardi's *Biblioteca matematica italiana* (I, 500–502). Ghaligai writes (fol. 71*B*): $x = cosa = c^{\circ}$, $x^2 = censo =$, $x^3 = cubo =$, $x^5 = relato =$, $x^7 = pronico =$ 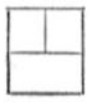, $x^{11} = tronico =$ 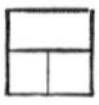, $x^{13} = dromico =$. He uses the m° for "minus" and the $\tilde{p}$ and *e* for "plus," but frequently writes in full *piu* and *meno*.

[1] *Pratica d'arithmetica di Francesco Ghaligai Fiorentino* (Nuouamente Riuista, & con somma Diligenza Ristampata. In Firenze. M.D.LII).

Equality is expressed by dashes (— — —); a single dash (—) is used also to separate factors. The repetition of a symbol, simply to fill up an interval, is found much later also in connection with the sign of equality (=). Thus, John Wallis, in his *Mathesis universalis* ([Oxford, 1657], p. 104) writes: $1+2-3= \;=\; =0$.

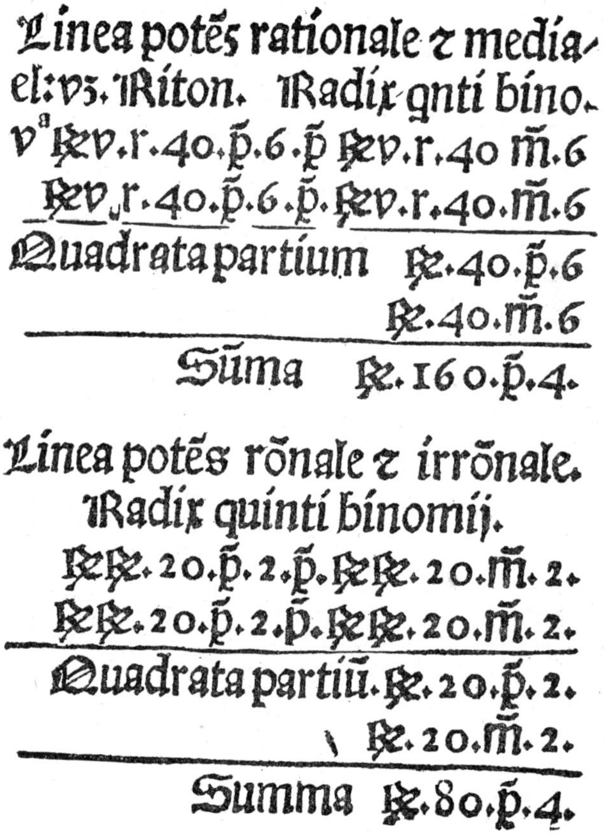
Linea potēs rationale z media-
le: vz. Riton. Radix qnti bino-
vª ℞v.r.40.p̄.6.p̄ ℞v.r.40 m̄.6
℞v.r.40.p̄.6.p̄.℞v.r.40.m̄.6
Quadrata partium ℞.40.p̄.6
℞.40.m̄.6
Sūma ℞.160.p̄.4.

Linea potēs rōnale z irrōnale.
Radix quinti binomij.
℞℞.20.p̄.2.p̄.℞℞.20.m̄.2.
℞℞.20.p̄.2.p̄.℞℞.20.m̄.2.
Quadrata partiū.℞.20.p̄.2.
℞.20.m̄.2.
Summa ℞.80.p̄.4.

FIG. 43.—Printed on the margin of fol. 123*B* of Pacioli's *Summa*, Part I (1523). The same occurs in the edition of 1494.

Ghaligai does not claim these symbols as his invention, but ascribes them to his teacher, Giovanni del Sodo, in the statement (folio 71*B*): "Dimostratione di 8 figure, le quale Giovanni del Sodo pratica la sua Arciba & perche in parte terro 'el suo stile le dimostreto.' "[1] The page shown (Fig. 45) contains the closing part of the

[1] *Op. cit.* (1552), fols. 2*B*, 65; Eneström, *Bibliotheca mathematica*, 3. S., Vol. VIII, 1907–8, p. 96.

solution of the problem to find three numbers, P, S, T, in continued proportion, such that $S^2 = P + T$, and, each number being multiplied

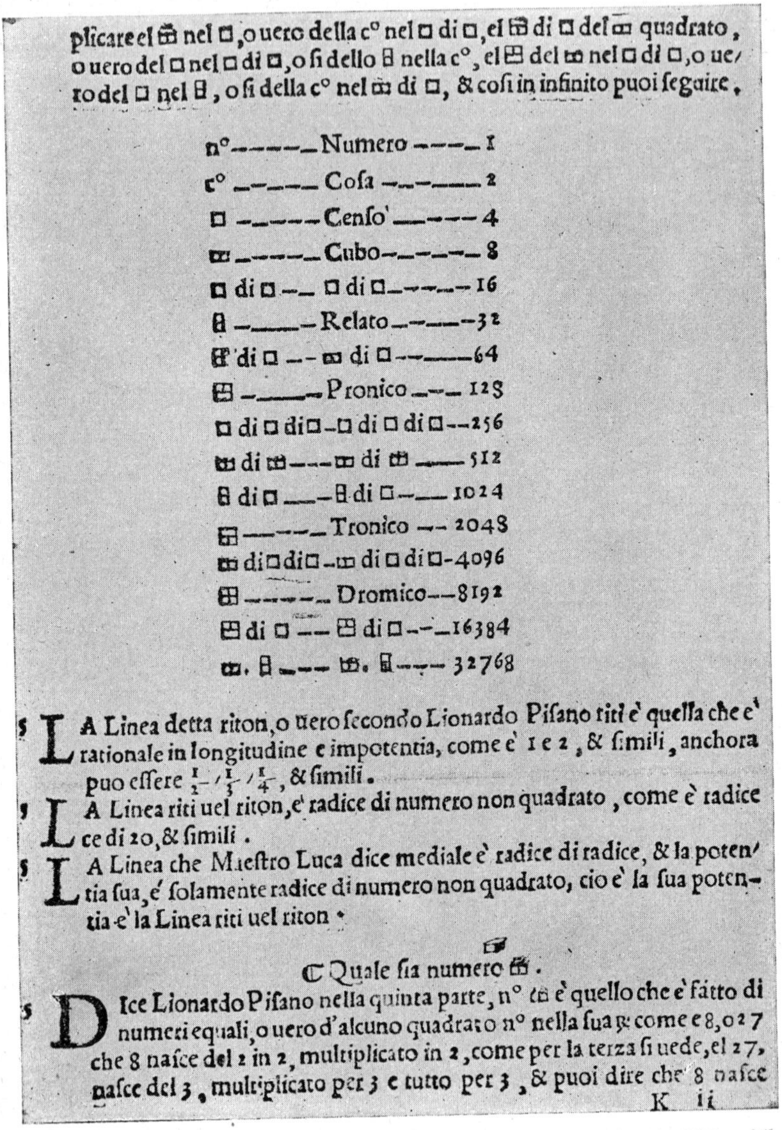

plicare el ⊞ nel ◻, ouero della c° nel ◻ di ◻, el ⊞ di ◻ del ⊞ quadrato, ouero del ◻ nel ◻ di ◻, o si dello 𝖡 nella c°, el ⊞ del ⊞ nel ◻ di ◻, o ue/ ro del ◻ nel 𝖡, o si della c° nel ⊞ di ◻, & cosi in infinito puoi seguire.

n°	Numero	1
c°	Cosa	2
◻	Censo	4
⊞	Cubo	8
◻ di ◻	◻ di ◻	16
𝖡	Relato	32
⊞ di ◻	⊞ di ◻	64
⊞	Pronico	128
◻ di ◻ di ◻	◻ di ◻ di ◻	256
⊞ di ⊞	⊞ di ⊞	512
𝖡 di ◻	𝖡 di ◻	1024
⊞	Tronico	2048
⊞ di ◻ di ◻	⊞ di ◻ di ◻	4096
⊞	Dromico	8192
⊞ di ◻	⊞ di ◻	16384
⊞. 𝖡	⊞. 𝖡	32768

¶ LA Linea detta riton, o uero secondo Lionardo Pisano titi e quella che e rationale in longitudine e impotentia, come e 1 e 2, & simili, anchora puo essere $\frac{1}{2}$ / $\frac{1}{3}$ / $\frac{1}{4}$, & simili.

¶ LA Linea riti uel riton, e radice di numero non quadrato, come e radice ce di 20, & simili.

¶ LA Linea che Maestro Luca dice mediale e radice di radice, & la poten/ tia sua, e solamente radice di numero non quadrato, cio e la sua poten_ tia e la Linea riti uel riton ·

¶ Quale sia numero ⊞ ·

¶ DIce Lionardo Pisano nella quinta parte, n° ⊞ e quello che e fatto di numeri equali, ouero d'alcuno quadrato n° nella sua ᵱ come e 8, o 27 che 8 nasce del 2 in 2, multiplicato in 2, come per la terza si uede, el 27, nasce del 3, multiplicato per 3 e tutto per 3, & puoi dire che 8 nasce

K ii

FIG. 44.—Part of fol. 72 of Ghaligai's *Pratica d'arithmetica* (1552). This exhibits more fully his designation of powers.

by the sum of the other two, the sum of these products is equal to twice the second number multiplied by the sum of the other two, plus 72. Ghaligai lets $S=3co$ or $3x$. He has found $x=2$, and the root of x^2 equal to $\sqrt{4}$.

The translation of the text in Figure 45 is as follows: "equal to $\sqrt{4}$, and the $\sqrt{x^4}$ is equal to $\sqrt{16}$, hence the first quantity was $18-\sqrt{288}$, and the second was 6, and the third $18+\sqrt{288}$.

$S.\quad 3x\quad P.$ and $T.\quad 9x^2$ $\qquad\qquad\qquad 18x^2+6x,\times 3x$.

$P.\quad 4\frac{1}{2}x^2-\sqrt{20\frac{1}{4}x^4-9x^2}$

$T.\quad 4\frac{1}{2}x^2+\sqrt{20\frac{1}{4}x^4-9x^2}$, $\qquad\qquad =54x^3+18x^2=54x^3+72$

$S.\quad 3x$ $\qquad\qquad\qquad\qquad\qquad\qquad 18x^2=72$

$\qquad\qquad P.\quad 4\frac{1}{2}x^2-\sqrt{20\frac{1}{4}x^4-9x^2}$ $\qquad\qquad \sqrt{4}$

$=9x^2+3x,\times 2$

$\qquad\qquad\quad \backslash -4 ---/$ $\qquad\qquad$ Value of x which is 2

$\qquad\qquad\quad 18\quad \sqrt{324}$ $\qquad\qquad P.$ was $18-\sqrt{288}$

$\qquad\qquad\qquad\quad 36$ $\qquad\qquad\qquad S.$ was 6

$\qquad\qquad\qquad \sqrt{288}$ $\qquad\qquad\quad T.$ was $18+\sqrt{288}$

Proof

$24+\sqrt{288}$ $\qquad\qquad\qquad 24-\sqrt{288}$

$18-\sqrt{288}$ $\qquad\qquad\qquad 18+\sqrt{288}$

$432+\sqrt{93,312}-288$ $\qquad 432+\sqrt{165,888}-288$

$288-\sqrt{165,888}$ $\qquad\qquad 288-\sqrt{93,312}$

144 $\qquad\qquad\qquad\qquad 144$

$\qquad\qquad\quad 144+\sqrt{93,312}$ $\qquad\qquad\qquad 18-\sqrt{288}$

$\qquad\qquad\qquad -\sqrt{165,888}$ $\qquad\qquad\qquad 18+\sqrt{288}$

$\qquad\qquad\qquad +\sqrt{165,888}$

$\qquad\qquad\quad 144-\sqrt{93,312}$ $\qquad\qquad\qquad =36,\times 6$

$\qquad\qquad\quad$ Gives 288 $\qquad\qquad\qquad 216,\times 2$

$\qquad\qquad\qquad\quad 216$

$\qquad\qquad\qquad\qquad\qquad\qquad\qquad\qquad 432$

$\qquad\qquad\quad$ Gives 504 $\qquad\qquad\qquad\qquad 72$

$\qquad\qquad\qquad\qquad\qquad$ As it should 504."

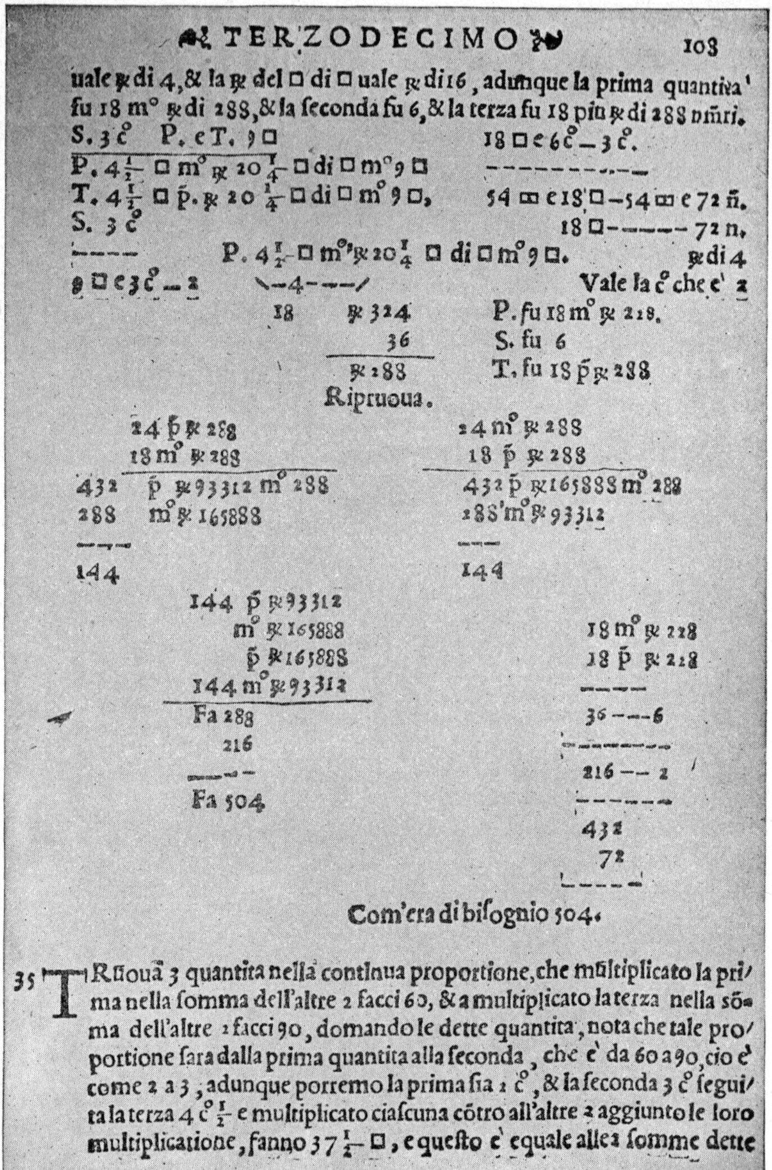

☙ TERZODECIMO ☙   103

uale ℞ di 4, & la ℞ del □ di □ uale ℞ di 16 , adunque la prima quantiva'
fu 18 m° ℞ di 288, & la seconda fu 6, & la terza fu 18 piu ℞ di 288 nmri.

S. 3 ç   P. e T. 9 □   18 □ e 6 ç — 3 ç.

P̄. 4½ □ m° ℞ 20¼ □ di □ m° 9 □   ----------
T. 4½ □ p̄. ℞ 20¼ □ di □ m° 9 □,   54 ▭ e 18 □ — 54 ▭ e 72 ñ.
S. 3 ç   18 □ — ---- — 72 n,

        P. 4½ □ m° ℞ 20¼ □ di □ m° 9 □.   ℞ di 4

θ □ e 3 ç — 2   \—4—--/   Vale la ç che e' 2

        18   ℞ 324   P. fu 18 m° ℞ 218.
                    36   S. fu 6
                 ℞ 288   T. fu 18 p̄ ℞ 288

            Ripruoua.

    24 p̄ ℞ 288   24 m° ℞ 288
    18 m° ℞ 288   18 p̄ ℞ 288
432   p̄ ℞ 93312 m° 288   432 p̄ ℞ 165888 m° 288
288   m° ℞ 165888   288 m° ℞ 93312

---   ---

144   144

        144 p̄ ℞ 93312
          m° ℞ 165888
            p̄ ℞ 165888   18 m° ℞ 218
        144 m° ℞ 93312   18 p̄ ℞ 218

        Fa 288   36 — 6
          216
                                          216 — 2
        Fa 504
                                            432
                                             72

        Com'era di bisognio 504.

35 **T**Rūoua 3 quantita nella continua proportione, che multiplicato la pri/
    ma nella somma dell'altre 2 facci 60, & a multiplicato la terza nella sō=
    ma dell'altre 2 facci 90, domando le dette quantita, nota che tale pro/
    portione sara dalla prima quantita alla seconda , che e' da 60 a 90, cio e'
    come 2 a 3 , adunque porremo la prima sia 2 ç , & la seconda 3 ç segui/
    ta la terza 4 ç ½ e multiplicato ciascuna cōtro all'altre 2 aggiunto le loro
    multiplicatione, fanno 37½ □ , e questo e' equale alle 2 somme dette

FIG. 45.—Ghaligai's *Pratica d'arithmetica* (1552), fol. 108

The following equations are taken from the same edition of 1552:

Translation

(Folio 110) $\frac{1}{4}$ □ di □ $\tilde{m}$ $\frac{1}{4}$ di □ -1 □ $\frac{1}{4}x^4 - \frac{1}{4}x^2 = x^2$

$\frac{1}{4}$ □ di □ $-1\frac{1}{4}$ di □ $\frac{1}{4}x^4 = 1\frac{1}{4}x^2$

$\frac{1}{4}$ □ $-----1\frac{1}{4}$ $\tilde{n}$ $\frac{1}{4}x^2 = 1\frac{1}{4}$

(Folio 113) $\frac{1}{4}$ □ □ □ $\overset{\circ}{m}$ 4 □ $----$ 4 □ $\frac{1}{4}x^4 - 4x^2 = 4x^2$

$\frac{1}{4}$ □ □ $--$ 8 □ $\frac{1}{4}x^4 = 8x^2$

Ghaligai uses his combinations of little squares to mark the orders of roots. Thus, folio 84B, R □ di 3600 — che $è$ 60, i.e., $\sqrt{3,600} = 60$; folio 72B, la R ⬜ di 8 $diciamo$ 2, i.e., $\sqrt[3]{8} = 2$; folio 73B, R ⊟ di 7776 for $\sqrt[5]{7,776}$; folio 73B, R ⬜ di □ di 262144 for $\sqrt[6]{262,144}$.

ITALIAN: HIERONYMO CARDAN
(1539, 1545, 1570)

140. Cardan uses $\tilde{p}$ and $\tilde{m}$ for "plus" and "minus" and R for "root." In his *Practica arithmeticae generalis* (Milano, 1539) he uses Pacioli's symbols *nu.*, *co.*, *ce.*, *cu.*, and denotes the successive higher powers, *ce.ce.*, *Rel. p.*, *cu.ce.*, *Rel. 2.*, *ce.ce.ce.*, *cu.cu.*, *ce. Rel.*[1] However, in his *Ars magna* (1545) Cardan does not use *co.* for x, *ce.* for x^2, etc., but speaks of "rem ignotam, quam vocamus positionem,"[2] and writes $60 + 20x = 100$ thus: "60. $\tilde{p}$. 20. positionibus aequalia 100." Farther on[3] he writes $x^2 + 2x = 48$ in the form "1. quad. $\tilde{p}$. 2. pos. aeq. 48.," x^4 in the form[4] "1. quadr. quad.," $x^5 + 6x^3 = 80$ in the form[5] "r. p^m $\tilde{p}$. 6. cub.   80," $x^5 = 7x^2 + 4$ in the form[6] "$r^m p^m$ 7. quad. $\tilde{p}$. 4." Observe that in the last two equations there is a blank space where we write the sign of equality ($=$). These equations appear in the text in separate lines; in the explanatory text is given *aequale* or *aequatur*. For the representation of a second unknown he follows Pacioli in using the word *quantitas*, which he abbreviates to *quan.* or *qua.* Thus[7] he writes $7x + 3y = 122$ in the form "7. pos. $\tilde{p}$. 3. qua. aequal. 122."

Attention should be called to the fact that in place of the $\tilde{p}$ and $\tilde{m}$, given in Cardan's *Opera*, Volume IV (printed in 1663), one finds in Cardan's original publication of the *Ars magna* (1545) the signs p:

[1] *Hieronymi Cardani operum tomvs quartvs* (Lvgdvni, 1663), p. 14.

[2] *Ibid.*, p. 227.

[3] *Ibid.*, p. 231.   [5] *Ibid.*

[4] *Ibid.*, p. 237.   [6] *Ibid.*, p. 239.

[7] *Ars magna* in *Operum tomvs quartvs*, p. 241, 242.

and m:. For example, in 1545 one finds $(5+\sqrt{-15})$ $(5-\sqrt{-15})=$ $25-(-15)=40$ printed in this form:

> " 5p: ℞ m: 15
> 5m: ℞ m: 15
>
> ──────────────
> 25m:m: 15 q̃d est 40 ,"

while in 1663 the same passage appears in the form:

> " 5. p̃. ℞. m̃. 15.
> 5. m̃. ℞. m. 15.
>
> ──────────────
> 25. m. m. 15. quad. est 40. "[1]

141. Cardan uses ℞ to mark square root. He employs[2] Pacioli's *radix vniversalis* to binomials and polynomials, thus "$R.V.7.\ \bar{p}\ ℞.\ 4.$ vel sic $(℞)$ $13.\bar{p}\ ℞.\ 9.$" for $\sqrt{7+\sqrt{4}}$ or $\sqrt{13+\sqrt{9}}$; "$℞.V.10.p.℞.16.p.3.p$ $℞.64.$" for $\sqrt{10+\sqrt{16}+3+\sqrt{64}}$. Cardan proceeds to new notations. He introduces the *radix ligata* to express the roots of each of the terms of a binomial; he writes: "$L℞.\ 7.\ \bar{p}R.\ 10.$"[3] for $\sqrt{7}+\sqrt{10}$. This L would seem superfluous, but was introduced to distinguish between the foregoing form and the *radix distincta*, as in "$℞.D.\ 9\ p.\ ℞.\ 4.$," which signified 3 and 2 taken separately. Accordingly, "$℞.D.\ 4.\ p.\ ℞.\ 9.$," multiplied into itself, gives $4+9$ or 13, while the "$R.L.\ 4.\ p.\ R.\ 9.$," multiplied into itself, gives $13+\sqrt{144}=25$. In later passages Cardan seldom uses the *radix ligata* and *radix distincta*.

In squaring binomials involving radicals, like "$℞.V.L.\ ℞.\ 5.\ \bar{p}.\ ℞.$ $1.\ \tilde{m}\ ℞.V.L.\ ℞.\ 5.\ \tilde{m}\ ℞.\ 1.$," he sometimes writes the binomial a second time, beneath the first, with the capital letter X between the two binomials, to indicate cross-multiplication.[4] Of interest is the following passage in the *Regula aliza* which Cardan brought out in 1570: "$℞p$: est p: ℞ m: quadrata nulla est iuxta usum communem" ("The square root of a positive number is positive; the square root of a negative number is not proper, according to the common acceptation").[5]

───────────

[1] See Tropfke, *op. cit.*, Vol. III (1922), p. 134, 135.

[2] Cardan, *op. cit.*, p. 14, 16, of the *Practica arithmeticae* of 1539.

[3] *Ibid.*, p. 16.

[4] *Ibid.*, p. 194.

[5] *Op. cit.* (Basel, 1570), p. 15. Reference taken from Eneström, *Bibliotheca mathematica*, Vol. XIII (1912–13), p. 163.

However, in the *Ars magna*[1] Cardan solves the problem, to divide **10** into two parts, whose product is 40, and writes (as shown above):

> " | 5. $\bar{p}$ ℞. $\bar{m}$. 15.
> 5. $\bar{m}$ ℞. $m..$ 15.
>
> 25 $m.m.$ 15. *quad. est* 40 . "

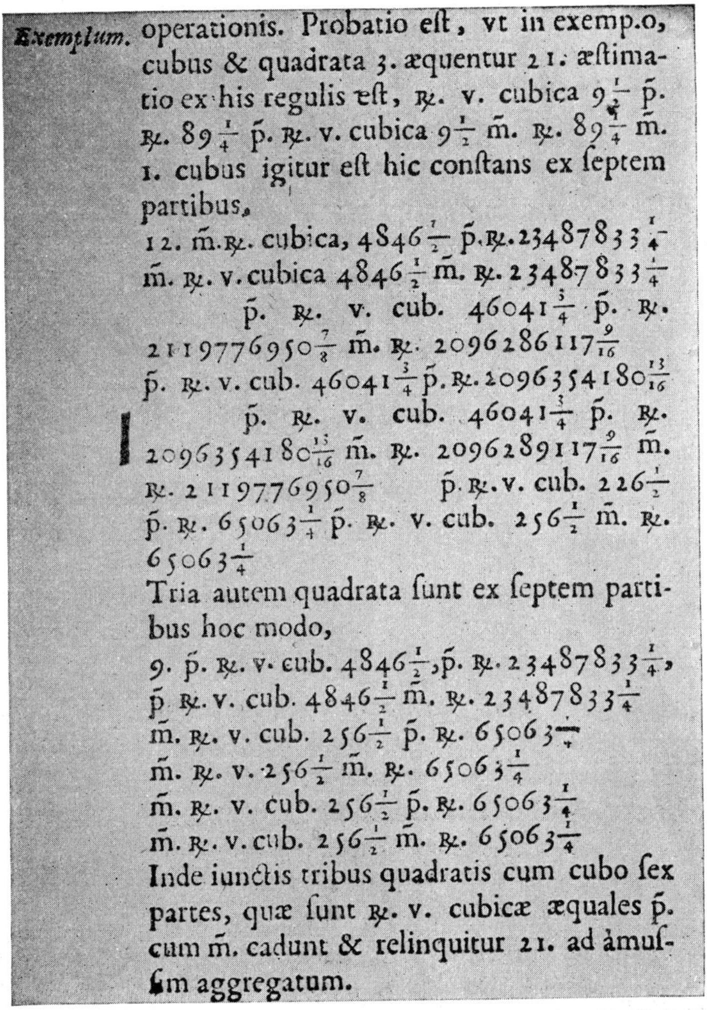

Exemplum. operationis. Probatio eft, vt in exemp.o, cubus & quadrata 3. æquentur 21. æftimatio ex his regulis eft, ℞. v. cubica $9\frac{1}{4}$ $\bar{p}$. ℞. $89\frac{1}{4}$ $\bar{p}$. ℞. v. cubica $9\frac{1}{4}$ $\bar{m}$. ℞. $89\frac{1}{4}$ $\bar{m}$. 1. cubus igitur eft hic conftans ex feptem partibus,

12. $\bar{m}$.℞. cubica, $4846\frac{1}{2}$ $\bar{p}$.℞.$23487833\frac{1}{4}$ $\bar{m}$. ℞. v. cubica $4846\frac{1}{4}$ $\bar{m}$. ℞. $23487833\frac{1}{4}$

$\bar{p}$. ℞. v. cub. $46041\frac{3}{4}$ $\bar{p}$. ℞. $2119776950\frac{7}{8}$ $\bar{m}$. ℞. $2096286117\frac{9}{16}$ $\bar{p}$. ℞. v. cub. $46041\frac{3}{4}$ $\bar{p}$.℞.$209635418 0\frac{13}{16}$

$\bar{p}$. ℞. v. cub. $46041\frac{1}{4}$ $\bar{p}$. ℞. $209635418 0\frac{13}{16}$ $\bar{m}$. ℞. $2096289117\frac{9}{16}$ $\bar{m}$. ℞. $2119776950\frac{7}{8}$ $\bar{p}$. ℞. v. cub. $226\frac{1}{2}$ $\bar{p}$. ℞. $65063\frac{1}{4}$ $\bar{p}$. ℞. v. cub. $256\frac{1}{2}$ $\bar{m}$. ℞. $65063\frac{1}{4}$

Tria autem quadrata funt ex feptem partibus hoc modo,

9. $\bar{p}$. ℞. v. cub. $4846\frac{1}{2}$,$\bar{p}$. ℞. $23487833\frac{1}{4}$, $\bar{p}$ ℞.v. cub. $4846\frac{1}{2}$ $\bar{m}$. ℞. $23487833\frac{1}{4}$ $\bar{m}$. ℞. v. cub. $256\frac{1}{2}$ $\bar{p}$. ℞. $65063\frac{1}{4}$ $\bar{m}$. ℞. v. $256\frac{1}{2}$ $\bar{m}$. ℞. $65063\frac{1}{4}$ $\bar{m}$. ℞. v. cub. $256\frac{1}{2}$ $\bar{p}$.℞. $65063\frac{1}{4}$ $\bar{m}$. ℞. v. cub. $256\frac{1}{2}$ $\bar{m}$. ℞. $65063\frac{1}{4}$

Inde iunctis tribus quadratis cum cubo fex partes, quæ funt ℞. v. cubicæ æquales $\bar{p}$. cum $\bar{m}$. cadunt & relinquitur 21. ad àmuffim aggregatum.

Fig. 46.—Part of a page (255) from the *Ars magna* as reprinted in H. Cardan's *Operum tomvs quartvs* (Lvgdvni, 1663). The *Ars magna* was first published in 1545.

[1] *Operum tomvs quartvs*, p. 287.

In one place Cardan not only designates known numbers by letters, but actually operates with them. He lets a and b stand for any given numbers and then remarks that $R\,\dfrac{a}{b}$ is the same as $\dfrac{Ra}{Rb}$, i.e., $\sqrt{\dfrac{a}{b}}$ is the same as $\dfrac{\sqrt{a}}{\sqrt{b}}$.[1]

Figure 46 deals with the cubic $x^3+3x^2=21$. As a check, the value of x, expressed in radicals, is substituted in the given equation. There are two misprints. The $226\frac{1}{2}$ should be $256\frac{1}{2}$. Second, the two lines which we have marked with a stroke on the left should be omitted, except the $\tilde{m}$ at the end. The process of substitution is unnecessarily complicated. For compactness of notation, Cardan's symbols rather surpass the modern symbols, as will be seen by comparing his passage with the following translation:

"The proof is as in the example $x^3+3x^2=21$. According to these rules, the result is $\sqrt[3]{9\frac{1}{2}+\sqrt{89\frac{1}{4}}}+\sqrt[3]{9\frac{1}{2}-\sqrt{89\frac{1}{4}}}.-1$. The cube [i.e., x^3] is made up of seven parts:

$$12-\sqrt[3]{4{,}846\tfrac{1}{2}+\sqrt{23{,}487{,}833\tfrac{1}{4}}}-\sqrt[3]{4{,}846\tfrac{1}{2}-\sqrt{23{,}487{,}833\tfrac{1}{4}}}$$
$$+\sqrt[3]{46{,}041\tfrac{3}{4}+\sqrt{2{,}119{,}776{,}950\tfrac{7}{8}}}-\sqrt{2{,}096{,}286{,}117\tfrac{9}{16}}$$
$$+\sqrt[3]{46{,}041\tfrac{3}{4}+\sqrt{2{,}096{,}354{,}180\tfrac{13}{16}}}-\sqrt{2{,}119{,}776{,}950\tfrac{7}{8}}$$
$$+\sqrt[3]{256\tfrac{1}{2}+\sqrt{65{,}063\tfrac{1}{4}}}+\sqrt[3]{256\tfrac{1}{2}-\sqrt{65{,}063\tfrac{1}{4}}}\ .$$

"The three squares [i.e., $3x^2$] are composed of seven parts in this manner:

$$9+\sqrt[3]{4{,}846\tfrac{1}{2}+\sqrt{23{,}487{,}833\tfrac{1}{4}}}$$
$$+\sqrt[3]{4{,}846\tfrac{1}{2}-\sqrt{23{,}487{,}833\tfrac{1}{4}}}$$
$$-\sqrt[3]{256\tfrac{1}{2}+\sqrt{65{,}063\tfrac{1}{4}}}$$
$$-\sqrt[3]{256\tfrac{1}{2}-\sqrt{65{,}063\tfrac{1}{4}}}$$
$$-\sqrt[3]{256\tfrac{1}{2}+\sqrt{65{,}063\tfrac{1}{4}}}$$
$$-\sqrt[3]{256\tfrac{1}{2}-\sqrt{65{,}063\tfrac{1}{4}}}\ .$$

Now, adding the three squares with the six parts in the cube, which are equal to the general cube root, there results 21, for the required aggregate."

[1] *De regula aliza* (1570), p. 111. Quoted by Eneström, *op. cit.*, Vol. VII (1906–7), p. 387.

In translation, Figure 47 is as follows:

"The *Quaestio VIII.*

"Divide 6 into three parts, in continued proportion, of which the sum of the squares of the first and second is 4. We let the first be the

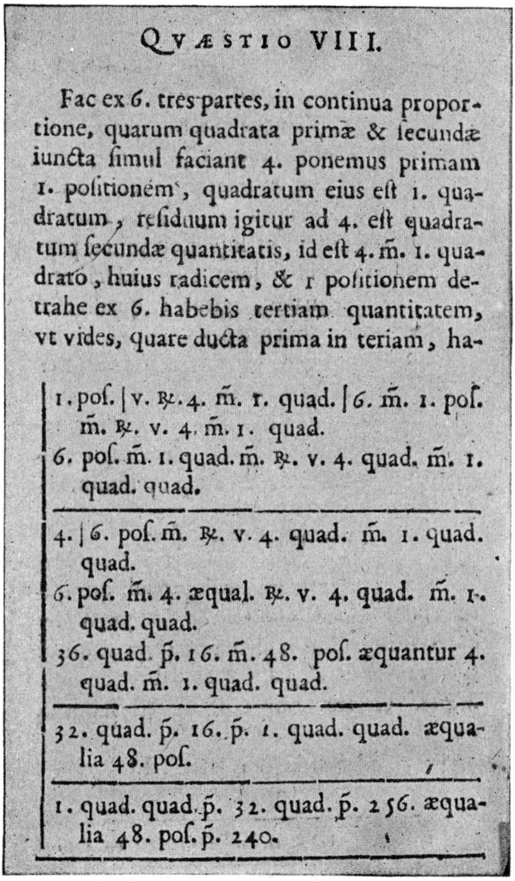

$Q_V \not\!E S T I O$ VIII.

Fac ex 6. tres partes, in continua proportione, quarum quadrata primæ & fecundæ
iuncta fimul faciant 4. ponemus primam
1. politioném, quadratum eius eft 1. quadratum, refiduum igitur ad 4. eft quadratum fecundæ quantitatis, id eft 4. m̄. 1. quadrato, huius radicem, & 1 politionem detrahe ex 6. habebis tertiam quantitatem,
vt vides, quare ducta prima in teriam, ha-

1. pof. | v. ℞. 4. m̄. 1. quad. | 6. m̄. 1. pof.
m̄. ℞. v. 4. m̄. 1. quad.

6. pof. m̄. 1. quad. m̄. ℞. v. 4. quad. m̄. 1.
quad. quad.

4. | 6. pof. m̄. ℞. v. 4. quad. m̄. 1. quad.
quad.

6. pof. m̄. 4. æqual. ℞. v. 4. quad. m̄. 1.
quad. quad.

36. quad. p̄. 16. m̄. 48. pof. æquantur 4.
quad. m̄. 1. quad. quad.

32. quad. p̄. 16. p̄. 1. quad. quad. æqualia 48. pof.

1. quad. quad. p̄. 32. quad. p̄. 256. æqualia 48. pof. p̄. 240.

Fig. 47.—Part of p. 297, from the *Ars magna*, as reprinted in H. Cardan's *Operum tomvs quartvs* (Lvgdvni, 1663).

1. position [i.e., x]; its square is 1. square [i.e., x^2]. Hence 4 minus this is the square of the second quantity, i.e., $4-1$. square [i.e., $4-x^2$]. Subtract from 6 the square root of this and also 1. position, and you will have the third quantity [i.e., $6-x-\sqrt{4-x^2}$], as you see, because the first multiplied by the third . . . . :

$$x \mid \sqrt{4-x^2} \mid 6-x-\sqrt{4-x^2} \mid$$
$$6x-x^2-\sqrt{4x^2-x^4}$$

$$4. \mid 6x-\sqrt{4x^2-x^4}$$
$$6x-4=\sqrt{4x^2-x^4}$$
$$36x^2+16-48x=4x^2-x^4$$

$$32x^2+16+x^4=48x$$

$$1x^4+32x^2+256=48x+240 \qquad \text{,,}$$

ITALIAN: NICOLO TARTAGLIA
(1537, 1543, 1546, 1556–60)

142. Nicolo Tartaglia's first publication, of 1537, contains little algebraic symbolism. He writes: "Radice .200. censi piu .10. cose" for $\sqrt{200x^2+10x}$, and "trouamo la cosa ualer Radice .200. men. 10." for "We find $x=\sqrt{200}-10$."[1] In his edition of Euclid's *Elements*[2] he writes "$R\!\!\!/ \ R\!\!\!/ \ R\!\!\!/ \ R\!\!\!/$" for the sixteenth root. In his *Qvesiti*[3] of 1546 one reads, "Sia .1. cubo de censo piu .48. equal à 14. cubi" for "Let $x^6+48=14x^3$," and "la $R\!\!\!/$. cuba de .8. ualera la cosa, cioe. 2." for "The $\sqrt[3]{8}$ equals x, which is 2."

More symbolism appeared ten years later. Then he used the $\bar{p}$ and $\tilde{m}$ of Pacioli to express "plus" and "minus," also the *co.*, *ce.*, *cu.*, etc., for the powers of numbers. Sometimes his abbreviations are less intense than those of Pacioli, as when he writes[4] *men* instead of $\tilde{m}$, or[5] *cen* instead of *ce*. Tartaglia uses $R\!\!\!/$ for *radix* or "root." Thus "la $R\!\!\!/ \ R\!\!\!/$ di $\frac{1}{16}$ è $\frac{1}{2}$,"[6] "la $R\!\!\!/$ cu. di $\frac{1}{8}$ è $\frac{1}{2}$,"[7] "la $R\!\!\!/$ rel. di $\frac{1}{32}$ è $\frac{1}{2}$,"[8] "la $R\!\!\!/$ cen. cu. di $\frac{1}{64}$ è $\frac{1}{2}$,"[9] "la $R\!\!\!/$ cu. cu. di $\frac{1}{512}$ è $\frac{1}{2}$,"[10] "la $R\!\!\!/$ terza rel. di $\frac{1}{2048}$ è $\frac{1}{2}$."[11]

143. Tartaglia writes proportion by separating the three terms which he writes down by two slanting lines. Thus,[12] he writes "9// 5//100," which means in modern notation $9:5=100:x$. For his occasional use of parentheses, see § 351.

---

[1] *Nova scientia* (Venice, 1537), last two pages of "Libro secondo."

[2] *Evclide Megarense* (Venice, 1569), fol. 229 (1st ed., 1543).

[3] *Qvesiti, et inventioni* (Venice, 1546), fol. 132.

[4] *Seconda parte del general trattato di nvmeri, et misvri de Nicolo Tartaglia* (Venice, 1556), fol. 88B.

[5] *Ibid.*, fol. 73.

[6] *Ibid.*, fol. 38.

[7] *Ibid.*, fol. 34.

[8] *Ibid.*, fol. 43.

[9] *Ibid.*, fol. 47B.

[10] *Ibid.*, fol. 60.

[11] *Ibid.*, fol. 68.

[12] *Ibid.*, fol. 162.

On the margin of the page shown in Figure 48 are given the symbols of powers of the unknown number, viz., *co.*, *ce.*, etc., up to the twenty-ninth power. In the illustrations of multiplication, the absolute number 5 is marked "5 n̊/0"; the 0 after the solidus indicates the *dignità* or power 0, as shown in the marginal table. His

11 MA per darte ad intendere vna regola, con la quale facilmente te conſeruarai in memoria le ſopraſcritte ripreſentationi, delli produtti delle ſopra ſcritte dignità multiplicate l'una fia l'altra, biſogna ſaper il numero ordinario di ciaſcuna di dette dignità, & ſe ben te aricordi, di ſopra ti ho detto qualmente il numero, conſiderato ſecondo ſe, non è dignità, ma ſolamẽte capo & principio di dette dignità, ſi come che anchora la vnità, conſiderata ſecondo ſe, non è numero, ma ſolamente principio del numero, adonque non eſſendo di nulla dignità il numero, gli daremo per ſuo ſegno. o. come che in margine ſi vede, & per che la. coſa. è la prima dignità, gli daremo per ſuo ſegno. 1. Et perche il cẽſo, è la ſecõda dignità, gli daremo per ſuo ſegno. 2. Et coſi perche il cubo, è la terza dignità, gli daremo per ſuo ſegno. 3. & coſi anchora per il ce. ce. è la quarta dignità, gli daremo per ſuo ſegno. 4. & coſi ſenza che piu oltra mi eſtenda andaremo procedendo di mano in mano nelle altre, come che in margine ſi vede annotato per fin alla 29 dignità, li quali ſegni de numeri, ſono ſituati nella continua progreſsione naturale arithmetica, & le dignità ſono ſituate nella continua proportionalità geometrica, Et ſe ben te aricordi nel 2° libro della 2ª parte a carte 131 nella ſeconda fazata, fu dichiarito nel primo Corellario della 8ª, che al multiplicare delle geometrice proportionalità, corriſponde il ſummare nelle arithmetice, E per tanto al multiplicare vna dignità fia vn'altra (che ſono nella proportionalità geometrica) corriſponde il ſommar di lor ſegni (che ſono nella progreſsione, ouer ꝓportionalità arithmetica) Eſſempi gratia volendo ſaper che coſa produca, ouer faza a multiplicar, cenſi fia pri. rel. ſumma il ſegno del primo relato (qual'è 5.) con il ſegno del cenſo (qual'è 2) & farà 7, & perche 7. è il ſegno del 2° relato, diremo che a multiplicar cenſi fia pri. relati farà ſecondi relati. Et coſi volendo ſaper che coſa faccia, ouer produca, a multiplicare cubi fia cenſi de cenſi. ſumma il ſegno di cubi (che è 3) con il ſegno di cece. (qual'è 4.) farà pur 7. & per che 7. è pur 2° relati, diremo che a multipl. cece. fia cu. farà 2° rel. & accio meglio apprẽdi quello che voglio dire, qua di ſotto te ne ponerò molti eſſẽpi ĩ figura.

a multipl.	5	n̊/o	a multipl.	4	n̊/o	a multipl.	3	co. 1	a multipl.	5	co. 1
fia	6	n̊/o	fia	8	co./1	fia	4	co. 1	fia	7	cu. 3
fara	30	n̊/o	fara	32	co./1	fara	12	ce. 2	fara	35	cece. 4
a multipli.	5	cu. 3	a multipli.	9	ce. 2	a multipl.	8	pri. rel. 5	a multipl.	3	cu. 3
fia	8	rel. 5	fia	6	rel. 5	fia	7	cu. ce. 6	fia	8	ce. cu. 6
fara	40	cecece. 8	fara	54	2° rel. 7	fara	56	3° rel. 11	fara	24	cu. cu. 9

Et coſi procederai in tutte le altre dignità ſi alte, come baſſe, vero è, che biſogna hauer in memoria i ſegni di dette dignità, & anchora le ſpecie della dignità corriſpondẽte a ciaſcun numero, cioè ſe la ſumma del ſegno delli cubi (quel'è 3.) con il ſegno delli ce. cu. (qual'è 6.) fa. 9. non hauendo in memoria che'l detto 9. è il ſegno di cubi de cubi, nõ ſapreſsimo che a multiplicar cubi fia cenſi cubi, faceſſe cubi de cubi, e però biſogna in cio aduertire.

Li ſotto ſcritti ſono li naturali ſegni per numero delle dette noſtre dignità algebratice.

	Segni p numeri
n	0
co	1
ce	2
cu	3
cece	4
pri. rel.	5
cecu	6
2° rel.	7
cecece	8
cucu	9
ce. pri. rel.	10
3° rel.	11
cucece	12
4° rel.	13
ce. 2° rel.	14
cu. pri. rel.	15
cececece	16
5° rel.	17
cucecu	18
6° rel.	19
cece. pri. rel.	20
cu. 2° rel.	21
ce. 3° rel.	22
7° rel.	23
cucecece	24
8° rel.	25
ce. 4° rel.	26
cucucu	27
cece. 2° rel.	28
9° rel.	29

Et coſi diſcorrẽdo in infinito.

A ii

FIG. 48.—Part of a page from Tartaglia's *La sesta parte del general trattato de nvmeri, et misvre* (Venice, 1560), fol. 2.

illustrations stress the rule that in multiplication of one *dignità* by another, the numbers expressing the *dignità* of the factors must be added.

ITALIAN: RAFAELE BOMBELLI (1572, 1579)

144. Bombelli's *L'algebra* appeared at Venice in 1572 and again at Bologna in 1579. He used *p.* and *m.* for "plus" and "minus."

Following Cardan, Bombelli used almost always *radix legata* for a root affecting only one term. To write two or more terms into one, Bombelli wrote an *L* right after the ℞ and an inverted ⅃ at the end of the expression to be radicated. Thus he wrote: ℞ L 7 *p*.℞ 14 ⅃ for our modern $\sqrt{7+\sqrt{14}}$, also[1] *Rq* L *Rc* L *Rq* 68 *p*.2 ⅃ *m* *Rc* L *R q* 68 *m* 2 ⅃⅃ for the modern $\sqrt{\left\{\sqrt[3]{(\sqrt{68}+2)}-\sqrt[3]{(\sqrt{68}-2)}\right\}}$.

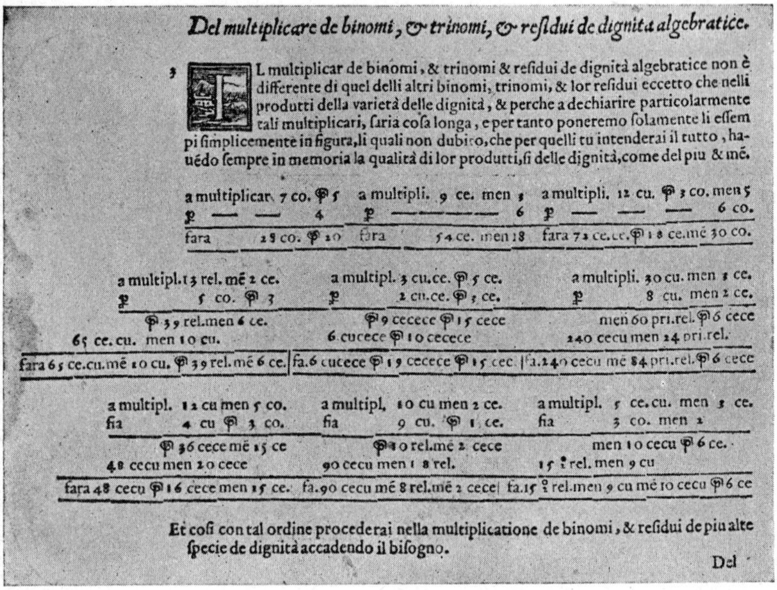

Del multiplicare de binomi, & trinomi, & residui de dignità algebratice.

3 IL multiplicar de binomi, & trinomi & residui de dignità algebratice non è differente di quel delli altri binomi, trinomi, & lor residui eccetto che nelli produtti della varietà delle dignità, & perche a dechiarire particolarmente tali multiplicari, saria cosa longa, e per tanto poneremo solamente li essempi simplicemente in figura, li quali non dubito, che per quelli tu intenderai il tutto, hauẽdo sempre in memoria la qualità di lor produtti, si delle dignità, come del piu & mẽ.

a multiplicar 7 co. p̃ 5 — p̃ 4 — fara 28 co. p̃ 20
a multipli. 9 ce. men 3 — p̃ 6 — fara 54 ce. men 18
a multipli. 12 cu. p̃ 3 co. men 5 — p̃ 6 co. — fara 72 ce.ce. p̃ 18 ce. mẽ 30 co.

a multipl. 13 rel. mẽ 2 ce. — p̃ 5 co. p̃ 3 — p̃ 39 rel. men 6 ce. — 65 ce.cu. men 10 cu. — fara 65 ce.cu. mẽ 10 cu. p̃ 39 rel. mẽ 6 ce.
a multipl. 3 cu.ce. p̃ 5 ce. — p̃ 2 cu.ce. p̃ 3 ce. — p̃ 9 cecece p̃ 15 cece — 6 cucece p̃ 10 cecece — fa. 6 cucece p̃ 19 cecece p̃ 15 cece
a multipli. 30 cu. men 3 ce. — p̃ 8 cu. men 2 ce. — men 60 pri.rel. p̃ 6 cece — 240 cecu men 24 pri.rel. — fa. 240 cecu mẽ 84 pri.rel. p̃ 6 cece

a multipl. 12 cu men 5 co. — fia 4 cu p̃ 3 co. — p̃ 36 cece mẽ 15 ce — 48 cecu men 20 cece — fara 48 cecu p̃ 16 cece men 15 ce.
a multipl. 10 cu men 2 ce. — fia 9 cu. p̃ 1 ce. — p̃ 10 rel. mẽ 2 cece — 90 cecu men 18 rel. — fa. 90 cecu mẽ 8 rel. mẽ 2 cece
a multipl. 5 ce.cu. men 3 ce. — fia 3 co. men 2 — men 10 cecu p̃ 6 ce. — 15 ? rel. men 9 cu — fa. 15 ? rel. men 9 cu mẽ 10 cecu p̃ 6 ce

Et cosi con tal ordine procederai nella multiplicatione de binomi, & residui de piu alte specie de dignità accadendo il bisogno.

Del

FIG. 49.—Part of a page from Tartaglia's *La sesta parte del general trattato de nvmeri, et misvre* (Venice, 1560), fol. 4. Shows multiplication of binomials. Observe the fancy .p̃. for "plus." For "minus" he writes here *mẽ* or *men*.

An important change in notation was made for the expression of powers which was new in Italian algebras. The change is along the line of what is found in Chuquet's manuscript of 1484. It is nothing less than the introduction of positive integral exponents, but without writing the base to which they belonged. As long as the exponents were applied only to the unknown x, there seemed no need of writing the x. The notation is shown in Figure 50.

[1] Copied by Cantor, *op. cit.*, Vol. II (2d ed., 1913), p. 624, from Bombelli's *L'algebra*, p. 99.

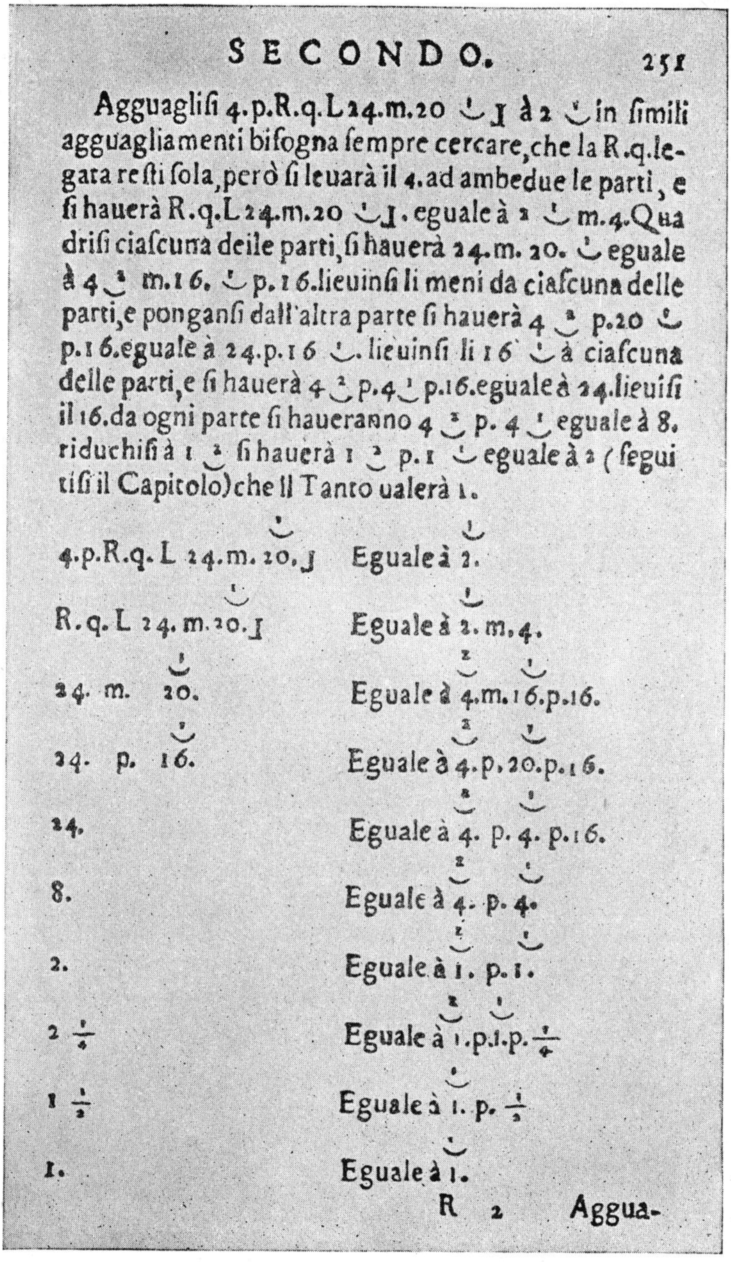

SECONDO. 251

Agguaglifi 4.p.R.q.L24.m.20 ‿ J à 2 ‿ in fimili
agguagliamenti bifogna fempre cercare, che la R.q.le-
gata refti fola, però fi leuarà il 4. ad ambedue le parti, e
fi hauerà R.q.L24.m.20 ‿ J. eguale à 2 ‿ m.4.Qua
drifi ciafcuna deile parti, fi hauerà 24.m. 20. ‿ eguale
à 4 ‿ m.16. ‿ p.16.lieuinfi li meni da ciafcuna delle
parti, e ponganfi dall'altra parte fi hauerà 4 ‿ p.20 ‿
p.16.eguale à 24.p.16 ‿. lieuinfi li 16 ‿ à ciafcuna
deile parti, e fi hauerà 4 ‿ p.4 ‿ p.16.eguale à 24.lieuifi
il 16.da ogni parte fi haueranno 4 ‿ p. 4 ‿ eguale à 8.
riduchifi à 1 ‿ fi hauerà 1 ‿ p.1 ‿ eguale à 2 (fegui
tifi il Capitolo)che il Tanto ualerà 1.

4.p.R.q.L 24.m. 20, J   Eguale à 2.

R.q.L 24. m.20. J   Eguale à 2. m, 4.

24. m. 20.   Eguale à 4.m.16.p.16.

24. p. 16.   Eguale à 4.p.20.p.16.

24.   Eguale à 4. p. 4. p.16.

8.   Eguale à 4. p. 4.

2.   Eguale à 1. p. 1.

$2 \frac{1}{2}$ Eguale à 1.p.1.p. $\frac{1}{2}$

$1 \frac{1}{2}$ Eguale à 1. p. $\frac{1}{2}$

1.   Eguale à 1.

R 2   Aggua-

Fig. 50.—From Bombelli's *L'algebra* (1572)

In Figure 50 the equations are:

$$4+\sqrt{24-20x}=2x \, ,$$
$$\sqrt{24-20x}=2x-4 \, ,$$
$$24-20x=4x^2-16x+16 \, ,$$
$$24+16x=4x^2+20x+16 \, ,$$
$$24=4x^2+4x+16 \, ,$$
$$8=4x^2+4x \, ,$$
$$2=x^2+x \, ,$$
$$2\tfrac{1}{4}=x^2+x+\tfrac{1}{4} \, ,$$
$$1\tfrac{1}{2}=x+\tfrac{1}{2} \, ,$$
$$1=x \, .$$

Bombelli expressed square root by $R.$ $q.$, cube root by $R.$ $c.$, fourth root by R $R.$ $q.$, fifth root (*Radice prima incomposta, ouer relata*) by $R.$ $p.$ $r.$, sixth root by $R.$ $q.$ $c.$, seventh root by $R.$ $s.$ $r.$, the square root of a polynomial (*Radice quadrata legata*) by $R.$ $q.$ L ⌐; the cube root of a polynomial (*Radice cubica legata*) by $R.$ $c.$ L ⌐. Some of these symbols are shown in Figure 51. He finds the sum of $\sqrt[3]{72-\sqrt{1,088}}$ and $\sqrt[3]{\sqrt{4,352}+16}$ to be $\sqrt[3]{232+\sqrt{53,312}}$.

The first part of the sentence preceding page 161 of Bombelli's *Algebra*, as shown in Figure 51, is "Sommisi $R.$ $c.$ L $R.$ $q.$ 4352 .$p.$ 16.⌐ con $R.$ $c.$ L 72. $m.$ $R.$ $q.$ 1088.⌐."

145. Bombelli's *Algebra* existed in manuscript about twenty years before it was published. The part of a page reproduced in Figure 52 is of interest as showing that the mode of expressing aggregation of terms is different from the mode in the printed texts. We have here the expression of the radicals representing x for the cubic $x^3=32x+24$. Note the use of horizontal lines with cross-bars at the ends; the lines are placed below the terms to be united, as was the case in Chuquet. Observe also that here a negative number is not allowed to stand alone: $--1069$ is written $0-1069$. The cube root is designated by R^3, as in Chuquet.

A manuscript, kept in the Library of the University of Bologna, contains data regarding the sign of equality ($=$). These data have been communicated to me by Professor E. Bortolotti and tend to show that ($=$) as a sign of equality was developed at Bologna independently of Robert Recorde and perhaps earlier.

The problem treated in Figure 53 is to divide 900 into two parts, one of which is the cube root of the other. The smaller part is desig-

R.q. 1088. ɪ Quefte due R.ſi poſſono ſommare, perche il lato di R. c. L 72. m. R. q. 1088. ch'è R. c. L
68. m. 2. ɪ è in proportione dupla à R. c. L R. q. 43-
52. m. 16. ɪ reſiduo di R.c. L R.q. 4352. p. 16. ɪ però ſi poſſono ſommare (com'è detto) partendo la mag
giore per la minore, cioè per R.c. L 72. m. R .q. 1088, ɪ
che moltiplicata uia il ſuo Binomio (come ſi uede nel-

R.c. L 72. m. R.q. 1088.   R.c. L R.q. 4352. p. 16. ɪ
R.c. L 72. p. R.q. 1088.   R.c. L 72. p. R.q. 1088. ɪ

R.c. 4096.   R.c. L. R.q. 18415616. p. 3328. ɪ
Lato 16. partitore.   Lato R.q. 172. p. 4.

$$\text{Auenimento } R.q. 1 \tfrac{1}{16} \ p. \tfrac{1}{4}, \text{ giontoli } 1,$$
$$\text{fa } R.q. 1 \tfrac{1}{16} \ p. 1 \tfrac{1}{4} \ \text{Via } R.q. 1 \tfrac{1}{16} \ p. 1 \tfrac{1}{4}$$

$$\text{Fà } R.q. 4. \tfrac{1}{6} \tfrac{1}{4} \ p. 2. \tfrac{1}{8}$$
$$R.q. 1 \tfrac{1}{16} \ p. 1 \tfrac{1}{4}$$

$$R.c. L \ 5 \tfrac{1}{4} \tfrac{1}{6} \ p. R.q. 35. \tfrac{2}{2} \tfrac{1}{6} \ ɪ \cdot$$
R.c. L 72. m. R.q. 1088. ɪ

Somma.   R.c. L 232. p R.q. 53312. ɪ

la figura) fà 16, e queſto è il partitore, e moltiplicato
R.c. L R.q. 4352. p. 16. ɪ uia Rad. c. L 72. p. Rad.
1088. ɪ Binomio del partitore fà R.c. L R.q. 18415616.
p. 3328. ɪ, che il ſuo lato è R. q. 272. p. 4, che parti-
to per 16. ne uiene R. q. 1 $\tfrac{1}{16}$ p. $\tfrac{1}{4}$, che aggiontoli 1
per regola fà 1 $\tfrac{1}{4}$ p. Rad. q. 1, $\tfrac{1}{6}$, e quello ſi ha da
moltiplicare uia R.c. L 72. m. R.q. 1088. ɪ però riducaſi à
L   R.c.

Fɪɢ. 51.—Bombelli's *Algebra*, p. 161 of the 1579 impression, exhibiting the
calculus of radicals. In the third line of the computation, instead of 18,415,616
there should be 27,852,800. Notice the broken fractional lines, indicating difficulty
in printing fractions with large numerators and denominators.

nated by a symbol consisting of c and a flourish (probably intended for co). Then follows the equation 900 $\tilde{m}$ $1co^①=1cu^③$. (our $900-x=x^3$). One sees here a mixture of two notations for x and x^3: the notation co and cu made familiar by Luca Pacioli, and Bombelli's exponential notation, with the 1 and 3, placed above the line, each exponent resting in a cup. It is possible that the part of the algebra here photographed may go back as far as about 1550. The cross-writing in the photograph begins: "in libro vecchio a carte 82: quella di far di 10 due parti: dice messer Nicolo che l'ona e R 43 p 5 $\tilde{m}$ R18: et l'altra il resto sino a 10, cioe 5 $\tilde{m}$ R 43 $\tilde{p}$. R 18." This Nicolo is supposed to be Nicolo Tartaglia who died in 1557. The phrasing "Messer Nicolo" implies, so Bortolotti argues, that Nicolo was a living contemporary. If these contentions are valid, then the manuscript in question was written in 1557 or earlier.[1]

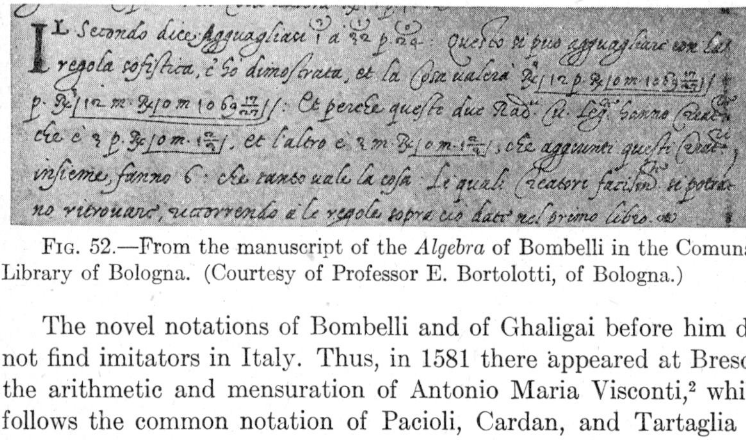

Fig. 52.—From the manuscript of the *Algebra* of Bombelli in the Comunale Library of Bologna. (Courtesy of Professor E. Bortolotti, of Bologna.)

The novel notations of Bombelli and of Ghaligai before him did not find imitators in Italy. Thus, in 1581 there appeared at Brescia the arithmetic and mensuration of Antonio Maria Visconti,[2] which follows the common notation of Pacioli, Cardan, and Tartaglia in designating powers of the unknown.

GERMAN: IOHANN WIDMAN
(1489, 1526)

146. Widman's *Behennde vnnd hübsche Rechnūg auff allen Kauff-manschafften* is the earliest printed arithmetic which contains the signs plus (+) and minus (−) (see §§ 201, 202).

[1] Since the foregoing was written, E. Bortolotti has published an article, on mathematics at Bologna in the sixteenth century, in the *Periodico di Matematiche* (4th ser., Vol. V, 1925), p. 147–84, which contains much detailed information, and fifteen facsimile reproductions of manuscripts exhibiting the notations then in use at Bologna, particularly the use of a dash (−) and the sign (=) to express equality.

[2] *Antonii Mariae Vicecomitis Civis Placentini practica numerorum & mensurarum* (Brixiae, 1581).

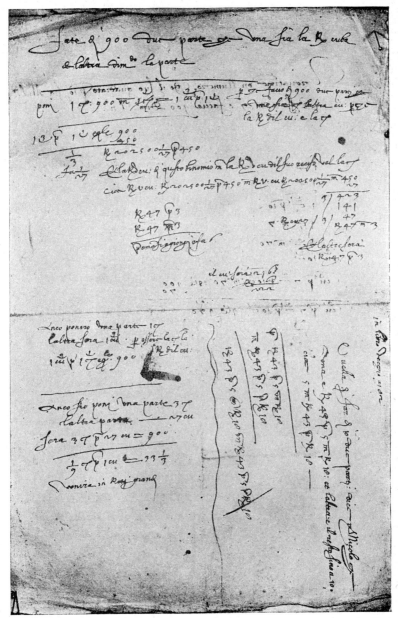

FIG. 53.—From a pamphlet (marked No. 595N, in the Library of the University of Bologna) containing studies and notes which Professor Bortolotti considers taken from the lessons of Pompeo Bolognetti ([Bologna?]–1568).

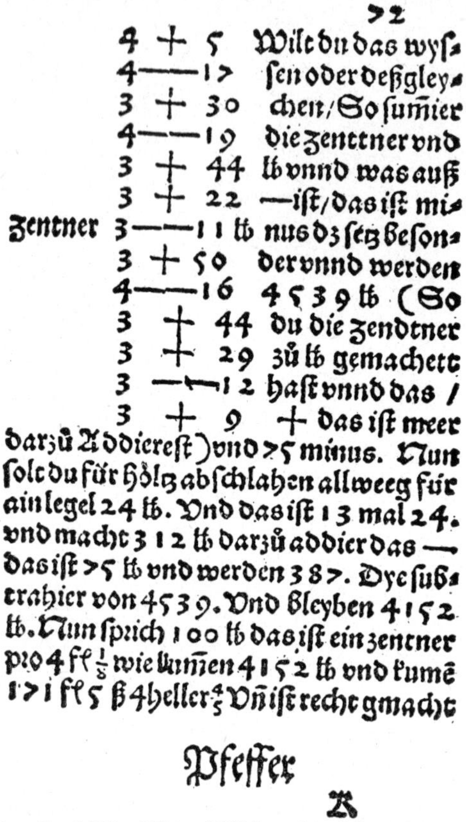

72

$$4 + 5 \quad \text{Wilt du das wyſ-}$$
$$4 \text{——} 17 \quad \text{ſen oder deßgley-}$$
$$3 + 30 \quad \text{chen/ So ſumier}$$
$$4 \text{——} 19 \quad \text{die zenttner vnd}$$
$$3 + 44 \quad \text{lb vnnd was auß}$$
$$3 + 22 \quad \text{—iſt/das iſt mi-}$$
$$\text{Zentner} \quad 3 \text{——} 11 \quad \text{lb nus dz ſetz beſon-}$$
$$3 + 50 \quad \text{der vnnd werden}$$
$$4 \text{——} 16 \quad 4539 \; \text{lb (So}$$
$$3 + 44 \quad \text{du die zendtner}$$
$$3 + 29 \quad \text{zů lb gemachett}$$
$$3 \text{—+} 12 \quad \text{haſt vnnd das /}$$
$$3 + 9 \quad + \text{das iſt meer}$$

darzů Addiereſt) vnd 75 minus. Nun
ſolt du für Höltz abſchlahen allweeg für
ain legel 24 lb. Vnd das iſt 13 mal 24.
vnd macht 312 lb darzů addier das —
das iſt 75 lb vnd werden 387. Dye ſub-
trahier von 4539. Vnd bleyben 4152
lb. Nun ſprich 100 lb das iſt ein zentner
pro 4 ff ⅛ wie kumen 4152 lb vnd kumē
171 ff 5 ß 4 heller ⅔ Vñ iſt recht gmacht

Pfeffer

Fig. 54.—From the 1526 edition of Widman's arithmetic. (Taken from D. E. Smith, *Rara arithmetica*, p. 40.)

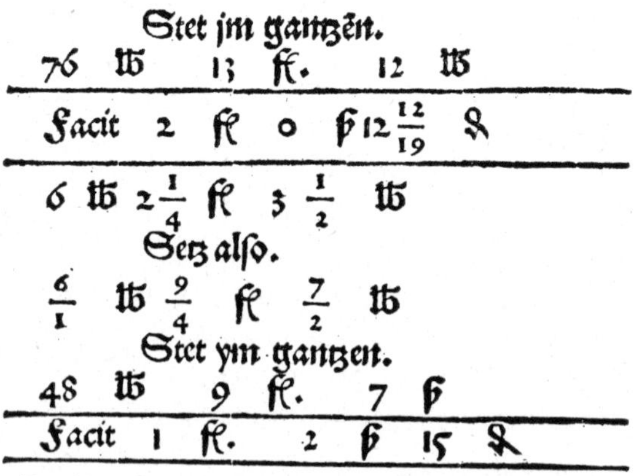

Stet im gantzen.			
76 lb	13 ff.	12 lb	
Facit 2 ff	0 f	$12\frac{12}{19}$ &	

6 lb $2\frac{1}{4}$ ff 3 $\frac{1}{2}$ lb

Setz alſo.

$\frac{6}{1}$ lb $\frac{9}{4}$ ff $\frac{7}{2}$ lb

Stet ym gantzen.

48 lb	9 ff.	7 f
Facit 1 ff.	2 f	15 &

Fig. 55.—From the arithmetic of Grammateus (1518)

AUSTRIAN: HEINRICH SCHREIBER (GRAMMATEUS) (1518, 1535)

147. Grammateus published an arithmetic and algebra, entitled *Ayn new Kunstlich Buech* (Vienna), printed at Nürnberg (1518), of which the second edition appeared in 1535. Grammateus used the

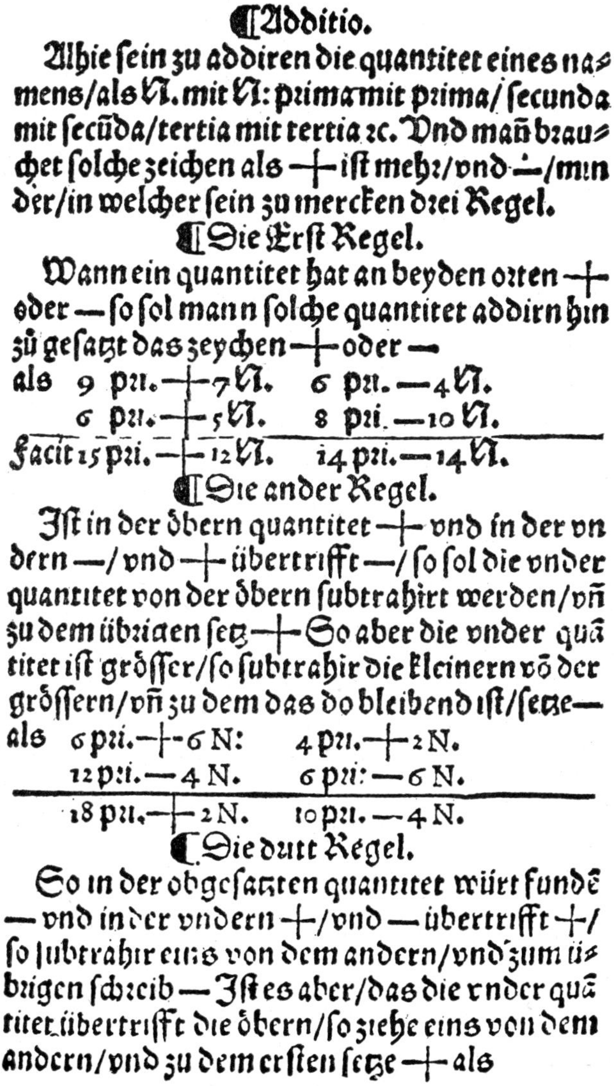

¶ Additio.

Alhie sein zu addiren die quantitet eines na-
mens/als N. mit N: prima mit prima/ secunda
mit secūda/tertia mit tertia ꝛc. Vnd man brauchet solche zeichen als + ist mehr/vnd ∸ /min
der/in welcher sein zu mercken drei Regel.

¶ Die Erst Regel.

Wann ein quantitet hat an beyden orten +
oder — so sol mann solche quantitet addirn hin
zu gesatzt das zeychen + oder —
als 9 pri. + 7 N. 6 pri. — 4 N.
6 pri. + 5 N. 8 pri. — 10 N.
facit 15 pri. + 12 N. 14 pri. — 14 N.

¶ Die ander Regel.

Ist in der obern quantitet + vnd in der vn
dern —/vnd + übertrifft —/so sol die vnder
quantitet von der obern subtrahirt werden/vñ
zu dem übrigen setz + So aber die vnder quã
titet ist grösser/so subtrahir die kleinern võ der
grössern/vñ zu dem das do bleibend ist/setze —
als 6 pri. + 6 N: 4 pri. + 2 N.
12 pri. — 4 N. 6 pri. — 6 N.
18 pri. + 2 N. 10 pri. — 4 N.

¶ Die dritt Regel.

So in der obgesatzten quantitet würt fundē
— vnd in der vndern +/vnd — übertrifft +/
so subtrahir eins von dem andern/vnd zum ü-
brigen schreib — Ist es aber/das die vnder quã
titet übertrifft die obern/so ziehe eins von dem
andern/vnd zu dem ersten setze + als

FIG. 56.—From the arithmetic of Grammateus (1535). (Taken from D. E. Smith, *Rara arithmetica*, p. 125.)

plus and minus signs in a technical sense for addition and subtraction. Figure 55 shows his mode of writing proportion: $76lb. : 13fl. = 12lb. : x$. He finds $x = 2fl.$ 0 s. $12\frac{1}{2}\frac{2}{3}\vartheta$. [$1fl. = 8s.$, $1s. = 30\vartheta$].

The unknown quantity x and its powers x^2, x^3, ... , were called, respectively, *pri* (*prima*), 2*a.* or *se.* (*seconda*), 3*a.* or *ter.* (*terza*), 4*a.*

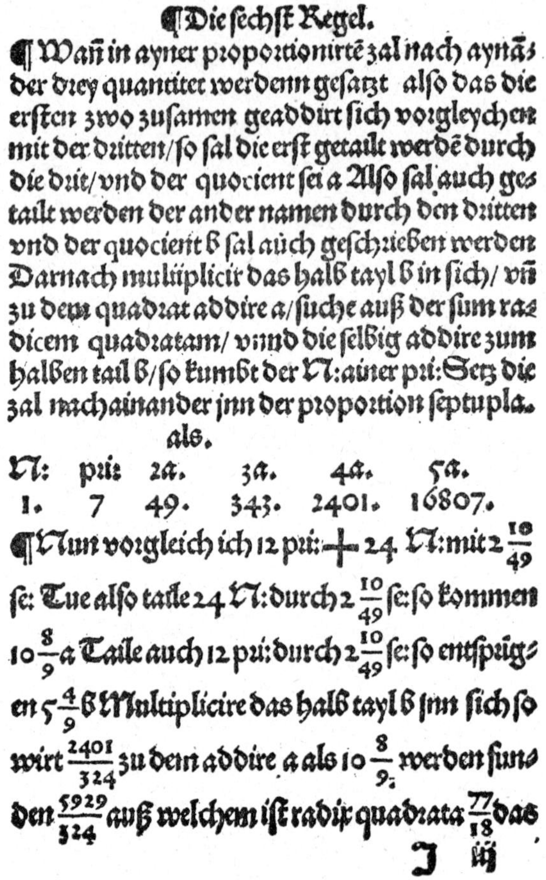

¶Die sechst Regel.

¶ Wañ in ayner proportionirte zal nach aynã der drey quantiter werdenn gesatzt also das die ersten zwo zusamen geaddirt sich vergleychen mit der dritten/so sal die erst getailt werdē durch die drit/vnd der quocient sei a Also sal auch ge taillt werden der ander namen durch den dritten vnd der quocient ß sal auch geschrieben werden Darnach multiplicir das halb tayl ß in sich/vñ zu dem quadrat addire a/suche auß der sum ra dicem quadratam/vnnd die selbig addire zum halben tail ß/so kumbt der N:ainer pri:Setz die zal nach ainander jnn der proportion septupla. als.

N: pri: 2*a.*   3*a.*   4*a.*   5*a.*
1.   7   49.   343.   2401. 16807.

¶Nun vergleich ich 12 pri:┼ 24 N:mit 2 $\frac{10}{49}$

se: Tue also taile 24 N:durch 2 $\frac{10}{49}$ se:so kommen

10 $\frac{8}{9}$ a Taile auch 12 pri:durch 2 $\frac{10}{49}$ se:so entspring

en 5 $\frac{4}{9}$ ß Multiplicire das halb tayl ß jnn sich so

wirt $\frac{2401}{324}$ zu dem addire a als 10 $\frac{8}{9}$ werden sind

den $\frac{5929}{324}$ auß welchem ist radix quadrata $\frac{77}{18}$ das

J iij

FIG. 57.—From the arithmetic of Grammateus (1518)

or *quart.* (*quarta*), 5*a.* or *quit.* (*quinta*), 6*a.* or *sex.* (*sexta*); *N.* stands for absolute number.

Fig. 56 shows addition of binomials. Figure 57 amounts to the solution of a quadratic equation. In translation: "The sixth rule: When in a proportioned number [i.e., in 1, x, x^2] three quantities are taken so that the first two added together are equal to the third [i.e.,

$d+ex=fx^2$], then the first shall be divided by [the coefficient of] the third and the quotient designated a. In the same way, divide the [coefficient of] the second by the [coefficient of] the third and the quotient designated b. Then multiply the half of b into itself and to the square add a; find the square root of the sum and add that to half of b. Thus is found the N. of 1 *pri.* [i.e., the value of x]. Place the number successively in the seven-fold proportion

N:	x	x^2	x^3	x^4	x^5
1.	7	49	343.	2,401.	16,807.

Now I equate $12x+24$ with $2\frac{10}{49}x^2$. Proceed thus: Divide 24 by $2\frac{10}{49}x^2$; there is obtained $10\frac{8}{9}a$. Divide also $12x$ by $2\frac{10}{49}x^2$; thus arises $5\frac{4}{9}b$. Multiplying the half of b by itself gives $\frac{2401}{324}$, to which adding a, i.e., $10\frac{8}{9}$, will yield $\frac{5929}{324}$, the square root of which is $\frac{77}{18}$; add this to half of the part b or $\frac{48}{18}$, and there results the number 7 as the number 1 *pri.* [i.e., x]."

The following example is quoted from Grammateus by Treutlein:[1]

" 6 *pri.*+8 N.	**Modern Symbols**
Durch	$6x +8$
5 *pri.*−7 N.	$5x -7$
30 *se.*+40 *pri.*	$30x^2+40x$
−42 *pri.*−56 N.	$-42x-56$
30 *se.*− 2 *pri.*−56 N. "	$30x^2-2x-56$.

In the notation of Grammateus, 9 *ter.*+30 *se.*−6 *pri.* +48N. stands for $9x^3+30x^2-6x+48$.[2]

We see in Grammateus an attempt to discard the old cossic symbols for the powers of the unknown quantity and to substitute in their place a more suitable symbolism. The words *prima, seconda*, etc., remind one of the nomenclature in Chuquet. His notation was adopted by Gielis van der Hoecke.

GERMAN: CHRISTOFF RUDOLFF
(1525)

148. Rudolff's *Behend vnnd Hubsch Rechnung durch die kunstreichen regeln Algebre so gemeincklich die Coss genent werden* (Strass-

---

[1] P. Treutlein, *Abhandlungen zur Geschichte der Mathematik*, Vol. II (Leipzig, 1879), p. 39.

[2] For further information on Grammateus, see C. I. Gerhardt, "Zur Geschichte der Algebra in Deutschland," *Monatsbericht d. k. Akademie der Wissenschaften zu Berlin* (1867), p. 51.

burg, 1525) is based on algebras that existed in manuscript (§ 203). Figure 58 exhibits the symbols for indicating powers up to the ninth. The symbol for *cubus* is simply the letter *c* with a final loop resembling the letter *e*, but is not intended as such. What appears below the symbols reads in translation: "*Dragma* or *numerus* is taken here as 1. It is no number, but assigns other numbers their kind. *Radix* is the

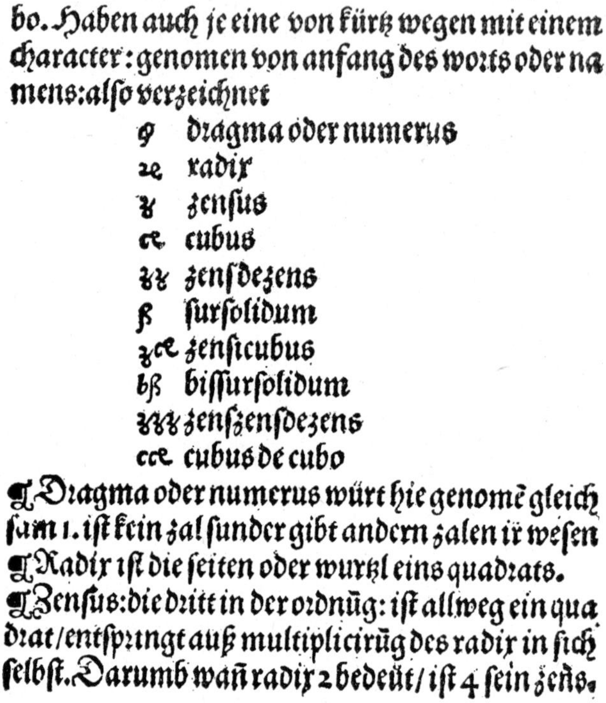
60. Haben auch je eine von kürtz wegen mit einem character: genomen von anfang des worts oder na mens: alſo verzeichnet

dragma oder numerus
radix
zenſus
cubus
zenſdezens
ſurſolidum
zenſicubus
biſſurſolidum
zenſzenſdezens
cubus de cubo

Dragma oder numerus würt hie genomē gleich ſam 1. iſt kein zal ſunder gibt andern zalen ir weſen
Radix iſt die ſeiten oder wurtzl eins quadrats.
Zenſus: die dritt in der ordnūg: iſt allweg ein qua drat/entſpringt auß multiplicirūg des radix in ſich ſelbſt. Darumb wañ radix 2 bedeüt/iſt 4 ſein zēns.

FIG. 58.—From Rudolff's *Coss* (1525)

side or root of a square. *Zensus*, the third in order, is always a square; it arises from the multiplication of the *radix* into itself. Thus, when *radix* means 2, then 4 is the *zensus*." Adam Riese assures us that these symbols were in general use ("zeichen ader benennung Di in gemeinen brauch teglich gehandelt werdenn").[1] They were adopted by Adam

[1] Riese's *Coss* was found, in manuscript, in the year 1855, in the Kirchen- und Schulbibliothek of Marienberg, Saxony; it was printed in 1892 in the following publication: *Adam Riese, sein Leben, seine Rechenbücher und seine Art zu rechnen. Die Coss von Adam Riese*, by Realgymnasialrektor Bruno Berlet, in Annaberg i. E., 1892.

Riese, Apian, Menher, and others. The addition of radicals is shown in Figure 59. Cube root is introduced in Rudolff's *Coss* of 1525 as follows: "Würt radix cubica in diesem algorithmo bedeut durch solchen character ⋎⋎⋎/ , als ⋏⋏⋏/ 8 is zu versteen radix cubica aufs 8."· ("In this algorithm the cubic root is expressed by this character ⋏⋏⋏/ , as ⋏⋏⋏/ 8 is to be understood to mean the cubic root of 8.") The fourth root Rudolff indicated by ⋏⋏/ ; the reader naturally wonders why two strokes should signify fourth root when three strokes indicate cube root. It is not at once evident that the sign for the fourth

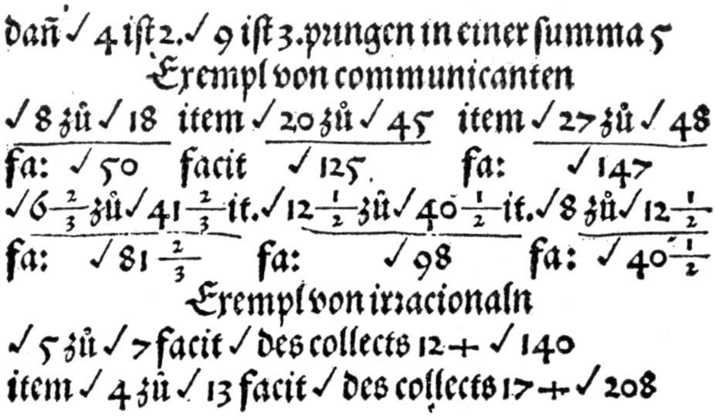

Fig. 59.—From Rudolff's *Coss* (1525)

root represented two successive square-root signs, thus, $\sqrt{\sqrt{}}$. This crudeness in notation was removed by Michael Stifel, as we shall see later.

The following example illustrates Rudolff's subtraction of fractions:[1]

$$\text{``} \frac{1\;\mathcal{X}-2}{12} \text{ von } \frac{12}{1\;\mathcal{X}+2} \text{ Rest } \frac{148-1\mathfrak{z}}{12\;\mathcal{X}+24}\text{ ,''}$$

On page 141 of his *Coss*, Rudolff indicates aggregation by a dot;[2] i.e., the dot in "$\sqrt{}.12+\sqrt{}140$" indicates that the expression is $\sqrt{12+\sqrt{140}}$, and not $\sqrt{12}+\sqrt{140}$. In Stifel sometimes a second dot appears at the end of the expression (§ 348). Similar use of the dot we shall find in Ludolph van Ceulen, P. A. Cataldi, and, in form of the colon (:), in William Oughtred.

When dealing with two unknown quantities, Rudolff represented

[1] Treutlein, "Die deutsche Coss," *op. cit.*, Vol. II, p. 40.

[2] G. Wertheim, *Abhandlungen zur Geschichte der Mathematik*, Vol. VIII (Leipzig, 1898), p. 153.

the second one by the small letter q, an abbreviation for *quantita*, which Pacioli had used for the second unknown.[1]

Interesting at this early period is the following use of the letters a, c, and d to represent ordinary numbers (folio Giij^a): "Nim $\frac{1}{2}$ solchs collects | setz es auff ein ort | dz werd von lere wegen c genennt. Darnach subtrahier das c vom a | das übrig werd gesprochen d. Nun sag ich dz $\sqrt{c}+\sqrt{d}$ ist quadrata radix des ersten binomij." ("Take $\frac{1}{2}$ this sum, assume for it a position, which, being empty, is called c. Then subtract c from a, what remains call d. Now I say that $\sqrt{c}+\sqrt{d}$ is the square root of the first binomial.")[2]

149. Rudolff was convinced that development of a science is dependent upon its symbols. In the Preface to the second part of Rudolff's *Coss* he states: "Das bezeugen alte bücher nit vor wenig jaren von der coss geschriben, in welchen die quantitetn, als dragma, res, substantia etc. nit durch character, sunder durch gantz geschribne wort dargegeben sein, vnd sunderlich in practicirung eines yeden exempels die frag gesetzt, ein ding, mit solchen worten, ponatur vna res." In translation: "This is evident from old books on algebra, written many years ago, in which quantities are represented, not by characters, but by words written out in full, 'drachm,' 'thing,' 'substance,' etc., and in the solution of each special example the statement was put, 'one thing,' in such words as *ponatur, una res*, etc."[3]

In another place Rudolff says: "Lernt die zalen der coss aussprechen vnnd durch ire charakter erkennen vnd schreiben."[4] ("Learn to pronounce the numbers of algebra and to recognize and write them by their characters.")

DUTCH: GIELIS VAN DER HOECKE

(1537)

150. An early Dutch algebra was published by Gielis van der Hoecke which appeared under the title, *In arithmetica een sonderlinge excellēt boeck* (Antwerp [1537]).[5] We see in this book the early appear-

---

[1] Chr. Rudolff, *Behend vnnd Hubsch Rechnung* (Strassburg, 1525), fol. R1^a. Quoted by Eneström, *Bibliotheca mathematica*, Vol. XI (1910–11), p. 357.

[2] Quoted from Rudolff by Eneström, *ibid.*, Vol. X (1909–10), p. 61.

[3] Quoted by Gerhardt, *op. cit.* (1870), p. 153. This quotation is taken from the second part of Gerhardt's article; the first part appeared in the same publication, for the year 1867, p. 38–54.

[4] *Op. cit.*, Buch I, Kap. 5, Bl. Dijr°; quoted by Tropfke, *op. cit.* (2. ed.), Vol. II, p. 7.

[5] On the date of publication, see Eneström, *op. cit.*, Vol. VII (1906–7), p. 211; Vol. X (1909–10), p. 87.

ance of the plus and minus signs in Holland. As the symbols for powers one finds here the notation of Grammateus, *N.*, *pri.*, *se.*, 3^a, 4^a, 5^a, etc., though occasionally, to fill out a space on a line, one en-

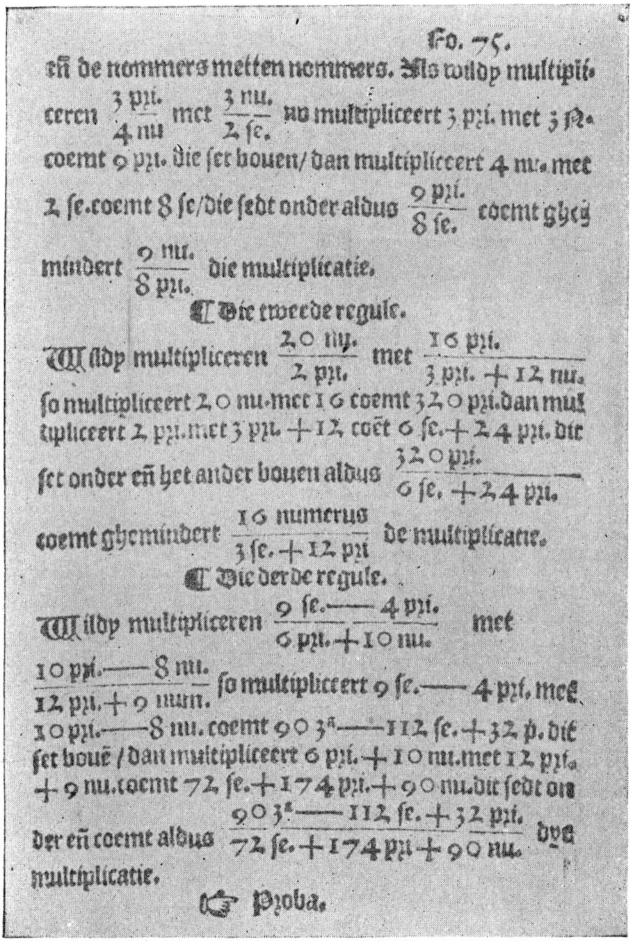
Fo. 75.
eñ de nommers metten nommers. Als wildy multipli-
ceren $\frac{3\ \text{pri.}}{4\ \text{nu}}$ met $\frac{3\ \text{nu.}}{2\ \text{se.}}$ so multipliceert 3 pri. met 3 N.
coemt 9 pri. die set bouen/ dan multipliceert 4 nu. met
2 se. coemt 8 se/die sedt onder aldus $\frac{9\ \text{pri.}}{8\ \text{se.}}$ coemt ghemindert $\frac{9\ \text{nu.}}{8\ \text{pri.}}$ die multiplicatie.
¶ Die tweede regule.
Wildy multipliceren $\frac{20\ \text{nu.}}{2\ \text{pri.}}$ met $\frac{16\ \text{pri.}}{3\ \text{pri.} + 12\ \text{nu.}}$
so multipliceert 20 nu. met 16 coemt 320 pri. dan mul-
tipliceert 2 pri. met 3 pri. + 12 coēt 6 se. + 24 pri. die
set onder eñ het ander bouen aldus $\frac{320\ \text{pri.}}{6\ \text{se.} + 24\ \text{pri.}}$
coemt ghemindert $\frac{16\ \text{numerus}}{3\ \text{se.} + 12\ \text{pri}}$ de multiplicatie.
¶ Die derde regule.
Wildy multipliceren $\frac{9\ \text{se.} - 4\ \text{pri.}}{6\ \text{pri.} + 10\ \text{nu.}}$ met
$\frac{10\ \text{pri.} - 8\ \text{nu.}}{12\ \text{pri.} + 9\ \text{num.}}$ so multipliceert 9 se. — 4 pri. met
10 pri. — 8 nu. coemt 90 3^a — 112 se. + 32 p. die
set bouē / dan multipliceert 6 pri. + 10 nu. met 12 pri.
+ 9 nu. coemt 72 se. + 174 pri. + 90 nu. die sedt on-
der eñ coemt aldus $\frac{90\ 3^a - 112\ \text{se.} + 32\ \text{pri.}}{72\ \text{se.} + 174\ \text{pri} + 90\ \text{nu.}}$ dye
multiplicatie.
☞ Proba.

FIG. 60.—From Gielis van der Hoecke's *In arithmetica* (1537). Multiplication of fractions by *regule cos.*

counters *numerus*, *num.*, or *nu.* in place of *N.;* also *secu.* in place of *se.* For *pri.* he uses a few times $\dot{p}$.

The translation of matter shown in Figure 60 is as follows: "[In order to multiply fractions simply multiply numerators by numerators] and denominators by denominators. Thus, if you wish to multiply $\frac{3x}{4}$ by $\frac{3}{2x^2}$, you multiply $3x$ by 3, this gives $9x$, which you write down. Then multiply 4 by $2x^2$, this gives $8x^2$, which you write under the other $\frac{9x}{8x^2}$. Simplified this becomes $\frac{9}{8x}$, the product. *Second rule:* If you wish to multiply $\frac{20}{2x}$ by $\frac{16x}{3x+12}$, multiply 20 by 16 [*sic*] which

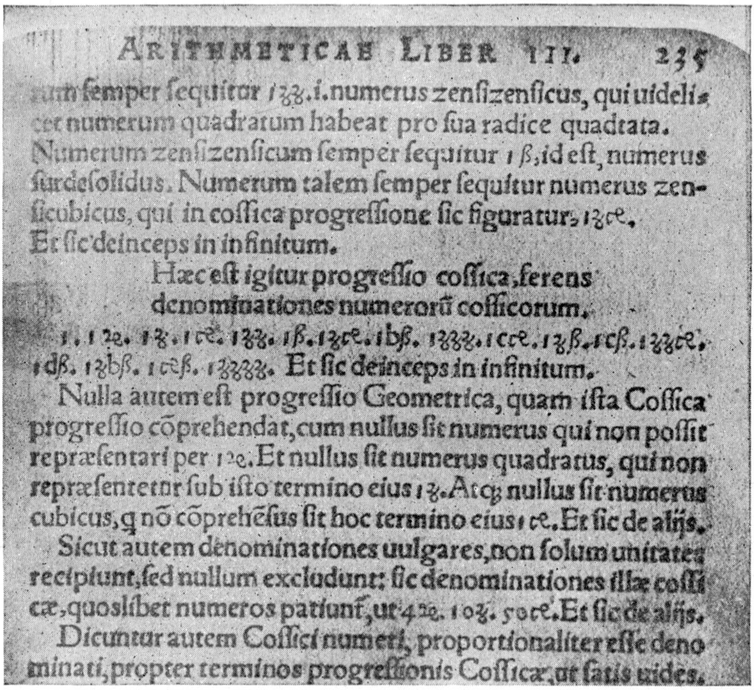
ARITHMETICAE LIBER III. 235

rum semper sequitur 1ʒʒ.i.numerus zensizensicus, qui uidelicet numerum quadratum habeat pro sua radice quadrata. Numerum zensizensicum semper sequitur 1ß, id est, numerus surdesolidus. Numerum talem semper sequitur numerus zensicubicus, qui in cossica progressione sic figuratur, 1ʒce. Et sic deinceps in infinitum.

Hæc est igitur progressio cossica, ferens denominationes numerorũ cossicorum.

1. 1𝓍. 1ʒ. 1ce. 1ʒʒ. 1ß. 1ʒce. 1bß. 1ʒʒʒ. 1cce. 1ʒß. 1cß. 1ʒʒce. 1dß. 1ʒbß. 1ceß. 1ʒʒʒʒ. Et sic deinceps in infinitum.

Nulla autem est progressio Geometrica, quam ista Cossica progressio cõprehendat, cum nullus sit numerus qui non possit repræsentari per 1𝓍. Et nullus sit numerus quadratus, qui non repræsentetur sub isto termino eius 1ʒ. Atq; nullus sit numerus cubicus, q nõ cõprehẽsus sit hoc termino eius 1ce. Et sic de alijs.

Sicut autem denominationes uulgares, non solum unitates recipiunt, sed nullum excludunt: sic denominationes illæ cossicæ, quoslibet numeros patiunt, ut 4𝓍. 10ʒ. 50ce. Et sic de alijs.

Dicuntur autem Cossici numeri, proportionaliter esse denominati, propter terminos progressionis Cossicæ, ut satis uides.

FIG. 61.—Part of a page from M. Stifel's *Arithmetica integra* (1544), fol. 235

gives $320x$, then multiply $2x$ by $3x+12$, which gives $6x^2+24x$. Place this under the other obtained above $\frac{320x}{6x^2+24x}$, this simplified gives:[1]

$\frac{16}{3x^2+12x}$, the product. . . . ."

As radical sign Gielis van der Hoecke does not use the German symbols of Rudolff, but the capital ℞ of the Italians. Thus he writes (fol. 90*B*) "6+℞8" for $6+\sqrt{8}$, "−℞ 32 *pri.*" for $-\sqrt{32x}$.

[1] The numerator should be 160, the denominator $3x+12$.

GERMAN: MICHAEL STIFEL
(1544, 1545, 1553)

151. Figure 61 is part of a page from Michael Stifel's important work on algebra, the *Arithmetica integra* (Nürnberg, 1544). From the ninth and the tenth lines of the text it will be seen that he uses the same symbols as Rudolff had used to designate powers, up to and including x^9. But Stifel carries here the notation as high as x^{16}. As Tropfke remarks,[1] the b in the symbol $b\beta$ of the seventh power leads Stifel to the happy thought of continuing the series as far as one may choose. Following the alphabet, his *Arithmetica integra* (1544) gives $c\beta = x^{11}$, $d\beta = x^{13}$, $e\beta = x^{17}$, etc.; in the revised *Coss* of Rudolff (1553), Stifel writes $\mathfrak{B}\beta$, $\mathfrak{C}\beta$, $\mathfrak{D}\beta$, $\mathfrak{G}\beta$. He was the first[2] who in print discarded the symbol for *dragma* and wrote a given number by itself. Where Rudolff, in his *Coss* of 1525 wrote 4ϕ, Stifel, in his 1553 edition of that book, wrote simply 4.

A multiplication from Stifel (*Arithmetica integra*, fol. 236v)[3] follows:

$$\text{[Concluding part} \quad \text{``}6\mathfrak{z} + 8\mathcal{R} - 6$$
$$\text{of a problem:]} \quad 2\mathfrak{z} - 4$$

$$12\mathfrak{z}\mathfrak{z} + 16\mathcal{d} - 12\mathfrak{z}$$
$$- 24\mathfrak{z} - 32\mathcal{R} + 24$$

$$12\mathfrak{z}\mathfrak{z} + 16\mathcal{d} - 36\mathfrak{z} - 32\mathcal{R} + 24 \text{ ''}$$

In Modern Symbols

$$6x^2 + 8x - 6$$
$$2x^2 - 4$$

$$12x^4 + 16x^3 - 12x^2$$
$$- 24x^2 - 32x + 24$$

$$12x^4 + 16x^3 - 36x^2 - 32x + 24$$

We give Stifel's treatment of the quartic equation, $1\mathfrak{z}\mathfrak{z} + 2\mathcal{d} + 6\mathfrak{z} + 5\mathcal{R} + 6$ *aequ.* 5550: "Quaeritur numerus ad quem additum suum quadratum faciat 5550. Pone igitur quod quadratum illud faciat $1A\mathfrak{z}$. tunc radix eius quadrata faciet $1A$. Et sic $1A\mathfrak{z} + 1A$. aequabitur

---

[1] *Op. cit.*, Vol. II, p. 120.

[2] Rudolff, *Coss* (1525), Signatur Hiiij (Stifel ed. [1553], p. 149); see Tropfke, *op. cit.*, Vol. II, p. 119, n. 651.

[3] Treutlein, *op. cit.*, p. 39.

5550. Itacq 1A_3. aequabit 5550−1A. Facit 1A. 74. Ergo cum. 1$_{33}$+ 2$cℓ$+6$_3$+5$\mathcal{R}$+6, aequetur. 5550. Sequitur quod. 74. aequetur 1$_3$+1$\mathcal{R}$+2. . . . . Facit itacq. 1$\mathcal{R}$.8."[1]

Translation:

$$``x^4+2x^3+6x^2+5x+6=5{,}550\ .$$

Required the number which, when its square is added to it, gives 5,550. Accordingly, take the square, which it makes, to be A^2. Then the square root of that square is A. Then $A^2+A=5{,}550$ and $A^2=$ 5,550−A. A becomes 74. Hence, since $x^4+2x^3+6x^2+5x+6=5{,}550$, it follows that $74=1x^2+x+2$. . . . . Therefore x becomes 8."

152. When Stifel uses more than the one unknown quantity $\mathcal{R}$, he at first follows Cardan in using the symbol q (abbreviation for *quantita*),[2] but later he represents the other unknown quantities by A, B, C. . . . . In the last example in the book he employs five un-knowns, $\mathcal{R}$, A, B, C, D. In the example solved in Figure 62 he repre-sents the unknowns by $\mathcal{R}$, A, B. The translation is as follows:

"Required three numbers in continued proportion such that the multiplication of the [sum of] the two extremes and the difference by which the extremes exceed the middle number gives 4,335. And the multiplications of that same difference and the sum of all three gives 6,069.

$A+x$ is the sum of the extremes,

$A-x$ the middle number,

$2A$ the sum of all three,

$2x$ the difference by which the extremes exceed the

middle. Then $2x$ multiplied into the sum of the extremes, i.e., in $A+x$, yields $2xA+2x^2=4{,}335$. Then $2x$ multiplied into $2A$ or the sum of all make $4xA=6{,}069$.

"Take these two equations together. From the first it follows that $xA=\dfrac{4{,}335-2x^2}{2}$. But from the second it follows that $1xA=$ $\dfrac{6{,}069}{4}$. Hence $\dfrac{4{,}335-2x^2}{2}=\dfrac{6{,}069}{4}$, for, since they are equal to one and the same, they are equal to each other. Therefore [by reduction] $17{,}340-8x^2=12{,}138$, which gives $x^2=650\frac{1}{4}$ and $x=25\frac{1}{2}$.

[1] *Arithmetica integra*, fol. 307 *B*.

[2] *Ibid.*, III, vi, 252A. This reference is taken from H. Bosmans, *Bibliotheca mathematica* (3d ser., 1906–7), Vol. VII, p. 66.

De Arithmet. Cardani 313

Quærantur tres numeri continue proportionales, ita ut multiplicatio duorum extremorum, per differentiam, quam habent extremi simul, ultra numerum medium, faciant 4335. Et multiplicatio eiusdēm differentiæ, in summam omnium trium faciat 6069.

1 A + 1 2℮. Est summa extremorum.

1 A — 1 2℮. Est summa medij.

2 A. Est summa omnium trium.

2 2℮ Est differentia quam habent extremi ultra mediū. Itacp 2 2℮ multiplicatæ in summam extremorum, id est, in 1 A + 1 2℮. faciunt, 2 2℮ A + 2 ȝ. æquata, 4335. Deinde 2 2℮ multiplicatæ in 2 A seu in summam omnium, faciunt 4 2℮ A æquata 6069.

Confer iam duas æquationes illas. Nam ex priore sequitur quòd 1 2℮ A faciat $4335\frac{—}{2}$ $2ȝ$ Ex posteriore autē sequitur quod 1 2℮ A. faciat $\frac{6069}{4}$. Sequitur ergo quod $4335\frac{—}{2}$ $2ȝ$ & $\frac{6069}{4}$ inter se æquentur. Quia quæ uni & eidem sunt æqualia, etiam sibi inuicem sunt æqualia. Ergo (per reductionem) 17340 — 8 ȝ æquantur. 12138. facit 1 ȝ. 650 ¼ Et 1 2℮. facit 25½.

Restat iam ut 1 A. etiam resoluatur, facit autem (ut paulo superius uidimus) 1 2℮ A. $\frac{6069}{4}$. Cum igitur duo illa inter se sint æqualia, Diuide utruncp per 1 2℮. tunc inuenies 1 A. æquari, seu facere. $\frac{6069}{42℮}$. Cum autem 1 2℮ faciat 25½. facient 4 2℮. 102. Itacp 6069. diuisa per 102. faciunt 59½. Et tantum facit 1 A. Quare 1 A — 1 2℮. id est, medius numerus facit 34. Et 1 A + 1 2℮. id est, summa duorum extremorum facit 85. Iam igitur oritur noua quæstio hæc.

Diuidantur 85 in duas partes, ita ut 34 mediet inter eas proportionaliter.

Sic stant numeri.

　　1 B.　　34.　　85 — 1 B.

Vnde 85 B — 1 Bȝ æquatur 1156. facit 1 B 17. Et sic stant numeri exempli. 17. 34. 68.

　　　　　　　　　KK　　　　Exem

Fig. 62.—From Stifel's *Arithmetica integra* (1544), fol. 313

"It remains to find also $1A$. One has [as we saw just above] $1xA = \dfrac{6,069}{4}$. Since these two are equal to each other, divide each by x, and there follows $A = \dfrac{6,069}{4x}$. But as $x = 25\frac{1}{2}$, one has $4x = 102$, and $6,069$ divided by 102 gives $59\frac{1}{2}$. And that is what A amounts to. Since $A - x$, i.e., the middle number equals 34, and $A + x$, i.e., the sum of the two extremes is 85, there arises this new problem:

"Divide 85 into two parts so that 34 is a mean proportional between them. These are the numbers:

$$B \, , \qquad 34 \, , \qquad 85 - B \, .$$

Since $85B - B^2 = 1,156$, there follows $B = 17$. And the numbers of the example are 17, 34, 68."

Observe the absence of a sign of equality in Stifel, equality being expressed in words or by juxtaposition of the expressions that are equal; observe also the designation of the square of the unknown B by the sign B_3. Notice that the fractional line is very short in the case of fractions with binomial (or polynomial) numerators—a singularity found in other parts of the *Arithmetica integra*. Another oddity is Stifel's designation of the multiplication of fractions.[1] They are written as we write ascending continued fractions. Thus

$$\frac{\frac{\frac{3}{4}}{\frac{2}{3}}}{\frac{1}{7}}$$

means "Tres quartae, duarum tertiarum, unius septimae," i.e., $\frac{3}{4}$ of $\frac{2}{3}$ of $\frac{1}{7}$.

The example in Fig. 62 is taken from the closing part of the *Arithmetica integra* where Cardan's *Ars magna*, particularly the solutions of cubic and quartic equations, receive attention. Of interest is Stifel's suggestion to his readers that, in studying Cardan's *Ars magna*, they should translate Cardan's algebraic statements into the German symbolic language: "Get accustomed to transform the signs used by him into our own. Although his signs are the older, ours are the more commodious, at least according to my judgment."[2]

[1] *Arithmetica integra* (1548), p. 7; quoted by S. Günther, *Vermischte Untersuchungen* (Leipzig, 1876), p. 131.

[2] *Arithmetica integra* (Nürnberg, 1544), Appendix, p. 306. The passage, as quoted by Tropfke, *op. cit.*, Vol. II (2. ed.), p. 7, is as follows: "Assuescas, signa eius, quibus ipse utitur, transfigurare ad signa nostra. Quamvis enim signa quibus ipse utitur, uetustiora sint nostris, tamen nostra signa (meo quidē iudicio) illis sunt commodiora."

153. Stifel rejected Rudolff's symbols for radicals of higher order and wrote $\sqrt{3}$ for $\sqrt{\ }$, $\sqrt{c\!\!\!/}$ for $\sqrt[3]{\ }$, etc., as will be seen more fully later.

But he adopts Rudolff's dot notation for indicating the root of a binomial:[1]

"$\sqrt{3} \cdot 12 + \sqrt{3}6 \cdot - \cdot \sqrt{3} \cdot 12 - \sqrt{3}6$ has for its square $12 + \sqrt{3}6 + 12 - \sqrt{3}6 - \sqrt{3}138 - \sqrt{3}138$"; i.e., "$\sqrt{12 + \sqrt{6}} - \sqrt{12 - \sqrt{6}}$ has for its square $12 + \sqrt{6} + 12 - \sqrt{6} - \sqrt{138} - \sqrt{138}$." Again:[2] "Tertio vide, utrũ $\sqrt{3} \cdot \sqrt{3}$ $12500 - 50$ addita ad $\sqrt{3} \cdot \sqrt{3}$ $12500 + 50$. faciat $\sqrt{3} \cdot \sqrt{3} 50000 + 200$" ("Third, see whether $\sqrt{\sqrt{12{,}500} - 50}$ added to $\sqrt{\sqrt{12{,}500} + 50}$ makes $\sqrt{\sqrt{50{,}000} + 200}$"). The dot is employed to indicate that the root of all the terms following is required.

154. Apparently with the aim of popularizing algebra in Germany by giving an exposition of it in the German language, Stifel wrote in 1545 his *Deutsche arithmetica*[3] in which the unknown x is expressed by *sum*, x^2 by "*sum: sum*," etc. The nature of the book is indicated by the following equation:

"Der Algorithmus meiner deutschen Coss braucht zum ersten schlecht vnd ledige zalē | wie der gemein Algorithmus | als da sind 1 2 3 4 5 etc. Zum audern braucht er die selbigen zalen vnder diesem namen | Sum̃a. Vnd wirt dieser nam Suma | also verzeichnet | Sum̄: Als hie | 1 sum̄: 2 sum̄: 3 sum̄ etc. . . . . So ich aber 2 sum: Multiplicir mit 3 sum: so kom̃en mir 6 sum: sum: Das mag ich also lesen | 6 summē summarum | wie man deñ im Deutschē offt findet | sum̃a sum̃arum. . . . . Soll ich multipliciren 6 sum: sum: sum: mit 12 sum: sum: sum: So sprich ich | 12 mal 6. macht 72 sum: sum: sum: sum: sum sum . . . ."[4] Translation: "The algorithm of my Deutsche Coss uses, to start with, simply the pure numbers of the ordinary algorithm, namely, 1, 2, 3, 4, 5, etc. Besides this it uses these same numbers under the name of *summa*. And this name *summa* is marked *sum̄:*, as in 1 *sum̄:* 2 *sum̄:* 3 *sum̄*, etc. . . . . But when I multiply 2 *sum:* by 3 *sum:* I obtain 6 *sum: sum:*. This I may read | 6 *summē summarum* | for in German one encounters often *sum̃a sum̃arum*. . . . . When I am to multiply 6 *sum: sum: sum:* by 12 *sum: sum: sum:*, I say | 12 times 6 makes 72 *sum: sum: sum: sum: sum: sum:* . . . ."

[1] *Op. cit.*, fol. 138a.   [2] *Ibid.*, fol. 315a.

[3] *Op. cit. Inhaltend. Die Hauszrechnung. Deutsche Coss. Rechnung* (1545).

[4] Treutlein, *op. cit.*, Vol. II, p. 34. For a facsimile reproduction of a page of Stifel's *Deutsche arithmetica*, see D. E. Smith, *Rara arithmetica* (1898), p. 234.

The inelegance of this notation results from an effort to render the subject easy; Stifel abandoned the notation in his later publications, except that the repetition of factors to denote powers reappears in 1553 in his "Cossische Progresʒ" (§ 156).

In this work of 1545 Stifel does not use the radical signs found in his *Arithmetica integra;* now he uses $\mathcal{z}\!\diagup$, $\overline{\mathcal{z}}\!\diagup$, $\overline{\overline{\mathcal{z}}}\!\diagup$, for square, cube, and fourth root, respectively. He gives (fol. 74) the German capital letter 𝔐 as the sign of multiplication, and the capital letter 𝔇 as the sign of division, but does not use either in the entire book.[1]

155. In 1553 Stifel brought out a revised edition of Rudolff's *Coss.* Interesting is Stifel's comparison of Rudolff's notation of radicals with his own, as given at the end of page 134 (see Fig. 63*a*), and his declaration of superiority of his own symbols. On page 135 we read: "How much more convenient my own signs are than those of Rudolff, no doubt everyone who deals with these algorithms will notice for himself. But I too shall often use the sign $\sqrt{}$ in place of the $\sqrt{3}$, for brevity.

"But if one places this sign before a simple number which has not the root which the sign indicates, then from that simple number arises a surd number.

"Now my signs are much more convenient and clearer than those of Christoff. They are also more complete for they embrace all sorts of numbers in the arithmetic of surds. They are [here he gives the symbols in the middle of p. 135, shown in Fig. 63*b*]. Such a list of surd numbers Christoff's symbols do not supply, yet they belong to this topic.

"Thus my signs are adapted to advance the subject by putting in place of so many algorithms a single and correct algorithm, as we shall see.

"In the first place, the signs (as listed) themselves indicate to you how you are to name or pronounce the surds. Thus, $\sqrt{\beta 6}$ means the sursolid root of 6, etc. Moreover, they show you how they are to be reduced, by which reduction the declared unification of many (indeed all such) algorithms arises and is established."

156. Stifel suggests on folio 61*B* also another notation (which, however, he does not use) for the progression of powers of x, which he calls "die Cossische Progresʒ." We quote the following:

"Es mag aber die Cossische Progresʒ auch also verzeychnet werden:

$$\begin{array}{ccccc} 0 & 1 & 2 & 3 & 4 \\ \end{array}$$
$$1 \cdot 1A \cdot 1AA \cdot 1AAA \cdot 1AAAA \cdot \text{etc.}$$

[1] Cantor, *op. cit.*, Vol. II (2. ed., 1913), p. 444.

Item auch also:

$$\begin{array}{ccccccccccc} 0 & & 1 & & 2 & & 3 & & 4 & & \\ 1 & \cdot & 1B & \cdot & 1BB & \cdot & 1BBB & \cdot & 1BBBB & \cdot & \text{etc.} \end{array}$$

Item auch also:

$$\begin{array}{ccccccccccc} 0 & & 1 & & 2 & & 3 & & 4 & & \\ 1 & \cdot & 1C & \cdot & 1CC & \cdot & 1CCC & \cdot & 1CCCC & \cdot & \text{etc.} \end{array}$$

Vnd so fort an von andern Buchstaben."[1]

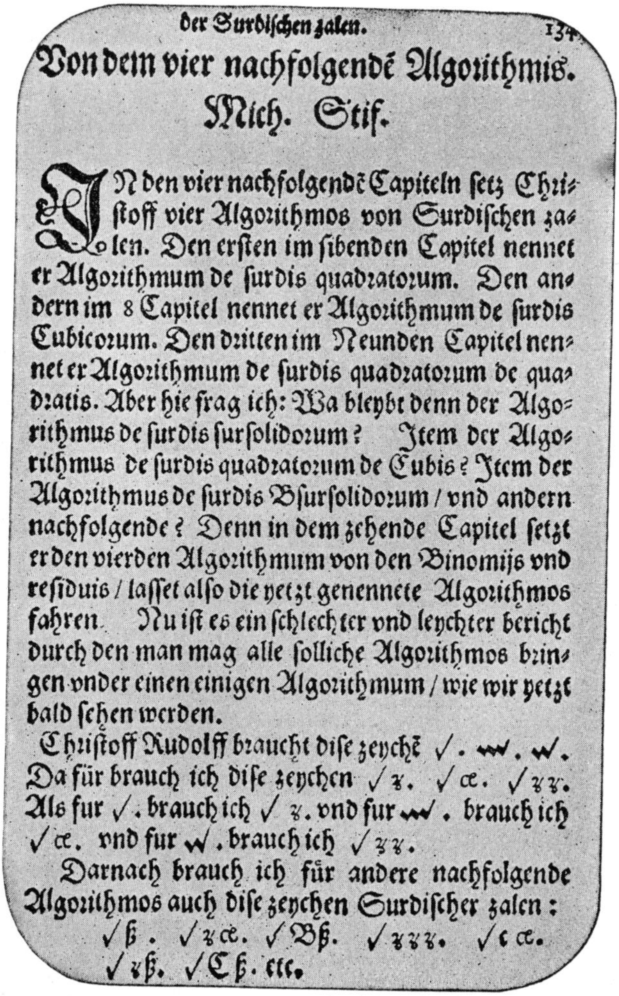

der Surdischen zalen. 134

Von dem vier nachfolgendẽ Algorithmis. Mich. Stif.

IN den vier nachfolgendẽ Capiteln setz Christoff vier Algorithmos von Surdischen zalen. Den ersten im sibenden Capitel nennet er Algorithmum de surdis quadratorum. Den andern im 8 Capitel nennet er Algorithmum de surdis Cubicorum. Den dritten im Neunden Capitel nennet er Algorithmum de surdis quadratorum de quadratis. Aber hie frag ich: Wa bleybt denn der Algorithmus de surdis sursolidorum? Item der Algorithmus de surdis quadratorum de Cubis? Item der Algorithmus de surdis Bsursolidorum / vnd andern nachfolgende? Denn in dem zehende Capitel setzt er den vierden Algorithmum von den Binomijs vnd residuis / lasset also die yetzt genennete Algorithmos fahren. Nu ist es ein schlechter vnd leychter bericht durch den man mag alle solliche Algorithmos bringen vnder einen einigen Algorithmum / wie wir yetzt bald sehen werden.

Christoff Rudolff braucht dise zeychẽ √. ⱳ. w. Da für brauch ich dise zeychen √ʒ. √ce. √ʒʒ. Als fur √. brauch ich √ʒ. vnd fur ⱳ. brauch ich √ce. vnd fur w. brauch ich √ʒʒ.

Darnach brauch ich für andere nachfolgende Algorithmos auch dise zeychen Surdischer zalen:
√ß. √ʒce. √Bß. √ʒʒʒ. √cce. √ʒß. √Cß. etc.

FIG. 63*a*.—This shows p. 134 of Stifel's edition of Rudolff's *Coss* (1553)

[1] Treutlein, *op. cit.*, Vol. II (1879), p. 34.

We see here introduced the idea of repeating a letter to designate powers, an idea carried out extensively by Harriot about seventy-five

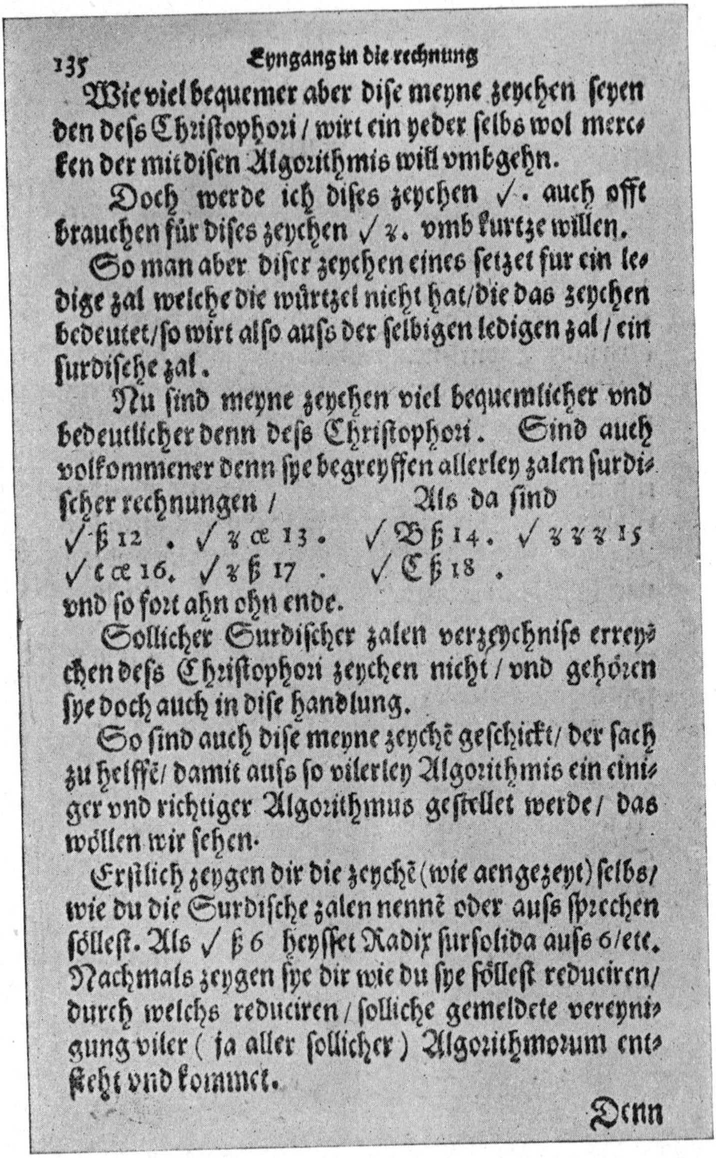

135 Eyngang in die rechnung

Wie viel bequemer aber diſe meyne zeychen ſeyen den deſs Chriſtophori / wirt ein yeder ſelbs wol mercken der mit diſen Algorithmis will vmbgehn.

Doch werde ich diſes zeychen √. auch offt brauchen für diſes zeychen √ℨ. vmb kurtze willen.

So man aber diſer zeychen eines ſetzet für ein ledige zal welche die würtzel nicht hat/die das zeychen bedeutet/ſo wirt alſo auſs der ſelbigen ledigen zal / ein ſurdiſche zal.

Nu ſind meyne zeychen viel bequemlicher vnd bedeutlicher denn deſs Chriſtophori. Sind auch volkommener denn ſye begreyffen allerley zalen ſurdiſcher rechnungen / Als da ſind

√ß 12. √ℨcœ 13. √𝔅ß 14. √ℨℨℨ 15

√ccœ 16. √ℨß 17. √ℭß 18.

vnd ſo fort ahn ohn ende.

Sollicher Surdiſcher zalen verzeychniſs erreychen deſs Chriſtophori zeychen nicht / vnd gehören ſye doch auch in diſe handlung.

So ſind auch diſe meyne zeychē geſchickt/ der ſach zu helffē/ damit auſs ſo vilerley Algorithmis ein einiger vnd richtiger Algorithmus geſtellet werde/ das wöllen wir ſehen.

Erſtlich zeygen dir die zeychē (wie aengezeyt) ſelbs/ wie du die Surdiſche zalen nennē oder auſs ſprechen ſölleſt. Als √ß 6 heyſſet Radix ſurſolida auſs 6/etc.

Nachmals zeygen ſye dir wie du ſye ſölleſt reduciren/ durch welchs reduciren / ſolliche gemeldete vereynigung viler (ja aller ſollicher) Algorithmorum entſteht vnd kommet.

Denn

FIG. 63b.—This shows p. 135 of Stifel's edition of Rudolff's *Coss* (1553) years later. The product of two quantities, of which each is represented by a letter, is designated by juxtaposition.

GERMAN: NICOLAUS COPERNICUS
(1566)

157. Copernicus died in 1543. The quotation from his *De revolutionibus orbium coelestium* (1566; 1st ed., 1543)[1] shows that the exposition is devoid of algebraic symbols and is almost wholly rhetorical. We find a curious mixture of modes of expressing numbers: Roman numerals, Hindu-Arabic numerals, and numbers written out in words. We quote from folio 12:

"Circulum autem communi Mathematicorum consensu in CCCLX. partes distribuimus. Dimetientem uero CXX. partibus asciscebant prisci. At posteriores, ut scrupulorum euitarent inuolutionem in multiplicationibus & diuisionibus numerorum circa ipsas lineas, quae ut plurimum incommensurabiles sunt longitudine, saepius etiam potentia, alij duodecies centena milia, alij uigesies, alij aliter rationalem constituerunt diametrum, ab eo tempore quo indicae numerorum figurae sunt usu receptae. Qui quidem numerus quemcunque alium, sine Graecum, sine Latinum singulari quadam promptitudine superat, & omni generi supputationum aptissime sese accommodat. Nos quoq, eam ob causam accepimus diametri 200000. partes tanquam sufficientes, que, possint errorem excludere patentem."

Copernicus does not seem to have been exposed to the early movements in the fields of algebra and symbolic trigonometry.

GERMAN: JOHANNES SCHEUBEL
(1545, 1551)

158. Scheubel was professor at the University of Tübingen, and was a follower of Stifel, though deviating somewhat from Stifel's notations. In Scheubel's arithmetic[2] of 1545 one finds the scratch method in division of numbers. The book is of interest because it does not use the + and − signs which the author used in his algebra; the + and − were at that time not supposed to belong to arithmetic proper, as distinguished from algebra.

[1] *Nicolai Copernici Torinensis de Revolvtionibus Orbium Coelestium, Libri VI. .. ... Item, de Libris Revolvtionvm Nicolai Copernici Narratio prima, per M. Georgium Ioachimum Rheticum ad D. Ioan. Schonerum scripta. Basileae* (date at the end of volume, M.D.LXVI).

[2] *De Nvmeris et Diversis Rationibvs seu Regulis computationum Opusculum, a Ioanne Scheubelio compositum* . . . . (1545).

Scheubel in 1550 brought out at Basel an edition of the first six books of Euclid which contains as an introduction an exposition of algebra,[1] covering seventy-six pages, which is applied to the working of examples illustrating geometric theorems in Euclid.

159. Scheubel begins with the explanation of the symbols for powers employed by Rudolff and Stifel, but unlike Stifel he retains a symbol for *numerus* or *dragma*. He explains these symbols, up to the twelfth power, and remarks that the list may be continued indefinitely. But there is no need, he says, of extending this unwieldy designation, since the ordinal natural numbers afford an easy nomenclature. Then he introduces an idea found in Chuquet, Grammateus, and others, but does it in a less happy manner than did his predecessors. But first let us quote from his text. After having explained the symbol for *dragma* and for x he says (p. 2): "The third of them ʒ, which, since it is produced by multiplication of the *radix* into itself, and indeed the first [multiplication], is called the *Prima* quantity and furthermore is noted by the syllable *Pri*. Even so the fourth ȼ, since it is produced secondly by the multiplication of that same *radix* by the square, i.e., by the *Prima* quantity, is called the *Second* quantity, marked by the syllable *Se*. Thus the fifth sign ʒʒ, which springs thirdly from the multiplication of the *radix*, is called the *Tertia* quantity, noted by the syllable *Ter*. . . . ."[2] And so he introduces the series of symbols, *N., Ra., Pri., Se., Ter., Quar., Quin., Sex., Sep.* . . . . , which are abbreviations for the words *numerus, radix, prima quantitas* (because it arises from one multiplication), *secunda quantitas* (because it arises from two multiplications), and so on. This scheme gives rise to the oddity of designating x^n by the number $n-1$, such as we have not hitherto encountered. In Pacioli one finds the contrary relation, i.e., the designation of x^{n-1} by x^n (§ 136). Scheubel's notation does not coincide with that of Grammateus, who more judiciously had used *pri., se.,* etc., to designate x, x^2, etc. (§ 147). Scheubel's singular notation is illustrated by

[1] *Evclidis Megarensis, Philosophi et Mathematici excellentissimi, sex libri priores de Geometricis principijs, Graeci et Latini* . . . . *Algebrae porro regvlae, propter nvmerorum exempla, passim propositionibus adiecta, his libris praemissae sunt, eadenque demonstratae. Authore Ioanne Schevbelio,* . . . . *Basileae* (1550). I used the copy belonging to the Library of the University of Michigan.

[2] "Tertius de, ʒ. qui cū ex multiplicatione radicis in se producatur, et primo quidem: Prima quantitas, et Pri etiam syllaba notata, appelletur. Quartus uerò ȼ quia ex multiplicatione eiusdem radicis cum quadrato, hoc est, cum prima quantitate, secundò producitur: Se syllaba notata, Secunda quantitas dicitur. Sic character quintus, ʒʒ, quia ex multiplicatione radicis cum secunda quantitate tertio nascitur: Ter syllaba notata, Tertia etiam quantitias dicitur. . . . ."

Figure 64, where he shows the three rules for solving quadratic equations. The first rule deals with the solution of $4x^2+3x=217$, the second with $3x+175=4x^2$, the third with $3x^2+217=52x$. These different cases arose from the consideration of algebraic signs, it being desired that the terms be so written as to appear in the positive form. Only positive roots are found.

ALIVD EXEMPLVM.

PRIMI CANONIS.   SECVNDI CANONIS.

Pri.   ra.   N   ra.   N   pri.

4   +   3   equales   217   3 + 175 æqu. 4

Hic, quia maximi characteris numerus non est unitas, diuisione, ut dictum est, ei succurri debet. Veniunt autem facta diuisione,

pri.   ra.   N   ra.   N   pri.

1   + $\frac{3}{4}$ æqu. $\frac{217}{4}$ $\frac{3}{4}$ + $\frac{175}{4}$ equ.   1

$\frac{3}{8}$ in se, $\frac{9}{64}$ + $\frac{217}{4}$ $\frac{3}{8}$ in se. $\frac{9}{64}$ + $\frac{175}{4}$

ueni. $\frac{3481}{64}$. Huius ra.   ueni. $\frac{2809}{64}$. Huius ra.

sunt 7$\frac{3}{8}$ minus $\frac{3}{8}$ sunt 6$\frac{5}{8}$ plus $\frac{3}{8}$

manent   7   ueniunt 7

radicis ualor.   radicis ualor.

ALIVD TERTII CANONIS EXEMPLVM.

3 pri.   +   217 N   æquales   52 ra.

Et hic, quia maximi characteris numerus non est unitas, diuisione ei succurrendum erit. Veniunt autem hoc facto,

1 pri. + $\frac{217}{3}$ N   æquales $\frac{52}{3}$ N

$\frac{26}{3}$ in se. $\frac{676}{9}$, minus $\frac{217}{3}$, manet $\frac{25}{9}$

Huius ra, qua. est 1$\frac{2}{3}$ $\begin{cases} de \\ ad \end{cases}$ 8$\frac{2}{3}$,   & manent 7, uel proueniunt 10$\frac{1}{3}$, Vterq; radicis ualor, quod examinari potest.

FIG. 64.—Part of p. 28 in Scheubel's Introduction to his *Euclid*, printed at Basel in 1550.

Under proportion we quote one example (p. 41):

" 3 ra.+4 N. ualent 8 se.+4 pri.
    quanti 8 ter.−4 ra.

$$Facit \; \frac{64\ sex.+32\ quin.-32\ ter.-16\ se.}{3\ ra.+4\ N.}$$ "

In modern notation:

$$3x+4 \text{ are worth } 8x^3+4x^2$$
$$\text{how much } 8x^4-4x \;.$$
$$\text{Result } \frac{64x^7+32x^6-32x^4-16x^3}{3x+4} \;.$$

In the treatment of irrationals or *numeri surdi* Scheubel uses two notations, one of which is the abbreviation Ra. or ra. for *radix*, or "square root," $ra.cu.$ for "cube root," $ra.ra.$ for "fourth root." Confusion from the double use of ra. (to signify "root" and also to signify x) is avoided by the following implied understanding: If ra. is followed by a number, the square root of that number is meant; if ra. is preceded by a number, then ra. stands for x. Thus "8 ra." means $8x$; "ra. 12" means $\sqrt{12}$.

Scheubel's second mode of indicating roots is by Rudolff's symbols for square, cube, and fourth roots. He makes the following statement (p. 35) which relates to the origin of $\sqrt{\ }$: "Many, however, are in the habit, as well they may, to note the desired roots by their points with a stroke ascending on the right side, and thus they prefix for the square root, where it is needed for any number, the sign $\sqrt{\ }$: for the cube root, $\wedge\!\!\wedge\!\!\diagup$; and for the fourth root $\wedge\!\!\diagup$."[1] Both systems of notation are used, sometimes even in the same example. Thus, he considers (p. 37) the addition of "ra. 15 ad ra. 17" (i.e., $\sqrt{15}+\sqrt{17}$) and gives the result "$ra.col.$ 32$+\sqrt{1020}$" (i.e. $\sqrt{32+\sqrt{1,020}}$). The $ra.col.$ (*radix collecti*) indicates the square root of the binomial. Scheubel uses also the $ra.re$ (*radix residui*) and *radix binomij*. For example (p. 55), he writes "$ra.re.$ $\sqrt{15}-\sqrt{12}$" for $\sqrt{\sqrt{15}-\sqrt{12}}$. Scheubel suggests a third notation for irrationals (p. 35), of which he makes no further use, namely, *radix se.* for "cube root," the abbreviation for *secundae quantitatis radix*.

The algebraic part of Scheubel's book of 1550 was reprinted in 1551 in Paris, under the title *Algebrae compendiosa facilisqve descriptio*.[2]

---

[1] "Solent tamen multi, et bene etiam, has desideratas radices, suis punctis cum lines quadam à dextro latere ascendente, notare, atque sic pro radice quidem quadrata, ubi haec in aliquo numero desideratur, notam $\sqrt{\ }$: pro cubica uerò, $\wedge\!\!\wedge\!\!\diagup$: ac radicis radice deinde, $\wedge\!\!\diagup$ praeponunt."

[2] Our information on the 1551 publication is drawn from H. Staigmüller, "Johannes Scheubel, ein deutscher Algebraiker des XVI. Jahrhunderts," *Abhandlungen zur Geschichte der Mathematik*, Vol. IX (Leipzig, 1899), p. 431–69; A. Witting and M. Gebhardt, *Beispiele zur Geschichte der Mathematik*, II. Teil

It is of importance as representing the first appearance in France of the symbols $+$ and $-$ and of some other German symbols in algebra.

Charles Hutton says of Scheubel's *Algebrae compendiosa* (1551): "The work is most beautifully printed, and is a very clear though succinct treatise; and both in the form and matter much resembles a modern printed book."[1]

MALTESE: WIL. KLEBITIUS
(1565)

160. Through the courtesy of Professor H. Bosmans, of Brussels, we are able to reproduce a page of a rare and curious little volume containing exercises on equations of the first degree in one unknown number, written by Wilhelm Klebitius and printed at Antwerp in 1565.[2] The symbolism follows Scheubel, particularly in the fancy form given to the plus sign. The unknown is represented by "$1R$."

The first problem in Figure 65 is as follows: Find a number whose double is as much below 30,000 as the number itself is below 20,000. In the solution of the second and third problems the notational peculiarity is that $\frac{1}{3}R. - \frac{1}{3}$ is taken to mean $\frac{1}{3}R. - \frac{1}{3}R.$, and $1R. - \frac{6}{8}$ to mean $1R. - \frac{6}{8}R.$

GERMAN: CHRISTOPHORUS CLAVIUS
(1608)

161. Though German, Christophorus Clavius spent the latter part of his life in Rome and was active in the reform of the calendar. His *Algebra*[3] marks the appearance in Italy of the German $+$ and $-$ signs, and of algebraic symbols used by Stifel. Clavius is one of the very first to use round parentheses to express aggregation. From his *Algebra* we quote (p. 15): "Pleriqve auctores pro signo $+$ ponunt literam *P*, vt significet plus: pro signo vero $-$ ponunt literam *M*, vt significet minus. Sed placet nobis vti nostris signis, vt à literis distinguantur, ne confusio oriatur." Translation: "Many authors put in place of the sign $+$ the letter *P*, which signifies "plus":

(Leipzig-Berlin, 1913), p. 25; Tropfke, *op. cit.*, Vol. I (1902), p. 195, 198; Charles Hutton, *Tracts on Mathematical and Philosophical Subjects*, Vol. II (London, 1812), p. 241–43; L. C. Karpinski, *Robert of Chester's* . . . . *Al-Khowarizmi*, p. 39–41.

[1] Charles Hutton, *op. cit.*, p. 242.

[2] The title is *Insvlae Melitensis, qvam alias Maltam vocant, Historia, quaestionib. aliquot Mathematicis reddita iucundior.* At the bottom of the last page: "*Avth. Wil. Kebitio.*"

[3] *Algebra Christophori Clavii Bambergensis e Societate Iesv.* (Romae, M.DC.VIII).

likewise, for the sign $-$ they put the letter M, which signifies "minus." But we prefer to use our signs; as they are different from letters, no confusion arises."

In his arithmetic, Clavius has a distinct notation for "fractions of fractional numbers," but strangely he does not use it in the ordinary

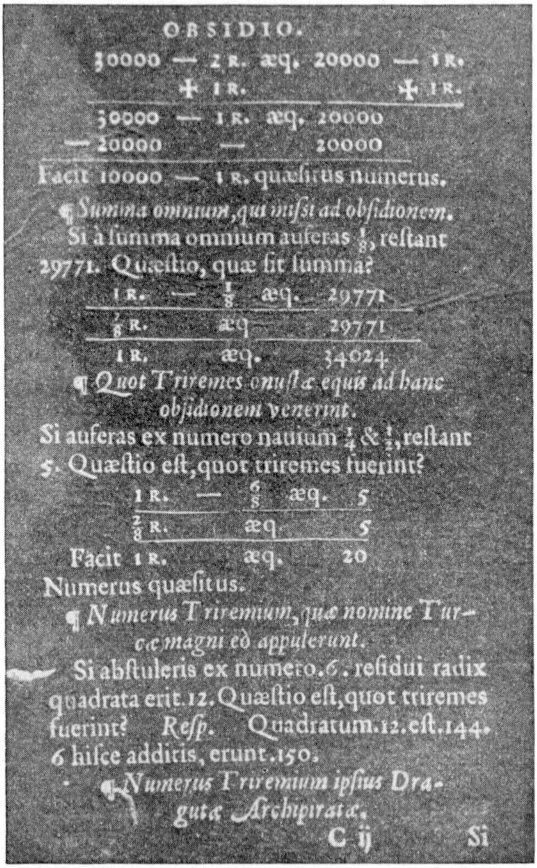
OBSIDIO.

30000 — 2 R. æq. 20000 — 1 R.
+ 1 R. + 1 R.

30000 — 1 R. æq. 20000
— 20000 — 20000

Facit 10000 — 1 R. quæſitus numerus.

¶ Summa omnium, qui inſit ad obſidionem.

Si à ſumma omnium auferas $\frac{1}{8}$, reſtant 29771. Quæſtio, quæ ſit ſumma?

1 R. — $\frac{1}{8}$ æq. 29771
$\frac{7}{8}$ R. æq 29771
1 R. æq. 34024

¶ Quot Triremes onuſtæ equis ad hanc obſidionem venerint.

Si auferas ex numero nauium $\frac{1}{4}$ & $\frac{1}{2}$, reſtant 5. Quæſtio eſt, quot triremes fuerint?

1 R. — $\frac{6}{8}$ æq. 5
$\frac{2}{8}$ R. æq 5
Facit 1 R. æq. 20

Numerus quæſitus.

¶ Numerus Triremium, quæ nomine Turcæ magni eò appulerunt.

Si abſtuleris ex numero .6. reſidui radix quadrata erit .12. Quæſtio eſt, quot triremes fuerint? Reſp. Quadratum .12. eſt .144. 6 hiſce additis, erunt .150.

¶ Numerus Triremium ipſius Dragutæ Archipiratæ.

C ij Si

FIG. 65.—Page from W. Klebitius (1565)

multiplication of fractions. His $\frac{3}{5}\cdot\frac{4}{7}\cdot$ means $\frac{3}{5}$ of $\frac{4}{7}$. He says: "Vt praedicta minutia minutiae ita scribenda est $\frac{3}{5}\cdot\frac{4}{7}\cdot$ pronuntiaturque sic. Tres quintae quatuor septimarũ vnius integri."[1] Similarly, $\frac{2}{3}\cdot\frac{3}{4}\cdot\frac{1}{6}\cdot\frac{1}{2}\cdot$ yields $\frac{6}{144}$. The distinctive feature in this notation is the

[1] *Epitome arithmeticae* (Rome, 1583), p. 68; see also p. 87.

omission of the fractional line after the first fraction.[1] The dot cannot be considered here as the symbol of multiplication. No matter what the operation may be, all numbers, fractional or integral, in the

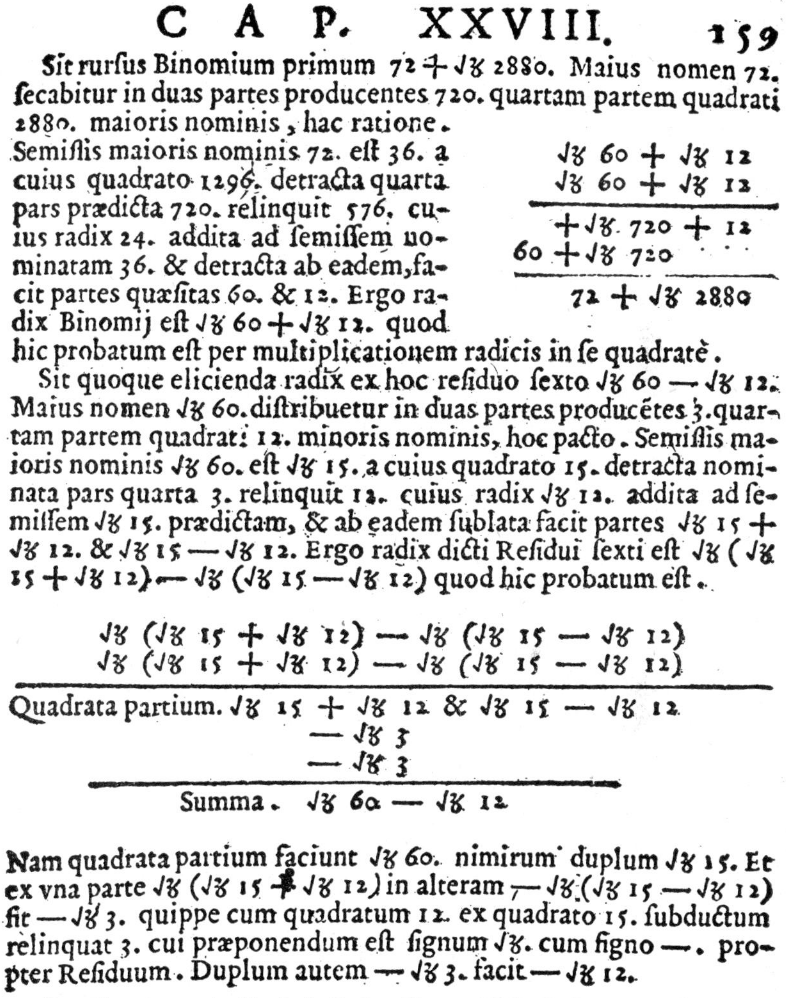

C A P. XXVIII. 159

Sit rurſus Binomium primum 72 + √⅋ 2880. Maius nomen 72. ſecabitur in duas partes producentes 720. quartam partem quadrati 2880. maioris nominis, hac ratione.

Semiſſis maioris nominis 72. eſt 36. a cuius quadrato 1296. detracta quarta pars prædicta 720. relinquit 576. cuius radix 24. addita ad ſemiſſem nominatam 36. & detracta ab eadem, facit partes quæſitas 60. & 12. Ergo radix Binomij eſt √⅋ 60 + √⅋ 12. quod

$$\begin{array}{r} \sqrt{⅋}\ 60\ +\ \sqrt{⅋}\ 12 \\ \sqrt{⅋}\ 60\ +\ \sqrt{⅋}\ 12 \\ \hline +\sqrt{⅋}\ 720\ +\ 12 \\ 60\ +\sqrt{⅋}\ 720 \\ \hline 72\ +\ \sqrt{⅋}\ 2880 \end{array}$$

hic probatum eſt per multiplicationem radicis in ſe quadratè.

Sit quoque elicienda radix ex hoc reſiduo ſexto √⅋ 60 — √⅋ 12. Maius nomen √⅋ 60. diſtribuetur in duas partes producétes 3. quartam partem quadrati 12. minoris nominis, hoc pacto. Semiſſis maioris nominis √⅋ 60. eſt √⅋ 15. a cuius quadrato 15. detracta nominata pars quarta 3. relinquit 12. cuius radix √⅋ 12. addita ad ſemiſſem √⅋ 15. prædictam, & ab eadem ſublata facit partes √⅋ 15 + √⅋ 12. & √⅋ 15 — √⅋ 12. Ergo radix dicti Reſidui ſexti eſt √⅋ (√⅋ 15 + √⅋ 12) — √⅋ (√⅋ 15 — √⅋ 12) quod hic probatum eſt.

$$\begin{array}{l} \sqrt{⅋}\ (\sqrt{⅋}\ 15\ +\ \sqrt{⅋}\ 12)\ -\ \sqrt{⅋}\ (\sqrt{⅋}\ 15\ -\ \sqrt{⅋}\ 12) \\ \sqrt{⅋}\ (\sqrt{⅋}\ 15\ +\ \sqrt{⅋}\ 12)\ -\ \sqrt{⅋}\ (\sqrt{⅋}\ 15\ -\ \sqrt{⅋}\ 12) \\ \hline \end{array}$$

Quadrata partium. √⅋ 15 + √⅋ 12 & √⅋ 15 — √⅋ 12
— √⅋ 3
— √⅋ 3
—————————————————
Summa. √⅋ 60 — √⅋ 12

Nam quadrata partium faciunt √⅋ 60. nimirum duplum √⅋ 15. Et ex vna parte √⅋ (√⅋ 15 + √⅋ 12) in alteram — √⅋ (√⅋ 15 — √⅋ 12) fit — √⅋ 3. quippe cum quadratum 12. ex quadrato 15. ſubductum relinquat 3. cui præponendum eſt ſignum √⅋. cum ſigno —. propter Reſiduum. Duplum autem — √⅋ 3. facit — √⅋ 12.

FIG. 66.—A page in Clavius' *Algebra* (Rome, 1608). It shows one of the very earliest uses of round parentheses to express aggregation of terms.

arithmetic of Clavius are followed by a dot. The dot made the numbers stand out more conspicuously.

[1] In the edition of the arithmetic of Clavius that appeared at Cologne in 1601, p. 88, 126, none of the fractional lines are omitted in the foregoing passages.

As symbol of the unknown quantity Clavius uses[1] the German $\mathfrak{X}$. In case of additional unknowns, he adopts $1A$, $1B$, etc., but he refers to the notation $1q$, $2q$, etc., as having been used by Cardan, Nonius, and others, to represent unknowns. He writes: $3\mathfrak{X}+4A$, $4B-3A$ for $3x+4y$, $4z-3y$.

Clavius' *Astrolabium* (Rome, 1593) and his edition of the last nine books of *Euclid* (Rome, 1589) contain no algebraic symbolism and are rhetorical in exposition.

BELGIUM: SIMON STEVIN
(1585)

162. Stevin was influenced in his notation of powers by Bombelli, whose exponent placed in a circular arc became with Stevin an exponent inside of a circle. Stevin's systematic development of decimal fractions is published in 1585 in a Flemish booklet, *La thiende*,[2] and also in French in his *La disme*. In decimal fractions his exponents may be interpreted as having the base one-tenth. Page 16 (in Fig. 67) shows the notation of decimal fractions and the multiplication of 32.57 by 89.46, yielding the product 2913.7122. The translation is as follows:

"III. *Proposition, on multiplication:* Being given a decimal fraction to be multiplied, and the multiplier, to find their product.

"*Explanation of what is given:* Let the number to be multiplied be 32.57, and the multiplier 89.46. *Required*, to find their product. *Process:* One places the given numbers in order as shown here and multiplies according to the ordinary procedure in the multiplication of integral numbers, in this wise: [see the multiplication].

"Given the product (by the third problem of our *Arithmetic*) 29137122; now to know what this means, one adds the two last of the given signs, one (2) and the other (2), which are together (4). We say therefore that the sign of the last character of the product is (4), the which being known, all the others are marked according to their successive positions, in such a manner that 2913.7122 is the required product. *Proof:* The given number to be multiplied 32.57 (according to the third definition) is equal to $32\frac{5}{10}\frac{7}{100}$, together $32\frac{57}{100}$. And for the same reason the multiplier 89.46 becomes $89\frac{46}{100}$. Multiplying the said $32\frac{57}{100}$ by the same, gives a product (by the twelfth problem of our *Arithmetic*) $2913\frac{7122}{10000}$; but this same value has also the said product 2913.7122; this is therefore the correct product, which we

[1] *Algebra*, p. 72.

[2] A facsimile edition of *La "thiende"* was brought out in 1924 at Anvers by H. Bosmans.

were to prove. But let us give also the reason why ② multiplied by ②, gives the product ④ (which is the sum of their numbers), also why ④ times ⑤ gives the product ⑨, and why ⓪ times ③ gives ③, etc. We take $\frac{2}{10}$ and $\frac{3}{100}$ (which by the third definition of this Disme are .2 and .03; their product is $\frac{6}{1000}$ which, according to our third definition, is equal to .006. Multiplying, therefore, ① by ② gives the

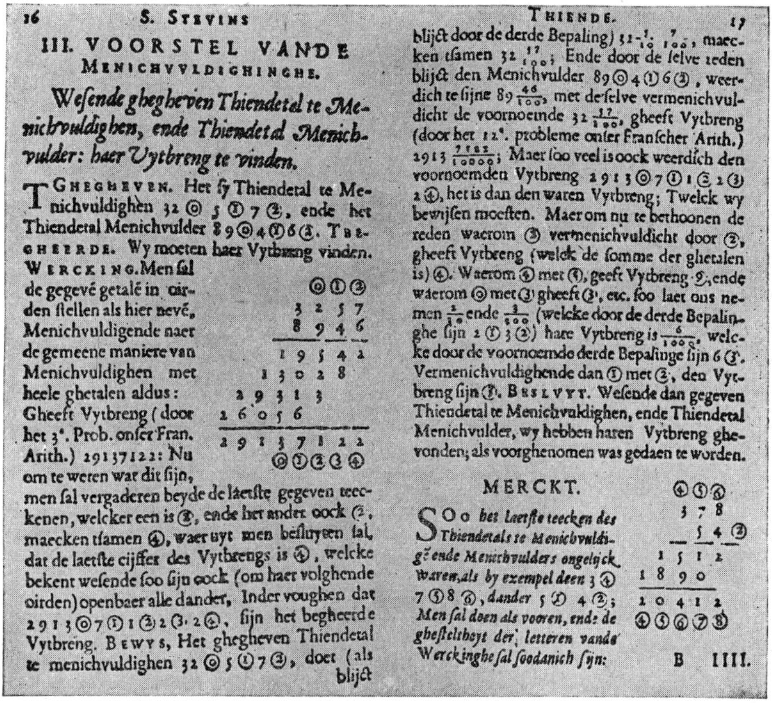

16 S. STEVINS

III. VOORSTEL VANDE MENICHVVLDIGHINGHE.

Wesende ghegheven Thiendetal te Menichvuldighen, ende Thiendetal Menichvulder: haer Vytbreng te vinden.

T GHEGHEVEN. Het sy Thiendetal te Menichvuldighen 32 ⓪ 5 ① 7 ②, ende het Thiendetal Menichvulder 89 ⓪ 4 ① 6 ②. TBEGHEERDE. Wy moeten haer Vytbreng vinden. WERCKING. Men sal de gegevẽ getalẽ in oirden stellen als hier nevẽ, Menichvuldigende naer de gemeene maniere van Menichvuldighen met heele ghetalen aldus: Gheeft Vytbreng (door het 3°. Prob. onser Fran. Arith.) 29137122: Nu om te weten wat dit sijn, men sal vergaderen beyde de laetste gegeven teeckenen, welcker een is ②, ende het ander oock ②, maecken tsamen ④, waer uyt men besluyten sal, dat de laetste cijffer des Vytbrengs is ④, welcke bekent wesende soo sijn oock (om haer volghende oirden) openbaer alle dander, Inder voughen dat 2913 ⓪ 7 ① 1 ② 2 ③ 2 ④, sijn het begheerde Vytbreng. BEWYS, Het ghegheven Thiendetal te menichvuldighen 32 ⓪ 5 ① 7 ②, doet (als blijct

					⓪	①	②
				3	2	5	7
				8	9	4	6
			1	9	5	4	2
		1	3	0	2	8	
	2	9	3	1	3		
2	6	0	5	6			
2	9	1	3	7	1	2	2
			⓪	①	②	③	④

THIENDE. 17

blijct door de derde Bepaling) 32 $\frac{5}{10}$ $\frac{7}{100}$, maecken tsamen 32 $\frac{57}{100}$; Ende door de selve reden blijct den Menichvulder 89 ⓪ 4 ① 6 ②, weerdich te sijne 89 $\frac{46}{100}$, met deselve vermenichvuldicht de voornoemde 32 $\frac{57}{100}$, gheeft Vytbreng (door het 12°. probleme onser Franscher Arith.) 2913 $\frac{7122}{10000}$; Maer soo veel is oock weerdich den voornoemden Vytbreng 2913 ⓪ 7 ① 1 ② 2 ③ 2 ④, het is dan den waren Vytbreng; Twelck wy bewijsen moesten. Maer om nu te bethoonen de reden waerom ② vermenichvuldicht door ②, gheeft Vytbreng (welck de somme der ghetalen is) ④. Waerom ④ met ⑤, geeft Vytbreng ⑨, ende waerom ⓪ met ③ gheeft ③, etc. soo laet ons nemen $\frac{2}{10}$ ende $\frac{3}{100}$ (welcke door de derde Bepalinghe sijn 2 ① 3 ②) haer Vytbreng is $\frac{6}{1000}$, welcke door de voornoemde derde Bepalinge sijn 6 ③. Vermenichvuldighende dan ① met ②, den Vytbreng sijn ③. BESLVYT. Wesende dan gegeven Thiendetal te Menichvuldighen, ende Thiendetal Menichvulder, wy hebben haren Vytbreng ghevonden; als voorghenomen was gedaen te worden.

MERCKT.

SOo het laetste teecken des Thiendetals te Menichvuldigen ende Menichvulders ongelijck waren, als by exempel deen 3 ④ 7 ⑤ 8 ⑥, dander 5 ① 4 ②; Men sal doen als vooren, ende de ghesteltheyt der letteren vande Werckinghe sal soodanich sijn:

		④	⑤	⑥
		3	7	8
		5	4	②
	1	5	1	2
1	8	9	0	
2	0	4	1	2
④	⑤	⑥	⑦	⑧

B IIII.

FIG. 67.—Two pages in S. Stevin's *Thiende* (1585). The same, in French, is found in *Les œuvres mathématiques de Simon Stevin* (ed. A. Girard; Leyden, 1634), p. 209.

product ③, a number made up of the sum of the numbers of the given signs. *Conclusion:* Being therefore given a decimal number as a multiplicand, and also a multiplier, we have found their product, as was to be done.

"*Note:* If the last sign of the numbers to be multiplied is not the same as the sign of the last number of the multiplier, if, for example, the one is 3④7⑤8⑥, and the other 5①4②, one proceeds as above and the disposition of the characters in the operation is as shown: [see process on p. 17]."

A translation of the *La disme* into English was brought out by Robert Norman at London in 1608 under the title, *Disme: The Art of*

Qvestion XX.

T Rouvons un ⊙ tel, que ſon quarré — 12 multiplié par la
ſomme du double d icelui ⊙ , & le quarré de — 2 & 4 , le
produict ſoit egal au quarré du produict de — 2 par icelui ⊙ requis.

Constrvction.

Soit le nombre requis   1 ① | 4
Son quarré 1 ② , auquel ajouſté — 12
   faict   1 ② — 12 | 4
Qui multiplié par la ſomme du double du nom-
   bre requis , & le quarré de — 2 & 4 , qui eſt
   par   2 ① + 8 , faict 2 ③ + 8 ② — 24 ① — 96 | 64
Egal au quarré du produict de — 2 , par 1 ①
   premier en l'ordre, qui eſt à   4 ② |
     Leſquels reduicts, 1 ③ ſera egale à — 2 ② + 12 ① +
48 ; Et 1 ① par le 71 probleme, vaudra 4.

    Je di que 4 eſt le nombre requis. *Demonſtration.* Le
quarré de 4 eſt 16, qui avec — 12 faict 4 , qui multiplié
par 16 (16 pour la ſomme du double d'iceluy 4, & le
quarré de — 2 & encore 4) faict 64, qui ſont egales au
quarré du produict de — 2, par le 4 trouvé, ſelon le requis ; ce qu'il falloit demonſtrer.

                               Qve-

Fig. 68.—From p. 98 of *L'arithmétique* in Stevin's *Œuvres mathématiques* (Leyden, 1634).

Tenths, or Decimall Arithmetike. Norman does not use circles, but round parentheses placed close together, the exponent is placed high, as in (²). The use of parentheses instead of circles was doubtless typographically more convenient.

Stevin uses the circles containing numerals also in algebra. Thus a circle with 1 inside means x, with 2 inside means x^2, and so on. In Stevin's *Œuvres* of 1634 the use of the circle is not always adhered to. Occasionally one finds, for x^4, for example,[1] the signs $(\overset{}{4})$ and (4).

The translation of Figure 68 is as follows: "To find a number such that if its square -12, is multiplied by the sum of double that number and the square of -2 or 4, the product shall be equal to the square of the product of -2 and the required number.

Solution

"Let the required number be . . . . x |4

Its square x^2, to which is added -12 gives x^2-12 |4

This multiplied by the sum of double the required number and the square of -2 or 4, i.e., by $2x+8$, gives $2x^3+8x^2-24x-96$ equal to the |64 square of the product of -2 and x, i.e., equal to. . . . $4x^2$ Which reduced, $x^3 = -2x^2+12x+48$; and x, by the problem 71, becomes 4. I say that 4 is the required number.

"*Demonstration:* The square of 4 is 16, which added to -12 gives 4, which multiplied by 16 (16 being the sum of double itself 4, and the square of -2 or 4) gives 64, which is equal to the square of the product of -2 and 4, as required; which was to be demonstrated."

If more than one unknown occurs, Stevin marks[2] the first unknown "1⊙," the second "1 *secund.* ⊙," and so on. In solving a Diophantine problem on the division of 80 into three parts, Stevin represents the first part by "1⊙," the second by "1 *secund.* ⊙," the third by "$-$① -1 *secund.* ①$+80$." The second plus $\frac{1}{5}$ the first $+$ 6 minus the binomial $\frac{1}{6}$ the second $+$ 7 yields him "$\frac{5}{6}$ *secund.* ⊙$+$ $\frac{1}{5}$⊙-1." The sum of the third and $\frac{1}{6}$ the second, $+$ 7, minus the binomial $\frac{1}{7}$ the third $+$ 8 yields him "$\frac{6}{7}$⊙$-\frac{29}{42}$ *secund.* ⊙$+\frac{473}{7}$." By the conditions of the problem, the two results are equal, and he obtains "1 *Secund.* ⊙ *Aequalem*$-\frac{111}{160}$⊙$+45$." In his *L'arithmétique*[3] one finds "12 *sec.* ④$+23$①M *sec.* ②$+10$②," which means $12y^4+23xy^2+10x^2$, the M signifying here "multiplication" as it had with Stifel (§ 154). Stevin uses also D for "division."

163. For radicals Stevin uses symbols apparently suggested by

<hr>

[1] *Les Œuvres mathématiques de Simon Stevin* (1634), p. 83, 85.

[2] Stevin, *Tomvs Qvintvs mathematicorvm Hypomnematvm de Miscellaneis* (Leiden, 1608), p. 516.

[3] Stevin, *Œuvres mathématiques* (Leyden, 1634), p. 60, 91, of "**Le II. livre** d'arith."

those of Christoff Rudolff, but not identical with them. Notice the shapes of the radicals in Figure 69. One stroke yields the usual square root symbol $\sqrt{}$, two strokes indicate the fourth root, three strokes the eighth root, etc. Cube root is marked by $\sqrt{}$ followed by a 3 inside a circle; $\sqrt{}$ followed by a 3 inside a circle means the cube root twice taken, i.e., the ninth root. Notice that $\sqrt{3}$)(②) means $\sqrt{3}$ times x^2, not $\sqrt{3x^2}$; the)(is a sign of separation of factors. In place of the u or v to express "universal" root, Stevin uses $bino$ ("binomial") root.

Stevin says that $\frac{3}{2}$ placed within a circle means $x^{\frac{3}{2}}$, but he does not actually use this notation. His words are (p. 6 of *Œuvres* [*Arithmetic*]), "$\frac{3}{2}$ en un circle seroit le charactere de racine quarrée de ③, par ce que telle $\frac{3}{2}$ en circle multipliée en soy donne produict ③, et ainsi des autres." A notation for fractional exponents had been suggested much earlier by Oresme (§ 123).

LORRAINE: ALBERT GIRARD
(1629)

164. Girard[1] uses $+$ and $-$, but mentions $\div$ as another sign used for "minus." He uses $=$ for "difference entre les quantitez où il se treuve." He introduces two new symbols: *ff*, *plus que*; §, *moins que*. In further explanation he says: "Touchant les lettres de l'Alphabet au lieu des nombres: soit A & aussi B deux grandeurs: la somme est $A+B$, leur différence est $A=B$, (ou bien si A est majeur on dira que c'est $A-B$) leur produit est AB, mais divisant A par B viendra $\dfrac{A}{B}$ comme és fractions: les voyelles se posent pour les choses incognues." This use of the vowels to represent the unknowns is in line with the practice of Vieta.

The marks (2), (3), (4), ... , indicate the second, third, fourth, ... , powers. When placed before, or to the left, of a number, they signify the respective power of that number; when placed after a number, they signify the power of the unknown quantity. In this respect Girard follows the general plan found in Schöner's edition of the *Algebra* of Ramus. But

BRIEFVE COLLECTION DES CHARACTERES QV'ON VSERA EN CESTE ARITHMETIQVE.

VEu que la cognoissance des characteres est de grande conséquence, par ce qu'on les use en l'Arithmetique au lieu des mots, nous les ajousterons icy, (combien qu'au precedent chascun a esté amplement declaré

[Continued on page 159]

[1] *Invention nouvelle en l'Algebre, A Amsterdam* (M.DC.XXIX); reimpression par Dr. D. Bierens de Haan (Leiden, 1884), fol. *B*.

Girard adopts the practice of Stevin in using fractional exponents. Thus, "$(\frac{3}{2})49$" means $(\sqrt{49})^3 = 343$, while "$49(\frac{3}{2})$" means $49x^{\frac{3}{2}}$. He points out that $18(0)$ is the same as 18, that $(1)18$ is the same as $18(0)$.

We see in Girard an extension of the notations of Chuquet, Bombelli, and Stevin; the notations of Bombelli and Stevin are only variants of that of Chuquet.

The conflict between the notation of roots by the use of fractional exponents and by the use of radical signs had begun at the time of Girard. "Or pource que $\sqrt{}$ est en usage, on le pourra prendre au lieu de $(\frac{1}{2})$ à cause aussi de sa facilité, signifiant racine seconde, ou racine quarée; que si on veut poursuivre la progression on pourra au lieu de $\sqrt{}$ marquer $\overset{2}{\sqrt{}}$; & pour la racine cubique, ou tierce, ainsi $\overset{3}{\sqrt{}}$ ou bien $(\frac{1}{3})$, ou bié $\mathcal{ct}$, ce qui peut estre au choix, mais pour en dire mon opinion les fractions sont plus expresses & plus propres à exprimer en perfection, & $\sqrt{}$ plus faciles et expedientes, comme $\overset{5}{\sqrt{}}32$ est à dire la racine de 32, & est 2. Quoy que ce soit l'un & l'autre sont facils

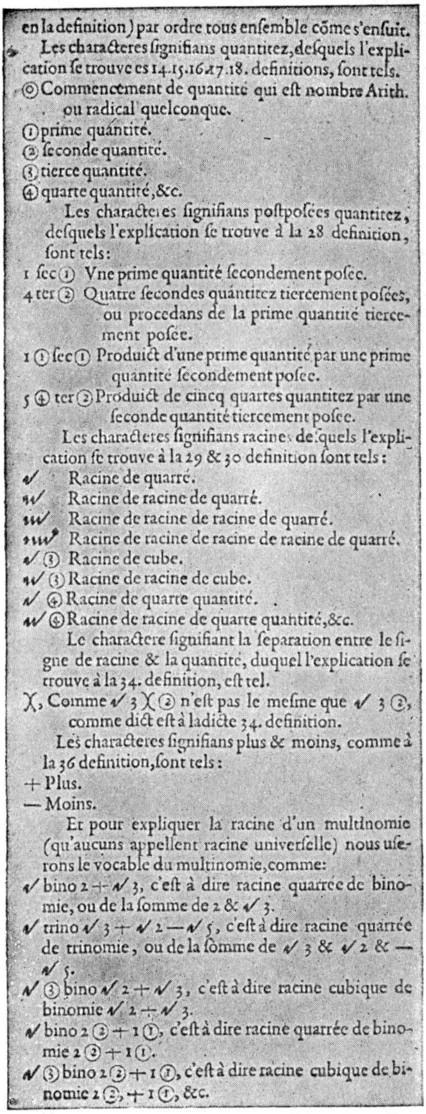

en la definition) par ordre tous enfemble côme s'enfuit.

Les characteres fignifians quantitez, defquels l'explication fe trouve es 14.15.16.17.18. definitions, font tels.

⊙ Commencement de quantité qui eft nombre Arith. ou radical quelconque.

① prime quantité.
② feconde quantité.
③ tierce quantité.
④ quarte quantité, &c.

Les characteres fignifians poftpofées quantitez, defquels l'explication fe trouve à la 28 definition, font tels.

1 fec① Vne prime quantité fecondement pofee.
4 ter② Quatre fecondes quantitez tiercement pofées, ou procedans de la prime quantité tiercement pofée.
1 ① fec① Produict d'une prime quantité par une prime quantité fecondement pofee.
5 ④ ter② Produict de cincq quartes quantitez par une feconde quantité tiercement pofee.

Les characteres fignifians racine, defquels l'explication fe trouve à la 29 & 30 definition font tels:

√ Racine de quarré.
√√ Racine de racine de quarré.
√√√ Racine de racine de racine de quarré.
√√√√ Racine de racine de racine de racine de quarré.
√③ Racine de cube.
√√③ Racine de racine de cube.
√④ Racine de quarte quantité.
√√④ Racine de racine de quarte quantité, &c.

Le character fignifiant la feparation entre le figne de racine & la quantité, duquel l'explication fe trouve à la 34. definition, eft tel.

χ, Comme √3 χ② n'eft pas le mefme que √3②, comme dict eft à ladicte 34. definition.

Les characteres fignifians plus & moins, comme à la 36 definition, font tels:

+ Plus.
— Moins.

Et pour expliquer la racine d'un multinomie (qu'aucuns appellent racine univerfelle) nous uferons le vocable du multinomie, comme:

√ bino 2 + √3, c'eft à dire racine quarrée de binomie, ou de la fomme de 2 & √3.
√ trino √3 + √2 — √5, c'eft à dire racine quarrée de trinomie, ou de la fomme de √3 & √2 & — √5.
√③ bino √2 + √3, c'eft à dire racine cubique de binomie √2 + √3.
√ bino 2②+1①, c'eft à dire racine quarrée de binomie 2②+1①.
√③ bino 2②+1①, c'eft à dire racine cubique de binomie 2②+1①, &c.

Fig. 69.—From S. Stevin's *L'arithmétique* in *Œuvres mathématiques* (ed. A. Girard; Leyden, 1634), p. 19.

à comprendre, mais $\sqrt{}$ et $\mathcal{ct}$ sont pris pour facilité." Girard appears to be the first to suggest placing the index of the root in the opening of the radical sign, as $\overset{\cdot}{\sqrt{}}$. Sometimes he writes $\sqrt{}\sqrt{}$ for $\overset{\cdot}{\sqrt{}}$.

The book contains other notations which are not specially explained. Thus the cube of $B+C$ is given in the form $B(B_q+C_q^3)+C(B_q^3+C_q)$.

We see here the use of round parentheses, which we encountered before in the *Algebra* of Clavius and, once, in Cardan. Notice also that C_q^3 means here $3C^2$.

Autre exemple	In Modern Symbols
"Soit 1(3) esgale à $-6(1)+20$	Let $x^3 = -6x+20$
Divisons tout par 1(1)	Divide all by x,
$1(2)$ esgale à $-6+\dfrac{20}{1(1)}$."	$x^2 = -6+\dfrac{20}{x}$.

Again (fol. F3): "Soit 1(3) esgale à 12(1)-18 (impossible d'estre esgal)

car le $\frac{1}{3}$ est 4   9 qui est $\frac{1}{2}$ de 18

son cube 64   81 son quarré .

Et puis que 81 est plus que 64, l'equation est impossible & inepte."

Translation: "Let $x^3 = 12x - 18$ (impossible to be equal)

because the $\frac{1}{3}$ is 4   9 which is $\frac{1}{2}$ of 18

its cube 64   81 its square

And since 81 is more than 64, the equation is impossible and inept."

A few times Girard uses parentheses also to indicate multiplication (see *op. cit.*, folios C_3^b, D_3^a, F_4^b).

GERMAN-SPANISH: MARCO AUREL

(1552)

165. Aurel states that his book is the first algebra published in Spain. He was a German, as appears from the title-page: *Libro primero de Arithmetica Algebratica ... por Marco Aurel, natural Aleman* (Valencia, 1552).[1] It is due to his German training that German algebraic symbols appear in this text published in Spain. There is hardly a trace in it of Italian symbolism. As seen in Figure 70, the plus $(+)$ and minus $(-)$ signs are used, also the German symbols for powers of the unknown, and the clumsy Rudolffian symbols for roots of different

---

[1] Aurel's algebra is briefly described by Julio Rey Pastor, *Los mathemáticos españoles del siglo XVI* (Oviedo, 1913), p. 36 n.; see *Bibliotheca mathematica*, Vol. IV (2d ser., 1890), p. 34.

orders. In place of the dot, used by Rudolff and Stifel, to express the root of a polynomial, Aurel employs the letter v, signifying universal root or *rayz vniuersal*. This v is found in Italian texts.

Declaracion de algunos characteres, que para las rayzes seran necessarios.

Para tratar de tales numeros, y otros semejantes, seria cosa larga, y no galana, poner los tales nõbres a la larga : mas desseando huyr esto, y euitar toda prolixidad, procure poner aqui algunos, que para en esta arte eran necessarios. Y son √. √. √√. √ v. √ v. √ v. +. —. Delos quales el p°, significa, y quiere dezir rayz quadrada: el 2°, rayz quadrada de rayz qua-drada, o rayz de rayz: el 3°, rayz cubica: el 4°, rayz vniuersal: el 5°, rayz de rayz vniuersal: el 6°, rayz cubica vniuersal: el 7°, mas: y el 8°, menos. Exẽplo, √ 4, quiere dezir rayz quadrada de 4, que es 2. √ 5, quiere dezir rayz de 5. &c. √ 20 + √ 7, quiere dezir, rayz de rayz de 20, y mas rayz cubica de 7. √ v. 8 — √ 3, quiere dezir, rayz quadrada de todo esto: q̃ es 8 — √ 3. Digo rayz vniuersal de 8 menos rayz de 3, q̃ es √ del residuo &c. Esto me ha parescido bueno, tome cada vno los q̃ querra, o escriualos a la larga, todo a su plazer: que por esto no pierde ni gana la sciencia, o arte.

L 3 La

Nota. Para ver el quociente, en partir vn caracter por otro, que sera, pornas los dichos caracteres en orden, como en el multiplicar heziste : y assi mesmo señalados cõ 1, 2, 3, &c. Y encima del 8, vn zero, assi.

0. 1. 2. 3. 4. 5. 6. 7. 8. 9.

Y assi como en el multiplicar summas las quantidades q̃ estan

Fig. 70.—From Aurel's *Arithmetica algebratica* (1552). (Courtesy of the Library of the University of Michigan.) Above is part of fol. 43, showing the + and —, and the radical signs of Rudolff, also the $\sqrt{v}$. Below is a part of fol. 73B, containing the German signs for the powers of the unknown and the sign for a given number.

166. Nuñez' *Libro de algebra* (1567)[1] bears in the Dedication the date December 1, 1564. The manuscript was first prepared in the Portuguese language some thirty years previous to Nuñez' preparation of this Spanish translation. The author draws entirely from Italian authors. He mentions Pacioli, Tartaglia, and Cardan.

The notation used by Nuñez is that of Pacioli and Tartaglia. He uses the terms *Numero, cosa, censo, cubo, censo de censo, relato primo, censo de cubo* or *cubo de censo, relato segundo, censo de censo de cẽso, cubo de cubo, censo de relato primo,* and their respective abbreviations *co., ce., cu., ce.ce., re.p°, ce.cu.* or *cu.ce., re.seg°. ce.ce.ce., cu.cu., ce.re.p°.* He uses $\bar{p}$ for *más* ("more"), and $\tilde{m}$ for *menos* ("les"). The only use made of the ✠ is in cross-multiplication, as shown in the following sentence (fol. 41): "... partiremos luego $\dfrac{12.}{1.co.}$ por $\dfrac{2.cu.\bar{p}.8.}{1.ce.}$ como si fuessen puros quebrados, multiplicãdo en ✠, y verna por quociente $\dfrac{12.ce.}{2.ce.ce.\bar{p}.8.co.}$ el qual quebrado abreuiado por numero y por dignidad verna a este quebrado $\dfrac{6.co.}{1.cu.\bar{p}.4.}$." This expression, *multiplicando en* ✠, occurs often.

Square root is indicated by *R.*, cube root by *R.cu.*, fourth root by *R.R.*, eighth root by *R.R.R.* (fol. 207). Following Cardan, Nuñez uses *L.R.* and *R.V.* to indicate, respectively, the *ligatura* ("combination") of roots and the *Raiz vniuersal* ("universal root," i.e., root of a binomial or polynomial). This is explained in the following passage (fol. 45*b*): "... diziendo assi: *L.R.7$\bar{p}$R.4.$\bar{p}$.3.* que significa vna quantidad sorda compuesta de .3. y 2. que son 5. con la *R.*7. o diziendo assi: *L.R.3$\bar{p}$2.co. Raiz vniuersal* es raiz de raiz ligada con numero o con otra raiz o dignidad. Como si dixessemos assi: *R.v.* 22 $\bar{p}$ *R*₉."

Singular notations are 2. *co.* ¼. for $2\frac{1}{4}x$ (fol. 32), and 2. *co.* ⅖ for $2\frac{2}{5}x$ (fol. 36*b*). Observe also that integers occurring in the running text are usually placed between dots, in the same way as was customary in manuscripts.

Although at this time our exponential notation was not yet invented and adopted, the notion of exponents of powers was quite well understood, as well as the addition of exponents to form the product

[1] *Libro de Algebra en arithmetica y Geometria. Compuesto por el Doctor Pedro Nuñez, Cosmographo Mayor del Rey de Portugal, y Cathedratico Jubilado en la Cathedra de Mathematicas en la Vniuersidad de Coymbra* (En Anvers, 1567).

of terms having the same base. To show this we quote from Nuñez the following (fol. 26b):

"... si queremos multiplicar .4. *co.* por .5. *ce.* diremos asi .4. por .5. hazen .20. y porque .1. denominaciõ de co. sũmado con .2. denominacion de censo hazen .3. que es denominaciõ de cubo. Diremos por tanto q .4. *co.* por .5. *ce.* hazen .20. *cu.* ... si multiplicamos .4. *cu.* por .8. *ce.ce.* diremos assi, la denominacion del cubo es .3. y la denominaciõ del censo de censo es .4. q̃ sũmadas hazẽ .7. q̃ sera la denominaciõ dela dignidad engẽdrada, y por que .4. por .8. hazen .32. diremos por tanto, que .4. *cu.* multiplicados por .8. *ce.ce.* hazen .32. dignida•les, que tienen .7. por denominacion, a que llaman relatos segundos."

Nuñez' division[1] of $12x^3+18x^2+27x+17$ by $4x+3$, yielding the quotient $3x^2+2\frac{1}{4}x+5\frac{1}{16}+\dfrac{1\frac{13}{16}}{4x+3}$, is as follows:

"Partidor .4.*co.p̄*.3 | 12.*cu.p̄*.18.*ce.p̄*.27.*co.p̄*.17.
$\quad\quad\quad\quad\quad\quad\quad$ | 12.*cu.p̄*. 9.*ce.*

$\quad\quad\quad\quad\quad\quad\quad\quad\quad\quad$ 9.*ce.p̄*.27.*co.p̄*.17.
$\quad\quad\quad\quad\quad\quad\quad\quad\quad\quad$ 9.*ce.p̄*. 6.*co*.$\frac{3}{4}$.

$\quad\quad\quad\quad\quad\quad\quad\quad\quad\quad\quad\quad$ 20.*co*.$\frac{1}{4}$.*p̄*.17.
$\quad\quad\quad\quad\quad\quad\quad\quad\quad\quad\quad\quad$ 20.*co*.$\frac{1}{4}$.*p̄*.15$\frac{3}{16}$.

$\quad\quad\quad\quad\quad\quad\quad\quad\quad\quad\quad\quad\quad\quad\quad\quad$ 1$\frac{13}{16}$

$\quad\quad\quad$ **3**.*ce.p̄*.2.*co*.$\frac{1}{4}$.*p̄*.5$\frac{1}{16}$.*p̄*.1$\frac{13}{16}$
$\quad\quad\quad\quad\quad\quad\quad\quad$ par .4.*co.p̄*.3."

Observe the "20.*co*.$\frac{1}{4}$" for $20\frac{1}{4}x$, the symbol for the unknown appearing between the integer and the fraction.

Cardan's solution of $x^3+3x=36$ is $\sqrt[3]{\sqrt{325}+18}-\sqrt[3]{\sqrt{325}-18}$, and is written by Nuñez as follows:

$$R.V.cu.R.325.\bar{p}.18.\tilde{m}.R.V.cu..R.325.\tilde{m}.18.$$

As in many other writers the V signifies *vniversal* and denotes, not the cube root of $\sqrt{325}$ alone, but of the binomial $\sqrt{325}+18$; in other words, the V takes the place of a parenthesis.

[1] See H. Bosmans, "Sur le 'Libro de algebra' de Pedro Nuñez," *Bibliotheca mathematica*, Vol. VIII (3d ser., 1908), p. 160–62; see also Tropfke, *op. cit.* (2d ed.), Vol. III, p. 136, 137.

167. Robert Recorde's arithmetic, the *Grovnd of Artes*, appeared in many editions. We indicate Recorde's singular notation for proportion:[1]

$$3 \diagdown 16s.$$
$$8 \diagup 42s.\ 8d.$$

(direct) $3:8 = 16s.:42s.\ 8d.$

$$\tfrac{3}{4} \diagdown {}^{7}_{13}$$
$$\tfrac{5}{12} \diagup$$

(reverse) $1\tfrac{5}{2}:\tfrac{3}{4} = \tfrac{7}{13}:x$

There is nothing in Recorde's notation to distinguish between the "rule of proportion direct" and the "rule of porportion reverse." The difference appears in the interpretation. In the foregoing "direct" proportion, you multiply 8 and 16, and divide the product by 3. In the "reverse" proportion, the processes of multiplication and division are interchanged. In the former case we have $8 \times 16 \div 3 = x$, in the second case we have $\tfrac{3}{4} \times {}^{7}_{13} \div {}^{5}_{12} = x$. In both cases the large strokes in Z serve as guides to the proper sequence of the numbers.

168. In Recorde's algebra, *The Whetstone of Witte* (London, 1557), the most original and historically important is the sign of equality ($=$), shown in Figure 71. Notice also the plus ($+$) and minus ($-$) signs which make here their first appearance in an English book.

In the designation of powers Recorde uses the symbols of Stifel and gives a table of powers occupying a page and ending with the eightieth power. The seventh power is denoted by $b\int\mathfrak{z}$; for the eleventh, thirteenth, seventeenth powers, he writes in place of the letter b the letters c, d, E, respectively. The eightieth power is denoted by $\mathfrak{zzzz}\int\mathfrak{z}$, showing that the Hindu multiplicative method of combining the symbols was followed.

Figure 72 shows addition of fractions. The fractions to be added are separated by the word "to." Horizontal lines are drawn above and below the two fractions; above the upper line is written the new numerator and below the lower line is written the new denominator. In "Another Example of Addition," there are added the fractions $\dfrac{5x^6 + 3x^5}{6x^9}$ and $\dfrac{20x^3 - 6x^5}{6x^9}$.

---

[1] *Op. cit.* (London, 1646), p. 175, 315. There was an edition in 1543 which was probably the first.

Square root Recorde indicates by $\sqrt{}$. or $\sqrt{3}$, cube root by
$\mathcal{w}$. or $\mathcal{w}$.cc. Following Rudolff, he indicates the fourth root by

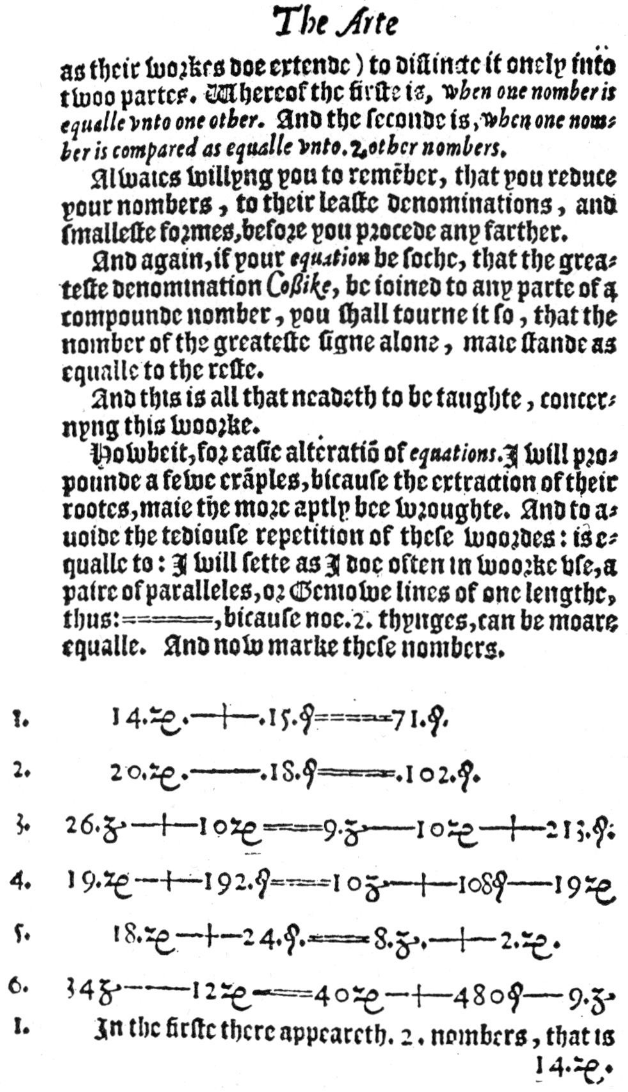

The Arte

as their woikes doe extende) to diftincte it onely into
twoo partes. Whereof the firfte is, *when one number is
equalle vnto one other.* And the feconde is, *when one nomber is compared as equalle vnto.2,other nombers.*

Alwaies willyng you to remēber, that you reduce
your nombers, to their leafte denominations, and
fmallefte foimes, befoie you piocede any farther.

And again, if your *equation* be foche, that the greatefte denomination Coßike, be ioined to any parte of a
compounde nomber, you ſhall tourne it fo, that the
nomber of the greatefte figne alone, maie ftande as
equalle to the refte.

And this is all that neadeth to be taughte, conccr-
nyng this wooike.

Howbeit, foi eafie alteratiō of *equations.*I will pio-
pounde a fewe crāples, bicaufe the extraction of their
rootes, maie the moie aptly bee wioughte. And to a-
uoide the tedioufe repetition of thefe wooides: is e-
qualle to: I will fette as I doe often in wooike vfe, a
paire of paralleles, oi Gemowe lines of one lengthe,
thus:=======, bicaufe noe.2. thynges, can be moare
equalle. And now marke thefe nombers.

1. $14.\not ze.--\!\!|--.15.\not q=====71.\not q.$

2. $20.\not ze.------.18.\not q=====.102.\not q.$

3. $26.\not z--\!\!|--10\not ze===9.\not z----10\not ze--\!\!|--213.\not q.$

4. $19.\not ze--\!\!|--192.\not q====10\not z--\!\!|--108\not q----19\not ze$

5. $18.\not ze--\!\!|--24.\not q.===8.\not z.--\!\!|--2.\not ze.$

6. $34\not z------12\not ze===40\not ze--\!\!|--480\not q----9.\not z.$

1. In the firfte there appeareth. 2. nombers, that is
    $14.\not ze.$

FIG. 71.—From Robert Recorde's *Whetstone of Witte* (1557)

$\mathcal{w}$., but Recorde writes it also $\mathcal{w}$₃₃. Instructive is the dialogue on
these signs, carried on between master and scholar:

"*Scholar:* It were againste reason, to take reason for those signes, whiche be set voluntarily to signifie any thyng; although some tymes there bee a certaine apte conformitie in soche thynges. And in these

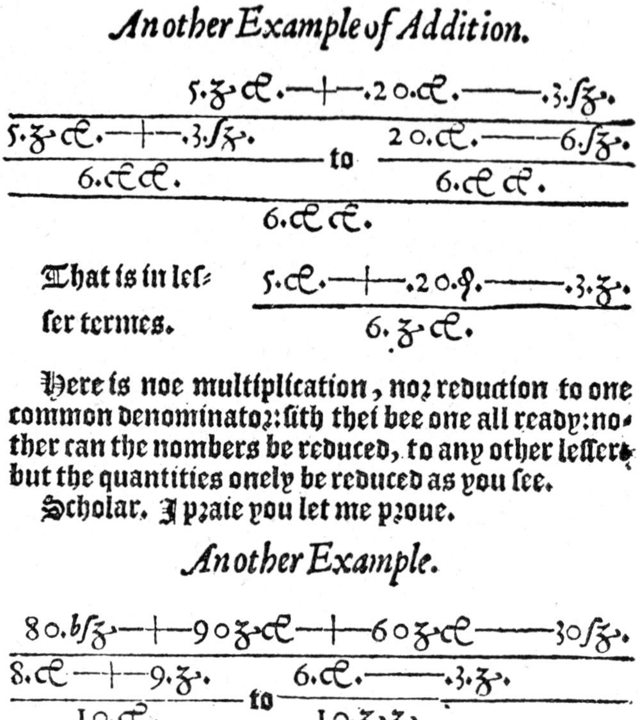

An other Example of Addition.

That is in lef-
fer termes.

Here is noe multiplication, nor reduction to one common denominator: fith thei bee one all ready: nother can the nombers be reduced, to any other letters but the quantities onely be reduced as you fee.

Scholar. I praie you let me proue.

An other Example.

Mafter. Marke your worke well, before you reduce it.

Scholar. I fee my faulte: I haue fette.2. nombers feuerally, with one figne *Cofsike* : by reafon I did not forefee, that. ℛ. multiplied with. ℛ. doeth make the like

Fig. 72.—Fractions in Recorde's *Whetstone of Witte* (1557)

figures, the number of their minomes, seameth disagreable to their order.

"*Master:* In that there is some reason to bee thewed: for as .√. declareth the multiplication of a nomber, ones by it self; so .⩗⩗. representeth that multiplication *Cubike,* in whiche the roote is represented thrise. And .√√. standeth for .√.√. that is .2. figures of *Square* multiplication: and is not expressed with .4. minomes. For so should it seme to expresse moare then .2. *Square* multiplications. But voluntarie signes, it is inoughe to knowe that this thei doe signifie.

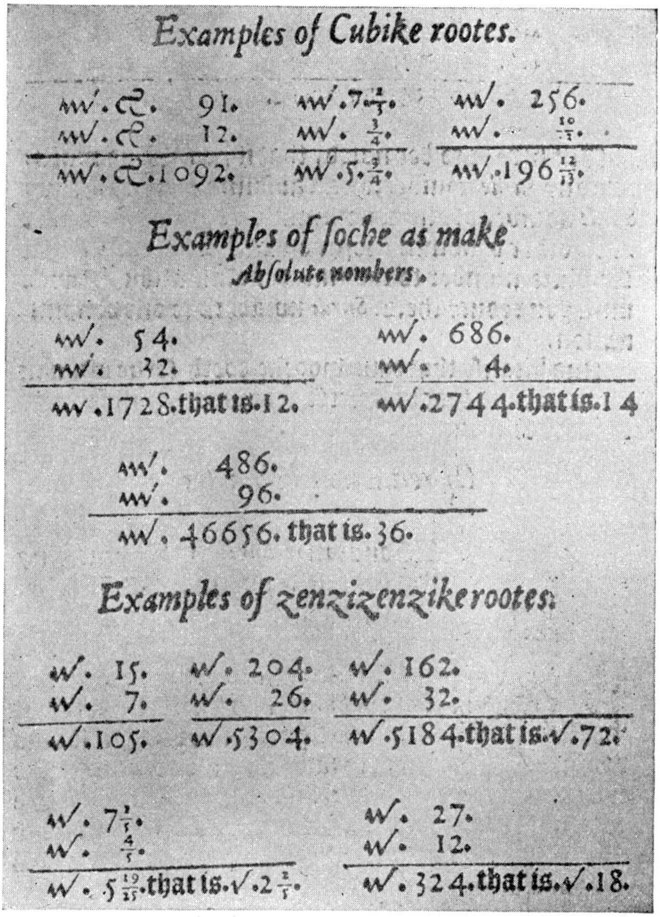

Examples of Cubike rootes.

√√√. 91. √√√. 12. — √√√. 1092.

√√√. 7⅔. √√√. ¾. — √√√. 5¾.

√√√. 256. √√√. 10/13. — √√√. 196 12/13.

Examples of soche as make Absolute nombers.

√√√. 54. √√√. 32. — √√√. 1728. that is. 12.

√√√. 686. √√√. 4. — √√√. 2744. that is. 14

√√√. 486. √√√. 96. — √√√. 46656. that is. 36.

Examples of zenzizenzike rootes.

√√. 15. √√. 7. — √√. 105.

√√. 204. √√. 26. — √√. 5304.

√√. 162. √√. 32. — √√. 5184. that is. √. 72.

√√. 7⅕. √√. ⅘. — √√. 5 19/25. that is. √. 2⅖.

√√. 27. √√. 12. — √√. 324. that is. √. 18.

FIG. 73.—Radicals in Recorde's *Whetstone of Witte* (1557)

And if any manne can diuise other, moare easie or apter in use, that maie well be received."

Figure 73 shows the multiplication of radicals. The first two exercises are $\sqrt[3]{91}\times\sqrt[3]{12}=\sqrt[3]{1,092}$, $\sqrt[3]{7\frac{2}{3}}\times\sqrt[3]{\frac{3}{4}}=\sqrt[3]{5\frac{3}{4}}$. Under fourth roots one finds $\sqrt[4]{15}\times\sqrt[4]{7}=\sqrt[4]{105}$.

ENGLISH: JOHN DEE

(1570)

169. John Dee wrote a Preface to Henry Billingsley's edition of *Euclid* (London, 1570). This Preface is a discussion of the mathematical sciences. The radical symbols shown in Figure 74 are those of Stifel. German influences predominated.

Fɪɢ. 74.—Radicals, John Dee's Preface to Billingsley's edition of *Euclid* (1570).

In Figure 75 Dee explains that if $a:b=c:d$, then also $a:a-b=c:c-d$. He illustrates this numerically by taking $9:6=12:8$. Notice Dee's use of the word "proportion" in the sense of "ratio." Attention is drawn to the mode of writing the two proportions $9.6:12.8$ and $9.3:12.4$, near the margin. Except for the use of a single colon (:),

Fɪɢ. 75.—Proportion in John Dee's Preface to Billingsley's edition of *Euclid* (1570).

in place of the double colon (: :), this is exactly the notation later used by Oughtred in his *Clavis mathematicae*. It is possible that Oughtred took the symbols from Dee. Dee's Preface also indicates the origin of these symbols. They are simply the rhetorical marks used in the text. See more particularly the second to the last line, "as 9. to 3: so 12. to 4:"

170. The *Stratioticos*[1] was brought out by Thomas Digges, the
son of Leonard Digges. It seems that the original draft of the book
was the work of Leonard; the enlargement of the manuscript and its
preparation for print were due to Thomas.

The notation employed for powers is indicated by the following
quotations (p. 33):

"In this Arte of Numbers Cossical, wae proceede from the Roote
by Multiplication, to create all Squares, Cubes, Zenzizenzike, and
Sur Solides, wyth all other that in this Science are used, the whyche
by Example maye best bee explaned.

1	2	3	4	5	6	7	8	9	10	11	12
Roo.	*Sq.*	*Cu.*	*SqS.*	*S∫o.*	*SqC.*	*B∫S.*	*SSSq.*	*CC.*	*S∫S.*	*C∫S.*	*SSC.* "
2	4	8	16	32	64	128	256	512	1024	2048	4096

Again (p. 32):

". . . . Of these [Roote, Square, Cube] are all the rest composed. For the Square being four, againe squared, maketh his
Squared square 16, with his Character ouer him. The nexte being not
made by the Square or Cubike, Multiplication of any of the former,
can not take his name from *Square* or *Cube*, and is therefore called a
Surd solide, and is onely created by Multiplicatiõ of 2 the Roote, in
16 the SqS. making 32 with his cõuenient Character ouer him & for
distinctiõ is tearmed $\overset{e}{y}$ first *Surd solide* . . . . the nexte being 128, is
not made of square or Cubique Multiplication of any, but only by the
Multiplication of the Squared Cube in his Roote, and therefore is
tearmed the *B.S.solide*, or seconde S. solide. . . . .

"This I have rather added for custome sake, bycause in all parts
of the world these *Characters* and names of *Sq.* and *Cu.* etc. are used,
but bycause I find another kinde of *Character* by my Father deuised,
farre more readie in *Multiplications, Diuisions*, and other *Cossical*
operations, I will not doubt, hauing Reason on my side, to dissent
from common custome in this poynt, and vse these Characters en-
suing: [What follows is on page 35 and is reproduced here in Fig. 76]."

[1] *An Arithmeticall Militare Treatise, named Stratioticos: compendiously teaching
the Science of Nũbers, as well in Fractions as Integers, and so much of the Rules and
Aequations Algebraicall and Arte of Numbers Cossicall, as are requisite for the Profes-
sion of a Soldiour. Together with the Moderne Militare Discipline, Offices, Lawes and
Dueties in euery wel gouerned Campe and Armie to be observed: Long since attẽpted
by Leonard Digges Gentleman, Augmented, digested, and lately finished, by Thomas
Digges, his Sonne* . . . . (At London, 1579).

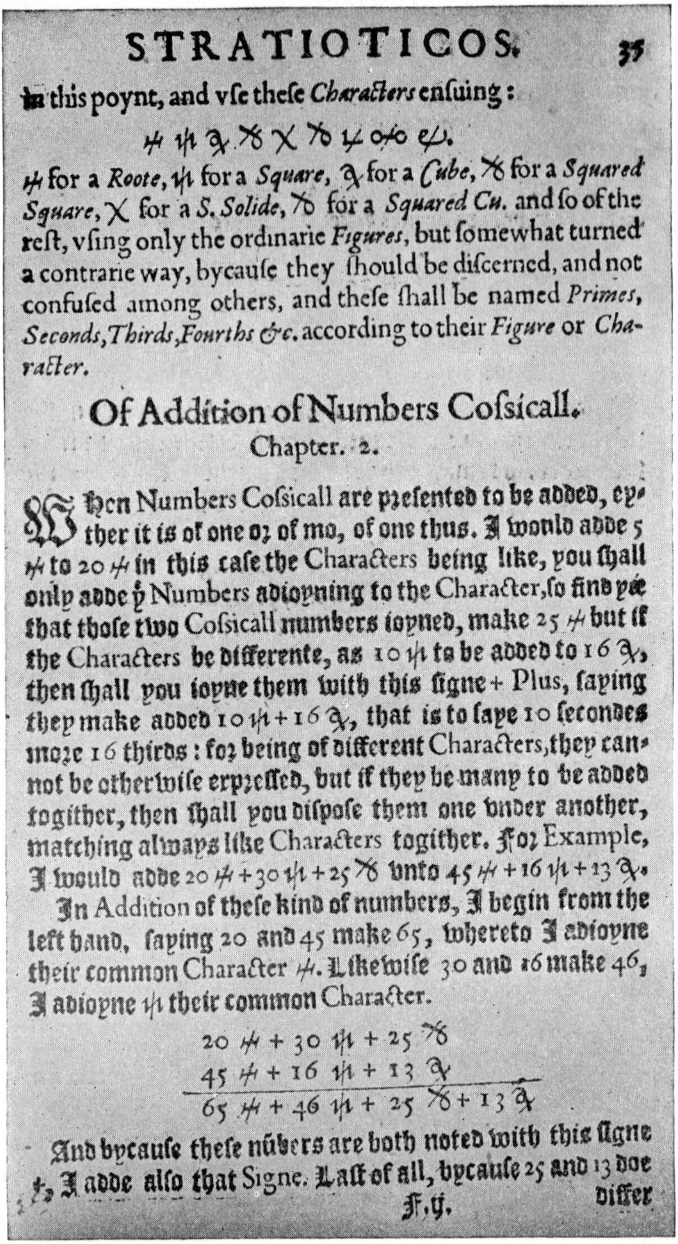

STRATIOTICOS.   35

in this poynt, and vſe theſe *Characters* enſuing :

$$\text{4 4 4 4 X 4 4 4 4}$$

4 for a *Roote*, 4 for a *Square*, 4 for a *Cube*, 4 for a *Squared Square*, X for a *S. Solide*, 4 for a *Squared Cu.* and ſo of the reſt, vſing only the ordinarie *Figures*, but ſomewhat turned a contrarie way, bycauſe they ſhould be diſcerned, and not confuſed among others, and theſe ſhall be named *Primes*, *Seconds*, *Thirds*, *Fourths &c.* according to their *Figure* or *Character*.

Of Addition of Numbers Coſsicall.
Chapter. 2.

When Numbers Coſsicall are preſented to be added, eyther it is of one or of mo, of one thus. I would adde 5 4 to 20 4 in this caſe the Characters being like, you ſhall only adde the Numbers adioyning to the Character, ſo find ye that thoſe two Coſsicall numbers ioyned, make 25 4 but if the Characters be differente, as 10 4 is be added to 16 4, then ſhall you ioyne them with this ſigne + Plus, ſaying they make added 10 4 + 16 4, that is to ſaye 10 ſecondes more 16 thirds : for being of different Characters, they cannot be otherwiſe expreſſed, but if they be many to be added together, then ſhall you diſpoſe them one vnder another, matching always like Characters together. For Example, I would adde 20 4 + 30 4 + 25 4 vnto 45 4 + 16 4 + 13 4.

In Addition of theſe kind of numbers, I begin from the left hand, ſaying 20 and 45 make 65, whereto I adioyne their common Character 4. Likewiſe 30 and 16 make 46, I adioyne 4 their common Character.

$$\begin{array}{r}
20 \; 4 + 30 \; 4 + 25 \; 4 \\
45 \; 4 + 16 \; 4 + 13 \; 4 \\
\hline
65 \; 4 + 46 \; 4 + 25 \; 4 + 13 \; 4
\end{array}$$

And bycauſe theſe nũbers are both noted with this ſigne +, I adde alſo that Signe. Laſt of all, bycauſe 25 and 13 doe differ

F. y.

Fig. 76.—Leonard and Thomas Digges, *Stratioticos* (1579), p. 35, showing the unknown and its powers to x^9.

As stated by the authors, the symbols are simply the numerals somewhat disfigured and crossed out by an extra stroke, to prevent confusion with the ordinary figures. The example at the bottom of page 35 is the addition of $20x+30x^2+25x^3$ and $45x+16x^2+13x^3$. It is noteworthy that in 1610 Cataldi in Italy devised a similar scheme for representing the powers of an unknown (§ 340).

The treatment of equations is shown on page 46, which is reproduced in Figure 77. Observe the symbol for zero in lines 4 and 7; this form is used only when the zero stands by itself.

A little later, on page 51, the authors, without explanation, begin to use a sign of equality. Previously the state of equality had been expressed in words, "equall to," "are." The sign of equality looks as if it were made up of two letters C in these positions $\Im C$ and crossed by two horizontal lines. See Figure 78.

This sign of equality is more elaborate than that previously devised by Robert Recorde. The Digges sign requires four strokes of the pen; the Recorde sign demands only two, yet is perfectly clear. The Digges symbol appears again on five or more later pages of the *Stratioticos*. Perhaps the sign is the astronomical symbol for *Pisces* ("the Fishes"), with an extra horizontal line. The top equation on page 51 is $x^2=6x+27$.

ENGLISH: THOMAS MASTERSON
(1592)

171. The domination of German symbols over English authors of the sixteenth century is shown further by the *Arithmeticke* of Thomas Masterson (London, 1592). Stifel's symbols for powers are used. We reproduce (in Fig. 79) a page showing the symbols for radicals.

FRENCH: JACQUES PELETIER
(1554)

172. Jacques Peletier du Mans resided in Paris, Bordeaux, Beziers, Lyon, and Rome. He died in Paris. His algebra, *De occvlta Parte Nvmerorvm, Quam Algebram vocant, Libri duo* (Paris, 1554, and several other editions),[1] shows in the symbolism used both German and Italian influences: German in the designation of powers and roots, done in the manner of Stifel; Italian in the use of p. and m. for "plus" and "minus."

[1] All our information is drawn from H. Bosmans, "L'algèbre de Jacques Peletier du Mans," *Extrait de la revue des questions scientifiques* (Bruxelles: January, 1907), p. 1–61.

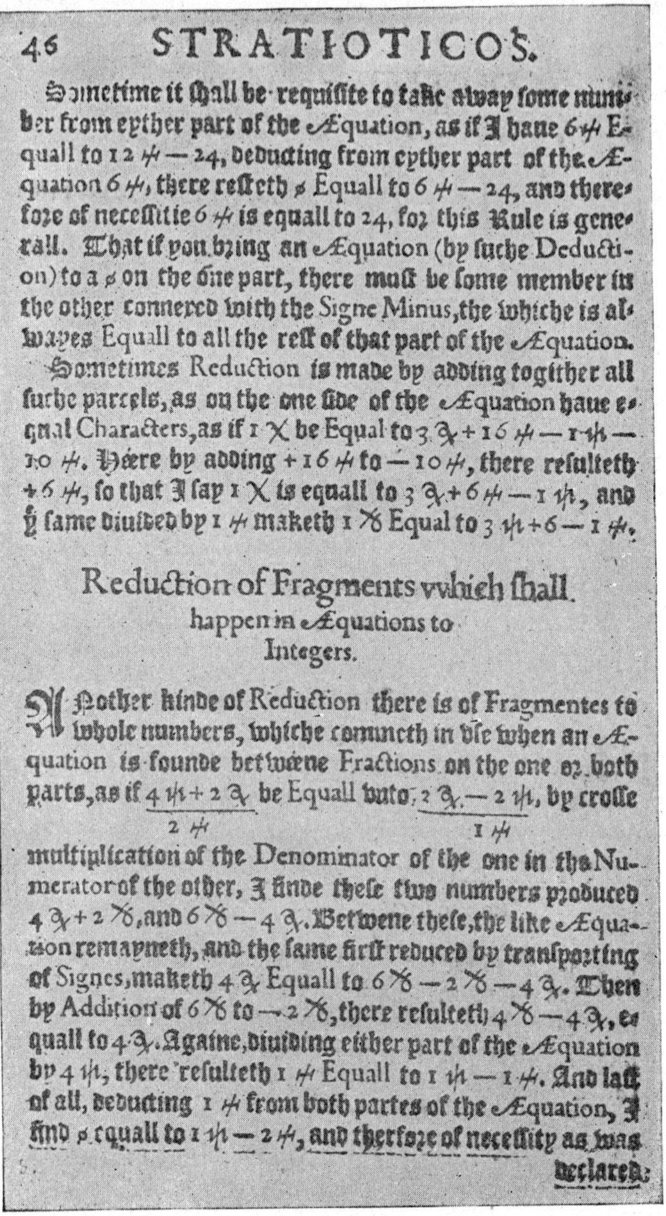

46 STRATIOTICOS.

Sometime it shall be requisite to take away some number from eyther part of the Æquation, as if I haue 6 ℳ Equall to 12 ℳ — 24, deducting from eyther part of the Æquation 6 ℳ, there resteth ∅ Equall to 6 ℳ — 24, and therefore of necessitie 6 ℳ is equall to 24, for this Rule is generall. That if you bring an Æquation (by suche Deduction) to a ∅ on the one part, there must be some member in the other connered with the Signe Minus, the whiche is alwayes Equall to all the rest of that part of the Æquation.

Sometimes Reduction is made by adding togither all suche parcels, as on the one side of the Æquation haue egual Characters, as if 1 ✗ be Equal to 3 ꒙ + 16 ℳ — 1 ℳ — 10 ℳ. Héere by adding + 16 ℳ to — 10 ℳ, there resulteth + 6 ℳ, so that I say 1 ✗ is equall to 3 ꒙ + 6 ℳ — 1 ℳ, and ẏ same diuided by 1 ℳ maketh 1 ✗ Equal to 3 ℳ + 6 — 1 ℳ.

Reduction of Fragments vvhich shall happen in Æquations to Integers.

Another kinde of Reduction there is of Fragmentes to whole numbers, whiche commeth in vse when an Æquation is founde betwéene Fractions on the one or both parts, as if $\frac{4\,ℳ + 2\,꒙}{2\,ℳ}$ be Equall vnto $\frac{2\,꒙ - 2\,ℳ}{1\,ℳ}$, by crosse multiplication of the Denominator of the one in the Numerator of the other, I finde these two numbers produced 4 ꒙ + 2 ✗, and 6 ✗ — 4 ꒙. Betwene these, the like Æquation remayneth, and the same first reduced by transporting of Signes, maketh 4 ꒙ Equall to 6 ✗ — 2 ✗ — 4 ꒙. Then by Addition of 6 ✗ to — 2 ✗, there resulteth 4 ✗ — 4 ꒙, equall to 4 ꒙. Againe, diuiding either part of the Æquation by 4 ℳ, there resulteth 1 ℳ Equall to 1 ℳ — 1 ℳ. And last of all, deducting 1 ℳ from both partes of the Æquation, I find ∅ equall to 1 ℳ — 2 ℳ, and therfore of necessity as was declared.

FIG. 77.—Equations in Digges, *Stratioticos* (1579)

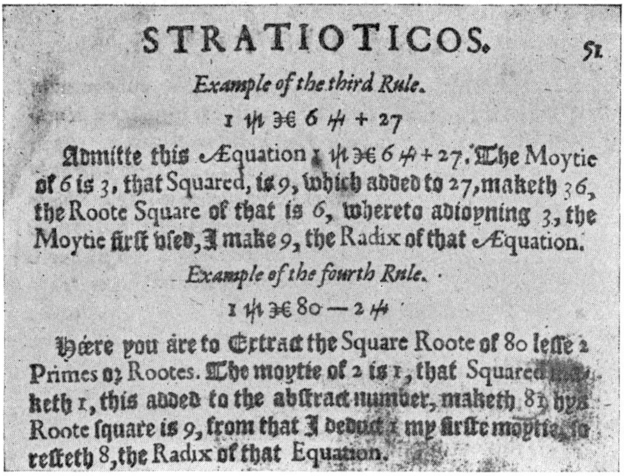

STRATIOTICOS. 51

Example of the third Rule.

1 ȝ ═ 6 ℞ + 27

Admitte this Æquation 1 ȝ ═ 6 ℞ + 27. The Moytie of 6 is 3, that Squared, is 9, which added to 27, maketh 36, the Roote Square of that is 6, whereto adioyning 3, the Moytie first vsed, I make 9, the Radix of that Æquation.

Example of the fourth Rule.

1 ȝ ═ 80 — 2 ℞

Here you are to Extract the Square Roote of 80 lesse 2 Primes or Rootes. The moytie of 2 is 1, that Squared maketh 1, this added to the abstract number, maketh 81, hys Roote square is 9, from that I deduct 1 my firste moytie, so resteth 8, the Radix of that Equation.

FIG. 78.—Sign of equality in Digges, *Stratioticos* (1579). This page exhibits also the solution of quadratic equations.

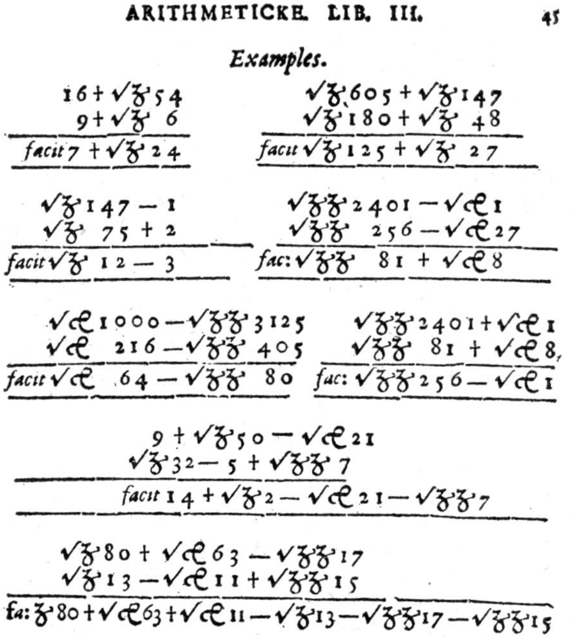

ARITHMETICKE. LIB. III. 45

Examples.

16 + √ℨ 54
9 + √ℨ 6
facit 7 + √ℨ 24

√ℨ 605 + √ℨ 147
√ℨ 180 + √ℨ 48
facit √ℨ 125 + √ℨ 27

√ℨ 147 — 1
√ℨ 75 + 2
facit √ℨ 12 — 3

√ℨℨ 2401 — √ℭ 1
√ℨℨ 256 — √ℭ 27
fac: √ℨℨ 81 + √ℭ 8

√ℭ 1000 — √ℨℨ 3125
√ℭ 216 — √ℨℨ 405
facit √ℭ 64 — √ℨℨ 80

√ℨℨ 2401 + √ℭ 1
√ℨℨ 81 + √ℭ 8,
fac: √ℨℨ 256 — √ℭ 1

9 + √ℨ 50 — √ℭ 21
√ℨ 32 — 5 + √ℨℨ 7
facit 14 + √ℨ 2 — √ℭ 21 — √ℨℨ 7

√ℨ 80 + √ℭ 63 — √ℨℨ 17
√ℨ 13 — √ℭ 11 + √ℨℨ 15
fa: ℨ 80 + √ℭ 63 + √ℭ 11 — √ℨ 13 — √ℨℨ 17 — √ℨℨ 15

FIG. 79.—Thomas Masterson, *Arithmeticke* (1592), part of p. 45

Page 8 (reproduced in Fig. 80) is in translation: "[The arithmetical progression, according to the natural order of counting,]
furnishes us successive terms for showing the Radicand numbers
and their signs, as you see from the table given here [here appears the
table given in Fig. 80].

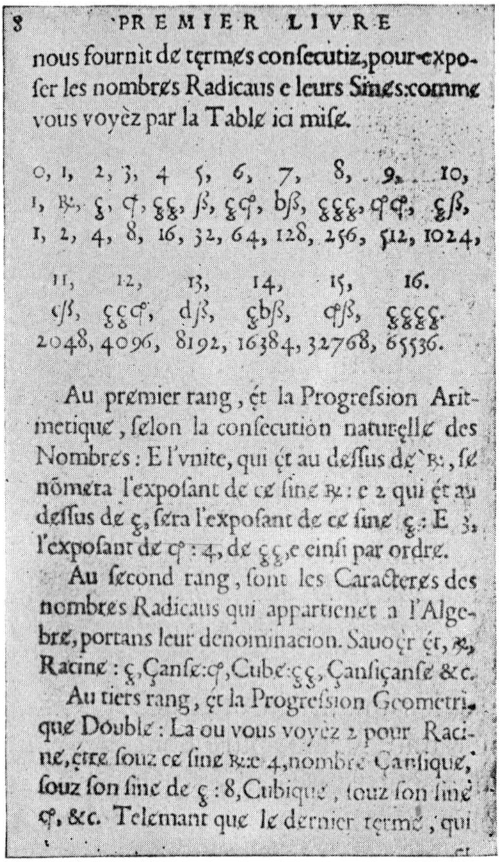

FIG. 80.—Designation of powers in J. Peletier's *Algebra* (1554)

"In the first line is the arithmetical progression, according to the
natural order of the numbers; and the one which is above the ℞
numbers the exponent of this sign ℞; the 2 which is above the ʒ is the
exponent of this sign ʒ; and 3 is the exponent of *c*, 4 of ʒʒ, and so on.

"In the second line are the characters of the Radicand numbers which pertain to algebra, marking their denomination." Then are explained the names of the symbols, as given in French, viz., ℞ *racine*, ȝ *çanse*, ꝯ *cube*, etc.

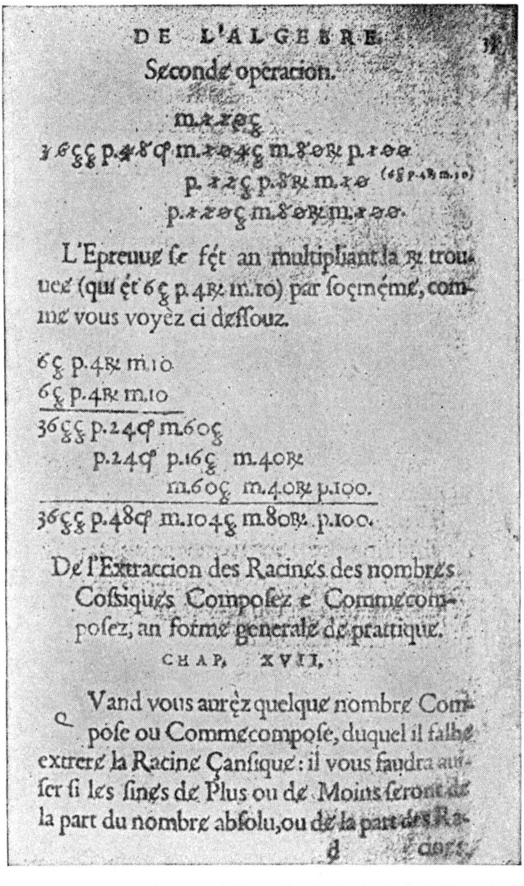
DE L'ALGEBRE. 33

Seconde operacion.

m.120ȝ
36ȝȝ p.48ꝯ m.104ȝ m.80R p.100
p.12ȝ p.8R m.10 (6ȝ p.4R m.10)
p.120ȝ m.80R m.100.

L'Epreuue se fét an multipliant la R trouuee (qui ét 6ȝ p.4R m.10) par soęmęme, comme vous voyez ci dessouz.

6ȝ p.4R m.10
6ȝ p.4R m.10
36ȝȝ p.24ꝯ m.60ȝ
p.24ꝯ p.16ȝ m.40R
m.60ȝ m.40R p.100.
36ȝȝ p.48ꝯ m.104ȝ m.80R p.100.

De l'Extraccion des Racines des nombres Cossiques Composez e Commecomposez, an forme generale de prattique.

CHAP. XVII.

QVand vous aurez quelque nombre Compose ou Commecompose, duquel il falhe extrere la Racine Çansique : il vous faudra auiser si les sines de Plus ou de Moins seront de la part du nombre absolu, ou de la part des Racines,

d

FIG. 81.—Algebraic operations in Peletier's *Algebra* (1554)

Page 33 (shown in Fig. 81) begins with the extraction of a square root and a "proof" of the correctness of the work. The root extraction is, in modern symbols:

$$\begin{array}{l} \qquad\qquad -120x^2 \\ 36x^4+48x^3-104x^2-80x+100 \\ \qquad\qquad +\ 12x^2+\ 8x-10(6x^2+4x-10) \\ \qquad\qquad +120x^2+80x-100\ . \end{array}$$

The "proof" is thus:

$$
\begin{array}{r}
6x^2 + 4x - 10 \\
6x^2 + 4x - 10 \\
\hline
36x^4 + 24x^3 - 60x^2 \\
+ 24x^3 + 16x^2 - 40x \\
- 60x^2 - 40x + 100 \\
\hline
36x^4 + 48x^3 - 104x^2 - 80x + 100
\end{array}
$$

Further on in this book Peletier gives:

$$\sqrt{3}\ 15\ p.\ \sqrt{3}8, \text{ signifying } \sqrt{15} + \sqrt{8}\ .$$

$$\sqrt{3}\ .\ 15\ p.\ \sqrt{3}8, \text{ signifying } \sqrt{15 + \sqrt{8}}\ .$$

FRENCH: JEAN BUTEON
(1559)

173. Deeply influenced by geometrical considerations was Jean Buteon,[1] in his *Logistica quae et Arithmetica vulgo dicitur* (Lugduni, 1559). In the part of the book on algebra he rejects the words *res*, *census*, etc., and introduces in their place the Latin words for "line," "square," "cube," using the symbols ρ, $\diamondsuit$, $\square$. He employs also P and M, both as signs of operation and of quality. Calling the sides of an equation *continens* and *contentum*, respectively, he writes between them the sign [as long as the equation is not reduced to the simplest form and the *contentum*, therefore, not in its final form. Later the *contentum* is inclosed in the completed rectangle [ ]. Thus Buteon writes $3\rho\ M\ 7$ [8 and then draws the inferences, 3ρ [15], 1ρ [5]. Again he writes $\frac{1}{4}\ \diamondsuit$ [100, hence $1\diamondsuit$ [400], 1ρ [20]. In modern symbols: $3x - 7 = 8$, $3x = 15$, $x = 5$; $\frac{1}{4}x^2 = 100$, $x^2 = 400$, $x = 20$. Another example: $\frac{1}{8}\ \square\ P\ 2$ [218, $\frac{1}{8}\ \square$ [216, $1\ \square$ [1728], 1ρ [12]; in modern form $\frac{1}{8}x^3 + 2 = 218$, $\frac{1}{8}x^3 = 216$, $x^3 = 1,728$, $x = 12$.

When more than one unknown quantity arises, they are represented by the capitals A, B, C. Buteon gives examples involving only positive terms and then omits the P. In finding three numbers subject to the conditions $x + \frac{1}{2}y + \frac{1}{2}z = 17$, $y + \frac{1}{3}x + \frac{1}{3}z = 17$, $z + \frac{1}{4}x + \frac{1}{4}y = 17$, he writes:

$$
\begin{array}{l}
1A\ ,\ \frac{1}{2}B\ ,\ \frac{1}{2}C\ [17 \\
1B\ ,\ \frac{1}{3}A\ ,\ \frac{1}{3}C\ [17 \\
1C\ ,\ \frac{1}{4}A\ ,\ \frac{1}{4}B\ [17
\end{array}
$$

---

[1] Our information is drawn from G. Werthheim's article on Buteon, *Bibliotheca mathematica*, Vol. II (3d ser., 1901), p. 213–19.

and derives from them the next equations in the solution:

$$2A . 1B . 1C \; [34$$
$$1A . 3B . 1C \; [51$$
$$1A . 1B . 4C \; [68, \text{ etc.}$$

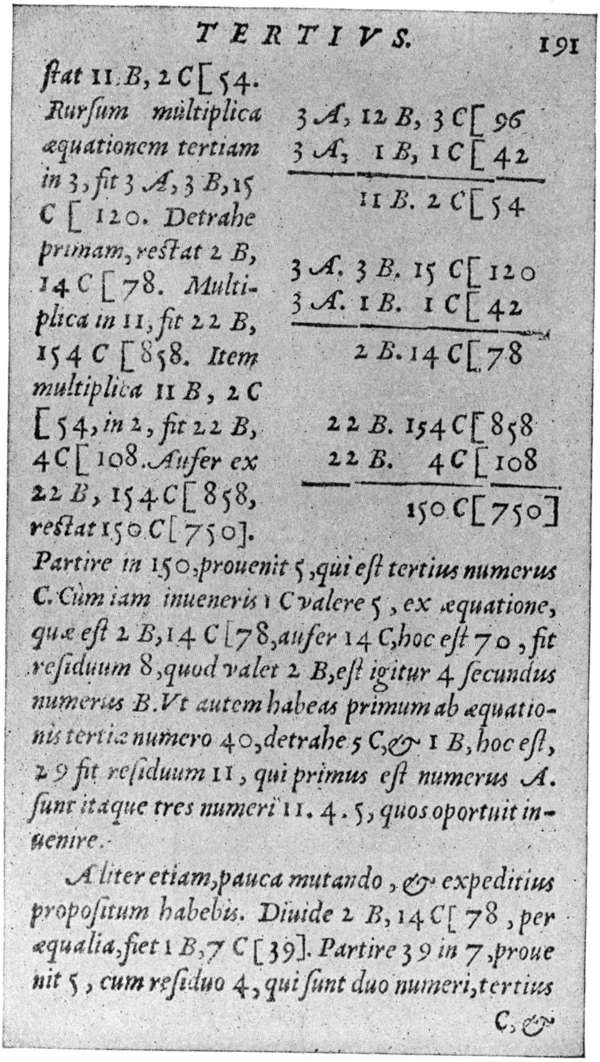

TERTIVS. 191

stat 11.*B*, 2 *C* [54.

Rurfum multiplica   3 *A*, 12 *B*, 3 *C* [96
æquationem tertiam   3 *A*, 1 *B*, 1 *C* [42
in 3, *fit* 3 *A*, 3 *B*, 15 ―――――――――
C [120. *Detrahe*   11 *B*. 2 *C* [54
primam, reſtat 2 *B*,
14 *C* [78. *Multi-*   3 *A*. 3 *B*. 15 *C* [120
plica in 11, *fit* 22 *B*,   3 *A*. 1 *B*. 1 *C* [42
154 *C* [858. *Item* ――――――――――
multiplica 11 *B*, 2 *C*   2 *B*. 14 *C* [78
[54, *in* 2, *fit* 22 *B*,
4 *C* [108. *Aufer ex*   22 *B*. 154 *C* [858
22 *B*, 154 *C* [858,   22 *B*.   4 *C* [108
reſtat 150 *C* [750]. ――――――――――
        150 *C* [750]

Partire in 150, *prouenit* 5, *qui eſt tertius numerus*
C. Cùm iam inueneris 1 *C valere* 5 , *ex æquatione,*
quæ eſt 2 *B*, 14 *C* [78, *aufer* 14 *C, hoc eſt* 7 0 , *fit*
refiduum 8, *quod valet* 2 *B, eſt igitur* 4 *fecundus*
numerus B. Vt autem habeas primum ab æquatio-
nis tertiæ numero 40, *detrahe* 5 *C, & * 1 *B, hoc eſt,*
2 9 *fit refiduum* 11 , *qui primus eſt numerus A.*
funt itaque tres numeri 11. 4. 5, *quos oportuit in-*
uenire.

 Aliter etiam, pauca mutando , & expeditius
propoſitum habebis. Diuide 2 *B*, 14 *C* [78 , *per*
æqualia, fiet 1 *B*, 7 *C* [39]. *Partire* 3 9 *in* 7, *proue-*
nit 5 , *cum refiduo* 4, *qui funt duo numeri, tertius*
            C, &

FIG. 82.—From J. Buteon, *Arithmetica* (1559)

In Figure 82 the equations are as follows:

$$3A+12B+\ 3C = \ 96$$
$$3A+\ 1B+\ 1C = \ 42$$

$$11B+\ 2C = \ 54$$

$$3A+\ 3B+\ 15C = 120$$
$$3A+\ 1B+\ 1C = \ 42$$

$$2B+\ 14C = \ 78$$

$$22B+154C = 858$$
$$22B+\ 4C = 108$$

$$150C = 750$$

FRENCH: GUILLAUME GOSSELIN
(1577)

174. A brief but very good elementary exposition of algebra was given by G. Gosselin in his *De arte magna*, published in Paris in 1577. Although the plus (+) and minus (−) signs must have been more or less familiar to Frenchmen through the *Algebra* of Scheubel, published in Paris in 1551 and 1552, nevertheless Gosselin does not use them. Like Peletier, Gosselin follows the Italians on this point, only Gosselin uses the capital letters P and M for "plus" and "minus," instead of the usual and more convenient small letters.[1] He defines his notation for powers by the following statement (chap. vi, fol. v):

$$L \cdot 2 \cdot Q \cdot 4 \cdot C \cdot 8 \cdot QQ \cdot 16 \cdot RP \cdot 32 \cdot QC \cdot 64 \cdot RS \cdot 128 \cdot CC \cdot 512 \ .$$

Here RP and RS signify, respectively, *relatum primum* and *relatum secundum*.

Accordingly,

" $12L\ M\ 1Q\ P\ 48$ aequalia $144\ M\ 24L\ P\ 2Q$ "

means

$$12x - x^2 + 48 = 144 - 24x + 2x^2 \ .$$

[1] Our information is drawn mainly from H. Bosmans' article on Gosselin, *Bibliotheca mathematica*, Vol. VII (1906–7), p. 44–66.

The translation of Figure 83 is as follows:

" . . . . Thus I multiply $4x - 6x^2 + 7$ by $3x^2$ and there results $12x^3 - 18x^4 + 21x^2$ which I write below the straight line; then I multi-

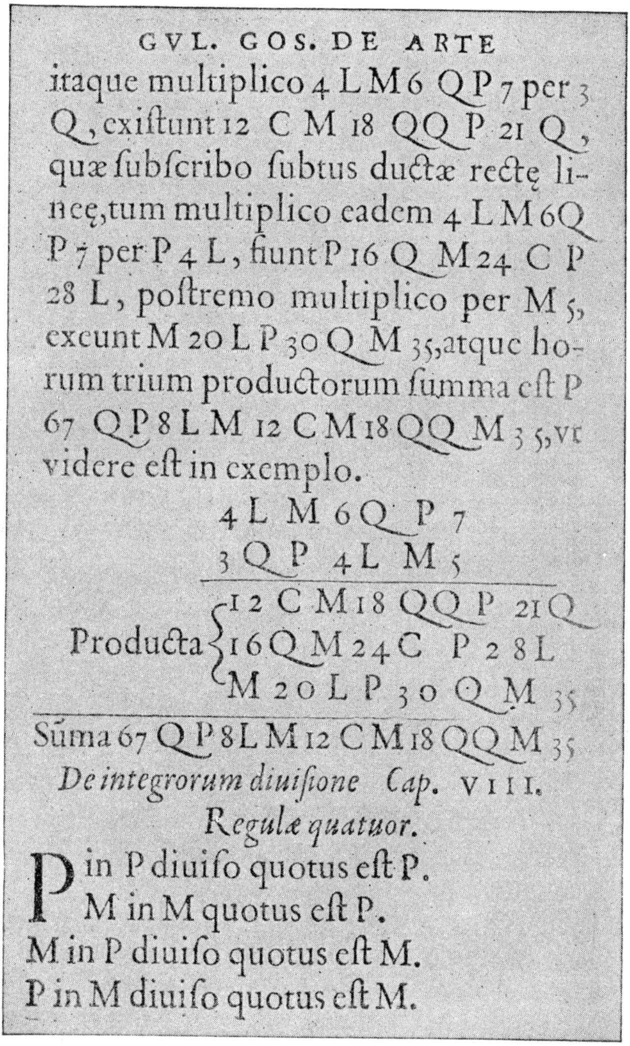

GVL. GOS. DE ARTE

itaque multiplico 4 L M 6 Q P 7 per 3 Q , exiſtunt 12 C M 18 QQ P 21 Q , quæ ſubſcribo ſubtus ductæ rectę li- ncę,tum multiplico eadem 4 L M 6 Q P 7 per P 4 L , fiunt P 16 Q M 24 C P 28 L , poſtremo multiplico per M 5, excunt M 20 L P 30 Q M 35,atquc ho- rum trium productorum ſumma eſt P 67 Q P 8 L M 12 C M 18 QQ M 3 5,vt videre eſt in exemplo.

4 L M 6 Q P 7

3 Q P 4 L M 5

Producta { 12 C M 18 QQ P 21 Q
16 Q M 24 C P 2 8 L
M 2 0 L P 3 0 Q M 35

Sŭma 67 Q P 8 L M 12 C M 18 QQ M 35

De integrorum diuiſione Cap. VIII.

Regulæ quatuor.

P in P diuiſo quotus eſt P.
M in M quotus eſt P.
M in P diuiſo quotus eſt M.
P in M diuiſo quotus eſt M.

Fig. 83.—Fol. 45v° of Gosselin's *De arte magna* (1577)

ply the same $4x - 6x^2 + 7$ by $+4x$, and there results $+16x^2 - 24x^3 + 28x$; lastly I multiply by -5 and there results $-20x + 30x^2 - 35$.

And the sum of these three products is $67x^2+8x-12x^3-18x^4-35$, as will be seen in the example.

$$4x - 6x^2 + 7$$
$$3x^2 + 4x - 5$$

$$\text{Products}\begin{cases} 12x^3-18x^4+21x^2 \\ 16x^2-24x^3+28x \\ -20x+30x^2-35 \end{cases}$$

$$\text{Sum } \overline{67x^2+8x-12x^3-18x^4-35} \,.$$

On the Division of Integers, chapter viii

Four Rules

$+$ divided in $+$ the quotient is $+$
$-$ divided in $-$ the quotient is $+$
$-$ divided in $+$ the quotient is $-$
$+$ divided in $-$ the quotient is $-$ "

175. Proceeding to radicals we quote (fol. 47*B*): "Est autem laterum duplex genus simplicium et compositorum. Simplicia sunt *L*9, *LC*8, *LL*16, etc. Composita vero ut *LV*24 *P L*29, *LV*6 *P L*8." In translation: "There are moreover two kinds of radicals, simple and composite. The simple are like $\sqrt{9}$, $\sqrt[3]{8}$, $\sqrt[4]{16}$, etc. The composite are like $\sqrt{24+\sqrt{29}}$, $\sqrt{6+\sqrt{8}}$." First to be noticed is the difference between *L*9 and 9*L*. They mean, respectively, $\sqrt{9}$ and $9x$. We have encountered somewhat similar conventions in Pacioli, with whom ℞ meant a power when used in the form, say, "℞ . 5ᵉ" (i.e., x^4), while ℞ meant a root when followed by a number, as in ℞ .200. (i.e., $\sqrt{200}$) (see § 135). Somewhat later the same principle of relative position occurs in Albert Girard, but with a different symbol, the circle. Gosselin's *LV* meant of course *latus universale*. Other examples of his notation of radicals are *LVL*10 *P L*5, for $\sqrt{\sqrt{10}+\sqrt{5}}$, and *LVCL*5 *P LC*10 for $\sqrt[3]{\sqrt{5}+\sqrt[3]{10}}$.

In the solution of simultaneous equations involving only positive terms Gosselin uses as the unknowns the capital letters $A, B, C, \ldots$ (similar to the notation of Stifel and Buteon), and omits the sign *P* for "plus"; he does this in five problems involving positive terms, following here an idea of Buteo. In the problem 5, taken from Buteo, Gosselin finds four numbers, of which the first, together with half of the remaining, gives 17; the second with the third of the remaining gives 12; and the third with a fourth of the remaining gives 13; and the fourth with a sixth of the remaining gives 13. Gosselin lets A, B, C, D be the four numbers and then writes:

Modern Notation

" $1A\frac{1}{2}B\frac{1}{2}C\frac{1}{2}D$ aequalia 17 , $x+\frac{1}{2}y+\frac{1}{2}z+\frac{1}{2}w=17$,
$1B\frac{1}{3}A\frac{1}{3}C\frac{1}{3}D$ aequalia 12, etc. " $y+\frac{1}{3}x+\frac{1}{3}z+\frac{1}{3}w=12$.

He is able to effect the solution without introducing negative terms.

In another place Gosselin follows Italian and German writers in representing a second unknown quantity by q, the contraction of *quantitas*. He writes (fols. 84B, 85A) "1L P 2q M 20 aequalia sunt 1L P 30" (i.e., $1x+2y-20=1x+30$) and obtains "2q aequales 50, fit 1q 25" (i.e., $2y=50$, $y=25$).

FRENCH: FRANCIS VIETA
(1591 and Later)

176. Sometimes, Vieta's notation as it appears in his early publications is somewhat different from that in his collected works, edited by Fr. van Schooten in 1646. For example, our modern $\dfrac{3BD^2-3BA^2}{4}$ is printed in Vieta's *Zeteticorum libri v* (Tours, 1593) as

$$\frac{\text{" } B \text{ in } D \text{ quadratum } 3-B \text{ in } A \text{ quadratum } 3 \text{ "}}{4},$$

while in 1646 it is reprinted[1] in the form

$$\frac{\text{" } B \text{ in } Dq\ 3-B \text{ in } Aq\ 3 \text{ "}}{4}.$$

Further differences in notation are pointed out by J. Tropfke:[2]

Zeteticorum libri v (1593)

Fol. 3B: $\text{" } \dfrac{B \text{ in } A}{D}+\left\{\begin{array}{c} B \text{ in } A \\ -B \text{ in } H \\ \hline F \end{array}\right\}$ aequabuntur B ."

Modern: $\dfrac{Bx}{D}+\dfrac{Bx-B\cdot H}{F}=B$.

Lib. II, 22:   " $l\dfrac{25}{3}-l\dfrac{5}{3}$."

---

[1] *Francisci Vietae Opera mathematica* (ed. Fr. à Schooten; Lvgdvni Batavorvm, 1646), p. 60. This difference in notation has been pointed out by H. Bosmans, in an article on Oughtred, in *Extrait des annales de la société scientifique de Bruxelles*, Vol. XXXV, fasc. 1 (2d part), p. 22.

[2] *Op. cit.*, Vol. III (2d ed., 1922), p. 139.

Lib. IV, 10: " B in $\left\{\begin{array}{l}D \text{ quadratum} \\ +B \text{ in } D\end{array}\right\}$."

Modern: $B(D^2+BD)$.

Lib. IV, 20: " D in $\left\{\begin{array}{l}B \text{ cubum } 2 \\ -D \text{ cubo}\end{array}\right\}$."

Modern: $D(2B^3-D^3)$.

Van Schooten edition of Vieta (1646)

P. 46:   " $\dfrac{B \text{ in } A}{D}+\dfrac{B \text{ in } A-B \text{ in } H}{F}$ aequabitur B ."

P. 56:   " $\sqrt{\dfrac{25}{3}}-\sqrt{\dfrac{5}{3}}$."

P. 70:   " B in $\overline{D \text{ quad.}+B \text{ in } D}$."

P. 74:   " $\overline{D}$ in $\overline{B \text{ cubum } 2-D \text{ cubo}}$."

Figure 84 exhibits defective typographical work. As in Stifel's *Arithmetica integra,* so here, the fractional line is drawn too short. In the translation of this passage we put the sign of multiplication ($\times$) in place of the word *in*: ". . . . Because what multiplication brings about above, the same is undone by division, as $\dfrac{B\times A}{B}$, i.e., A; and $\dfrac{B\times A^2}{B}$ is A^2.

Thus in additions, required, to $\dfrac{A^2}{B}$ to add Z. The sum is $\dfrac{A^2+Z\times B}{B}$; or required, to $\dfrac{A^2}{B}$ to add $\dfrac{Z^2}{G}$. The sum is $\dfrac{G\times A^2+B\times Z^2}{B\times G}$.

In subtraction, required, from $\dfrac{A^2}{B}$ to subtract Z. The remainder is $\dfrac{A^2-Z\times B}{B}$. Or required, from $\dfrac{A^2}{B}$ to subtract $\dfrac{Z^2}{G}$. The remainder is $\dfrac{A^2\times G-Z^2\times B}{B\times G}$,"

Observe that Vieta uses the signs plus (+) and minus (−), which had appeared at Paris in the *Algebra* of Scheubel (1551). Outstanding in the foregoing illustrations from Vieta is the appearance of capital letters as the representatives of general magnitudes. Vieta was the first to do this systematically. Sometimes, Regiomontamus, Rudolff, Adam Riese, and Stifel in Germany, and Cardan in Italy, used letters at an earlier date, but Vieta extended this idea and first made it an essential part of algebra. Vieta's words,[1] as found in his *Isagoge*, are: "That this work may be aided by a certain artifice, given magnitudes are to be distinguished from the uncertain required ones by a symbolism, uniform and always readily seen, as is possible by designating the required quantities by letter A or by other vowel letters $A, I, O, V, Y,$ and the given ones by the letters B, G, D or by other consonants."[2]

Vieta's use of letters representing known magnitudes as coefficients of letters representing unknown magnitudes is altogether new. In discussing Vieta's designation of unknown quantities by vowels,

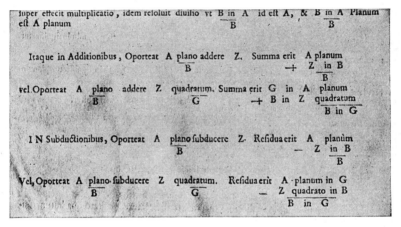

Fig. 84.—From Vieta's *In artem analyticam Isagoge* (1591). (I am indebted to Professor H. Bosmans for this photograph.)

C. Henry remarks: "Thus in a century which numbers fewer Orientalists of eminence than the century of Vieta, it may be difficult not to regard this choice as an indication of a renaissance of Semitic languages; every one knows that in Hebrew and in Arabic only the consonants are given and that the vowels must be recovered from them."[3]

177. Vieta uses = for the expression of arithmetical difference. He says: "However when it is not stated which magnitude is the greater and which is the less, yet the subtraction must be carried out,

---

[1] Vieta, *Opera mathematica* (1646), p. 8.

[2] "Quod opus, ut arte aliqua juvetur, symbolo constanti et perpetuo ac bene conspicuo date magnitudines ab incertis quaesititiis distinguantur, ut pote magnitudines quaesititias elemento A aliave litera vocali, *E, I, O, V, Y* datas elementis *B, G, D,* aliisve consonis designando."

[3] "Sur l'origine de quelques notations mathématiques," *Revue archéologique,* Vol. XXXVIII (N.S., 1879), p. 8.

the sign of difference is $=\!=$, i.e., an uncertain minus. Thus, given A^2 and B^2, the difference is $A^2=\!=B^2$, or $B^2=\!=A^2$."[1]

We illustrate Vieta's mode of writing equations in his *Isagoge:* "*B* in *A* quadratum plus *D* plano in *A* aequari *Z* solido," i.e., $BA^2+D^2A=Z^3$, where *A* is the unknown quantity and the consonants are the known magnitudes. In Vieta's *Ad Logisticen speciosam notae priores* one finds: "*A* cubus, $+A$ quadrato in *B* ter, $+A$ in *B* quadratum ter, $+B$ cubo," for $A^3+3A^2B+3AB^2+B^3$.[2]

We copy from Vieta's *De emendatione aequationum tractatus secundus* (1615),[3] as printed in 1646, the solution of the cubic $x^3+3B^2x=2Z^3$:

"Proponatur *A* cubus $+$ *B* plano 3 in *A*, aequari *Z* solido 2. Oportet facere quod propositum est. *E* quad. $+A$ in *E*, aequetur *B* plano. Vnde *B* planum ex hujus modi aequationis constitutione, intelligitur rectangulum sub duobus lateribus quorum minus est *E*,

differentia à majore *A*. igitur $\dfrac{B \text{ planum} - E \text{ quad.}}{E}$ erit *A*. Quare

$\dfrac{B \text{ plano-plano-planum} - E \text{ quad. in } B \text{ plano-planum } 3 + E \text{ quad.}}{E \text{ cubo}}\, -$

$\dfrac{\text{quad. in } B \text{ planum } 3 - E \text{ cubo-cubo}}{E} + \dfrac{B \text{ pl. pl. } 3. - B \text{ pl. in Eq. } 3}{E}$ aequa-

bitur *Z* solido 2 .

"Et omnibus per *E* cubum ductis et ex arte concinnatis, *E* cubi quad. $+Z$ solido 2 in *E* cubum, aequabitur *B* plani-cubo.[4]

"Quae aequatio est quadrati affirmate affecti, radicem habentis solidam. Facta itaque reductio est quae imperabatur.

"*Confectarium:* Itaque si *A* cubus $+$ *B* plano 3 in *A*, aequetur *Z* solido 2, $\&\ \sqrt{B}$ plano-plano-plani $+$ *Z* solido-solido $-$ *Z* solido,

aequetur *D* cubo. Ergo $\dfrac{B \text{ planum} -- D \text{ quad.}}{D}$, sit *A* de qua quaeritur."

Translation: "Given $x^3+3B^2x=2Z^3$. To solve this, let $y^2+yx=B^2$. Since B^2 from the constitution of such an equation is understood to be a rectangle of which the less of the two sides is y, and the difference between it and the larger side is x. Therefore $\dfrac{B^2-y^2}{y}=x$. Whence

$$\frac{B^6-3B^4y^2+3B^2y^4-y^6}{y^3}+\frac{3B^4-3B^2y^2}{y}=2Z^3 \ .$$

---

[1] "Cum autem non proponitur utra magnitudo sit major vel minor, et tamen subductio facienda est, nota differentiae est $=\!=$ id est, minus incerto: ut propositis *A* quadrato et *B* plano, differentia erit *A* quadratum $=\!=B$ plano, vel *B* planum $A=\!=$quadrato" (Vieta, *Opera mathematica* [1646], p. 5).

[2] *Ibid.*, p. 17.   [3] *Ibid.*, p. 149.

[4] "*B* plani-cubo" should be "*B* cubo-cubo," and "*E* cubi quad." should be "*E* cubo-cubo."

All terms being multiplied by y^3, and properly ordered, one obtains $y^6 + 2Z^3y^3 = B^6$. As this equation is quadratic with a positive affected term, it has also a cube root. Thus the required reduction is effected.

"*Conclusion:* If therefore $x^3 + 3B^2x = 2Z^3$, and $\sqrt{B^6 + Z^6} - Z^3 = D^3$, then $\dfrac{B^2 - D^2}{D}$ is x, as required."

The value of x in $x^3 + 3B^2x = 2Z^3$ is written on page 150 of the 1646 edition thus:

$$\text{``} \sqrt{C.\sqrt{B} \text{ plano-plano-plani} + Z \text{ solido-solido} + Z \text{ solido} -}$$
$$\sqrt{C.\sqrt{B} \text{ plano-plano-plani} + Z \text{ solido-solido.} - Z \text{ solido .''}}$$

The combining of vinculum and radical sign shown here indicates the influence of Descartes upon Van Schooten, the editor of Vieta's collected works. As regards Vieta's own notations, it is evident that compactness was not secured by him to the same degree as by earlier writers. For powers he did not adopt either the Italian symbolism of Pacioli, Tartaglia, and Cardan or the German symbolism of Rudolff and Stifel. It must be emphasized that the radical sign, as found in the 1646 edition of his works, is a modification introduced by Van Schooten. Vieta himself rejected the radical sign and used, instead, the letter l (*latus*, "the side of a square") or the word *radix*. The l had been introduced by Ramus (§ 322); in the *Zeteticorum*, etc., of 1593 Vieta wrote l. 121 for $\sqrt{121}$. In the 1646 edition (p. 400) one finds $\sqrt{2 + \sqrt{2 + \sqrt{2 + \sqrt{2}}}}$, which is Van Schooten's revision of the text of Vieta; Vieta's own symbolism for this expression was, in 1593,[1]

"Radix binomiae 2

$+$Radix binomiae $\left\{\begin{array}{l} 2 \\ +\text{radix binomiae} \end{array}\right. \left\{\begin{array}{l} 2 \\ +\text{radice 2 ,''} \end{array}\right.$

and in 1595,[2]

"R. bin. $2 + R$. bin. $2 + R$. bin. $2 + R$. 2. ,"

a notation employed also by his contemporary Adrian Van Roomen.

178. Vieta distinguished between number and magnitude even in his notation. In numerical equations the unknown number is no longer represented by a vowel; the unknown number and its powers are represented, respectively, by N (*numerus*), Q (*quadratus*), C (*cubus*), and

---

[1] *Variorum de rebus mathem. Responsorum liber VIII* (Tours, 1593), corollary to Caput XVIII, p. 12v°. This and the next reference are taken from Tropfke, *op. cit.*, Vol. II (1921), p. 152, 153.

[2] *Ad Problema quod omnibus mathematicis totius orbis construendum proposuit Adrianus Romanus, Francisci Vietae responsum* (Paris, 1595), Bl. *A* IV°.

combinations of them. Coefficients are now written to the left of the letters to which they belong.

Thus,[1] "Si $65C-1QQ$, aequetur 1,481,544, fit $1N57$," i.e., if $65x^3-x^4=1,481,544$, then $x=57$. Again,[2] the "$B3$ in A quad." occurring in the regular text is displaced in the accompanying example by "$6Q$," where $B=2$.

Figure 85 further illustrates the notation, as printed in 1646.

Vieta died in 1603. The *De emendatione aeqvationvm* was first printed in 1615 under the editorship of Vieta's English friend, Alexander Anderson, who found Vieta's manuscript incomplete and con-

THEOREMA I.

Sɪ A cubus -+ B in A quadr. 3 -+ D plano in A, æquetur B cubo 2—D plano in B. A quad. -+ B in A 2, æquabitur B quad. 2 — D plano.

Quoniam enim A quadr. -+ B in A 2, æquatur B quadr. 2 — D plano. Ductis igitur omnibus in A. A cubus -+ B in A quad. 2, æquabitur B quad. in A 2 — D plano in A.

Et iifdem ductis in B. B in A quad. -+ B quadr. in A 2, æquabitur B cubo 2—D plano in B. Iungatur ducta æqualia æqualibus. A cubus + B in A quad. 3 + B quad. in A 2, æquabitur B quad. in A 2 — D plano in A + B cubo 2 — D plano in B.

Et deleta utrinque adfectione B quad. in A 2, & ad æqualitatis ordinationem, tranflata per antithefin D plani in A adfectione. A cubus + B in A quadr. 3 + D plano in A, æquabitur B cubo 2 — D plano in B. Quod quidem ita fe habet.

1 C + 30 Q + 44 N, *æquatur* 1560. *Igitur* 1 Q + 20 N, *æquabitur* 156. *& fit* 1 N 6.

FIG. 85.—From Vieta's *De emendatione aeqvationvm*, in *Opera mathematica* (1646), p. 154.

taining omissions which had to be supplied to make the tract intelligible. The question arises, Is the notation N, Q, C due to Vieta or to Anderson?[3] There is no valid evidence against the view that Vieta did use them. These letters were used before Vieta by Xylander in his edition of Diophantus (1575) and in Van Schooten's edition[4] of the *Ad problema, quod omnibus mathematicis totius orbis construendum proposuit Adrianus Romanus*. It will be noticed that the letter N stands here for x, while in some other writers it is used in the designation of absolute number as in Grammateus (1518), who writes our $12x^3-24$ thus: "12 ter. mi. 24N." After Vieta N appears as a mark for absolute number in the *Sommaire de l'algebre* of Denis Henrion[5]

[1] Vieta, *Opera mathematica* (1646), p. 223.   [2] *Op. cit.*, p. 130.

[3] See Eneström, *Bibliotheca mathematica*, Vol. XIII (1912–13), p. 166, 167.

[4] Vieta, *Opera mathematica* (1646), p. 306, 307.

[5] Denis Henrion, *Les qvinze livres des elemens d'Evclide* (4th ed.; Paris, 1631), p. 675–788. First edition, Paris, 1615. (Courtesy of Library of University of Michigan.)

which was inserted in his French edition of Euclid. Henrion did not adopt Vieta's literal coefficients in equations and further showed his conservatism in having no sign of equality, in representing the powers of the unknown by R, q, c, qq, β, qc, $b\beta$, qqq, cc, $q\beta$, $c\beta$, qqc, etc., and in using the "scratch method" in division of algebraic polynomials, as found much earlier in Stifel.[1] The one novel feature in Henrion was his regular use of round parentheses to express aggregation.

ITALIAN: BONAVENTURA CAVALIERI
(1647)

179. Cavalieri's *Geometria indivisibilibvs* (Bologna, 1635 and 1653) is as rhetorical in its exposition as is the original text of Euclid's *Elements*. No use whatever is made of arithmetical or algebraic signs, not even of $+$ and $-$, or p and m.

An invasion of German algebraic symbolism into Italy had taken place in Clavius' *Algebra*, which was printed at Rome in 1608. That German and French symbolism had gained ground at the time of Cavalieri appears from his *Exercitationes geometriae sex* (1647), from which Figure 86 is taken. Plus signs of fancy shape appear, also Vieta's *in* to indicate "times." The figure shows the expansion of $(a+b)^n$ for $n = 2, 3, 4$. Observe that the numerical coefficients are written after the literal factors to which they belong.

ENGLISH: WILLIAM OUGHTRED
(1631, 1632, 1657)

180. William Oughtred placed unusual emphasis upon the use of mathematical symbols. His symbol for multiplication, his notation for proportion, and his sign for difference met with wide adoption in Continental Europe as well as Great Britain. He used as many as one hundred and fifty symbols, many of which were, of course, introduced by earlier writers. The most influential of his books was the *Clavis mathematicae*, the first edition[2] of which appeared in 1631, later Latin editions of which bear the dates of 1648, 1652, 1667, 1693.

---

[1] M. Stifel, *Arithmetica integra* (1544), fol. 239A.

[2] The first edition did not contain *Clavis mathematicae* as the leading words in the title. The exact title of the 1631 edition was: *Arithmeticae in|numeris et speci-| ebvs institvtio:|Qvae tvm logisticae, tvm analyti|cae, atqve adeo|totivs mathematicae, qvasi|clavis|est.|—Ad nobilissimvm spe|ctatissimumque iuvenem Dn. Gvilel|mvm Howard, Ordinis, qui dici|tur, Balnei Equitem, honoratissimi Dn.| Thomae, Comitis Arvndeliae & | Svrriae, Comitis Mareschal|li Angliae, &c. filium.—|Lon- dini,|Apud Thomam Harpervm,| M. DC. xxxi.*

A second impression of the 1693 or fifth edition appeared in 1698.
Two English editions of this book came out in 1647 and 1694.

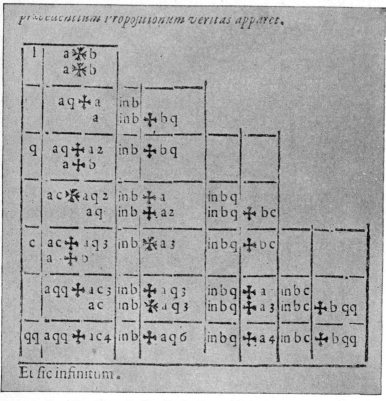

FIG. 86.—From B. Cavalieri's *Exercitationes* (1647), p. 268

We shall use the following abbreviations for the designation of
tracts which were added to one or another of the different editions of
the *Clavis mathematicae*:

Eq. = *De Aequationum affectarvm resolvtione in numeris*
Eu. = *Elementi decimi Euclidis declaratio*
So. = *De Solidis regularibus, tractatus*
An. = *De Anatocismo, sive usura composita*
Fa. = *Regula falsae positionis*
Ar. = *Theorematum in libris Archimedis de sphaera &*
     cylindro declaratio
Ho. = *Horologia scioterica in plano, Geometricè delineandi*
     modus

In 1632 there appeared, in London, Oughtred's *The Circles of Proportion*, which was brought out again in 1633 with an *Addition vnto the Vse of the Instrvment called the Circles of Proportion.*[1] Another edition bears the date 1660. In 1657 was published Oughtred's *Trigonometria,*[2] in Latin, and his *Trigonometrie*, an English translation.

We have arranged Oughtred's symbols, as found in his various works, in tabular form.[3] The texts referred to are placed at the head of the table, the symbols in the column at the extreme left. Each number in the table indicates the page of the text cited at the head of the column containing the symbol given on the left. Thus, the notation :: in geometrical proportion occurs on page 7 of the *Clavis* of 1648. The page assigned is not always the first on which the symbol occurs in that volume.

[1] In our tables this *Addition* is referred to as *Ad.*

[2] In our tables *Ca.* stands for *Cænones sinuum tangentium*, etc., which is the title for the tables in the *Trigonometria.*

[3] These tables were first published, with notes, in the *University of California Publications in Mathematics*, Vol. I, No. 8 (1920), p. 171–86.

181.   OUGHTRED'S MATHEMATICAL SYMBOLS

Symbols	Meanings of Symbols	Clavis mathematica							Circ. of Prop. 1632, 1633	Trigono. (Latin), 1657	Opusc. Posth., 1677	Oughtr. Explic., 1682
		1631	1647	1648	1652	1667	1693	1694				
=	Equal to	38	34	53	30	15	16	73	20	3	3	29
0|56	Separatrix²	1	1	1	1	1	1	2	3	13	63	1
0.56	Separatrix	......	......	......	......	......	......	......	......	235	......	......
.|56	Separatrix³	......	......	......	......	......	......	17	......	......	......	5
0,56	Separatrix	......	......	......	......	......	......	......	......	......	221	......
0|00005	.00005	......	......	3	3	3	3	......	......	......	......	2
a.b	Ratio a:b, or ⁴a—b	5	8	7	12	7	7	25	7	o	3	27
2.314	⎱ Separating⁵	......	......	......	......	......	......	......	4	235	......	......
2.314	⎰ the mantissa	......	Eq.136	158	150	113	113	175	......			
2̄.314	— Characteristic	......	Eq.167	158	150	150	150	207	......			
:	Arithm. proportion⁶	......	......	22	21	21	21	32	......			
:	a:b, ratio⁷	......	......	An.162	......	......	......	......	24	10	36	140	......
R.S	Given ratio	21	28	30	32	25	25	49	19	......	87	42
::	Geomet. proportion⁸	5	8	7	7	7	7	11	i	3	3	27
÷	Contin. proportion	13	18	16	16	16	16	25	......	34	142	29
⋮	Contin. proportion	......	......	......	......	......	......	......	......	......	......	114
=	Geom.⁹ proportion	......	......	......	......	......	......	......	......	......	101	8o
: :	(,10	45	57	107	104	52	53	149	96	......	101	......
:	(	40	58	99	92	56	92	119	35	32	101	75
: .	( )	......	115	106	104	104	104	95	......	......	102	53
:	( )	......	58	......	95	95	63	122	......	......	......	93
. :	( )	......	65	58	57	57	63	95	......	......	......	97
.	( )11	......	......	......	......	......	An. 42	97	......	......	......	116
. .	( )11	......	......	......	......	......	......	89	......	......	......	101
(	(,12	......	......	......	......	......	......	......	......	......	......	81
∵	Therefore	......	......	......	......	......	......	......	......	......	151	......
+	Addition13	2	3	3	57	3	3	4	99	3	3	5
pl	Addition	49	3	3	57	3	3	4	96	......	112	5
mo	Addition¹	......	......	......	......	......	......	4	......			
—	Subtraction	2	3	3	10	3	3	4	21	3	4	5
±	Plus or minus	51	57	106	56	53	17	140	......	16	......	97
mi	Subtraction	......	3	66	57	3	3	4	96	......	130	......
e	Less¹⁴	......	......	......	......	......	......	4	......			
2	Negative z	......	1	9	1	1	5	16	......	......	......	8
×	Multiplication¹⁵	7	10	10	10	10	10	13	37	32	143	......

182. OUGHTRED'S MATHEMATICAL SYMBOLS—*Cont.*

Symbols	Meanings of Symbols	Clavis mathematicae							Circ. of Prop.[1] 1632, 1633	Trigono. (Latin), 1657	Opusc. Posth., 1677	Oughtr. Explic., 1682
		1631	1647	1648	1652	1667	1693	1694				
$Hq\,bq$	× By juxtaposition	7	11	10	11	10	37	13	......	5	87	17
in	Multiplication[16]	7	10	10	10	10	10	......	96	......	219	59
$\frac{3}{4}$	Fraction, division	8	12	11	11	11	11	23	21	16	9	5
$a)b(c$	$b \div a = c$	10	14	13	14	14	13	21	.....	.....	99	50
$\frac{4}{3}]\frac{3}{2}[\frac{9}{8}$	$\frac{3}{2} \div \frac{4}{3} = \frac{9}{8}$	......									156	......
Aq	AA	7	11	10	10	10	10	14			104	17
Ac	AAA	7	11	10	10	10	10	14			105	25
Aqq	$AAAA$	7	11	10	10	10	10	14			106	41
Aqc	$AAAAA$	7	11	10	10	10	10	55			......	67
Acc	$AAAAAA$	7	11	10	10	10	10	55			......	41
ABq	$\overline{AB}^2$ [17]	......	11	11	11	11	11	15			......	......
$\boxed{4} \dots \boxed{10}$	4th .... 10th power	23	37	35	34	34	......	53				55
$|4| \dots [10]$	4th .... 10th power				35	35	34	52				
$a^2 \dots a^7$	$a^2 \dots a^7$										205	24
Q	*Quaesitum*	......	17	16	16	16	16	25				......
Q	Square[18]	38	33	31	57	30	30	47	28	5	100	75
Qu	Square											105
C	Cube	38	33	31	30	30	30	47	28	......	62	53
Cu	Cube	......	136	128	123	123	123	175				
QQ	4th power	45	33	61	30	30	30	47			210	
QC	5th power	......	33	31	30	30	30	47			210	
D	Diameter	......		187	Eu. 21	Eu. 21	Eu. 20	......	37			
L, l	*Latus, radix*[19]	......	121	113	110	110	110	158	37			139
$\angle$	Angle									19	192	
$\angle\angle$	Angles									16		
P	Perimeter								37			
P	$ZA - Aq$	41										
R	Radius	......	120	111	109	109	109	154	37	32	211	......
R, R	Remainder	......	152	134	126	128	142					45
R	Rational			166	Eu. 1	Eu. 1	Eu. 1					
$\cap$	*Superficies curva*				Ar. I	Ar. 1	Ar. 1					
$\sqrt{}$	Root	......	33	31	30	30	30	47			102	......
$\sqrt{}$	Square root	......	53	48	47	47	47	70			134	......
$\sqrt{q}$	Square root	35	49	48	46	46	46	65	96			......
$\sqrt{b}$	*Latus binomii*	......	33	31	30	30	30	47				

183.   OUGHTRED'S MATHEMATICAL SYMBOLS—*Cont.*

Symbols	Meanings of Symbols	Clavis mathematicae							Circ. of Prop.[1] 1632, 1633	Trigono. (Latin), 1657	Opusc. Posth., 1677	Oughtr. Explic., 1682
		1631	1647	1648	1652	1667	1693	1694				
$\sqrt{}\,r$	*Latus residui*	......	34	31	30	30	30	47				
$\sqrt{}\,u$	Sq. rt. of polyno.[20]	......	55	53	53	52	52	96				
$\sqrt{}\,qq$	4th root	35	52	47	46	46	48	69				
$\sqrt{}\,c$	Cube root	35	52	49	46	46	46	69				
$\sqrt{}\,qc$	5th root	35	49	47	46	46	46	65				
$\sqrt{}\,cc$	6th root	37	52	49	48	48	48	69				
$\sqrt{}\,ccc$	9th root	......										
$\sqrt{}\,cccc$	12th root	37	52	49	48	48	48	69				
$\sqrt{}\,qu$	Square root	49	......									
$\sqrt{}\,\boxed{12}$ or $\sqrt{}\,\boxed{12}$ }	12th root	37	52	50	49	49	49	69				
rq, rc	$\sqrt{}$, $\sqrt[3]{}$	......										73
r, ru	Square root	......										74, 96
A, E	Nos., $A > E$	21	33	31	30	30	30	47		......	87	53
Z	$A + E$[21]	21	33	31	30	30	30	47	19	16	87	53
X	$A - E$	21	33	31	30	30	30	47	......	16	87	53
Z	$A^2 + E^2$	41	33	31	30	30	30	47			98	54
X	$A^2 - E^2$	41	33	31	30	30	30	47			99	54
Z	$A^3 + E^3$	44	33	31	30	30	30	47				94
X	$A^3 - E^3$	44	33	31	30	30	30	47				94
	$a + e$	......	......	167	*Eu.* 1	*Eu.* 1	*Eu.* I					
	$a - e$	......	......		*Eu.* 1	*Eu.* 1	*Eu.* I					
	$a^2 + b^2$	......	......	167	*Eu.* 2	*Eu.* 2	*Eu.* I					
	$a^2 - b^2$	......	......	167	*Eu.* 2	*Eu.* 2	*Eu.* I					
⌐	*Majus*[22]	......	*Ho.* 17	166	145	*Eu.* 1	*Eu.* I					
⌐	*Minus*	......	*Ho.* 17	166	*Eu.* 1	*Eu.* 1	*Eu.* I					
⌐	*Non majus*	......	......	166	*Eu.* 1	*Eu.* 1	*Eu.* I					
⌐	*Non minus*	......	......	166	*Eu.* 1	*Eu.* 1	*Eu.* I					
⌐	*Minus*[23]	......	*Ho.* 30	......								
⌐	*Minus*[23]	......	......	......	*Ho.* 31	*Ho.* 29						
∷	Major ratio	......	......	166	*Eu.* 1	*Eu.* 1	*Eu.* I			11	......
∷	Minor ratio	......	......	166	*Eu.* 1	*Eu.* 1	*Eu.* I			6	......
<	Less than[22]	......	......							4	......
>	Greater than	......	......							4	......
⊐⊐	*Commensurabilia*	......	......	166	*Eu.* 1	*Eu.* 1	*Eu.* I				
⊐⊐	*Incommensurabilia*	......	......	166	*Eu.* 1	*Eu.* 1	*Eu.* I				

184. OUGHTRED'S MATHEMATICAL SYMBOLS—*Cont.*

Symbols	Meanings of Symbols	Clavis mathematicae							Circ. of Prop.[1] 1632, 1633	Trigono. (Latin), 1657	Opusc. Posth., 1677	Oughtr. Explic., 1682
		1631	1647	1648	1652	1667	1693	1694				
	Commens. potentia			166	Eu. 1	Eu. 1	Eu. I					
	Incommens. potentia			166	Eu. 1	Eu. 1	Eu. I					
	Rationale			166	Eu. 1	Eu. 1	Eu. I					
	Irrationale			166	Eu. 1	Eu. 1	Eu. I					
	Medium			166	Eu. 1	Eu. 1	Eu. I					
	Line, cut extr. and mean ratio			166	Eu. 1	Eu. 1	Eu. I					
	Major ejus portio			166	Eu. 1	Eu. 1	Eu. I					
	Minor ejus portio			166	Eu. 1	Eu. 1	Eu. I					
sim	Simile			166	Eu. 1	Eu. 1	Eu. I			33		
	Proxime majus			166	Eu. 1	Eu. 1	Eu. I					
	Proxime minus			166	Eu. 1	Eu. 1	Eu. I					
	Aequale vel minus			166	Eu. 1	Eu. 1	Eu. I					
	Aequale vel majus			166	Eu. 1	Eu. 1	Eu. I					
	Rectangulum	51		167	Eu. 2	Eu. 2	Eu. I			17	149	
	Quadratum			167	Eu. 2	Eu. 2	Eu. I					
△	Triangulum			167	Eu. 2	Eu. 2	Eu. I				147	
	Latus, radix			167	Eu. 2	Eu. 2	Eu. I					
	Media proportion			167	Eu. 2	Eu. 2	Eu. I					
	Differentia[24]				Eu. 2	Eu. 2	Eu. I					
‖	Parallel										197	
log	Logarithm		172	158	150	150	122	207		17		
log:Q :	Log. of square		135	127	122	122	122	174				
S	Sine[25]		Ho. 29						96	5	172	
t	Tangent		Ho. 29						96	3	174	
se	Secant									14		
sv	Sinus versus	76	107	99	98	98	98	140				
s ver	Sinus versus[26]									5		
sin : com	Sine complement								Ad. 69			
s co	Cosine								96	3	174	
t co	Cotangent								96	3		
se co	Cosecant									4		
sin	Sine				Ho. 41	Ho. 41	Ho. 42		Ad. 69	35		37
tan	Tangent								Ad. 69	Ca. 3		
sec	Secant								Ad. 41			
sec:parall	Sum of secants								Ad. 41			

185.   OUGHTRED'S MATHEMATICAL SYMBOLS—*Cont.*

Symbols	Meanings of Symbols	Clavis Mathematicae							Circ. of Prop. [1] 1632, 1633	Trigono. (Latin), 1657	Opusc. Posth. 1677	Oughtr. Explic. 1632
		1631	1647	1648	1652	1667	1693	1694				
tang	Tangent	......	*Ho.* 29	......	*Ho.* 41	*Ho.* 41	*Ho.* 42	......	12	235	......	......
C	.01 of a degree									236		
Cent	.01 of a degree									235		
' '' '''	Degr., min., sec.	......	21	20	21	20	21	32	66		......	36
Ho. ' ''	Hours, min., sec.								67			
,	180 $\stackrel{\circ}{-}$ angle									2		
$\overline{V}$	Equal in no. of degr.									6		
$\frac{\pi}{\delta}$	$\pi=3.1416$	..\...	72	69	66	66	66	99				
——	Canceled[27]	68	100	94	90	90	90	131				
M	Mean proportion				*Ar.* 1	*Ar.* 1	*Ar.* 1					
m	Minus	∠										
(figure)	$-\times-=2,\ -\div-=-$ 3 4 3 4 9[28] / 2 3 2 3 8	20	32	30	29	29	29	45				
Gr.	Degree		20	19	*Ho.* 23	*Ho.* 23	19	29		235		
min.	Minute								*Ad.* 19	......		
⌐	*Differentia*										134	
⊢⊣	*Aequalia tempore*										68	
Lo	Logarithm									*Ca.* 2		
I	Separatrix									244		
D	*Differentia*								19	237		
Tri, tri	Triangle		76	191	*Eu.* 26	70	69			24		
M	Cent. minute of arc									*Ca.* 2		
X	Multiplication[29]									5	101	16
Z *cru*	Z sum, X diff.									17		
Z *crur*	of sides of									16		
X *cru*	rectangle[30]									17		
X *crur*	or triangle									16		
A	Unknown	38	53	51	50	50	50	72			113	84
L	Altit. frust. of pyramid or cone	77	109	101	99	99	99	141				
T	Altit. of part cut off	77	109	101	99	99	99	142				
α	First term	13	85, 18	80, 17	78, 16	78, 16	78, 16	116, 26	19			30, 116
ω	Last term		85, 18	80, 17	78, 16	78, 16	78, 16	116, 26				30. 116
T	No. of terms		85	80	78	78	78	116				11
X	Common differ.		85	80	78	78	78	116				116
Z	Sum of all terms		85, 18	80, 17	78, 16	78, 16	78, 16	116, 26	19			30, 116

(The last six rows are bracketed as "o progressions.")

186. Historical notes[1] to the tables in §§ 181–85:

1. All the symbols, except "Log," which we saw in the 1660 edition of the *Circles of Proportion*, are given in the editions of 1632 and 1633.

2. In the first half of the seventeenth century the notation for decimal fractions engaged the attention of mathematicians in England as it did elsewhere (see §§ 276–89). In 1608 an English translation of Stevin's well-known tract was brought out, with some additions, in London by Robert Norton, under the title, *Disme: The Art of Tenths, or, Decimall Arithmetike* (§ 276). Stevin's notation is followed also by Henry Lyte in his *Art of Tens or Decimall Arithmetique* (London, 1619), and in *Johnsons Arithmetick* (2d ed.; London, 1633), where 3576.725 is written $3576|725$. William Purser in his *Compound Interest and Annuities* (London, 1634), p. 8, uses the colon (:) as the separator, as did Adrianus Metius in his *Geometriae practicae pars I et II* (Lvgd., 1625), p. 149, and Rich. Balam in his *Algebra* (London, 1653), p. 4. The decimal point or comma appears in John Napier's *Rabdologia* (Edinburgh, 1617). Oughtred's notation for decimals must have delayed the general adoption of the decimal point or comma.

3. This mixture of the old and the new decimal notation occurs in the *Key* of 1694 (*Notes*) and in Gilbert Clark's *Oughtredus explicatus*[2] only once; no reference is made to it in the table of errata of either book. On Oughtred's *Opuscula mathematica hactenus inedita*, the mixed notation $128,\underline{57}$ occurs on p. 193 fourteen times. Oughtred's regular notation $128|\underline{57}$ hardly ever occurs in this book. We have seen similar mixed notations in the *Miscellanies: or Mathematical Lucubrations, of Mr. Samuel Foster, Sometime publike Professor of Astronomie in Gresham Colledge, in London*, by John Twysden (London, 1659), p. 13 of the "Observationes eclipsium"; we find there $32.\underline{466}$, $31.\underline{008}$.

4. The dot (.), used to indicate ratio, is not, as claimed by some writers, used by Oughtred for division. Oughtred does not state in his book that the dot (.) signifies division. We quote from an early and a late edition of the *Clavis*. He says in the *Clavis* of 1694, p. 45, and in the one of 1648, p. 30, "to continue ratios is to multiply them *as if they were* fractions." Fractions, as well as divisions, are indicated by a horizontal line. Nor does the statement from the *Clavis* of 1694, p. 20, and the edition of 1648, p. 12, "In Division, as the Divisor is to Unity, so is the Dividend to the Quotient," prove that he looked upon ratio as an indicated division. It does not do so any more than the sentence from the *Clavis* of 1694, and the one of 1648, p. 7, "In Multiplication, as 1 is to either of the factors, so is the other to the Product," proves that he considered ratio an indicated multiplication. Oughtred says (*Clavis* of 1694, p. 19, and the one of 1631, p. 8): "If Two Numbers stand one above another with a Line drawn between them, 'tis as much as to say, that the upper is to be divided by the under; as $\frac{12}{4}$ and $\frac{5}{12}$."

---

[1] N. 1 refers to the *Circles of Proportion*. The other notes apply to the superscripts found in the column, "Meanings of Symbols."

[2] This is not a book written by Oughtred, but merely a commentary on the *Clavis*. Nevertheless, it seemed desirable to refer to its notation, which helps to show the changes then in progress.

In further confirmation of our view we quote from Oughtred's letter to W. Robinson: "Division is wrought by setting the divisor under the dividend with a line between them."[1]

5. In Gilbert Clark's *Oughtredus explicatus* there is no mark whatever to separate the characteristic and mantissa. This is a step backward.

6. Oughtred's language (*Clavis* of 1652, p. 21) is: "Ut $7.4 : 12.9$ vel $7.7 - 3 : 12.12$ -3. Arithmeticè proportionales sunt." As later in his work he does not use arithmetical proportion in symbolic analysis, it is not easy to decide whether the symbols just quoted were intended by Oughtred as part of his algebraic symbolism or merely as punctuation marks in ordinary writing. Oughtred's notation is adopted in the article "Caractere" of the *Encyclopédie méthodique* (*mathématiques*), Paris: Liège, 1784 (see § 249).

7. In the publications referred to in the table, of the years 1648 and 1694, the use of : to signify ratio has been found to occur only once in each copy; hence we are inclined to look upon this notation in these copies as printer's errors. We are able to show that the colon (:) was used to designate geometric ratio some years before 1657, by at least two authors, Vincent Wing the astronomer, and a schoolmaster who hides himself behind the initials "R.B." Wing wrote several works.

8. Oughtred's notation $A.B::C.D$, is the earliest serviceable symbolism for proportion. Before that proportions were either stated in words as was customary in rhetorical modes of exposition, or else was expressed by writing the terms of the proportion in a line with dashes or dots to separate them. This practice was inadequate for the needs of the new symbolic algebra. Hence Oughtred's notation met with ready acceptance (see §§ 248–59).

9. We have seen this notation only once in this book, namely, in the expression $R.S. = 3.2$.

10. Oughtred says (*Clavis* of 1694, p. 47), in connection with the radical sign, "If the Power be included between two Points at both ends, it signifies the universal Root of all that Quantity so included; which is sometimes also signified by b and r, as the $\sqrt{b}$ is the Binomial Root, the $\sqrt{r}$ the Residual Root." This notation is in no edition strictly adhered to; the second : is often omitted when all the terms to the end of the polynomial are affected by the radical sign or by the sign for a power. In later editions still greater tendency to a departure from the original notation is evident. Sometimes one dot takes the place of the two dots at the end; sometimes the two end dots are given, but the first two are omitted; in a few instances one dot at both ends is used, or one dot at the beginning and no symbol at the end; however, these cases are very rare and are perhaps only printer's errors We copy the following illustrations:

$Q : A - E$: est $Aq - 2AE + Eq$, for $(A-E)^2 = A^2 - 2AE + E^2$ (from *Clavis* of 1631, p. 45)

$\frac{1}{2}BCq \pm \sqrt{q} : \frac{1}{4}BCqq - CMqq. = \left.\begin{matrix}BAq\\CAq\end{matrix}\right\}$, for $\frac{1}{2}\overline{BC^2} \pm \sqrt{(\frac{1}{4}\overline{BC^4} - \overline{CM^4})} = \overline{BA^2}$ or $\overline{CA^2}$ (from *Clavis* of 1648, p. 106)

$\sqrt{q} : BA + CA = BC + D$, for $\sqrt{(BA + CA)} = BC + D$ (from *Clavis* of 1631, p. 40)

$\frac{AB}{2} + \sqrt{q}\frac{ABq}{4} - \frac{C \times S}{R} : = A.$, for $\frac{\overline{AB}}{2} + \sqrt{\left(\frac{\overline{AB^2}}{4} - \frac{\overline{C \times S}}{R}\right)} = A.$ (from *Clavis* of 1652, p. 95)

[1] Rigaud, *Correspondence of Scientific Men of the Seventeenth Century*, Vol. I (1841), Letter VI, p. 8.

$Q.Hc+Ch$: for $(Hc+Ch)^2$ (from *Clavis* of 1652, p. 57)

$Q.A-X=$, for $(A-X)^2=$ (from *Clavis* of 1694, p. 97)

$\dfrac{B}{2}+r.u.\ \dfrac{Bq}{4}-CD.=A$, for $\dfrac{B}{2}+\sqrt{\left(\dfrac{B^2}{4}-CD\right)}=A$ (from *Oughtredus explicatus* [1682], p. 101)

11. These notations to signify aggregation occur very seldom in the texts referred to and may be simply printer's errors.

12. Mathematical parentheses occur also on p. 75, 80, and 117 of G. Clark's *Oughtredus explicatus*.

13. In the *Clavis* of 1631, p. 2, it says, "Signum additionis siue affirmationis, est+plus" and "Signum subductionis, siue negationis est−minus." In the edition of 1694 it says simply, "The Sign of Addition is + more" and "The Sign of Subtraction is − less," thereby ignoring, in the definition, the double function played by these symbols.

14. In the errata following the Preface of the 1694 edition it says, for "*more* or *mo.* r. [ead] *plus* or *pl.*"; for *less* or *le.* r.[ead] *minus* or *mi.*"

15. Oughtred's *Clavis mathematicae* of 1631 is not the first appearance of $\times$ as a symbol for multiplication. In Edward Wright's translation of John Napier's *Descriptio*, entitled *A Description of the Admirable Table of Logarithms* (London, 1618), the letter "X" is given as the sign of multiplication in the part of the book called "An Appendix to the Logarithms, shewing the practise of the calculation of Triangles, etc."

The use of the letters x and X for multiplication is not uncommon during the seventeenth and beginning of the eighteenth centuries. We note the following instances: Vincent Wing, *Doctrina theorica* (London, 1656), p. 63; John Wallis, *Arithmetica infinitorum* (Oxford, 1655), p. 115, 172; *Moore's Arithmetick in two Books*, by Jonas Moore (London, 1660), p. 108; *Antoine Arnauld, Novveavx elemens de geometrie* (Paris, 1667), p. 6; Lord Brounker, *Philosophical Transactions*, Vol. II (London, 1668), p. 466; *Exercitatio geometrica, auctore Laurentio Lorenzinio, Vincentii Viviani discipulo* (Florence, 1721). John Wallis used the $\bowtie$ in his *Elenchus geometriae Hobbianae* (Oxoniae, 1655), p. 23.

16. *in* as a symbol of multiplication carries with it also a collective meaning; for example, the *Clavis* of 1652 has on p. 77, "Erit $\frac{1}{2}Z+\frac{1}{2}B$ in $\frac{1}{2}Z-\frac{1}{2}B=\frac{1}{4}Zq-\frac{1}{4}Bq.$"

17. That is, the line AB squared.

18. These capital letters precede the expression to be raised to a power. Seldom are they used to indicate powers of monomials. From the *Clavis* of 1652, p. 65, we quote:

$$\text{“}Q:A+E:+Eq=2Q:\tfrac{1}{2}A+E:+2Q.\tfrac{1}{2}A\text{ ,”}$$

$$\text{i.e., } (A+E)^2+E^2=2(\tfrac{1}{2}A+E)^2+2\left(\frac{A}{2}\right)^2.$$

19. L and l stand for the same thing, "side" or "root," l being used generally when the coefficients of the unknown quantity are given in Hindu-Arabic numerals, so that all the letters in the equation, viz., l, q, c, qq, qc, etc., are small letters. The *Clavis* of 1694, p. 158, uses L in a place where the Latin editions use l.

20. The symbol $\sqrt{u}$ does not occur in the *Clavis* of 1631 and is not defined in the later editions. The following throws light upon its significance. In the 1631 edition, chap. xvi, sec. 8, p. 40, the author takes $\sqrt{qBA}+B=CA$, gets from it $\sqrt{qBA}=CA-B$, then squares both sides and solves for the unknown A. He passes

next to a radical involving two terms, and says: "Item $\sqrt{q}$ vniuers : $BA+CA$: $-$ $D=BC$: vel per transpositionem $\sqrt{q}$: $BA+CA=BC+D$"; he squares both sides and solves for A. In the later editions he writes "$\sqrt{u}$" in place of "$\sqrt{q}$ vniuers : "

21. The sum $Z=A+E$ and the difference $X=A-E$ are used later in imitation of Oughtred by Samuel Foster in his *Miscellanies* (London, 1659), "Of Projection," p. 8, and by Sir Jonas Moore in his *Arithmetick* (3d ed.; London, 1688), p. 404; John Wallis in his *Operum mathematicorum pars prima* (Oxford, 1657), p. 169, and other parts of his mathematical writings.

22. Harriot's symbols $>$ for "greater" and $<$ for "less" were far superior to the corresponding symbols used by Oughtred.

23. This notation for "less than" in the *Ho.* occurs only in the explanation of "Fig. *EE.*" In the text (chap. ix) the regular notation explained in *Eu.* is used.

24. The symbol ∞ so closely resembles the symbol ∞ which was used by John Wallis in his *Operum mathematicorum pars prima* (Oxford, 1657), p. 208, 247, 334, 335, that the two symbols were probably intended to be one and the same. It is difficult to assign a good reason why Wallis, who greatly admired Oughtred and was editor of the later Latin editions of his *Clavis mathematicae*, should purposely reject Oughtred's ∞ and intentionally introduce ∞ as a substitute symbol.

25. Von Braunmühl, in his *Geschichte der Trigonometrie* (2. Teil; Leipzig, 1903), p. 42, 91, refers to Oughtred's *Trigonometria* of 1657 as containing the earliest use of abbreviations of trigonometric functions and points out that a half-century later the army of writers on trigonometry had hardly yet reached the standard set by Oughtred. This statement must be modified in several respects (see §§ 500–526).

26. This reference is to the English edition, the *Trigonometrie* of 1657. In the Latin edition there is printed on p. 5, by mistake, *s* instead of *s versus*. The table of errata makes reference to this misprint.

27. The horizontal line was printed beneath the expression that was being crossed out. Thus, on p. 68 of the *Clavis* of 1631 there is:

$$BGqq - \underline{BGq \times 2BK \times BD} + BKq \times BDq$$
$$= BGq \times BDq + BGq \times BKq - \underline{BGq \times 2BK \times BD} + BGq \times 4CAq.$$

28. This notation, says Oughtred, was used by ancient writers on music, who "are wont to connect the terms of ratios, either to be continued" as in $\frac{3}{2} \times \frac{4}{3} = 2$, "or diminish'd" as in $\frac{3}{2} \div \frac{4}{3} = \frac{9}{8}$.

29. See n. 15.

30. *Cru* and *crur* are abbreviations for *crurum*, side of a rectangle or right triangle. Hence Z *cru* means the sum of the sides, X *cru*, the difference of the sides.

187. Oughtred's recognition of the importance of notation is voiced in the following passage:

". . . . Which Treatise being not written in the usuall synthetical manner, nor with verbous expressions, but in the inventive way of Analitice, and with symboles or notes of things instead of words, seemed unto many very hard; though indeed it was but their owne diffidence, being scared by the newness of the delivery; and not any

difficulty in the thing it selfe. For this specious and symbolicall manner, neither racketh the memory with multiplicity of words, nor
chargeth the phantasie with comparing and laying things together;
but plainly presenteth to the eye the whole course and processe of
every operation and argumentation."[1]

Again in his *Circles of Proportion* (1632), p. 20:

"This manner of setting downe theoremes, whether they be Proportions, or Equations, by Symboles or notes of words, is most excellent, artificiall, and doctrinall. Wherefore I earnestly exhort every
one, that desireth though but to looke into these noble Sciences
Mathematicall, to accustome themselves unto it: and indeede it is
easie, being most agreeable to reason, yea even to sence. And out of
this working may many singular consectaries be drawne: which
without this would, it may be, for ever lye hid."

ENGLISH: THOMAS HARRIOT

(1631)

188. Thomas Harriot's *Artis analyticae praxis* (London, 1631)
appeared as a posthumous publication. He used small letters in place
of Vieta's capitals, indicated powers by the repetition of factors, and
invented $>$ and $<$ for "greater" and "less."

Harriot used a very long sign of equality $=$. The following quotation shows his introduction of the now customary signs for "greater"
and "smaller" (p. 10):

"Comparationis signa in sequentibus vsurpanda.

Aequalitatis $=$ ut $a = b$. significet a aequalem ipi b.

Maioritatis $>$ ut $a > b$. significet a maiorem quam b.

Minoritatis $<$ ut $a < b$ significet a minorem quam b."

Noteworthy is the notation for multiplication, consisting of a
vertical line on the right of two expressions to be multiplied together
of which one is written below the other; also the notation for complex
fractions in which the principal fractional line is drawn double. Thus
(p. 10):

$$\left.\begin{array}{c} \dfrac{ac}{b} \\[4pt] b \end{array}\right| = \frac{acb}{b} = ac \;.$$

$$\dfrac{\dfrac{aaa}{b}}{d} = \frac{aaa}{bd} \;.$$

---

[1] William Oughtred, *The Key of the Mathematicks* (London, 1647), Preface.

Harriot places a dot between the numerical coefficient and the other factors of a term. Excepting only a very few cases which seem to be printer's errors, this notation is employed throughout. Thus (p. 60):

"Aequationis $aaa - 3.baa + 3.bba \Longrightarrow\; + 2.bbb$ est $2.b$. radix
$\qquad\qquad\qquad\qquad\qquad\qquad$ radici quaesititiae a. aequalis ."

Probably this dot was not intended as a sign of multiplication, but simply a means of separating a numeral from what follows, according to a custom of long standing in manuscripts and early printed books.

On the first twenty-six pages of his book, Harriot frequently writes all terms on one side of an equation. Thus (p. 26):

"Posito igitur $cdf = aaa$. est $aaa - cdf \Big|_{a+b} \Longrightarrow 0$

Est autem ex genesi $aaa - cdf \Big|_{a+b} \Longrightarrow aaaa + baaa - cdfa - bcdf$.

$\qquad$ quae est aequatio originalis hic designata.

Ergo . . . . $aaaa + baaa - cdfa - bcdf. \Longrightarrow 0$."

Sometimes Harriot writes underneath a given expression the result of carrying out the indicated operations, using a brace, but without using the regular sign of equality. This is seen in Figure 87. The first equation is $52 = -3a + aaa$, where the vowel a represents the unknown. Then the value of a is given by Tartaglia's formula, as $\sqrt[3]{26 + \sqrt{675}} + \sqrt[3]{26 - \sqrt{675}} = 4$. Notice that "$\sqrt{3}$.)" indicates that the cube root is taken of the binomial $26 + \sqrt{675}$.

In Figure 88 is exhibited Harriot's use of signs of equality placed vertically and expressing the equality of a polynomial printed above a horizontal line with a polynomial printed below another horizontal line. This exhibition of the various algebraic steps is clever.

FRENCH: PIERRE HÉRIGONE

(1634, 1644)

189. A full recognition of the importance of notation and an almost reckless eagerness to introduce an exhaustive set of symbols is exhibited in the *Cursus mathematicus* of Pierre Hérigone, in six volumes, in Latin and French, published at Paris in 1634 and, in a second edition, in 1644. At the beginning of the first volume is given

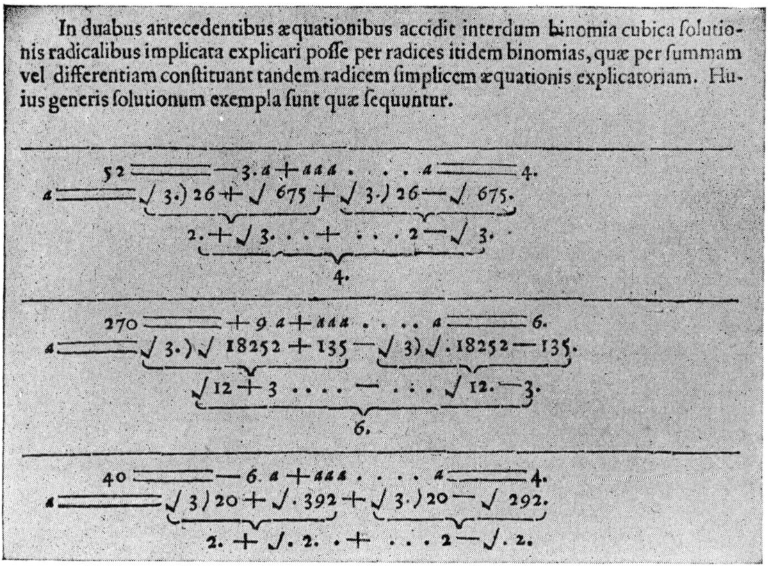

In duabus antecedentibus æquationibus accidit interdum binomia cubica folutionis radicalibus implicata explicari poffe per radices itidem binomias, quæ per fummam vel differentiam conftituant tandem radicem fimplicem æquationis explicatoriam. Huius generis folutionum exempla funt quæ fequuntur.

$$52 \quad\quad -3.a + aaa \ldots . a \quad\quad 4.$$
$$a \quad\quad \sqrt{3.)\,26} + \sqrt{675} + \sqrt{3.)\,26} - \sqrt{675.}$$
$$2. + \sqrt{3}. \ldots + \ldots 2 - \sqrt{3}.$$
$$4.$$

$$270 \quad\quad + 9.a + aaa \ldots .. a \quad\quad 6.$$
$$a \quad\quad \sqrt{3.)}\sqrt{18252 + 135} - \sqrt{3)}\sqrt{.18252 - 135.}$$
$$\sqrt{12 + 3} \ldots - \ldots \sqrt{12.} - 3.$$
$$6.$$

$$40 \quad\quad -6.a + aaa \ldots . a \quad\quad 4.$$
$$a \quad\quad \sqrt{3)\,20} + \sqrt{.392} + \sqrt{3.)\,20} - \sqrt{292.}$$
$$2. + \sqrt{.2.} . + \ldots 2 - \sqrt{.2.}$$

Fig. 87.—From Thomas Harriot's *Artis analyticae praxis* (1631), p. 101

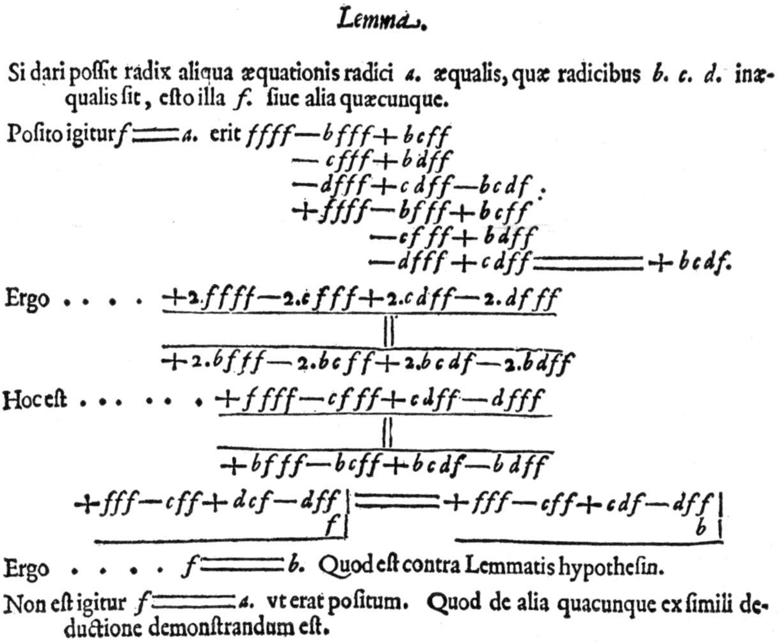

Lemma.

Si dari poffit radix aliqua æquationis radici *a.* æqualis, quæ radicibus *b. c. d.* inæqualis fit, efto illa *f.* fiue alia quæcunque.

Pofito igitur $f = a.$ erit $ffff - bfff + bcff$
$\quad\quad - cfff + bdff$
$\quad\quad - dfff + cdff - bcdf$:
$\quad\quad + ffff - bfff + bcff$
$\quad\quad - cfff + bdff$
$\quad\quad - dfff + cdff = + bcdf.$

Ergo . . . . $+2.ffff - 2.cfff + 2.cdff - 2.dfff$
$$\|$$
$$+2.bfff - 2.bcff + 2.bcdf - 2.bdff$$

Hoc eft . . . . . . $+ffff - cfff + cdff - dfff$
$$\|$$
$$+bfff - bcff + bcdf - bdff$$

$$+fff - cff + dcf - dff \,\Big|\, = + fff - cff + cdf - dff \,\Big|$$
$$f b\Big|$$

Ergo . . . . . $f = b.$ Quod eft contra Lemmatis hypothefin.

Non eft igitur $f = a.$ vt erat pofitum. Quod de alia quacunque ex fimili deductione demonftrandum eft.

Fig. 88.—From Thomas Harriot's *Artis analyticae praxis* (1631), p. 65

an explanation of the symbols. As found in the 1644 edition, the list
is as follows:

+ plus

∼ minus

·∼: differentia

◁ inter se, *entrélles*

◁ *n* in, *en*

◁ *ntr.* inter, *entre*

11 vel, *ou*

π, ad, *à*

5< pentagonum,   penta-
    gone

6< hexagonum

√·4< latus quadrati

√·5< latus pentagoni

a2 *A* quadratum

a3 *A* cubus

a4 *A* quadrato-quadratū.
    et sic infinitum.

= parallela

⊥ perpendicularis

·· est nota genitini, *sig-
    nifie (de)*

; est nota numeri plural-

is, signifie le plurier

2|2 aequalis

3|2 maior

2|3 minor

⅓ tertia pars

¼ quarta pars

⅔ duae tertiae

a,b, 11 *ab* rectangulum quod sit
    ductu *A* in *B*

· est punctum

— est recta linea

<, ∠ est angulus

⌐ est angulus rectus

⊙ est circulus

↻ ☾ est   pars   circumfer-
    entiae circuli

⌒, ⌣ est segmentu circuli

△ est triangulum

□ est quadratum

▭ est rectangulum

◇ est parallelogrammum

◇ piped. est parallelepipedum

In this list the symbols that are strikingly new are those for equality
and inequality, the ∼ as a minus sign, the — being made to represent
a straight line. Novel, also, is the expression of exponents in Hindu-
Arabic numerals and the placing of them to the right of the base, but
not in an elevated position. At the beginning of Volume VI is given a
notation for the aggregation of terms, in which the comma plays a
leading rôle:

"□· *a2∼5a+6, a∼4:* virgula, la virgule, dis-
    tinguit multiplicatorem *a∼4 à* multiplicãdo
    a2∼5a+6.
Ergo ₒ□ 5+4+3, 7∼3:∼10, est 38."

Modern: The rectangle $(a^2 - 5a + 6)(a - 4)$,
    Rectangle $(5+4+3)(7-3) - 10 = 38$.

"*hg* π *ga* 2|2 *hb* π *bd,* signifi. *HG* est *ad GA,* vt
    HB ad *BD .*"

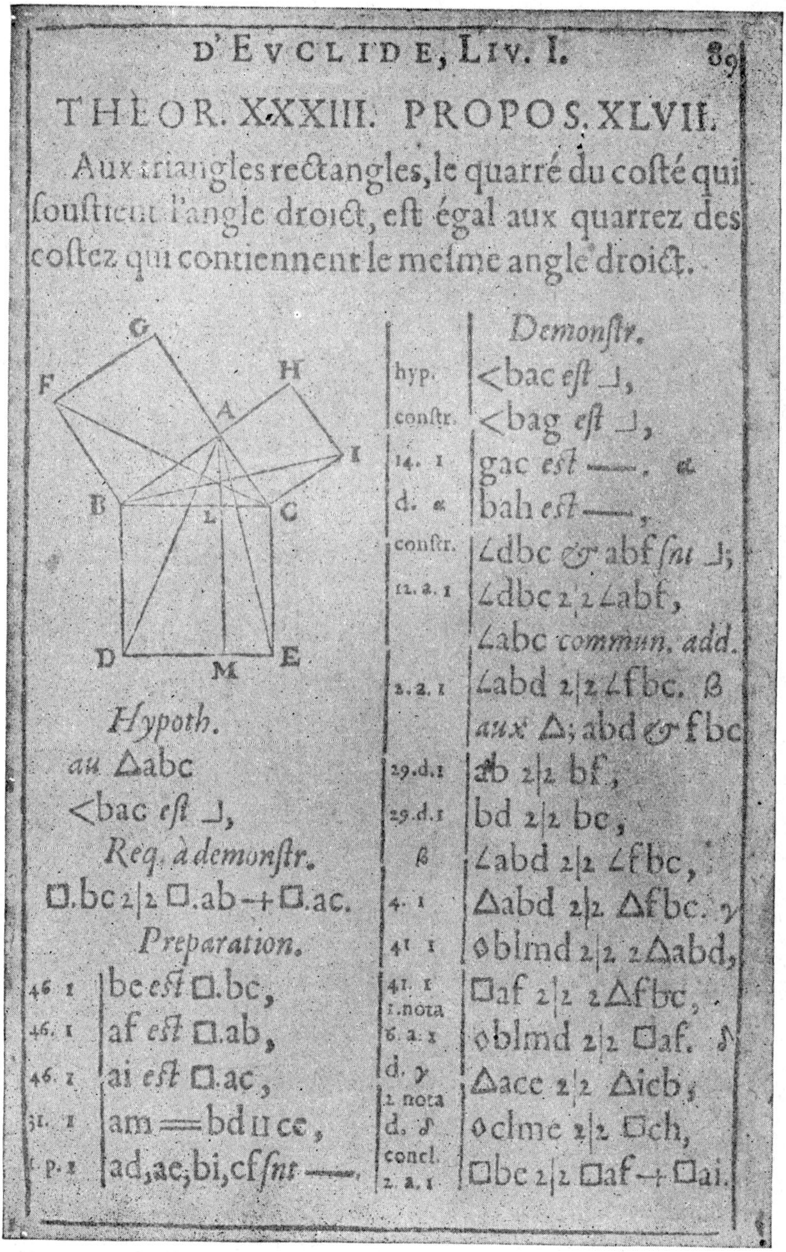

D'EVCLIDE, LIV. I. 89

THEOR. XXXIII. PROPOS. XLVII.

Aux triangles rectangles, le quarré du costé qui soustient l'angle droict, est égal aux quarrez des costez qui contiennent le mesme angle droict.

Hypoth.
au △abc
<bac *est* ⊥,

Req. à demonstr.
□.bc 2|2 □.ab + □.ac.

Preparation.

46. 1	be *est* □.bc,
46. 1	af *est* □.ab,
46. 1	ai *est* □.ac,
31. 1	am ═ bd ц ce,
1 p. 1	ad, ae, bi, cf *snt* ——.

Demonstr.

hyp.	<bac *est* ⊥,
constr.	<bag *est* ⊥,
14. 1	gac *est* ——. α
d. α	bah *est* ——,
constr.	∠dbc & abf *snt* ⊥;
12. a. 1	∠dbc 2|2 ∠abf,
	∠abc *commun. add.*
2. a. 1	∠abd 2|2 ∠fbc. β
	aux △; abd & fbc
29. d. 1	ab 2|2 bf,
29. d. 1	bd 2|2 bc,
β	∠abd 2|2 ∠fbc,
4. 1	△abd 2|2 △fbc. γ
41. 1	▱blmd 2|2 2△abd,
41. 1, 1. nota	□af 2|2 2△fbc,
6. a. 1	▱blmd 2|2 □af. δ
d. γ, 2 nota	△ace 2|2 △icb,
d. δ	▱clme 2|2 □ch,
concl. 2. a. 1	□bc 2|2 □af + □ai.

FIG. 89.—From P. Herigone, *Cursus mathematicus*, Vol. VI (1644); proof of the Pythagorean theorem.

Modern: $hg : ga = hb : bd$.

> "$\sqrt{\cdot}16+9$ est 5, se pormoit de serire plus dis-
> tinctement ainsi ,
> $\sqrt{\cdot}(16+9)$ $11\sqrt{\cdot16+9}$, est $5:\sqrt{\cdot}9$, $+4$, sont
> $7:\sqrt{\cdot}9$, $+\sqrt{\cdot}4$ sont 5: "

Modern: $\sqrt{\cdot}16+9$ is 5, can be written more clearly thus,
$\sqrt{\cdot}(16+9)$ or $\sqrt{\cdot16+9}$, is 5; $\sqrt{\cdot}9$, $+4$, are 7;
$\sqrt{\cdot}9$, $+\sqrt{\cdot}4$ are 5 .

FRENCH: JAMES HUME

(1635, 1636)

190. The final development of the modern notation for positive integral exponents took place in mathematical works written in French. Hume was British by birth. His *Le traité d'algèbre* (Paris, 1635) contains exponents and radical indexes expressed in Roman numerals. In Figure 90 we see that in 1635 the plus ($+$) and minus ($-$) signs were firmly established in France. The idea of writing exponents without the bases, which had been long prevalent in the writings of Chuquet, Bombelli, Stevin, and others, still prevails in the 1635 publication of Hume. Expressing exponents in Roman symbols made it possible to write the exponent on the same line with the coefficient without confusion of one with the other. The third of the examples in Figure 90 exhibits the multiplication of $8x^2+3x$ by $10x$, yielding the product $80x^3+30x^2$.

The translation of part of Figure 91 is as follows: "*Example:* Let there be two numbers $\sqrt{9}$ and $\sqrt[i]{8}$, to reduce them it will be necessary to take the square of $\sqrt[i]{8}$, because of the II which is with 9, and the square of the square of $\sqrt{9}$ and you obtain $\sqrt[i]{6561}$ and $\sqrt[i]{64}$. . . . .

$\sqrt[iii]{8}$ to $\sqrt[i]{64}$ | $\sqrt[iii]{3}$ to $\sqrt[i]{8}$ [should be $\sqrt[i]{9}$]

$\sqrt{9}$ to $\sqrt[i]{729}$ | $\sqrt{2}$ to $\sqrt[i]{9}$ [should be $\sqrt[i]{8}$]

$\sqrt[i]{3}$ to $\sqrt[i]{9}$

$\sqrt{2}$ to $\sqrt[i]{32}$."

The following year, Hume took an important step in his edition of *L'algèbre de Viète* (Paris, 1636), in which he wrote A^{iii} for A^3. Except for the use of the Roman numerals one has here the notation used by Descartes in 1637 in his *La géométrie* (see § 191).

191. Figure 92 shows a page from the first edition of Descartes'
La géométrie. Among the symbolic features of this book are: (1) the
use of small letters, as had been emphazised by Thomas Harriot;

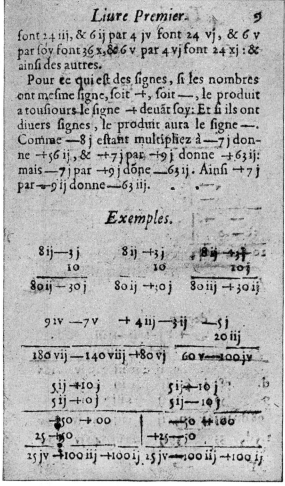

Fig. 90.—Roman numerals for unknown numbers in James Hume, *Algèbre*
(Paris, 1635).

(2) the writing of the positive integral exponents in Hindu-Arabic
numerals and in the position relative to the base as is practiced today,

except that *aa* is sometimes written for a^2; (3) the use of a new sign of equality, probably intended to represent the first two letters in the word *aequalis*, but apparently was the astronomical sign, ♉ taurus,

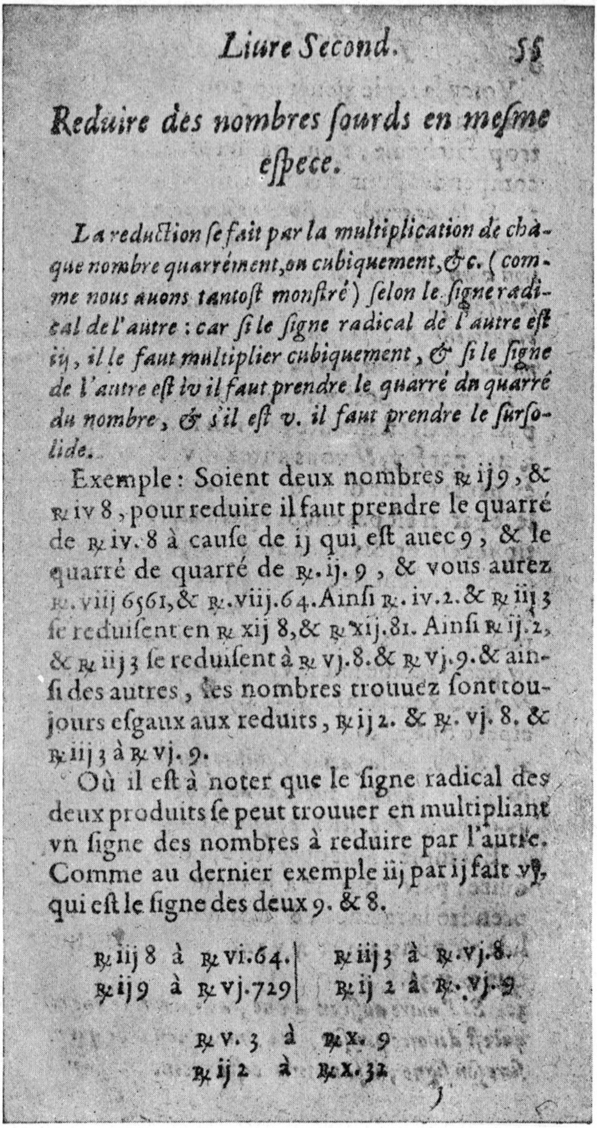

Liure Second. 55

Reduire des nombres sourds en mesme espece.

La reduction se fait par la multiplication de chaque nombre quarréement, ou cubiquement, &c. (comme nous auons tantost monstré) selon le signe radical de l'autre : car si le signe radical de l'autre est iij, il le faut multiplier cubiquement, & si le signe de l'autre est iv il faut prendre le quarré du quarré du nombre, & s'il est v. il faut prendre le sursolide.

Exemple : Soient deux nombres ℞ ij 9, & ℞ iv 8, pour reduire il faut prendre le quarré de ℞ iv. 8 à cause de ij qui est auec 9, & le quarré de quarré de ℞. ij. 9, & vous aurez ℞. viij 6561, & ℞.viij.64. Ainsi ℞. iv. 2. & ℞ iiij 3 se reduisent en ℞ xij 8, & ℞ xij. 81. Ainsi ℞ ij. 2, & ℞ iij 3 se reduisent à ℞ vj. 8. & ℞ vj. 9. & ainsi des autres, les nombres trouuez sont toujours esgaux aux reduits, ℞ ij 2. & ℞. vj. 8. & ℞ iiij 3 à ℞ vj. 9.

Où il est à noter que le signe radical des deux produits se peut trouuer en multipliant vn signe des nombres à reduire par l'autre. Comme au dernier exemple iij par ij fait vj, qui est le signe des deux 9. & 8.

℞ iij 8 à ℞ vi.64. | ℞ iiij 3 à ℞.vj.8.
℞ ij 9 à ℞ vj.729 | ℞ ij 2 à ℞.vj.9.

℞ v. 3 à ℞ x. 9
℞ ij 2 à ℞ x.32.

FIG. 91.—Radicals in James Hume, *Algèbre* (1635)

LIVRE PREMIER.   *303*

angle, iufques a O, en forte qu'N O foit efgale a N L, la toute O M eft z la ligne cherchée. Et elle s'exprime en cete forte

$$z \infty \tfrac{1}{2} a + \sqrt{\tfrac{1}{4} a a + b b}.$$

Que fi iay $y y \infty - a y + b b$, & qu'y foit la quantité qu'il faut trouuer, ie fais le mefme triangle rectangle N L M, & de fa baze M N i'ofte N P efgale a N L, & le refte P M eft y la racine cherchée. De façon que iay $y \infty - \tfrac{1}{2} a + \sqrt{\tfrac{1}{4} a a + b b}$. Et tout de mefme fi i'a-uois $x \infty - a x + b$. P M feroit x, & i'aurois

$$x \infty \sqrt{-\tfrac{1}{2} a + \sqrt{\tfrac{1}{4} a a + b b}} : \text{& ainfi des autres.}$$

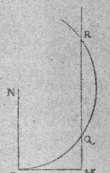

Enfin fi i'ay

$$z \infty a z - b b :$$

ie fais N L efgale à $\tfrac{1}{2} a$, & L M efgale à b côme deuãt, puis, au lieu de ioindre les poins M N, ie tire M Q R parallele a L N. & du cen-tre N par L ayant defcrit vn cer-cle qui la couppe aux poins Q & R, la ligne cherchée z eft M Q, oubiē M R, car en ce cas elle s'ex-prime en deux façons, a fçauoir $z \infty \tfrac{1}{2} a + \sqrt{\tfrac{1}{4} a a - b b}$, & $z \infty \tfrac{1}{2} a - \sqrt{\tfrac{1}{4} a a - b b}$.

Et fi le cercle, qui ayant fon centre au point N, paffe par le point L, ne couppe ny ne touche la ligne droite M Q R, il n'y a aucune racine en l'Equation, de façon qu'on peut affurer que la conftruction du problefme propofé eft impoffible.

Au

FIG. 92.—A page from René Descartes, *La géométrie* (1637)

placed horizontally, with the opening facing to the left; (4) the uniting of the vinculum with the German radical sign $\sqrt{}$, so as to give $\sqrt{}$, an adjustment generally used today.

The following is a quotation from Descartes' text (ed., Paris, 1886, p. 2): "Mais souvent on n'a pas besoin de tracer ainsi ces lignes sur le papier, et il suffit de les désigner par quelques lettres, chacune par une seule. Comme pour ajouter le ligne BD à GH, je nomme l'une a et l'autre b, et écris $a+b$; et $a-b$ pour soustraire b de a; et ab pour les multiplier l'une par l'autre; et $\dfrac{a}{b}$ pour diviser a par b; et aa ou a^2 pour multiplier a par soi-même; et a^3 pour le multiplier encore une fois par a, et ainsi à l'infini."

The translation is as follows: "But often there is no need thus to trace the lines on paper, and it suffices to designate them by certain letters, each by a single one. Thus, in adding the line BD to GH, I designate one a and the other b, and write $a+b$; and $a-b$ in subtracting b from a; and ab in multiplying the one by the other; and $\dfrac{a}{b}$ in dividing a by b; and aa or a^2 in multiplying a by itself; and a^3 in multiplying it once more again by a, and thus to infinity."

ENGLISH: ISAAC BARROW
(1655, 1660)

192. An enthusiastic admirer of Oughtred's symbolic methods was Isaac Barrow,[1] who adopted Oughtred's symbols, with hardly any changes, in his Latin (1655) and his English (1660) editions of *Euclid*. Figures 93 and 94 show pages of Barrow's *Euclid*.

ENGLISH: RICHARD RAWLINSON
(1655–68)

193. Sometime in the interval 1655–68 Richard Rawlinson, of Oxford, prepared a pamphlet which contains a collection of lithographed symbols that are shown in Figure 95, prepared from a crude freehand reproduction of the original symbols. The chief interest lies in the designation of an angle of a triangle and its opposite side by the same letter—one a capital letter, the other letter small. This simple device was introduced by L. Euler, but was suggested many years earlier by Rawlinson, as here shown. Rawlinson designated spherical

[1] For additional information on his symbols, see §§ 456, 528.

triangles by conspicuously rounded letters and plane triangles by letters straight in part.

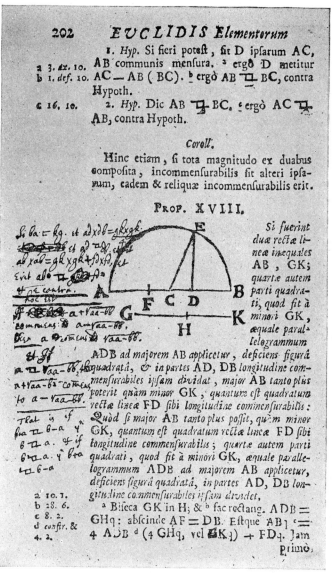

Fig. 93.—Latin edition (1655) of Barrow's *Euclid*. Notes by Isaac Newton. (Taken from *Isaac Newton: A Memorial Volume* [ed. W. J. Greenstreet; London, 1927], p. 168.)

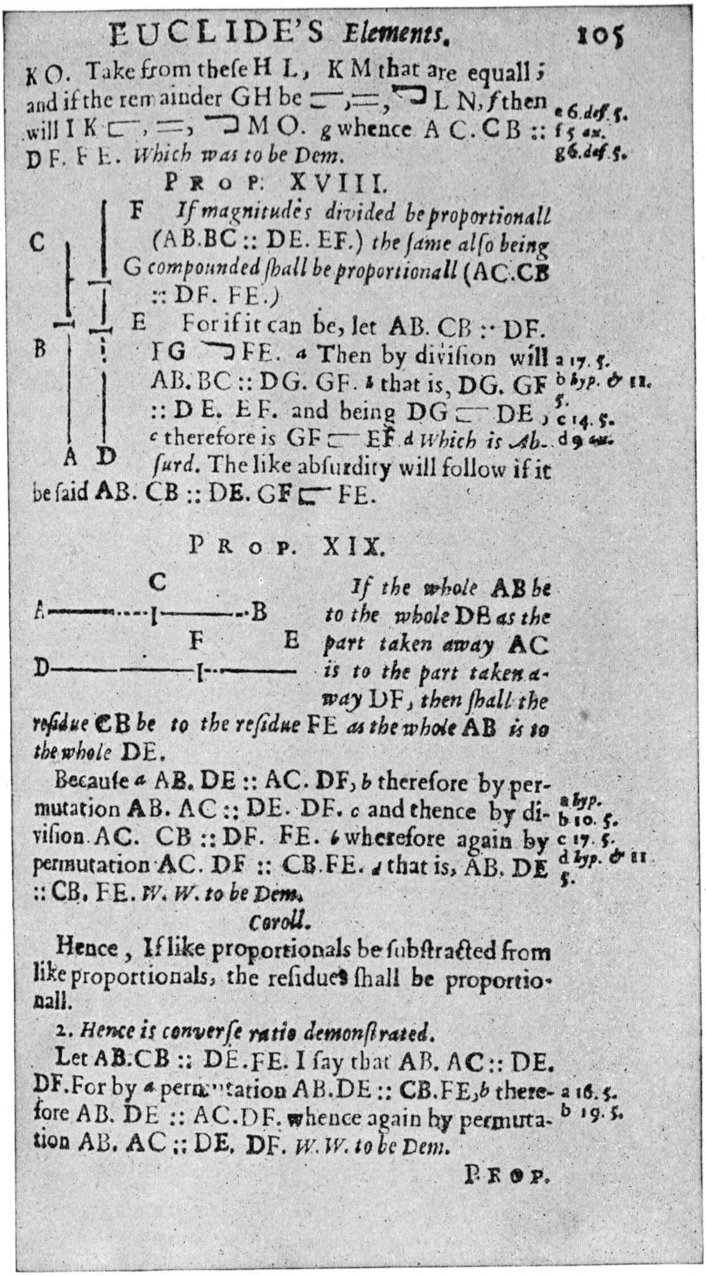

EUCLIDE'S *Elements.*   105

K O. Take from thefe H L, K M that are equall ;
and if the remainder GH be ⊏, ═, ⊐ L N, f then
will I K ⊏, ═, ⊐ M O. g whence A C. CB ::
DF. FE. *Which was to be Dem.*

a 6.def.5.
f 5 ax.
g6.def.5.

P R o p: XVIII.

F   *If magnitudes divided be proportionall*
(AB.BC :: DE. EF.) the fame alfo being
G *compounded fhall be proportionall (AC.CB*
:: DF. FE.)

E   For if it can be, let AB. CB :· DF.
IG ⊐FE. *a* Then by divifion will
AB. BC :: DG. GF. *b* that is, DG. GF
:: D E. EF. and being DG ⊏ DE,
c therefore is GF ⊏ EF. *d which is Ab.*
furd. The like abfurdity will follow if it
befaid AB. CB :: DE. GF ⊏ FE.

a 17.5.
b hyp. & 11.
5.
c 14.5.
d 9 ax.

P R o p. XIX.

C   *If the whole* AB *be*
A————·I———·B   *to the whole* DE *as the*
    F   E   *part taken away* AC
D————·I·———   *is to the part taken a-*
                   way DF, *then fhall the*
refidue CB *be   to the refidue* FE *as the whole* AB *is to*
the whole DE.

Becaufe *a* AB. DE :: AC. DF, *b* therefore by permutation AB. AC :: DE. DF. *c* and thence by divifion. AC. CB :: DF. FE. *b* wherefore again by
permutation AC. DF :: CB.FE. *d* that is, AB. DE
:: CB. FE. *W. W. to be Dem.*

a hyp.
b 10.5.
c 17.5.
d hyp. & 11
5.

Coroll.

Hence, If like proportionals be fubftracted from
like proportionals, the refidue fhall be proportionall.

2. *Hence is converfe ratio demonftrated.*

Let AB.CB :: DE.FE. I fay that AB. AC :: DE.
DF. For by *a* permutation AB.DE :: CB.FE, *b* therefore AB. DE :: AC.DF. whence again by permutation AB. AC :: DE. DF. *W. W. to be Dem.*

a 16.5.
b 19.5.

P R o P.

Fig. 94.—English edition of Isaac Barrow's *Euclid*

SWISS: JOHANN HEINRICH RAHN
(1659)

194. Rahn published in 1659 at Zurich his *Teutsche Algebra,*
which was translated by Thomas Brancker and published in 1668 at
London, with additions by John Pell. There were some changes in
the symbols as indicated in the following comparison:

Meaning	German Edition, 1659		English Edition, 1668	
1. Multiplication..............	*	(p. 7)	Same	(p. 6)
2. $a+b$ times $a-b$...........	$\begin{vmatrix} a+b \\ a-b \end{vmatrix}$	(p. 14)	Same	(p. 12)
3. Division..................	÷	(p. 8)	Same	(p. 7)
4. Cross-multiplication........	*×	(p. 25)	*X	(p. 23)
5. Involution................	Archimedean spiral (Fig. 96)	(p. 10)	Ligature of omicron and sigma (Fig. 97)	(p. 9)
6. Evolution................	Ligature of two epsilons (Fig.96)	(p. 11)	Same	(p. 9)
7. *Erfüll ein quadrat* Compleat the square }	$E\,\square$	(p 16)	$C\,\square$	(p. 14)
8. Sixth root................	$\sqrt{qc.}\begin{cases} aaa=\sqrt{a} \\ aa=\sqrt{c.a} \end{cases}$	(p. 34)	*cubo-cubick* $\sqrt{}$ of $aaa=\sqrt{a}$ *cubo-cubick* $\sqrt{}$ of $aa=\sqrt{c.a}$	(p. 32)
9. Therefore.................	∴ (usually)	(p. 53)	∴ (usually)	(p. 37)
10. Impossible (absurd)........	⊅	(p. 61)	⊃I	(p. 48)
11. Equation expressed in another way..............	,	(p. 67)	Same	(p. 54)
12. Indeterminate, "liberty of assuming an equation".......	(*	(p. 89)	Same	(p. 77)
13. Nos. in outer column referring to steps numbered in middle column............	1, 2, 3, etc.	(p. 3)	1, 2, 3, etc.	(p. 3)
14. Nos in outer column *not* referring to numbers in middle column................	1, 2, 3, etc.	(p. 3)	1̄, 2̄, 3̄, etc.	(p. 3)

REMARKS ON THESE SYMBOLS

No. 1.—Rahn's sign * for multiplication was used the same year as Brancker's translation, by
N. Mercator, in his *Logarithmotechnia* (London, 1668), p. 28.

No. 4.—If the lowest common multiple of abc and ad is required, Rahn writes $\dfrac{abc}{ad}=\dfrac{bc}{d}$; then

$\dfrac{abc}{ad}*\times\dfrac{bc}{d}$ yields $abcd$ in each of the two cross-multiplications.

No 8.—Rahn's and Brancker's modes of indicating the higher powers and roots differ in
principle and represent two different procedures which had been competing for supremacy for several
centuries. Rahn's $\sqrt{qc}$. means the sixth root, $2\times3=6$, and represents the Hindu idea. Brancker's
cubo-cubick root means the "sixth root," $3+3=6$, and represents the Diophantine idea.

No. 9.—In both editions occur both ∴ and ∵, but ∴ prevails in the earlier edition: ∵ prevails in
the later.

No. 10.—The symbols indicate that the operation is impossible or, in case of a root, that it is
imaginary.

No. 11.—The use of the comma is illustrated thus: The marginal column (1668, p. 54) gives
"6, 1," which means that the sixth equation "$Z=A$" and the first equation "$A=6$" yield $Z=6$.

No. 12.—For example, if in a right triangle h, b, c. we know only $b-c$, then one of the three
sides, say c, is indeterminate.

Page 73 of Rahn's *Teutsche Algebra* (shown in Fig. 96) shows:
(1) the first use of ÷ in print, as a sign of division; (2) the Archimedean spiral for involution; (3) the double epsilon for evolution; (4) the

use of capital letters B, D, E, for given numbers, and small letters a, b, for unknown numbers; (5) the ✳ for multiplication; (6) the first use of ∴ for "therefore"; (7) the three-column arrangement of which the left column contains the directions, the middle the numbers of

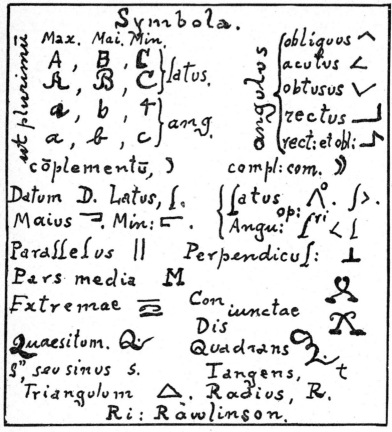

Fig. 95.—Freehand reproduction of Richard Rawlinson's symbols

the lines, the right the results of the operations. Thus, in line 3, we have "line 1, raised to the second power, gives $aa+2ab+bb=DD$."

ENGLISH: JOHN WALLIS
(1655, 1657, 1685)

195. Wallis used extensively symbols of Oughtred and Harriot, but of course he adopted the exponential notation of Descartes (1637). Wallis was a close student of the history of algebra, as is illustrated

by the exhibition of various notations of powers which Wallis gave in 1657. In Figure 98, on the left, are the names of powers. In the first column of symbols Wallis gives the German symbols as found in Stifel, which Wallis says sprang from the letters *r*, *z*, *c*, ſ, the first

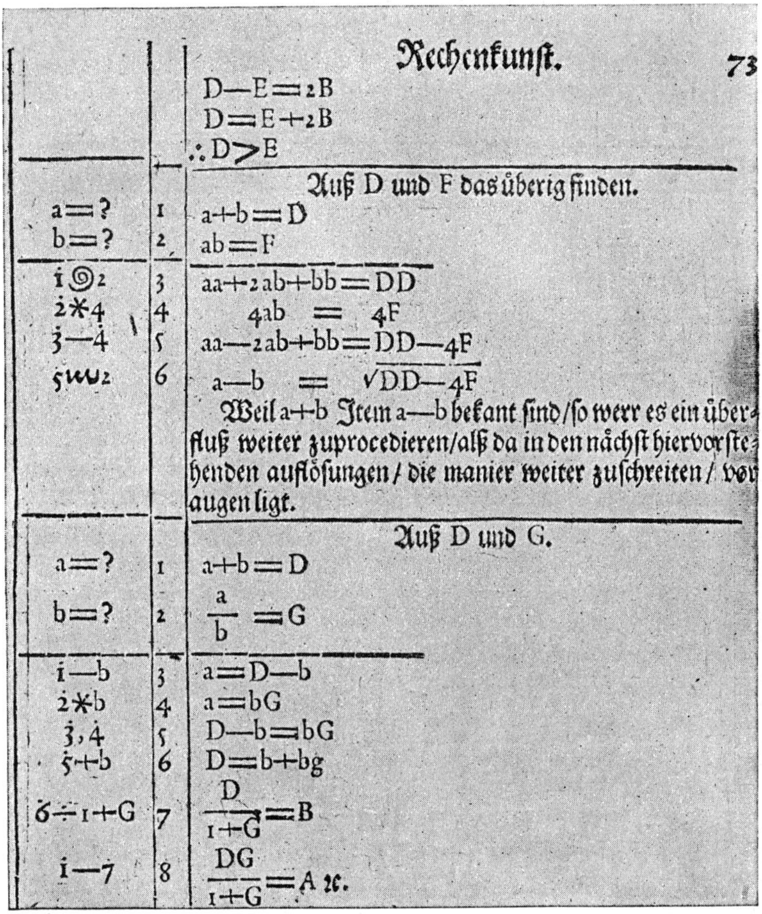
Rechenkunst. 73

$D - E = 2B$
$D = E + 2B$
$\therefore D > E$

Auß D und F das übrig finden.

$a = ?$	1	$a + b = D$
$b = ?$	2	$ab = F$
1 ⊚ 2	3	$aa + 2ab + bb = DD$
2 * 4	4	$4ab = 4F$
3 — 4	5	$aa - 2ab + bb = DD - 4F$
5 ωω 2	6	$a - b = \sqrt{DD - 4F}$

Weil a+b Item a—b bekant sind/so werr es ein überfluß weiter zuprocedieren/alß da in den nächst hiervorstehenden auflösungen/ die manier weiter zuschreiten/ vor augen ligt.

Auß D und G.

$a = ?$	1	$a + b = D$
$b = ?$	2	$\frac{a}{b} = G$
1 — b	3	$a = D - b$
2 * b	4	$a = bG$
3, 4	5	$D - b = bG$
5 + b	6	$D = b + bg$
6 ÷ 1 + G	7	$\frac{D}{1+G} = B$
1 — 7	8	$\frac{DG}{1+G} = A$ &c.

FIG. 96.—From Rahn, *Teutsche Algebra* (1659)

letters of the words *res, zensus, cubus, sursolidus.* In the second column are the letters *R*, *Q*, *C*, *S* and their combinations, Wallis remarking that for *R* some write *N;* these were used by Vieta in numerical equations. In the third column are Vieta's symbols in literal algebra, as abbreviated by Oughtred; in the fourth column Harriot's procedure is indicated; in the fifth column is Descartes' exponential notation.

In his *Arithmetica infinitorum*[1] he used the colon as a symbol for aggregation, as $\sqrt{}:a^2+1$ for $\sqrt{a^2+1}$, $\sqrt{}:aD-a^2:$ for $\sqrt{aD-a^2}$; Oughtred's notation for ratio and proportion, $\because$ for continued proportion. As the sign for multiplication one finds in this book X and ✕, both signs occurring sometimes on one and the same page (for instance, p. 172). In a table (p. 169) he puts □ for a given number: "Verbi gratiâ; si numerus hâc notâ □ designatus supponatur cognitus, reliqui omnes etiam cognoscentur." It is in this book and in his *De*

Resolution of Problemes.   61

By D and T, &c.

$a=?$
$b=?$

1 ☉·2	1	$a+b=D$
3—2	2	$aa+bb=T$
4✱2	3	$aa+bb+2ab=DD$
3—5	4	$2ab=DD-T$
6 ω 2	5	$4ab=2DD-2T$
	6	$aa+bb-2ab=2T-DD$
	7	$a-b=\sqrt{2T-DD},$ &c.

Fig. 97.—From Brancker's translation of Rahn (1668). The same arrangement of the solution as in 1659, but the omicron-sigma takes the place of the Archimedean spiral; the ordinal numbers in the outer column are not dotted, while the number in that column which does not refer to steps in the middle column carries a bar, $\bar{2}$. Step 5 means "line 4, multiplied by 2, gives $4ab=2DD-2T$."

sectionibus conicis that Wallis first introduces ∞ for infinity. He says (p. 70): "Cum enim primus terminus in serie Primanorum sit 0, primus terminus in serie reciproca erit ∞ vel infinitus : (sicut, in divisione, si diviso sit 0, quotiens erit infinitus)"; on pages 152, 153: " .... quippe $\frac{1}{\infty}$ (pars infinite parva) habenda erit pro nihilo," "$\infty \times \frac{1}{\infty} B = B$," "Nam ∞, ∞+1 ∞ −1, perinde sunt"; on page 168: "Quamvis enim ∞ ✕0 non aliquem determinate numerum designet. ...." An imitation of Oughtred is Wallis' "$\pi\!\!\!\!\!\!\!\frown : 1|\frac{3}{2}$," which occurs in his famous determination by interpolation of $\frac{4}{\pi}$ as the ratio of two infinite products. At this place he represents our $\frac{4}{\pi}$ by the symbol □.

---

[1] *Johannis Wallisii Arithmetica infinitorum* (Oxford, 1655).

He says also (p. 175): "Si igitur ut $\sqrt{}:3\times6$: significat terminum medium inter 3 et 6 in progressione Geometrica aequabili 3, 6, 12, etc. (continue multiplicando $3\times2\times2$ etc.) ita $m:1|\frac{3}{2}$: significet terminum medium inter 1 et $\frac{3}{2}$ in progressione Geometrica decrescente 1, $\frac{3}{2}$, $\frac{15}{8}$, etc. (continue multiplicando $1\times\frac{3}{2}\times\frac{5}{4}$, etc.) erit $\square=m:1|\frac{3}{2}$: Et propterea circulus est ad quadratum diametri, ut 1 ad $m:1|\frac{3}{2}$." He uses this symbol again in his *Treatise of Algebra* (1685), pages 296, 362.

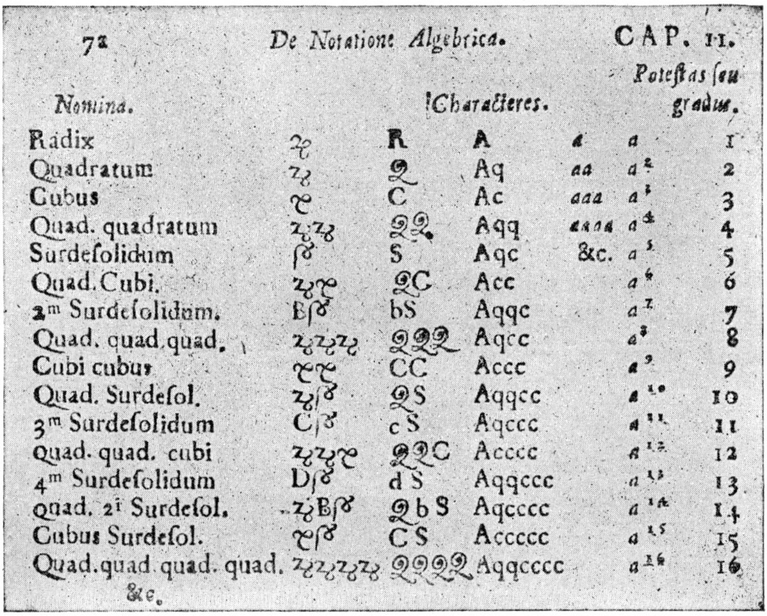

72	*De Notatione Algebrica.*				C A P. 11.	
					Poteſtas ſeu	
Nomina.			*Characteres.*		*gradus.*	
Radix	γ	R	A	a	a	1
Quadratum	$\gamma\gamma$	Q	Aq	aa	a^2	2
Cubus	γ	C	Ac	aaa	a^3	3
Quad. quadratum	$\gamma\gamma$	QQ	Aqq	$aaaa$	a^4	4
Surdeſolidum	$\int\delta$	S	Aqc	&c.	a^5	5
Quad. Cubi.	$\gamma\gamma$	QC	Acc		a^6	6
2m Surdeſolidum.	$\mathrm{B}\int\delta$	bS	Aqqc		a^7	7
Quad. quad. quad.	$\gamma\gamma\gamma$	QQQ	Aqcc		a^8	8
Cubi cubus	$\gamma\gamma$	CC	Accc		a^9	9
Quad. Surdeſol.	$\gamma\int\delta$	QS	Aqqcc		a^{10}	10
3m Surdeſolidum	$\mathrm{C}\int\delta$	cS	Aqccc		a^{11}	11
Quad. quad. cubi	$\gamma\gamma\gamma$	QQC	Accccc		a^{12}	12
4m Surdeſolidum	$\mathrm{D}\int\delta$	dS	Aqqccc		a^{13}	13
quad. 2i Surdeſol.	$\gamma\mathrm{B}\int\delta$	QbS	Aqcccc		a^{14}	14
Cubus Surdeſol.	$\gamma\int\delta$	CS	Acccccc		a^{15}	15
Quad. quad. quad. quad.	$\gamma\gamma\gamma\gamma$	QQQQ	Aqqcccc		a^{16}	16
&c.						

Fig. 98.—From John Wallis, *Operum mathematicorum pars prima* (Oxford, 1657), p. 72.

The absence of a special sign for division shows itself in such passages as (p. 135): "Ratio rationis hujus $\dfrac{1}{2\square}$ ad illam $\frac{1}{2}$, puta $\dfrac{1}{2}\Big)\dfrac{1}{2\square}\Big(\dfrac{1}{\square}$, erit......" He uses Oughtred's clumsy notation for decimal fractions, even though Napier had used the point or comma in 1617. On page 166 Wallis comes close to the modern radical notation; he writes "$\sqrt{}{}^{6}R$" for $\sqrt[6]{R}$. Yet on that very page he uses the old designation "$\sqrt{}qqR$" for $\sqrt[4]{R}$.

His notation for continued fractions is shown in the following quotation (p. 191):

"Esto igitur fractio ejusmodi continue fracta quaelibet, sic designata, $\dfrac{a}{a}\,\dfrac{b}{\beta}\,\dfrac{c}{\gamma}\,\dfrac{d}{\delta}\,\dfrac{e}{\epsilon}$, etc.,"

where

$$\frac{a}{a}\,\frac{b}{\beta}\equiv\frac{a\beta}{a\beta+b}\,.$$

The suggestion of the use of negative exponents, introduced later by Isaac Newton, is given in the following passage (p. 74): "Ubi autem series directae indices habent 1, 2, 3, etc. ut quae supra seriem Aequalium tot gradibus ascendunt; habebunt hae quidem (illis reciprocae) suos indices contrarios negativos -1, -2, -3, etc. tanquam tot gradibus infra seriem Aequalium descendentes."

In Wallis' *Mathesis universalis*,[1] the idea of positive and negative integral exponents is brought out in the explanation of the Hindu-Arabic notation. The same principle prevails in the sexagesimal notation, "hoc est, minuta prima, secunda, tertia, etc. ad dextram descendendo," while ascending on the left are units "quae vocantur Sexagena prima, secunda, tertia, etc. hoc modo.

$$\overset{\backslash\backslash\backslash\backslash}{4}\,\overset{}{9},\ \overset{\backslash\backslash\backslash}{3}\,\overset{}{6},\ \overset{\backslash\backslash}{2}\,\overset{}{5},\ \overset{\backslash}{1}\,\overset{}{5},\ \overset{\circ}{1},\ \overset{/}{1}\,\overset{}{5},\ \overset{//}{2}\,\overset{}{5},\ \overset{///}{3}\,\overset{}{6},\ \overset{////}{4}\,\overset{}{9}\,."$$

That the consideration of sexagesimal integers of denominations of higher orders was still in vogue is somewhat surprising.

On page 157 he explains both the "scratch method" of dividing one number by another and the method of long division now current, except that, in the latter method, he writes the divisor underneath the dividend. On page 240: "$A, M, V \,\overset{..}{\frown}$" for arithmetic proportion, i.e., to indicate $M-A=V-M$. On page 292, he introduces a general root d in this manner: "$\sqrt{^dR^d}=R$." Page 335 contains the following interesting combination of symbols:

"Si $\overbrace{A\,\cdot\,B\,\cdot\,C : \underbrace{a\,\cdot\,\beta\,\cdot\,\gamma}_{::}}^{::}$ in Modern Symbols
                              If $A:B=a:\beta$,
                              and $B:C=\beta:\gamma$,
Erit $A\,\cdot\,C :: a\,\cdot\,\gamma$."   then $A:C=a:\gamma$.

196. In the *Treatise of Algebra*[2] (p. 46), Wallis uses the decimal point, placed at the lower terminus of the letters, thus: 3.14159,

[1] *Johannis Wallisii Mathesis universalis: sive, Arithmeticum opus integrum* (Oxford, 1657), p. 65–68.

[2] *Op. cit.* (London, 1685).

26535. . . . . , but on page 232 he uses the comma, "12,756," ",3936."
On page 67, describing Oughtred's *Clavis mathematicae*, Wallis says:
"He doth also (to very great advantage) make use of several Ligatures,
or Compendious Notes, to signify the *Summs, Differences*, and *Rectangles* of several Quantities. As for instance, Of two quantities *A*
(the Greater, and *E* (the Lesser,) the Sum he calls *Z*, the Difference
X, the Rectangle Æ. . . . ." On page 109 Wallis summarizes various
practices: "The Root of such Binomial or Residual is called a Root
universal; and thus marked $\sqrt{u}$, (Root universal,) or $\sqrt{b}$, (Root of a
Binomial,) or $\sqrt{r}$, (Root of a Residual,) or drawing a Line over the
whole Compound quantity; or including it (as Oughtred used to do)
within two colons; or by some other distinction, whereby it may appear, that the note of Radicality respects, not only the single quantity
next adjoining, but the whole Aggregate. As $\sqrt{b}:2+\sqrt{3}\cdot\sqrt{r}:2-$
$\sqrt{3}\cdot\sqrt{u}:2\pm\sqrt{3}\cdot\sqrt{2\pm\sqrt{3}}\cdot\sqrt{}:2\pm\sqrt{3}$; etc."

On page 227 Wallis uses Rahn's sign $\div$ for division; along with the
colon as the sign of aggregation it gives rise to oddities in notation
like the following: "$ll-2laa+a^4:\div bb$."

On page 260, in a geometric problem, he writes "$\square AE$" for the
square of the line *AE;* he uses $\stackrel{\frown}{\div}$ for the absolute value of the
difference.

On page 317 his notation for infinite products and infinite series is
as follows:

$$\text{"}1\times1\tfrac{1}{3}\times1\tfrac{1}{24}\times1\tfrac{1}{48}\times1\tfrac{1}{80}\times1\tfrac{1}{120}\times1\tfrac{1}{168}\times\text{ etc."}$$
$$\text{"}1+\tfrac{1}{8}A+\tfrac{1}{24}B+\tfrac{1}{48}C+\tfrac{1}{80}D+\tfrac{1}{120}E+\tfrac{1}{168}F+\text{ etc." ;}$$

on page 322:

$$\text{"}\sqrt{}:2-\sqrt{}:2+\sqrt{}:2+\sqrt{2}\text{" for }\sqrt{2-\sqrt{2+\sqrt{2+\sqrt{2}}}}.$$

On page 332 he uses fractional exponents (Newton having introduced the modern notation for negative and fractional exponents in
1676) as follows:

$$\text{"}\sqrt{}^5:c^5+c^4x-x^5:\qquad\text{or}\qquad\overline{c^5+c^4x-x^5}\big|^{\frac{1}{5}}\text{ ."}$$

The difficulties experienced by the typesetter in printing fractional
exponents are exhibited on page 346, where we find, for example,
"$d\tfrac{1}{2}\ x\tfrac{1}{2}$" for $d^{\frac{1}{2}}x^{\frac{1}{2}}$. On page 123, the factoring of 5940 is shown as
follows:

"11)5)3)3)3)2)2) 5940 (2970(1485(495(165(55(11(1 ."

In a letter to John Collins, Wallis expresses himself on the sign of multiplication: "In printing my things, I had rather you make use of Mr. Oughtred's note of multiplication, $\times$, than that of $*$; the other being the more simple. And if it be thought apt to be mistaken for X, it may [be] helped by making the upper and lower angles more obtuse ⋈."[1] "I do not understand why the sign of multiplication $\times$ should more trouble the convenient placing of the fractions than the other signs $+ - = > ::$."[2]

Wallis, in presenting the history of algebra, stressed the work of Harriot and Oughtred. John Collins took some exception to Wallis' attitude, as is shown in the following illuminating letter. Collins says:[3] "You do not like those words of Vieta in his theorems, ex adjunctione plano solidi, plus quadrato quadrati, etc., and think Mr. Oughtred the first that abridged those expressions by symbols; but I dissent, and tell you 'twas done before by Cataldus, Geysius, and Camillus Gloriosus,[4] who in his first decade of exercises, (not the first tract,) printed at Naples in 1627, which was four years before the first edition of the Clavis, proposeth this equation just as I here give it you, viz., $1ccc+16qcc+41qqc-2304cc-18364qc-133000qq-54505c+3728q+8064N$ aequatur 4608, finds N or a root of it to be 24, and composeth the whole out of it for proof, just in Mr. Oughtred's symbols and method. Cataldus on Vieta came out fifteen years before, and I cannot quote that, as not having it by me. . . . . And as for Mr. Oughtred's method of symbols, this I say to it; it may be proper for you as a commentator to follow it, but divers I know, men of inferior rank that have good skill in algebra, that neither use nor approve it. . . . . Is not A^5 sooner wrote than Aqc? Let A be 2, the cube of 2 is 8, which squared is 64: one of the questions between Magnet Grisio and Gloriosus is whether $64 = A_{cc}$ or A_{qc}. The Cartesian method tells you it is A^6, and decides the doubt."

EXTRACT FROM ACTA ERUDITORUM[5]

197. "Monendum denique, nos in posterum in his Actis usuros esse Signis *Leibnitianis*, ubi cum *Algebraicis* res nobis fuerit, ne typothetis

---

[1] John Wallis to John Collins, July 21, 1668 (S. P. Rigaud, *Correspondence of Scientific Men of the Seventeenth Century*, Vol. II [Oxford, 1841], p. 492).

[2] Wallis to Collins, September 8, 1668 (*ibid.*, p. 494).

[3] Letter to John Wallis, about 1667 (*ibid.*, p. 477-80).

[4] *"Exercitationum Mathematicarum Decas prima*, Nap. 1627, and probably Cataldus' *Transformatio Geometrica*, Bonon. 1612."

[5] Taken from *Acta eruditorum* (Leipzig, 1708), p. 271.

taedia & molestias gratis creemus, utque ambiguitates evitemus. Loco igitur lineolae characteribus supraducendae parenthesin adhibebimus, immo in multiplicatione simplex comma, ex. gr. loco $\overline{\text{V}aa+bb}$ scribemus V($aa+bb$) & pro $\overline{aa+bb}\times c$ ponemus $aa+bb$, c. Divisionem designabimus per duo puncta, nisi peculiaris quaedam circumstantia morem vulgarem adhiberi suaserit. Ita nobis erit $a:b=\dfrac{a}{b}$. Et hinc peculiaribus signis ad denotandam proportionem nobis non erit opus. Si enim fuerit ut a ad b ita c ad d, erit $a:b=c:d$. Quod potentias attinet, $\overline{aa+bb}^{m}$ designabimus per $(aa+bb)^{m}$: unde & $\overset{m}{\text{V}}\overline{aa+bb}$ erit$=(aa+bb)^{1:m}$ & $\overset{m}{V}\overline{aa+bb^{n}}=(aa+bb)^{n:m}$. Nulli vero dubitamus fore, ut Geometrae omnes Acta haec legentes Signorum Leibnitianorum praestantiam animadvertant, & nobiscum in eadem consentiant."

The translation is as follows: "We hereby issue the reminder that in the future we shall use in these *Acta* the Leibnizian signs, where, when algebraic matters concern us, we do not choose the typographically troublesome and unnecessarily repugnant, and that we avoid ambiguity. Hence we shall prefer the parenthesis to the characters consisting of lines drawn above, and in multiplication by all means simply the comma; for example, in place of $\sqrt{aa+bb}$ we write $\sqrt{(aa+bb)}$ and for $\overline{aa+bb}\times c$ we take $aa+bb$, c. Division we mark with two dots, unless indeed some peculiar circumstance directs adherence to the usual practice. Accordingly, we have $a:b=\dfrac{a}{b}$. And it is not necessary to denote proportion by any special sign. For, if a is to b as c is to d, we have $a:b=c:d$. As regards powers, $\overline{aa+bb}^{m}$, we designate them by $(aa+bb)^{m}$; whence also $\overset{m}{V}\overline{aa+bb}$ becomes $=(aa+bb)^{1:m}$ and $\overset{m}{V}\overline{aa+bb^{n}}=(aa+bb)^{n:m}$. We do not doubt that all geometers who read the *Acta* will recognize the excellence of the Leibnizian symbols and will agree with us in this matter."

EXTRACT FROM MISCELLANEA BEROLINENSIA[1]

198. "*Monitum De Characteribus Algebraicis.*—Quoniam variant Geometrae in characterum usu, nova praesertim Analysi inventa; quae res legentibus non admodum provectis obscuritatem parit; ideo è re visum est exponere, quomodo Characteres adhibeantur Leibnitiano more, quem in his Miscellaneis secuturi sumus. *Literae*

---

[1] Taken from *Miscellanea Berolinensia* (1710), p. 155. Article due to G. W. Leibniz.

minusculae a, b, x, y solent significare magnitudines, vel quod idem
est, numeros indeterminatos: Majusculae verô, ut A, B, X, Y puncta
figurarum; ita ab significat factum ex a in b, sed AB rectam à puncto A
ad punctum B ductam. Huic tamen observationi adeo alligati non
sumus, ut non aliquando minusculas pro punctis, majusculas pro
numeris vel magnitudinibus usurpemus, quod facile apparebit ex
modo adhibendi. Solent etiam literae priores, ut a, b, pro quantitatibus cognitis vel saltem determinatis adhiberi, sed posteriores, ut
x, y, pro incognitis vel saltem pro variantibus.

"Interdum pro literis adhibentur Numeri, sed qui idem significant
quod literae, utiliter tamen usurpantur relationis exprimendae gratia.
Exempli causa: Sint binae aequationes generales secundi gradus pro
incognita, x; eas sic exprimere licebit: $10xx \,{\succ}\!\!{+}\!\cdot\, 11x \,{\succ}\!\!{+}\!\cdot\, 12 = 0$ &
$20xx \,{\succ}\!\!{+}\!\cdot\, 21x \,{\succ}\!\!{+}\!\cdot\, 22 = 0$ ita in progressu calculi ex ipsa notatione
apparet quantitatis cujusque relatio; nempe 21 (ex. gr.) per notam
dextram, quae est 1 agnoscitur esse coëfficiens ipsius x simplicis, at
per notam sinistram 2 agnoscitur esse ex. aeq. secunda: sed et servatur
lex quaedam homogeneorum. Et ope harum duarum aequationum
tollendo x, prodit aequatio, in qua similiter se habere oportet 10, 11,
12 et 12, 11, 10; item 20, 21, 22 et 22, 21, 20; et denique 10, 11, 12 se
habent ut. 20, 21, 22. id est si pro 10, 11, 12 substituas 20, 21, 22 et
vice versa manet eadem aequatio; idemque est in caeteris. Tales
numeri tractantur ut literae, veri autem numeri, discriminis causa,
parenthesibus includuntur vel aliter discernuntur. Ita in tali sensu
11.20. significat numeros indefinitos 11 et 20 in se invicem ductos, non
vero significat 220 quasi essent Numeri veri. Sed hic usus ordinarius
non est, rariusque adhibetur.

"*Signa, Additionis* nimirum et *Subtractionis*, sunt ${\succ}\!\!{+}\!\cdot$ plus, —
minus, ${\succ}\!\!{+}\!\cdot$ plus vel minus, $\overline{{\succ}\!\!{+}\!\cdot}$ priori oppositum minus vel plus. At
$(\overline{{\dashv}\!\cdot})$ vel $(\overline{{\dashv}\!\cdot})$ est nota ambiguitatis signorum, independens à
priori; et $((\overline{{\dashv}\!\cdot})$ vel $((\overline{{\dashv}\!\cdot})$ alia independens ab utraque; Differt
autem *Signum ambiguum a Differentia* quantitatum, quae etsi aliquan-
do incerta, non tamen ambigua est. . . . . Sed differentia inter a et
b, significat $a-b$, si a sit majus, et $b-a$ si b sit majus, quod etiam ap-
pellari potest moles ipsius $a-b$, intelligendo (exempli causa) ipsius
${\dashv}\!\cdot 2$ et ipsius $-\!-2$ molem esse eandem, nempe ${\dashv}\!\cdot 2$; ita si $a-b$
vocemus c utique *mol. c*, seu moles ipsius c erit ${\dashv}\!\cdot 2$, quae est quantitas affirmativa sive c sit affirmativa sive negativa, id est, sive sit c
idem quod ${\dashv}\!\cdot 2$, sive c sit idem quod $-\!-2$. Et quantitates duae
diversae eandem molem habentes semper habent idem quadratum.

"*Multiplicationem* plerumque signifare contenti sumus per nudam appositionem: sic ab significat a multiplicari per b, Numeros multiplicantes solemus praefigere, sic $3a$ significat triplum ipsius a interdum tamen punctum vel comma interponimus inter multiplicans et multiplicandum, velut cum 3, 2 significat 3 multiplicari per 2, quod facit 6, si 3 et 2 sunt numeri veri; et AB, CD significat rectam AB duci in rectam CD, atque inde fieri rectangulum. Sed et commata interdum hoc loco adhibemus utiliter, velut a, b✚c, vel AB, CD —┼•EF, id est, a duci in b —┼•c, vel AB in CD —┼•EF; sed de his mox, ubi de vinculis. Porro propria Nota Multiplicationis non solet esse necessaria, cum plerumque appositio, qualem diximus, sufficiat. Si tamen utilis aliquando sit, adhibebitur potius ⌢ quam ⋈, quia hoc ambiguitatem parit, et ita $AB⌢CD$ significat AB duci in CD.

"*Diviso* significatur interdum more vulgari per subscriptionem diuisoris sub ipso dividendo, intercedente linea, ita a dividi per b, significatur vulgo per $\dfrac{a}{b}$; plerumque tamen hoc evitare praestat, efficereque, ut in eadem linea permaneatur, quod sit interpositis duobus punctis; ita ut $a:b$ significat a dividi per b. Quod si $a:b$ rursus dividi debeat per c, poterimus scribere $a:b$, $:c$, vel $(a:b):c$. Etsi enim res hoc casu (sane simplici) facile aliter exprimi posset, fit enim $a:(bc)$ vel $a:bc$ non tamen semper divisio actu ipse facienda est, sed saepe tantum indicanda, et tunc praestat operationis dilatae processum per commata vel parentheses indicari. . . . . Et exponens interdum lineolis includitur hac modo $\boxed{3}(AB$ —┼•$BC)$ quo significatur cubus rectae AB —┼•BC. . . . . a^{e+n} et utiliter interdum lineola subducitur, ne literae exponentiales aliis confundantur; posset etiam scribi $\boxed{e+n}$ a. . . . .

" . . . . ita $\sqrt[3]{(a^3)}$ vel $\sqrt{\boxed{3}}(a^3)$ rursus est a, ... sed $\sqrt[3]{}2$ vel $\sqrt{\boxed{3}}2$ significat radicem cubicam ex eodem numero, et $\sqrt[e]{}2$ vel $\sqrt{\boxed{e}}2$ significat, radicem indeterminati gradus e ex 2 extrahendam. . . . .

"Pro vinculis vulgo solent adhiberi ductus linearum; sed quia lineis una super alia ductis, saepe nimium spatii occupatur, aliasque ob causas commodius plerumque adhibentur commata et parentheses. Sic a, $\overline{b ⋅┼⋅ c}$ idem est quod a, $b⋅┼⋅c$ vel $a(b⋅┼⋅c)$; et $\overline{a⋅┼⋅b}$, $\overline{c⋅┼⋅d}$ idem quod $a⋅┼⋅b$, $c⋅┼⋅d$ vel $(a⋅┼⋅)$ $(c⋅┼⋅)$, id est, $⋅┼⋅a⋅┼⋅b$ multiplicatum per $c⋅┼⋅d$. Et similiter vincula in vinculis exhibentur. Ita a, $\overline{bc⋅┼⋅ef⋅┼⋅g}$ etiam sic exprimetur, $a(bc⋅┼⋅e(f⋅┼⋅g))$ Et a, $\overline{bc⋅┼⋅ef⋅┼⋅g⋅┼⋅hlm}$, n potest etiam

sic exprimi: ⊣⟨ $(a(bc$⟩⊣•$e(f+g))+hlm)n$. Quod de vinculis multiplicationis, idem intelligi potest *de vinculis divisionis*, exempli gratia

$$\dfrac{\dfrac{a}{\dfrac{b}{c}\ \substack{e\\ f\,\dashv\,g}}\ \dashv\ \dfrac{h}{\dfrac{l}{m}}}{n}$$ sic scribetur in una linea

$$(a:((b:c)\dashv\!\!\!-\!\!\!-(e:,f\dashv\!\!\!-\!\!\!-g))\dashv\!\!\!-\!\!\!-h:(l:m)):n$$

nihilque in his difficultatis, modo teneamus, quicquid parenthesin
aliquam implet pro una quantitate haberi, ... Idemque igitur
locum habet in vinculis extractionis radicalis.

Sic $\sqrt{a^4\dashv\!\!\!-\!\!\!-\sqrt{e,f\dashv\!\!\!-\!\!\!-g}}$ idem est quod $\sqrt{(a^4\dashv\!\!\!-\!\!\!-\sqrt{(e(f\dashv\!\!\!-\!\!\!-g)))}}$
vel $\sqrt{(a^4\dashv\!\!\!-\!\!\!-\sqrt{(e,f\dashv\!\!\!-\!\!\!-g))}}$.

Et pro $\dfrac{\sqrt{aa\dashv\!\!-b\sqrt{cc\dashv\!\!-dd}}}{e\dashv\!\!-\sqrt{f\sqrt{gg\dashv\!\!-hh\dashv\!\!-kk}}}$
scribi poterit $\sqrt{(aa\dashv\!\!-b\sqrt{(cc\dashv\!\!-dd))}}:,$
$$e\dashv\!\!-\sqrt{(f\sqrt{(gg\dashv\!\!-hh)\dashv\!\!-kk)}} \ . \ . \ . \ .$$

itaque $a=b$ significat, a, esse equale ipsi b, et a═━b significat a esse
majus quam b, et a ══b significat a esse minus quam b.

"Sed et *proportionalitas* vel analogia de quantitatibus enunciatur,
id est, rationis identitas, quam possumus in Calculo exprimere per
notam aequalitatis, ut non sit opus peculiaribus notis. Itaqua a
esse ad b, sic ut l ad m, sic exprimere poterimus $a:b=l:m$, id est $\dfrac{a}{b}=\dfrac{l}{m}$.
Nota continue proportionalium erit ∺, ita ut ∺ $a.b.c$. etc. sint con-
tinuè proportionales. Interdum nota *Similitudinis* prodest, quae est
$\backsim$, item nota similitudinis aequalitatis simul, seu nota *congruitatis* ≌,
Sic $DEF \backsim PQR$ significabit Triangula haec duo esse similia; at DEF ≌
PQR significabit congruere inter se. Huic si tria inter se habeant
eandem rationem quam tria alia inter se, poterimus hoc exprimere
nota similitudinis, ut $a; b; \ \backsim l; m; $ n quod significat esse a ad b, ut l ad
m, et a ad c ut l ad n, et b ad c ut m ad n. . . . ."

The translation is as follows:

"*Recommendations on algebraic characters.*—Since geometers differ
in the use of characters, especially those of the newly invented analysis, a situation which perplexes those followers who as yet are not
very far advanced, it seems proper to explain the manner of using the
characters in the Leibnizian procedure, which we have adopted in the

Miscellanies. The small letters a, b, x, y, signify magnitudes, or what is the same thing, indeterminate numbers. The capitals on the other hand, as A, B, X, Y, stand for points of figures. Thus ab signifies the result of a times b, but AB signifies the right line drawn from the point A to the point B. We are, however, not bound to this convention, for not infrequently we shall employ small letters for points, capitals for numbers or magnitudes, as will be easily evident from the mode of statement. It is customary, however, to employ the first letters a, b, for known or fixed quantities, and the last letters x, y, for the unknowns or variables.

"Sometimes numbers are introduced instead of letters, but they signify the same as letters; they are convenient for the expression of relations. For example, let there be two general equations of the second degree having the unknown x. It is allowable to express them thus: $10xx + 11x + 12 = 0$ and $20xx + 21x + 22 = 0$. Then, in the progress of the calculation the relation of any quantity appears from the notation itself; thus, for example, in 21 the right digit which is 1 is recognized as the coefficient of x, and the left digit 2 is recognized as belonging to the second equation; but also a certain law of homogeneity is obeyed. And eliminating x by means of these two equations, an equation is obtained in which one has similarity in 10, 11, 12 and 12, 11, 10; also in 20, 21, 22 and 22, 21, 20; and lastly in 10, 11, 12 and 20, 21, 22. That is, if for 10, 11, 12, you substitute 20, 21, 22 and vice versa, there remains the same equation, and so on. Such numbers are treated as if letters. But for the sake of distinction, they are included in parentheses or otherwise marked. Accordingly, $11 \cdot 20$. signifies the indefinite numbers 11 and 20 multiplied one into the other; it does not signify 220 as it would if they were really numbers. But this usage is uncommon and is rarely applied.

"The *signs of addition* and subtraction are commonly $+$ plus, $-$ minus, $\pm$ plus or minus, $\mp$ the opposite to the preceding, minus or plus. Moreover $(\pm)$ or $(\mp)$ is the mark of ambiguity of signs that are independent at the start; and $((\pm))$ or $((\mp))$ are other signs independent of both the preceding. Now the symbol of ambiguity differs from the difference of quantities which, although sometimes undetermined, is not ambiguous. . . . . But $a - b$ signifies the difference between a and b when a is the greater, $b - a$ when b is the greater, and this absolute value (moles) may however be called itself $a - b$, by understanding that the absolute value of $+2$ and -2, for example, is the same, namely, $+2$. Accordingly, if $a - b$ is called c, then *mol. c* or the absolute value of c is $+2$, which is an affirmative quantity whether

c itself is positive or negative; i.e., either c is the same as $+2$, or c is the same as -2. Two different quantities having the same absolute value have always the same square.

"*Multiplication* we are commonly content to indicate by simple apposition: thus, ab signifies a multiplied by b. The multiplier we are accustomed to place in front; thus $3a$ means the triple of a itself. Sometimes, however, we insert a point or a comma between multiplier and multiplicand; thus, for example, $3,2$ signifies that 3 is multiplied by 2, which makes 6, when 3 and 2 are really numbers; and AB,CD signifies the right line AB multiplied into the right line CD, producing a rectangle. But we also apply the comma advantageously in such a case, for example,[1] as $a,b+c$, or $AB,CD+EF;$ i.e., a multiplied into $b+c$, or AB into $CD+EF;$ we speak about this soon, under vinculums. Formerly no sign of multiplication was considered necessary for, as stated above, commonly mere apposition sufficed. If, however, at any time a sign seems desirable use $\frown$ rather than $\times$, because the latter leads to ambiguity; accordingly, $AB\frown CD$ signifies AB times CD.

"*Division* is commonly marked by writing the divisor beneath its dividend, with a line of separation between them. Thus a divided by b is ordinarily indicated by $\dfrac{a}{b}$; often, however, it is preferable to avoid this notation and to arrange the signs so that they are brought into one and the same line; this may be done by the interposition of two points; thus $a:b$ signifies a divided by b. If $a:b$ in turn is to be divided by c, we may write $a:b$, $:c$, or $(a:b):c$. However, this should be expressed more simply in another way, namely, $a:(bc)$ or $a:bc$, for the division cannot always be actually carried out, but can be only indicated, and then it becomes necessary to mark the delayed process of the operation by commas or parentheses. . . . . Exponents are frequently inclosed by lines in this manner $\boxed{3}$ $(AB+BC)$, which means the cube of the line $AB+BC$ . . . . ; the exponents of a^{l+n} may also be advantageously written between the lines, so that the literal exponents will not be confounded with other letters; thus it may be written $\boxed{l+n}$ a. . . . . From $\sqrt[3]{}(a^3)$ or $\sqrt{}\boxed{3}(a^3)$ arises a . . . . ; but $\sqrt[3]{}2$ or $\sqrt{}\boxed{3}$ 2 means the cube root of the same number, and $\sqrt{}2$ or $\sqrt{}\boxed{e}2$ signifies the extraction of a root of the indeterminate order e. . . . .

"For aggregation it is customary to resort to the drawing of

---

[1] A similar use of the comma to separate factors and at the same time express aggregation occurs earlier in Hérigone (see § 189).

lines, but because lines drawn one above others often occupy too much space, and for other reasons, it is often more convenient to introduce commas and parentheses. Thus a, $\overline{b+c}$ is the same as a, $b+c$ or $a(b+c)$; and $\overline{a+b}$, $\overline{c+d}$ is the same as $a+b$, $c+d$, or $(a+b)$ $(c+d)$, i.e., $+a+b$ multiplied by $c+d$. And, similarly, vinculums are placed under vinculums. For example, a, $\overline{bc+e\overline{f+g}}$ is expressed also thus, $a(bc+e(f+g))$, and a, $\overline{bc+e\overline{f+g}+hlm,n}$ may be written also $+(a(bc+e(f+g))+hlm)n$. What relates to vinculums in multiplication applies to vinculums in division. For example,

$$\frac{\dfrac{a}{\dfrac{b}{c}+\dfrac{e}{f+g}}+\dfrac{h}{\dfrac{l}{m}}}{n} \qquad \text{may be written in one line thus:}$$

$$(a:\ ((b:c)+(e:,f+g))+h:(l:m)):n\ ,$$

and there is no difficulty in this, as long as we observe that whatever fills up a given parenthesis be taken as one quantity. . . . . The same is true of vinculums in the extraction of roots. Thus $\sqrt{a^4+\sqrt{e,\overline{f+g}}}$ is the same as $\sqrt{(a^4+\sqrt{(e(f+g))})}$ or $\sqrt{(a^4+\sqrt{(e,\ f+g)})}$. And for $\dfrac{\sqrt{aa+b\sqrt{cc+dd}}}{e+\sqrt{f\sqrt{gg+hh}+kk}}$ one may write $\sqrt{(aa+b\sqrt{(cc+dd)}):,e+}$

$\sqrt{(f\sqrt{(gg+hh)}+kk)}$. Again $a=b$ signifies that a is equal to b, and $a\rightleftharpoons b$ signifies that a is greater than b, and $a\rightleftharpoons b$ that a is less than b. Also proportionality or analogia of quantities, i.e., the identity of ratio, may be represented; we may express it in the calculus by the sign of equality, for there is no need of a special sign. Thus, we may indicate that a is to b as l is to m by $a:b=l:m$, i.e., $\dfrac{a}{b}=\dfrac{l}{m}$. The sign for continued proportion is $\div\div$, so that $\div\div$ a, b, c, and d are continued proportionals.

"There is adopted a sign for similitude; it is $\backsim$; also a sign for both similitude and equality, or a sign of congruence, $\eqsim$ accordingly, $DEF \backsim PQR$ signifies that the two triangles are similar; but $DEF \eqsim PRQ$ marks their congruence. Hence, if three quantities have to one another the same ratio that three others have to one another, we may mark this by a sign of similitude, as $a; b; c \backsim l; m; n$ means that a is to b as l is to m, and a is to c as l is to n, and b is to c as m is to n. . . . ."

In the second edition of the *Miscellanea Berolinensia*, of the year

1749, the typographical work is less faulty than in the first edition of
1710; some slight errors are corrected, but otherwise no alterations
are made, except that Harriot's signs for "greater than" and "less
than" are adopted in 1749 in place of the two horizontal lines of unequal length and thickness, given in 1710, as shown above.

199. *Conclusions.*—In a letter to Collins, John Wallis refers to a
change in algebraic notation that occurred in England during his
lifetime: "It is true, that as in other things so in mathematics, fashions
will daily alter, and that which Mr. Oughtred designed by great
letters may be now by others be designed by small; but a mathematician will, with the same ease and advantage, understand Ac, and a^3
or aaa."[1] This particular diversity is only a trifle as compared with
what is shown in a general survey of algebra in Europe during the
fifteenth, sixteenth, and seventeenth centuries. It is discouraging to
behold the extreme slowness of the process of unification.

In the latter part of the fifteenth century $\tilde{p}$ and $\tilde{m}$ became symbols
for "plus" and "minus" in France (§ 131) and Italy (§ 134). In Germany the Greek cross and the dash were introduced (§ 146). The two
rival notations competed against each other on European territory
for many years. The $\tilde{p}$ and $\tilde{m}$ never acquired a foothold in Germany.
The German $+$ and $-$ gradually penetrated different parts of Europe.
It is found in Scheubel's *Algebra* (§ 158), in Recorde's *Whetstone of
Witte*, and in the *Algebra* of Clavius. In Spain the German signs occur
in a book of 1552 (§ 204), only to be superseded by the $\tilde{p}$ and $\tilde{m}$ in
later algebras of the sixteenth century. The struggle lasted about
one hundred and thirty years, when the German signs won out everywhere except in Spain. Organized effort, in a few years, could have
ended this more than a century competition.

If one takes a cross-section of the notations for radical expressions
as they existed in algebra at the close of the sixteenth century, one
finds four fundamental symbols for indicating roots, the letters R and l,
the radical sign $\sqrt{}$ proper and the fractional exponent. The letters
R and l were sometimes used as capitals and sometimes as small
letters (§§ 135, 318–22). The student had to watch his step, for at times
these letters were used to mark, not roots, but the unknown quantity
x and, perhaps, also its powers (§ 136). When R stood for "root," it
became necessary to show whether the root of one term or of several
terms was meant. There sprang up at least seven different symbols
for the aggregation of terms affected by the R, namely, one of Chuquet
(§ 130), one of Pacioli (§ 135), two of Cardan (§ 141), the round paren-

---

[1] See Rigaud, *op. cit.*, Vol. II, p. 475.

thesis of Tartaglia (§ 351), the upright and inverted letter L of Bombelli
(§ 144), and the r $bin.$ and r $trinomia$ of A. V. Roomen (§ 343). There
were at least five ways of marking the orders of the root, those of
Chuquet (§ 130), De la Roche (§ 132), Pacioli (§ 135), Ghaligai
(§ 139), and Cardan (Fig. 46). With A. M. Visconti[1] the signs
$R.ce$ $cu.$ meant the "sixth root"; he used the multiplicative principle,
while Pacioli used the additive one in the notation of radicals. Thus
the letter R carried with it at least fifteen varieties of usage. In connection with the letter l, signifying $latus$ or "root," there were at least
four ways of designating the orders of the roots and the aggregation
of terms affected (§§ 291, 322). A unique line symbolism for roots of
different orders occurs in the manuscripts of John Napier (§ 323).

The radical signs for cube and fourth root had quite different
shapes as used by Rudolff (§§ 148, 326) and Stifel (§ 153). Though
clumsier than Stifel's, the signs of Rudolff retained their place in some
books for over a century (§ 328). To designate the order of the roots,
Stifel placed immediately after the radical sign the German abbreviations of the words $zensus$, $cubus$, $zensizensus$, $sursolidus$, etc. Stevin
(§ 163) made the important innovation of numeral indices. He placed
them within a circle. Thus he marked cube root by a radical sign
followed by the numeral 3 coraled in a circle. To mark the root of an
aggregation of terms, Rudolff (§§ 148, 348) introduced the dot placed
after the radical sign; Stifel sometimes used two dots, one before the
expression, the other after. Stevin (§§ 163, 343) and Digges (§§ 334,
343) had still different designations. Thus the radical sign carried
with it seven somewhat different styles of representation. Stevin
suggested also the possibility of fractional exponents (§ 163), the
fraction being placed inside a circle and before the radicand.

Altogether there were at the close of the sixteenth century twenty-
five or more varieties of symbols for the calculus of radicals with which
the student had to be familiar, if he desired to survey the publications
of his time.

Lambert Lincoln Jackson makes the following historical observations: "For a hundred years after the first printed arithmetic many
writers began their works with the line-reckoning and the Roman
numerals, and followed these by the Hindu arithmetic. The teaching
of numeration was a formidable task, since the new notation was so
unfamiliar to people generally."[2] In another place (p. 205) Jackson

---

[1] "Abbreviationes," *Practica numerorum, et mensurarum* (Brescia, 1581).

[2] *The Educational Significance of Sixteenth Century Arithmetic* (New York,
1906), p. 37, 38.

states: "Any phase of the growth of mathematical notation is an interesting study, but the chief educational lesson to be derived is that notation always grows too slowly. Older and inferior forms possess remarkable longevity, and the newer and superior forms appear feeble and backward. We have noted the state of transition in the sixteenth century from the Roman to the Hindu system of characters, the introduction of the symbols of operation, $+$, $-$, and the slow growth toward the decimal notation. The moral which this points for twentieth-century teachers is that they should not encourage history to repeat itself, but should assist in hastening new improvements."

The historian Tropfke expresses himself as follows: "How often has the question been put, what further achievements the patriarchs of Greek mathematics would have recorded, had they been in possession of our notation of numbers and symbols! Nothing stirs the historian as much as the contemplation of the gradual development of devices which the human mind has thought out, that he might approach the truth, enthroned in inaccessible sublimity and in its fullness always hidden from earth. Slowly, only very slowly, have these devices become what they are to man today. Numberless strokes of the file were necessary, many a chink, appearing suddenly, had to be mended, before the mathematician had at hand the sharp tool with which he could make a successful attack upon the problems confronting him. The history of algebraic language and writing presents no uniform picture. An assemblage of conscious and unconscious innovations, it too stands subject to the great world-law regulating living things, the principle of selection. Practical innovations make themselves felt, unsuitable ones sink into oblivion after a time. The force of habit is the greatest opponent of progress. How obstinate was the struggle, before the decimal division met with acceptation, before the proportional device was displaced by the equation, before the Indian numerals, the literal coefficients of Vieta, could initiate a world mathematics."[1]

Another phase is touched by Treutlein: "Nowhere more than in mathematics is intellectual content so intimately associated with the form in which it is presented, so that an improvement in the latter may well result in an improvement of the former. Particularly in arithmetic, a generalization and deepening of concept became possible only after the form of presentation had been altered. The history of our science supplies many examples in proof of this. If the Greeks had been in possession of our numeral notation, would their

---

[1] Tropfke, *Geschichte der Elementar-Mathematik*, Vol. II (Leipzig, 1921), p. 4, 5.

mathematics not present a different appearance? Would the binomial theorem have been possible without the generalized notation of powers? Indeed could the mathematics of the last three hundred years have assumed its degree of generality without Vieta's pervasive change of notation, without his introduction of general numbers? These instances, to which others from the history of modern mathematics could be added, show clearly the most intimate relation between substance and form."[1]

B. SPECIAL SURVEY OF THE USE OF NOTATIONS

SIGNS FOR ADDITION AND SUBTRACTION

200. *Early symbols.*—According to Hilprecht,[2] the early Babylonians had an ideogram, which he transliterates *LAL*, to signify "minus." In the hieratic papyrus of Ahmes and, more clearly in the hieroglyphic translation of it, a pair of legs walking forward is the sign of addition; away, the sign of subtraction.[3] In another Egyptian papyrus kept in the Museum of Fine Arts in Moscow,[4] a pair of legs walking forward has a different significance; there it means to square a number.

Figure 99, translated, is as follows (reading the figure from right to left):

"$\frac{2}{3}$ added and $\frac{1}{3}$ [of this sum] taken away, 10 remains.
Make $\frac{1}{10}$ of this 10: the result is 1, the remainder 9.
$\frac{2}{3}$ of it, namely, 6, added to it; the total is 15. $\frac{1}{3}$ of it is 5.
When 5 is taken away, the remainder is 10."

In the writing of unit fractions, juxtaposition meant addition, the unit fraction of greatest value being written first and the others in descending order of magnitude.

While in Diophantus addition was expressed merely by juxtaposition (§ 102), a sporadic use of a slanting line / for addition, also a semi-elliptical curve Ɔ for subtraction, and a combination of the two

[1] Treutlein, "Die deutsche Coss," *Abhandlungen z. Geschichte der Mathematik*, Vol. II (Leipzig, 1879), p. 27, 28.

[2] H. V. Hilprecht, *Babylonian Expedition: Mathematical etc. Tablets* (Philadelphia, 1906), p. 23.

[3] A. Eisenlohr, *op. cit.* (2d ed.), p. 46 (No. 28), 47, 237. See also the improved edition of the Ahmes papyrus, *The Rhind Mathematical Papyrus*, by T. Eric Peet (London, 1923), Plate J, No. 28; also p. 63.

[4] Peet, *op. cit.*, p. 20, 135: *Ancient Egypt* (1917), p. 101.

⋂ for the total result has been detected in Greek papyri.[1] Diophantus' sign for subtraction is well known (§ 103). The Hindus had no mark for addition (§ 106) except that, in the Bakhshali *Arithmetic, yu* is used for this purpose (§ 109). The Hindus distinguished negative quantities by a dot (§§ 106, 108), but the Bakhshali *Arithmetic* uses the sign + for subtraction (§ 109). The Arab al-Qalasâdî in the fifteenth century indicated addition by juxtaposition and had a special sign for subtraction (§ 124). The Frenchman Chuquet (1484), the Italian Pacioli (1494), and the sixteenth-century mathematicians in Italy used $\tilde{p}$ or p: for plus and $\tilde{m}$ or m: for "minus" (§§ 129, 134).

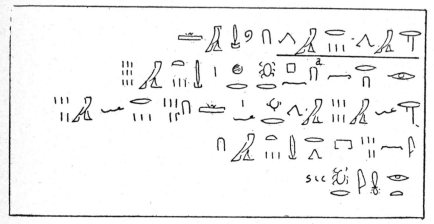

Fig. 99.—From the hieroglyphic translation of the Ahmes papyrus, Problem 28, showing a pair of legs walking forward, to indicate addition, and legs walking away, to indicate subtraction. (Taken from T. E. Peet, *The Rhind Mathematical Papyrus*, Plate J, No. 28.)

201. *Origin and meanings of the signs* + *and* −.—The modern algebraic signs + and − came into use in Germany during the last twenty years of the fifteenth century. They are first found in manuscripts. In the Dresden Library there is a volume of manuscripts, C. 80. One of these manuscripts is an algebra in German, written in the year 1481,[2] in which the minus sign makes its first appearance in

[1] H. Brugsch, *Numerorum apud veteres Aegyptios demoticorum doctrina. Ex papyris* (Berlin, 1849), p. 31; see also G. Friedlein, *Zahlzeichen und das elementare Rechnen* (Erlangen, 1869), p. 19 and Plate I.

[2] E. Wappler, *Abhandlungen zur Geschichte der Mathematik*, Vol. IX (1899), p. 539, n. 2; Wappler, *Zur Geschichte der deutschen Algebra im 15. Jahrhundert, Zwickauer Gymnasialprogramm von 1887*, p. 11–30 (quoted by Cantor, *op. cit.*, Vol. II [2d ed., 1900], p. 243, and by Tropfke, *op. cit.*, Vol. II [2d ed., 1921], p. 13).

algebra (Fig. 100); it is called *minnes*. Sometimes the − is placed after the term affected. In one case −4 is designated "4 das ist −." Addition is expressed by the word *vnd*.

In a Latin manuscript in the same collection of manuscripts, C. 80, in the Dresden Library, appear both symbols + and − as signs of operation (Fig. 101), but in some rare cases the + takes the place of *et* where the word does not mean addition but the general "and."[1] Repeatedly, however, is the word *et* used for addition.

It is of no little interest that J. Widman, who first used the + and − in print, studied these two manuscripts in the manuscript volume C. 80 of the Dresden Library and, in fact, annotated them. One of his marginal notes is shown in Figure 102. Widman lectured at the University of Leipzig, and a manuscript of notes taken in 1486 by a pupil is preserved in the Leipzig Library (Codex Lips. 1470).[2] These notes show a marked resemblance to the two Dresden manuscripts.

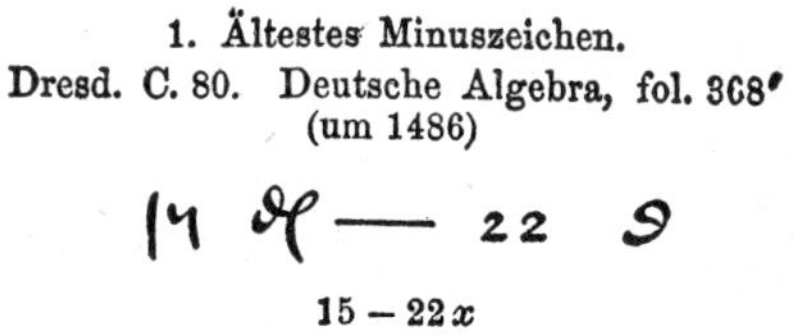

1. Ältestes Minuszeichen.
Dresd. C. 80. Deutsche Algebra, fol. 3C8′
(um 1486)

$$15 - 22x$$

Fig. 100.—Minus sign in a German MS, C. 80, Dresden Library. (Taken from J. Tropfke, *op. cit.*, Vol. II [1921], p. 14.)

The view that our + sign descended from one of the florescent forms for *et* in Latin manuscripts finds further support from works on

2. Ältestes Pluszeichen.
Dresd. C. 80. Lat. Algebra, fol. 350′
(um 1486)

$$x^3 + 2x^2$$

4. Dresd. C. 80.
Lateinische Algebra, fol. 352ᵗ

$$10 - x$$

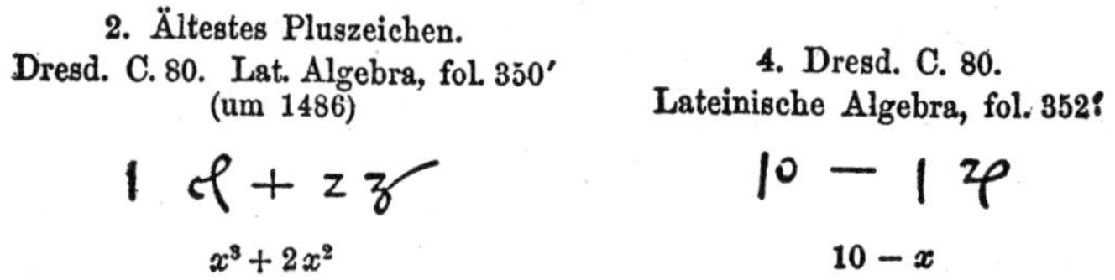

Fig. 101.—Plus and minus signs in a Latin MS, C. 80, Dresden Library. (Taken from Tropfke, *op. cit.*, Vol. II [2d ed., 1921], p. 14.)

paleography. J. L. Walther[3] enumerates one hundred and two different abbreviations found in Latin manuscripts for the word *et;* one of these, from a manuscript dated 1417, looks very much like the modern

[1] Wappler, *Programm* (1887), p. 13, 15.

[2] Wappler, *Zeitschrift Math. u. Physik*, Vol. XLV (Hist. lit. Abt., 1900), p. 7–9.

[3] *Lexicon diplomaticvm abbreviationes syllabarvm et vocvm in diplomatibvs et codicibvs a secvlo VIII. ad XVI. . . . . . Studio Joannis Lvdolfi VValtheri . . . .* (Ulmae, 1756), p. 456–59.

+. The downward stroke is not quite at right angles to the horizontal stroke, thus +.

Concerning the origin of the minus sign ($-$), we limit ourselves to the quotation of a recent summary of different hypotheses: "One knows nothing certain of the origin of the sign $-$; perhaps it is a simple bar used by merchants to separate the indication of the tare, for a long time called *minus*, from that of the total weight of merchandise; according to L. Rodet (*Actes Soc. philol. Alençon*, Vol. VIII [1879], p. 105) this sign was derived from an Egyptian hieratic sign. One has also sought the origin of our sign $-$ in the sign employed by Heron and Diophantus and which changed to $\top$ before it became $-$. Others still have advanced the view that the sign $-$ has its origin in the ὀβελός of the Alexandrian grammarians. None of these hypotheses is supported by plausible proof."[1]

Zusatz von WIDMANN.
5. Dresd. C. 80, fol. 349'
(um 1486)

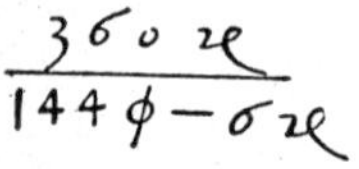

$$\frac{360\,x}{144 - 6x}$$

FIG. 102.—Widman's marginal note to MS C. 80, Dresden Library. (Taken from Tropfke.)

202. The sign $+$ first occurs in print in Widman's book in the question: "Als in diesē exēpel 16 ellñ pro 9 fl $\frac{1}{3}$ vñ $\frac{1}{4}+\frac{1}{5}$ eynss fl wy kūmē 36 ellñ máchss alsso Addir $\frac{1}{3}$ vñ $\frac{1}{4}$ vñ $\frac{1}{5}$ zu sāmen kumpt $\frac{47}{60}$ eynss fl Nu secz vñ machss nach der regl vñ kūmē 22 fl $\frac{1}{80}$ eynsz fl dz ist gerad 3 hlr in gold."[2] In translation: "Thus in this example, 16 ells [are bought] for 9 florins [and] $\frac{1}{3}$ and $\frac{1}{4}+\frac{1}{5}$ of a florin, what will 36 ells cost? Proceed thus: Add $\frac{1}{3}$ and $\frac{1}{4}$ and $\frac{1}{5}$ obtaining $\frac{47}{60}$ of a florin. Now put down and proceed according to the rule and there results 22 florin, and $\frac{1}{80}$ of a florin which is exactly 3 heller in gold." The $+$ in this passage stands for "and." Glaisher considers this $+$ a misprint for *vñ* (the contraction for *vnnd*, our "and"), but there are other places in Widman where $+$ clearly means "and," as we shall see later. There is no need of considering this $+$ a misprint.

On the same leaf Widman gives a problem on figs. We quote from the 1498 edition (see also Fig. 54 from the 1526 edition):

[1] *Encyclopédie des scien. math.*, Tome I, Vol. I (1904), p. 31, 32, n. 145.

[2] Johann Widman, *Behēde vnd hubsche Rechenung auff allen Kauffmanschafft* (Leipzig, 1489), unnumbered p. 87. Our quotation is taken from J. W. L. Glaisher's article, "On the Early History of Signs + and − and on the Early German Arithmeticians," *Messenger of Mathematics*, Vol. LI (1921–22), p. 6. Extracts from Widman are given by De Morgan, *Transactions of the Cambridge Philosophical Society*, Vol. XI, p. 205, and by Boncompagni, *Bulletino*, Vol. IX, p. 205.

"*Veygen.*—Itm̄ Eyner Kaufft 13 lagel veygen vn̄ nympt ye 1 ct pro 4 fl ½ ort Vnd wigt itliche lagel als dan hye nochuolget. vn̄ ich wolt wissen was an der sum brecht

	4+ 5	Wiltu dass
	4−17	wyssen der
	3+36	dess gleichn̄
	4−19	Szo sum −
	3+44	mir die ct
	3+22	Vnd lb vn̄
Czentner	3−11 lb	was − ist
	3+50	dz ist mi⁹
	4−16	dz secz besū
	3+44	der vn̄ wer
	3+29	dē 4539
	3−12	lb (So du
	3+ 9	die ct zcu lb

gemacht hast Vnnd das + das ist mer dar zu addirest) vnd 75 min⁹ Nu solt du fur holcz abschlahn̄ albeg fur eyn lagel 24 lb vn̄ dz ist 13 mol 24· vn̄ macht 312 lb dar zu addir dz − dz ist 75 lb vnnd werden 387 Die subtrahir vonn 4539 Vnnd pleybn̄ 4152 lb Nu sprich 100 lb das ist 1 ct pro 4 fl ½ wie kummen 4152 lb vnd kūmen 171 fl 5 ss 4 hlr ⅘ Vn̄ ist recht gemacht."[1]

In free translation the problem reads: "*Figs.*—Also, a person buys 13 barrels of figs and receives 1 centner for 4 florins and ½ ort (4⅛ florins), and the weight of each barrel is as follows: 4 ct+5 lb, 4 ct−17 lb, 3 ct+36 lb, 4 ct−19 lb, 3 ct+44 lb, 3 ct+22 lb, 3 ct−11 lb, 3 ct+50 lb, 4 ct−16 lb, 3 ct+44 lb, 3 ct+29 lb, 3 ct−12 lb, 3 ct+9 lb; and I would know what they cost. To know this or the like, sum the ct and lb and what is −, that is minus, set aside, and they become 4539 lb (if you bring the centners to lb and thereto add the +, that is more) and 75 minus. Now you must subtract for the wood 24 lb for each barrel and 13 times 24 is 312 to which you add the −, that is 75 lb and it becomes 387 which subtract from 4,539 and there remains 4152 lb. Now say 100 lb that is 1 ct for 4⅛ fl, what do 4152 lb come to, and they come to 171 fl 5 ss 4⅘ hlr which is right."

Similar problems are given by Widman, relating to pepper and soap. The examination of these passages has led to divergent opinions on the original significance of the + and −. De Morgan suspected

[1] The passage is quoted and discussed by Eneström, *Bibliotheca mathematica*, Vol. IX (3d ser., 1908–9), p. 156, 157, 248; see also *ibid.*, Vol. VIII, p. 199.

that they were warehouse marks, expressing excess or deficiency in weights of barrels of goods.[1] M. W. Drobisch,[2] who was the first to point out the occurrence of the signs + and − in Widman, says that Widman uses them in passing, as if they were sufficiently known, merely remarking, "Was − ist das ist minus vnd das + das ist mer." C. I. Gerhardt,[3] like De Morgan, says that the + and − were derived from mercantile practice.

But Widman assigned the two symbols other significations as well. In problems which he solved by false position the error has the + or − sign prefixed.[4] The − was used also to separate the terms of a proportion. In "11630−198   4610−78" it separates the first and second and the third and fourth terms. The "78" is the computed term, the fractional value of the fourth term being omitted in the earlier editions of Widman's arithmetic. The sign + occurs in the heading "Regula augmenti + decrementi" where it stands for the Latin *et* ("and"), and is not used there as a mathematical symbol. In another place Widman gives the example, "Itm̄ eyner hat kaufft 6 eyer−2 ℒ pro 4 ℒ+1 ey" ("Again, someone has bought 6 eggs− 2 ℒ for 4 ℒ+1 egg"), and asks for the cost of one egg. Here the − is simply a dash separating the words for the goods from the price. From this and other quotations Glaisher concludes that Widman used + and − "in all the ways in which they are used in algebra." But we have seen that Widman did not restrict the signs to that usage; the + was used for "and" when it did not mean addition; the − was used to indicate separation. In other words, Widman does not restrict the use of + and − to the technical meanings that they have in algebra.

203. In an anonymous manuscript,[5] probably written about the time when Widman's arithmetic appeared, use is made of symbolism in the presentation of algebraic rules, in part as follows:

"Conditiones circa + vel − in additione

$$\left.\begin{array}{l}+\ et\ +\\ -\ et\ -\end{array}\right\rangle facit\ \left.\begin{array}{l}+\\ -\end{array}\right\rangle \begin{array}{l} addatur\ non\ sumendo\ respectum\ quis\ numerus\ sit \\ superior. \end{array}$$

---

[1] De Morgan, *op. cit.*, Vol. XI, p. 206.

[2] *De Joannis Widmanni . . . . compendio* (Leipzig, 1840), p. 20 (quoted by Glaisher, *op. cit.*, p. 9).

[3] *Geschichte der Mathematik in Deutschland* (1877), p. 36: ". . . . dass diese Zeichen im kaufmännischen Verkehr üblich waren."

[4] Glaisher, *op. cit.*, p. 15.

[5] *Regulae Cosae vel Algebrae*, a Latin manuscript, written perhaps about 1450, but "surely before 1510," in the Vienna Library.

Si fuerit $\begin{cases} + \text{ et } - \\ - \text{ et } + \end{cases}$ simpliciter subtrahatur minor numerus a majori et residuo sua ascribatur nota,"[1] and similarly for subtraction. This manuscript of thirty-three leaves is supposed to have been used by Henricus Grammateus (Heinrich Schreiber) in the preparation of his *Rechenbuch* of 1518 and by Christoff Rudolff in his *Coss* of 1525.

Grammateus[2] in 1518 restricts his use of + and − to technical algebra: "Vnd man braucht solche zaichen als + ist vnnd, − mynnder" ("And one uses such signs as + [which] is 'and,' − 'less' "). See Figure 56 for the reproduction of this passage from the edition of 1535. The two signs came to be used freely in all German algebras, particularly those of Grammateus, Rudolff (1525), Stifel (1544), and in Riese's manuscript algebra (1524). In a text by Eysenhut[3] the + is used once in the addition of fractions; both + and − are employed many times in the *regula falsi* explained at the end of the book.

Arithmetics, more particularly commercial arithmetics, which did not present the algebraic method of solving problems, did not usually make use of the + and − symbols. L. L. Jackson says: "Although the symbols + and − were in existence in the fifteenth century, and appeared for the first time in print in Widman (1489), as shown in the illustration (p. 53), they do not appear in the arithmetics as signs of operation until the latter part of the sixteenth century. In fact, they did not pass from algebra to general use in arithmetic until the nineteenth century."[4]

204. *Spread of the + and − symbols.*—In Italy the symbols $\bar{p}$ and $\bar{m}$ served as convenient abbreviations for "plus" and "minus" at the end of the fifteenth century and during the sixteenth. In 1608 the German Clavius, residing in Rome, used the + and − in his algebra brought out in Rome (see Fig. 66). Camillo Gloriosi adopted them in his *Ad theorema geometricum* of 1613 and in his *Exercitationes mathematicae, decas I* (Naples, 1627) (§ 196). The + and − signs were used by B. Cavalieri (see Fig. 86) as if they were well known. The +

---

[1] C. I. Gerhardt, "Zur Geschichte der Algebra in Deutschland," *Monatsberichte der k. pr. Akademie d. Wissenschaften z. Berlin* (1870), p. 147.

[2] Henricus Grammateus, *Ayn New Kunstlich Buech* (Nürnberg: Widmung, 1518; publication probably in 1521). See Glaisher, *op. cit.*, p. 34.

[3] *Ein künstlich rechenbuch auff Zyffern / Lini vnd Wälschen Practica* (Augsburg, 1538). This reference is taken from Tropfke, *op. cit.*, Vol. I (2d ed., 1921), p. 58.

[4] *The Educational Significance of Sixteenth Century Arithmetic* (New York, 1906), p. 54.

and — were used in England in 1557 by Robert Recorde (Fig. 71) and in Holland in 1637 by Gillis van der Hoecke (Fig. 60). In France and Spain the German + and —, and the Italian $\tilde{p}$ and $\tilde{m}$, came in sharp competition. The German Scheubel in 1551 brought out at Paris an algebra containing the + and — (§ 158); nevertheless, the $\tilde{p}$ and $\tilde{m}$ (or the capital letters P, M) were retained by Peletier (Figs. 80, 81), Buteo (Fig. 82), and Gosselin (Fig. 83). But the adoption of the German signs by Ramus and Vieta (Figs. 84, 85) brought final victory for them in France. The Portuguese P. Nuñez (§ 166) used in his algebra (published in the Spanish language) the Italian $\tilde{p}$ and $\tilde{m}$. Before this, Marco Aurel,[1] a German residing in Spain, brought out an algebra at Valencia in 1552 which contained the + and — and the symbols for powers and roots found in Christoff Rudolff (§ 165). But ten years later the Spanish writer Pérez de Moya returned to the Italian symbolism[2] with its $\tilde{p}$ and $\tilde{m}$, and the use of $n.$, $co.$, ce, cu, for powers and r, rr, rrr for roots. Moya explains: "These characters I am moved to adopt, because others are not to be had in the printing office."[3] Of English authors[4] we have found only one using the Italian signs for "plus" and "minus," namely, the physician and mystic, Robert Fludd, whose numerous writings were nearly all published on the Continent. Fludd uses P and M for "plus" and "minus."

The + and —, and the $\tilde{p}$ and $\tilde{m}$, were introduced in the latter part of the fifteenth century, about the same time. They competed with each other for more than a century, and $\tilde{p}$ and $\tilde{m}$ finally lost out in the early part of the seventeenth century.

205. *Shapes of the plus sign.*—The plus sign, as found in print, has had three principal varieties of form: (1) the Greek cross +, as it is found in Widman (1489); (2) the Latin cross, $\dagger$ more frequently placed horizontally, —|— or —+—; (3) the form ⊁, or occasionally some form still more fanciful, like the eight-pointed Maltese cross ⊰⊱, or a cross having four rounded vases with tendrils drooping from their edges.

The Greek cross, with the horizontal stroke sometimes a little

---

[1] *Libro primero de Arithmetica Algebratica* . . . . *por Marco Aurel, natural Aleman* (Valencia, 1552).

[2] J. Rey Pastor, *Los mathemáticos españoles del siglo XVI* (Oviedo, 1913), p. 38.

[3] "Estos characteres me ha parecido poner, porque no auia otros en la imprenta" (*Ad theorema geometricvm, á nobilissimo viro propositum, Joannis Camilli Gloriosi responsum* [Venetiis, 1613], p. 26).

[4] See C. Henry, *Revue archeologique*, N.S., Vol. XXXVII, p. 329, who quotes from Fludd, *Utriusque cosmi* . . . . *Historia* (Oppenheim, 1617).

longer than the vertical one, was introduced by Widman and has been the prevailing form of plus sign ever since. It was the form commonly used by Grammateus, Rudolff, Stifel, Recorde, Digges, Clavius, Dee, Harriot, Oughtred, Rahn, Descartes, and most writers since their time.

206. The Latin cross, placed in a horizontal position, thus ─┼─, was used by Vieta[1] in 1591. The Latin cross was used by Romanus,[2] Hunt,[3] Hume,[4] Hérigone,[5] Mengoli,[6] Huygens,[7] Fermat,[8] by writers in the *Journal des* Sçavans,[9] Dechales,[10] Rolle,[11] Lamy,[12] L'Hospital,[13] Swedenborg,[14] Pardies,[15] Kresa,[16] Belidor,[17] De Moivre,[18] and Michelsen.[19] During the eighteenth century this form became less common and finally very rare.

Sometimes the Latin cross receives special ornaments in the form of a heavy dot at the end of each of the three shorter arms, or in the form of two or three prongs at each short arm, as in H. Vitalis.[20] A very ostentatious twelve-pointed cross, in which each of the four equal

[1] Vieta, *In artem analyticam isagoge* (Turonis, 1591).

[2] *Adriani Romani Canon triangvlorvm sphaericorum* . . . . (Mocvntiae, 1609).

[3] Nicolas Hunt, *The Hand-Maid to Arithmetick* (London, 1633), p. 130.

[4] James Hume, *Traité de l'algebre* (Paris, 1635), p. 4.

[5] P. Herigone, "Explicatis notarvm," *Cvrsvs mathematicvs*, Vol. I (Paris, 1634).

[6] Petro Mengoli, *Geometriae speciosae elementa* (Bologna, 1659), p. 33.

[7] *Christiani Hvgenii Holorogivm oscillatorivm* (Paris, 1673), p. 88.

[8] P. de Fermat, *Diophanti Alexandrini Arithmeticorum libri sex* (Toulouse, 1670), p. 30; see also Fermat, *Varia opera* (1679), p. 5.

[9] *Op. cit.* (Amsterdam, 1680), p. 160; *ibid.* (1693), p. 3, and other places.

[10] K. P. Claudii Francisci Milliet Dechales, *Mundus mathematicus*, Vol. I (Leyden, 1690), p. 577.

[11] M. Rolle, *Methode pour resoudre les egalitez de tous les degreez* (Paris, 1691) p. 15.

[12] Bernard Lamy, *Elemens des mathematiques* (3d ed.; Amsterdam, 1692), p. 61.

[13] L'Hospital, *Acta eruditorum* (1694), p. 194; *ibid.* (1695), p. 59; see also other places, for instance, *ibid.* (1711), Suppl., p. 40.

[14] Emanuel Swedenborg, *Daedalus Hyperborens* (Upsala, 1716), p. 5; reprinted in *Kungliga Vetenskaps Societetens i Upsala Tvåhundr aårsminne* (1910).

[15] *Œuvres du R. P. Pardies* (Lyon, 1695), p. 103.

[16] J. Kresa, *Analysis speciosa trigonometriae sphericae* (Prague, 1720), p. 57.

[17] B. F. de Belidor, *Nouveau cours de mathématique* (Paris, 1725), p. 10.

[18] A. de Moivre, *Miscellanea analytica* (London, 1730), p. 100.

[19] J. A. C. Michelsen, *Theorie der Gleichungen* (Berlin, 1791).

[20] "Algebra," *Lexicon mathematicum authore Hieronymo Vitali* (Rome, 1690).

arms has three prongs, is given by Carolo Renaldini.[1] In seventeenth-
and eighteenth-century books it is not an uncommon occurrence to
have two or three forms of plus signs in one and the same publication,
or to find the Latin cross in an upright or horizontal position, according to the crowded condition of a particular line in which the symbol
occurs.

207. The cross of the form ✠ was used in 1563 and earlier by the
Spaniard De Hortega,[2] also by Klebotius,[3] Romanus,[4] and Descartes.[5] It occurs not infrequently in the *Acta eruditorum*[6] of Leipzig,
and sometimes in the *Miscellanea Berolinensia*.[7] It was sometimes
used by Halley,[8] Weigel,[9] Swedenborg,[10] and Wolff.[11] Evidently this
symbol had a wide geographical distribution, but it never threatened
to assume supremacy over the less fanciful Greek cross.

A somewhat simpler form, ✚, consists of a Greek cross with four
uniformly heavy black arms, each terminating in a thin line drawn
across it. It is found, for example, in a work of Hindenburg,[12] and
renders the plus signs on a page unduly conspicuous.

Occasionally plus signs are found which make a "loud" display
on the printed page. Among these is the eight-pointed Maltese cross,

[1] *Caroli Renaldini Ars analytica mathematicvm* (Florence, 1665), p. 80, and
throughout the volume, while in the earlier edition (Anconnae, 1644) he uses both
the heavy cross and dagger form.

[2] Fray Juã de Hortega, *Tractado subtilissimo d'arismetica γ geometria* (Granada, 1563), leaf 51. Also (Seville, 1552), leaf 42.

[3] Guillaume Klebitius, *Insvlae Melitensis, quam alias Maltam vocant, Historia,
Quaestionib. aliquot Mathematicis reddita incundior* (Diest [Belgium], 1565). I
am indebted to Professor H. Bosmans for information relating to this book.

[4] Adr. Romanus. "Problema," *Ideae mathematicae pars prima* (Antwerp, 1593).

[5] René Descartes, *La géométrie* (1637), p. 325. This form of the plus sign is in-
frequent in this publication; the ordinary form (+) prevails.

[6] See, for instance, *op. cit.* (1682), p. 87; *ibid.* (1683), p. 204; *ibid.* (1691),
p. 179; *ibid.* (1694), p. 195; *ibid.* (1697), p. 131; *ibid.* (1698), p. 307; *ibid.* (1713),
p. 344.

[7] *Op. cit.*, p. 156. However, the Latin cross is used more frequently than the
form now under consideration. But in Vol. II (1723), the latter form is prevalent.

[8] E. Halley, *Philosophical Transactions*, Vol. XVII (London, 1692–94), p. 963;
ibid. (1700–1701), Vol. XXII, p. 625.

[9] *Erhardi Weigelii Philosophia Mathematica* (Jena, 1693), p. 135.

[10] E. Swedenborg, *op. cit.*, p. 32. The Latin cross is more prevalent in this
book.

[11] Christian Wolff, *Mathematisches Lexicon* (Leipzig, 1716), p. 14.

[12] Carl Friedrich Hindenburg, *Infinitinomii dignitatum . . . . leges ac Formulae*
(Göttingen, 1779).

of varying shape, found, for example, in James Gregory,[1] Corachan,[2] Wolff,[3] and Hindenburg.[4]

Sometimes the ordinary Greek cross has the horizontal stroke very much heavier or wider than the vertical, as is seen, for instance, in Fortunatus.[5] A form for plus —/— occurs in Johan Albert.[6]

208. *Varieties of minus signs.*—One of the curiosities in the history of mathematical notations is the fact that notwithstanding the extreme simplicity and convenience of the symbol − to indicate subtraction, a more complicated symbol of subtraction ÷ should have been proposed and been able to maintain itself with a considerable group of writers, during a period of four hundred years. As already shown, the first appearance in print of the symbols + and − for "plus" and "minus" is found in Widman's arithmetic. The sign − is one of the very simplest conceivable; therefore it is surprising that a modification of it should ever have been suggested.

Probably these printed signs have ancestors in handwritten documents, but the line of descent is usually difficult to trace with certainty (§ 201). The following quotation suggests another clue: "In the west-gothic writing before the ninth century one finds, as also Paoli remarks, that a short line has a dot placed above it ∸, to indicate *m*, in order to distinguish this mark from the simple line which signifies a contraction or the letter *N*. But from the ninth century down, this same west-gothic script always contains the dot over the line even when it is intended as a general mark."[7]

In print the writer has found the sign ∸ for "minus" only once. It occurs in the 1535 edition of the *Rechenbüchlin* of Grammateus (Fig. 56). He says: "Vnd man̄ brauchet solche zeichen als + ist mehr / vnd ∸ / minder."[8] Strange to say, this minus sign does not occur in the first edition (1518) of that book. The corresponding passage of the earlier edition reads: "Vnd man braucht solche zaichen

---

[1] *Geometriae pars vniversalis* (Padua, 1668), p. 20, 71, 105, 108.

[2] Juan Bautista Corachan, *Arithmetica demonstrada* (Barcelona, 1719), p. 326.

[3] Christian Wolff, *Elementa matheseos universae*, Tomus I (Halle, 1713), p. 252.

[4] *Op. cit.*

[5] P. F. Fortunatus, *Elementa matheseos* (Brixia, 1750), p. 7.

[6] Johan Albert, *New Rechenbüchlein auff der federn* (Wittemberg, 1541); taken from Glaisher, *op. cit.*, p. 40, 61.

[7] Adriano Cappelli, *Lexicon abbreviaturam* (Leipzig, 1901), p. xx.

[8] Henricus Grammateus, *Eyn new Künstlich behend and gewiss Rechenbüchlin* (1535; 1st ed., 1518). For a facsimile page of the 1535 edition, see D. E. Smith, *Rara arithmetica* (1908), p. 125.

als + ist vnnd / — mynnder." Nor does Grammateus use ÷ in other parts of the 1535 edition; in his mathematical operations the minus sign is always —.

The use of the dash and two dots, thus ÷, for "minus," has been found by Glaisher to have been used in 1525, in an arithmetic of Adam Riese,[1] who explains: "Sagenn sie der warheit zuuil so bezeychenn sie mit dem zeychen + plus wu aber zu wenigk so beschreib sie mit dem zeychen ÷ minus genant."[2]

No reason is given for the change from — to ÷. Nor did Riese use ÷ to the exclusion of —. He uses ÷ in his algebra, *Die Coss*, of 1524, which he did not publish, but which was printed[3] in 1892, and also in his arithmetic, published in Leipzig in 1550. Apparently, he used — more frequently than ÷.

Probably the reason for using ÷ to designate — lay in the fact that — was assigned more than one signification. In Widman's arithmetic — was used for subtraction or "minus," also for separating terms in proportion,[4] and for connecting each amount of an article (wool, for instance) with the cost per pound (§ 202). The symbol — was also used as a rhetorical symbol or dash in the same manner as it is used at the present time. No doubt, the underlying motive in introducing ÷ in place of — was the avoidance of confusion. This explanation receives support from the German astronomer Regiomontanus,[5] who, in his correspondence with the court astronomer at Ferrara, Giovanni Bianchini, used — as a sign of equality; and used for subtraction a different symbol, namely, $\overline{ig}$ (possibly a florescent form of $\tilde{m}$). With him $1 \ \overline{ig} \ r^e$ meant $1-x$.

Eleven years later, in 1546, Gall Splenlin, of Ulm, had published at Augsburg his *Arithmetica künstlicher Rechnung*, in which he uses ÷, saying: "Bedeut das zaichen + züuil, und das ÷ zü wenig."[6] Riese and Splenlin are the only arithmetical authors preceding the middle of the sixteenth century whom Glaisher mentions as using ÷ for subtraction or "minus."[7] Caspar Thierfeldern,[8] in his *Arithmetica*

---

[1] *Rechenung auff der linihen vnd federn in zal, masz, vnd gewicht* (Erfurt, 1525; 1st ed., 1522).

[2] This quotation is taken from Glaisher, *op. cit.*, p. 36.

[3] See Bruno Berlet, *Adam Riese* (Leipzig, Frankfurt am Main, 1892).

[4] Glaisher, *op. cit.*, p. 15.

[5] M. Curtze, *Abhandlungen zur Geschichte der mathematischen Wissenschaften*, Vol. XII (1902), p. 234; Karpinski, *Robert of Chester, etc.*, p. 37.

[6] See Glaisher, *op. cit.*, p. 43.

[7] *Ibid.*, Vol. LI, p. 1–148.   [8] See Jackson, *op. cit.*, p. 55, 220.

(Nuremberg, 1587), writes the equation (p. 110), "18 fl. $\div$ 85 gr. gleich 25 fl. $\div$ 232 gr."

With the beginning of the seventeenth century $\div$ for "minus" appears more frequently, but, as far as we have been able to ascertain only in German, Swiss, and Dutch books. A Dutch teacher, Jacob Vander Schuere, in his *Arithmetica* (Haarlem, 1600), defines $+$ and $-$, but lapses into using $\div$ in the solution of problems. A Swiss writer, Wilhelm Schey,[1] in 1600 and in 1602 uses both $\div$ and $\vcentcolon\vcentcolon$ for "minus." He writes $9+9$, $5\div12$, $6\div28$, where the first number signifies the weight in *centner* and the second indicates the excess or deficiency of the respective "pounds." In another place Schey writes "9 fl. $\vcentcolon\vcentcolon$ 1 ort," which means "9 florins less 1 ort or quart." In 1601 Nicolaus Reymers,[2] an astronomer and mathematician, uses regularly $\div$ for "minus" or subtraction; he writes

$$\text{"XXVIII} \quad \text{XII} \quad \text{X} \quad \text{VI} \quad \text{III} \quad \text{I} \quad \text{O}$$
$$\text{1 gr.} \quad 65532+18 \quad \div 30 \quad \div 18 \quad +12 \quad \div 8 \text{"}$$
$$\text{for } x^{28} = 65{,}532x^{12}+18x^{10}-30x^6-18x^3+12x-8 \ .$$

Peter Roth, of Nürnberg, uses $\vcentcolon\vcentcolon$ in writing[3] $3x^2-26x$. Johannes Faulhaber[4] at Ulm in Württemberg used $\div$ frequently. With him the horizontal stroke was long and thin, the dots being very near to it. The year following, the symbol occurs in an arithmetic of Ludolf van Ceulen,[5] who says in one place: "Subtraheert $\sqrt{7}$ van, $\sqrt{13}$, rest $\sqrt{13}$, weynigher $\sqrt{7}$, daerom stelt $\sqrt{13}$ voren en $\sqrt{7}$ achter, met een sulck teecken $\div$ tusschen beyde, vvelck teecmin beduyt, comt alsoo de begeerde rest $\sqrt{13}\div\sqrt{7}-$." However, in some parts of the book $-$ is used for subtraction. Albert Girard[6] mentions $\div$ as the symbol for "minus," but uses $-$. Otto Wesellow[7] brought out a book in which

---

[1] *Arithmetica oder die Kunst zu rechnen* (Basel, 1600–1602). We quote from D. E. Smith, *op. cit.*, p. 427, and from Matthäus Sterner, *Geschichte der Rechenkunst* (München and Leipzig, 1891), p. 280, 291.

[2] *Nicolai Raimari Ursi Dithmarsi . . . . arithmetica analytica, vulgo Cosa, oder Algebra* (zu Frankfurt an der Oder, 1601). We take this quotation from Gerhardt, *Geschichte der Mathematik in Deutschland* (1877), p. 85.

[3] *Arithmetica philosophica* (1608). We quote from Treutlein, "Die deutsche Coss," *Abhandlungen zur Geschichte der Mathematik*, Vol. II (Leipzig, 1879), p. 28, 37, 103.

[4] *Numerus figuratus sive arithmetica analytica* (Ulm, 1614), p. 11, 16.

[5] *De arithmetische en geometrische Fondamenten* (1615), p. 52, 55, 56.

[6] *Invention nouvelle en l'algebre* (Amsterdam, 1629), no paging. A facsimile edition appeared at Leiden in 1884.

[7] *Flores arithmetici* (drüdde vnde veerde deel; Bremen, 1617), p. 523.

$+$ and $\div$ stand for "plus" and "minus," respectively. These signs are used by Follinus,[1] by Stampioen (§ 508), by Daniel van Hovcke[2] who speaks of $+$ as signifying "mer en $\div$ min.," and by Johann Ardüser[3] in a geometry. It is interesting to observe that only thirteen years after the publication of Ardüser's book, another Swiss, J. H. Rahn, finding, perhaps, that there existed two signs for subtraction, but none for division, proceeded to use $\div$ to designate division. This practice did not meet with adoption in Switzerland, but was seized upon with great avidity as the symbol for division in a far-off country, England. In 1670 $\div$ was used for subtraction once by Huygens[4] in the *Philosophical Transactions*. Johann Hemelings[5] wrote $\div$ for "minus" and indicated, in an example, $14\frac{1}{2}$ legions less 1250 men by "14 1/2 Legion $\div$ 1250 Mann." The symbol is used by Tobias Beutel,[6] who writes "$81 \div 1R6561 \div 162.\ R.+1.\ zenss$" to represent our $81 - \sqrt{6561 - 162x + x^2}$. Kegel[7] explains how one can easily multiply by 41, by first multiplying by 6, then by 7, and finally subtracting the multiplicand; he writes "$7 \div 1$." In a set of seventeenth-century examination questions used at Nürnberg, reference is made to cossic operations involving quantities, "durch die Signa $+$ und $\div$ connectirt."[8]

The vitality of this redundant symbol of subtraction is shown by its continued existence during the eighteenth century. It was employed by Paricius,[9] of Regensburg. Schlesser[10] takes $\div$ to represent

---

[1] Hermannus Follinus, *Algebra sive liber de rebus occultis* (Coloniae, 1622), p. 113, 185.

[2] *Cyffer-Boeck* . . . . (den tweeden Druck: Rotterdam, 1628), p. 129–33.

[3] *Geometriae theoricae et practicae. Oder von dem Feldmässen* (Zürich, 1646), fol. 75.

[4] In a reply to Slusius, *Philosophical Transactions*, Vol. V (London, 1670), p. 6144.

[5] *Arithmetisch-Poetisch-u. Historisch-Erquick Stund* (Hannover, 1660); *Selbst-lehrendes Rechen-Buch* . . . . *durch Johannem Hemelingium* (Frankfurt, 1678). Quoted from Hugo Grosse, *Historische Rechenbücher des 16. and 17. Jahrhunderts* (Leipzig, 1901), p. 99, 112.

[6] *Geometrische Gallerie* (Leipzig, 1690), p. 46.

[7] Johann Michael Kegel, *New vermehrte arithmetica vulgaris et practica italica* (Frankfurt am Main, 1696). We quote from Sterner, *op. cit.*, p. 288.

[8] Fr. Unger, *Die Methodik der praktischen Arithmetik in historischer Entwickelung* (Leipzig, 1888), p. 30.

[9] Georg Heinrich Paricius, *Praxis arithmetices* (1706). We quote from Sterner, *op. cit.*, p. 349.

[10] Christian Schlesser, *Arithmetisches Haupt-Schlüssel* . . . . *Die Coss—oder Algebra* (Dresden and Leipzig, 1720).

"minus oder weniger." It was employed in the *Philosophical Transactions* by the Dutch astronomer N. Cruquius;[1] ÷ is found in Hübsch[2] and Crusius.[3] It was used very frequently as the symbol for subtraction and "minus" in the *Maandelykse Mathematische Liefhebbery*, Purmerende (1754–69). It is found in a Dutch arithmetic by Bartjens[4] which passed through many editions. The vitality of the symbol is displayed still further by its regular appearance in a book by van Steyn,[5] who, however, uses − in 1778.[6] Halcke states, "÷ of − het teken van *substractio minus* of min.,"[7] but uses − nearly everywhere. Praalder, of Utrecht, uses ordinarily the minus sign −, but in one place[8] he introduces, for the sake of clearness, as he says, the use of ÷ to mark the subtraction of complicated expressions. Thus, he writes "$= \div \overline{9\frac{1}{2} + 2\sqrt{26}}$." The ÷ occurs in a Leipzig magazine,[9] in a Dresden work by Illing,[10] in a Berlin text by Schmeisser,[11] who uses it also in expressing arithmetical ratio, as in "$2 \div 6 \div 10$." In a part of Klügel's[12] mathematical dictionary, published in 1831, it is stated that ÷ is used as a symbol for division, "but in German arithmetics is employed also to designate subtraction." A later use of it for "minus," that we have noticed, is in a Norwegian arithmetic.[13] In fact, in Scandinavian

[1] *Op. cit.*, Vol. XXXIII (London, 1726), p. 5, 7.

[2] J. G. G. Hübsch, *Arithmetica portensis* (Leipzig, 1748).

[3] David Arnold Crusius, *Anweisung zur Rechen-Kunst* (Halle, 1746), p. 54.

[4] *De vernieuwde Cyfferinge van Mr. Willem Bartjens, ... vermeerdert—ende verbetert, door Mr. Jan van Dam. . . . . en van alle voorgaande Fauten gezuyvert door Klaas Bosch* (Amsterdam, 1771), p. 174–77.

[5] Gerard van Steyn, *Liefhebbery der Reekenkonst* (eerste deel; Amsterdam, 1768), p. 3, 11, etc.

[6] *Ibid.* (2e Deels, 2e Stuk, 1778), p. 16.

[7] *Mathematisch Zinnen-Confect . . . . door Paul Halcken . . . . Uyt het Hoogduytsch vertaald . . . . dor Jacob Oostwoud* (Tweede Druk, Te Purmerende, 1768), p. 5.

[8] *Mathematische Voorstellen . . . . door . . . . Ludolf van Keulen . . . . door Laurens Praalder* (Amsterdam, 1777), p. 137.

[9] J. A. Kritter, *Leipziger Magazin für reine and angewandte Mathematik* (herausgegeben von J. Bernoulli und C. F. Hindenburg, 1788), p. 147–61.

[10] Carl Christian Illing, *Arithmetisches Handbuch für Lehrer in den Schulen* (Dresden, 1793), p. 11, 132.

[11] Friedrich Schmeisser, *Lehrbuch der reinen Mathesis* (1. Theil, Berlin, 1817), p. 45, 201.

[12] G. S. Klügel, "Zeichen," *Mathematisches Wörterbuch*. This article was written by J. A. Grunert.

[13] G. C. Krogh, *Regnebog for Begyndere* (Bergen, 1869), p. 15.

countries the sign ÷ for "minus" is found occasionally in the twentieth century. For instance, in a Danish scientific publication of the year 1915, a chemist expresses a range of temperature in the words "fra+18° C. til ÷ 18° C."[1]. In 1921 Ernst W. Selmer[2] wrote "0,72÷ 0,65 = 0,07." The difference in the dates that have been given, and the distances between the places of publication, make it certain that this symbol ÷ for "minus" had a much wider adoption in Germany, Switzerland, Holland, and Scandinavia than the number of our citations would indicate. But its use seems to have been confined to Teutonic peoples.

Several writers on mathematical history have incidentally called attention to one or two authors who used the symbol ÷ for "minus," but none of the historians revealed even a suspicion that this symbol had an almost continuous history extending over four centuries.

209. Sometimes the minus sign − appears broken up into two or three successive dashes or dots. In a book of 1610 and again of 1615, by Ludolph van Ceulen,[3] the minus sign occasionally takes the form − −. Richard Balam[4] uses three dots and says "3 · · · 7, 3 from 7"; he writes an arithmetical proportion in this manner: "2 · · · 4 = 3 · · · 5." Two or three dots are used in René Descartes' *Géométrie*, in the writings of Marin Mersenne,[5] and in many other seventeenth-century books, also in the *Journal des Sçavans* for the year 1686, printed in Amsterdam, where one finds (p. 482) "1 − − − R − − − 11" for $1 - \sqrt{-11}$, and in volumes of that *Journal* printed in the early part of the eighteenth century. Hérigone used ∼ for "minus" (§ 189), the − being pre-empted for *recta linea*.

From these observations it is evident that in the sixteenth and seventeenth centuries the forms of type for "minus" were not yet standardized. For this reason, several varieties were sometimes used on the same page.

This study emphasizes the difficulty experienced even in ordinary

[1] Johannes Boye Petersen, *Kgl. Danske Vidensk. Selskabs Skrifter, Nat. og. Math. Afd., 7. Raekke,* Vol. XII (Kopenhagen, 1915), p. 330; see also p. 221, 223, 226, 230, 238.

[2] *Skrifter utgit av Videnskapsselskapet i Kristiania* (1921)," Historisk-filosofisk Klasse" (2. Bind; Kristiania, 1922), article by Ernst W. Selmer, p. 11; see also p. 28, 29, 39, 47.

[3] *Circvlo et adscriptis liber. . . . . Omnia e vernacûlo Latina fecit et annotationibus illustravit Willebrordus Snellius* (Leyden, 1610), p. 128.

[4] *Algebra* (London, 1653), p. 5.

[5] *Cogitata Physico-Mathematica* (Paris, 1644), Praefatio generalis, "De Rationibus atque Proportionibus," p. xii, xiii.

arithmetic and algebra in reaching a common world-language. Centuries slip past before any marked step toward uniformity is made. It appears, indeed, as if blind chance were an uncertain guide to lead us away from the Babel of languages. The only hope for rapid approach of uniformity in mathematical symbolism lies in international co-operation through representative committees.

210. *Symbols for "plus or minus."*—The $\pm$ to designate "plus or minus" was used by Albert Girard in his *Tables*[1] of 1626, but with the interpolation of *ou*, thus "$o\overset{+}{\underset{-}{u}}$." The $\pm$ was employed by Oughtred in his *Clavis mathematicae* (1631), by Wallis,[2] by Jones[3] in his *Synopsis*, and by others. There was considerable experimentation on suitable notations for cases of simultaneous double signs. For example, in the third book of his *Géométrie*, Descartes uses a dot where we would write $\pm$. Thus he writes the equation "$+y^6 \cdot 2py^4 \overset{+pp}{4ryy} - qq \backsim 0$" and then comments on this: "Et pour les signes $+$ ou $-$ que iay omis, s'il y a eu $+p$ en la precedente Equation, il faut mettre en celle $-$ cy $+ 2p$, ou s'il ya eu $- p$, il faut mettre $- 2p$; & au contraire s'il ya eu $+ r$, il faut mettre $- 4r$, ..." The symbolism which in the *Miscellanea Berolinensia* of 1710 is attributed to Leibniz is given in § 198.

A different notation is found in Isaac Newton's *Universal Arithmetick:* "I denoted the Signs of b and c as being indeterminate by the Note $\perp$, which I use indifferently for $+$ or $-$, and its opposite $\top$ for the contrary."[4] These signs appear to be the $+$ with half of the vertical stroke excised. William Jones, when discussing quadratic equations, says: "Therefore if $\vee$ be put for the Sign of any Term, and $\wedge$ for the contrary, all *Forms of Quadratics* with their *Solutions*, will be reduc'd to this one. If $xx \vee ax \vee b = 0$ then $\wedge \frac{1}{2}a \pm aa \wedge b\big|^{\frac{1}{2}}$."[5] Later in the book (p. 189) Jones lets two horizontal dots represent any sign: "Suppose any *Equation* whatever, as $x^n \mathinner{..} ax^{n-1} \mathinner{..} bx^{n-2} \mathinner{..} cx^{n-3} \mathinner{..} dx^{n-4}$, etc. $\mathinner{...} A = 0$."

A symbol $\aleph$ standing for $\pm$ was used in 1649 and again as late as 1695, by van Schooten[6] in his editions of Descartes' geometry, also

---

[1] See *Bibliotheca mathematica* (3d ser., 1900), Vol. I, p. 66.

[2] J. Wallis, *Operum mathematicorum pars prima* (Oxford, 1657), p. 250.

[3] William Jones, *Synopsis Palmariorum matheseos* (London, 1706), p. 14.

[4] *Op. cit.* (trans. Mr. Ralphson . . . . rev. by Mr. Cunn; London, 1728), p. 172; also *ibid.* (rev. by Mr. Cunn . . . . expl. by Theaker Wilder; London, 1769), p. 321.

[5] *Op. cit.*, p. 148.

[6] *Renati Descartes Geometria* (Leyden, 1649), Appendix, p. 330; *ibid.* (Frankfurt am Main, 1695), p. 295, 444, 445.

by De Witt.[1] Wallis[2] wrote ႘ for + or −, and ႘ for the contrary.
The sign ႘ was used in a restricted way, by James Bernoulli;[3] he
says, "႘ significat + in pr. e − in post. hypoth.," i.e., the symbol
stood for + according to the first hypothesis, and for −, according to
the second hypothesis. He used this same symbol in his *Ars conjectandi* (1713), page 264. Van Schooten wrote also ႘ for ∓. It
should be added that ႘ appears also in the older printed Greek books
as a ligature or combination of two Greek letters, the omicron *o* and
the upsilon *v*. The ႘ appears also as an astronomical symbol for the
constellation Taurus.

Da Cunha[4] introduced ±' and ±', or ±' and ∓', to mean that
the upper signs shall be taken simultaneously in both or the lower
signs shall be taken simultaneously in both. Oliver, Wait, and Jones[5]
denoted positive or negative N by $\pm N$.

211. The symbol $[a]$ was introduced by Kronecker[6] to represent
0 or +1 or −1, according as a was 0 or +1 or −1. The symbol "sgn"
has been used by some recent writers, as, for instance, Peano,[7] Netto,[8]
and Le Vavasseur, in a manner like this: "sgn $A = +1$" when $A > 0$,
"sgn $A = -1$" when $A < 0$. That is, "sgn A" means the "sign of
A." Similarly, Kowalewski[9] denotes by "sgn $\mathfrak{P}$" +1 when $\mathfrak{P}$ is an
even, and −1 when $\mathfrak{P}$ is an odd, permutation.

The symbol $\sqrt{a^2}$ is sometimes taken in the sense[10] $\pm a$, but in equations involving $\sqrt{}$, the principal root $+a$ is understood.

212. *Certain other specialized uses of* + *and* −.—The use of each
of the signs + and − in a double sense—first, to signify addition and
subtraction; second, to indicate that a number is positive and negative—has met with opposition from writers who disregarded the advantages resulting from this double use, as seen in $a-(-b)=a+b$,

---

[1] Johannis de Witt, *Elementa Cvrvarvm Linearvm. Edita Operâ Francisci à
Schooten* (Amsterdam, 1683), p. 305.

[2] John Wallis, *Treatise of Algebra* (London, 1685), p. 210, 278.

[3] *Acta eruditorum* (1701), p. 214.

[4] J. A. da Cunha, *Principios mathematicos* (Lisbon, 1790), p. 126.

[5] *Treatise on Algebra* (2d ed.; Ithaca, 1887), p. 45.

[6] L. Kronecker, *Werke*, Vol. II (1897), p. 39.

[7] G. Peano, *Formulario mathematico*, Vol. V (Turin, 1908), p. 94.

[8] E. Netto and R. le Vavasseur, *Encyclopédie des scien. math.*, Tome I, Vol. II
(1907), p. 184; see also A. Voss and J. Molk, *ibid.*, Tome II, Vol. I (1912), p. 257,
n. 77.

[9] Gerhard Kowalewski, *Einführung in die Determinantentheorie* (Leipzig, 1909),
p. 18.

[10] See, for instance, *Encyclopédie des scien. math.*, Tome II, Vol. I, p. 257, n. 77.

and who aimed at extreme logical simplicity in expounding the elements of algebra to young pupils. As a remedy, German writers proposed a number of new symbols which are set forth by Schmeisser as follows:

"The use of the signs $+$ and $-$, not only for opposite magnitudes . . . . but also for Addition and Subtraction, frequently prevents clearness in these matters, and has even given rise to errors. For that reason other signs have been proposed for the positive and negative. Wilkins (*Die Lehre von d. entgegengesetzt. Grössen etc.*, Brschw., 1800) puts down the positive without signs $(+a=a)$ but places over the negative a dash, as in $-a=\bar{a}$. v. Winterfeld (*Anfangsgr. d. Rechenk.*, 2te Aufl. 1809) proposes for positive the sign ⊢ or Γ, for negative ⊣ or ¬. As more scientific he considers the inversion of the letters and numerals, but unfortunately some of them as i, r, o, x, etc., and 0, 1, 8, etc., cannot be inverted, while others, by this process, give rise to other letters as b, d, p, q, etc. Better are the more recent proposals of Winterfeld, to use for processes of computation the signs of the waxing and waning moon, namely for Addition), for Subtraction (, for Multiplication), for Division (, but as he himself acknowledges, even these are not perfectly suitable. . . . . Since in our day one does not yet, for love of correctness, abandon the things that are customary though faulty, it is for the present probably better to stress the significance of the concepts of the positive and additive, and of the negative and subtractive, in instruction, by the retention of the usual signs, or, what is the same thing, to let the qualitative and quantitative significance of $+$ and $-$ be brought out sharply. This procedure has the advantage moreover of more fully exercising the understanding."[1]

Wolfgang Bolyai[2] in 1832 draws a distinction between $+$ and $-$, and $+\!\!\!\!+$ and $\rightarrowtail$; the latter meaning the (intrinsic) "positive" and "negative." If A signifies $\rightarrowtail B$, then $-A$ signifies $+\!\!\!\!+B$.

213. In more recent time other notations for positive and negative numbers have been adopted by certain writers. Thus, Spitz[3] uses $\leftarrow a$ and $\rightarrow a$ for positive a and negative a, respectively. Méray[4] prefers $\vec{a}$, $\overleftarrow{a}$; Padé,[5] a_ρ, a_n; Oliver, Wait, and Jones[6] employ an ele-

[1] Friedrich Schmeisser, *op. cit.*, p. 42, 43.

[2] *Tentamen* (2d ed., T. I.; Budapestini, 1897), p. xi.

[3] C. Spitz, *Lehrbuch der alg. Arithmetik* (Leipzig, 1874), p. 12.

[4] Charles Méray, *Leçons nouv. de l'analyse infin.*, Vol. I (Paris, 1894), p. 11.

[5] H. Padé, *Premières leçons d'algèbre élém.* (Paris, 1892), p. 5.

[6] *Op. cit.*, p. 5.

vated $+$ or $-$ (as in $+10$, -10) as signs of "quality"; this practice has been followed in developing the fundamental operations in algebra by a considerable number of writers; for instance, by Fisher and Schwatt,[1] and by Slaught and Lennes.[2] In elementary algebra the special symbolisms which have been suggested to represent "positive number" or "negative number" have never met with wide adoption. Stolz and Gmeiner[3] write a, $\bar{a}$, for positive a and negative a. The designation .... -3, -2, -1, 0, $+1$, $+2$, $+3$, ..., occurs in Huntington's *Continuum* (1917), page 20.

214. A still different application of the sign $+$ has been made in the theory of integral numbers, according to which Peano[4] lets $a+$ signify the integer immediately following a, so that $a+$ means the integer $(a+1)$. For the same purpose, Huntington[5] and Stolz and Gmeiner[6] place the $+$ in the position of exponents, so that $5+=6$.

215. *Four unusual signs.*—The Englishman Philip Ronayne used in his *Treatise of Algebra* (London, 1727; 1st ed., 1717), page 4, two curious signs which he acknowledged were "not common," namely, the sign $\ominus$ to denote that "some Quantity indefinitely Less than the Term that next precedes it, is to be added," and the sign $\ominus$ that such a quantity is "to be subtracted," while the sign ϕ may mean "either $\ominus$ or $\ominus$ when it matters not which of them it is." We have not noticed these symbols in other texts.

How the progress of science may suggest new symbols in mathematics is illustrated by the composition of velocities as it occurs in Einstein's addition theorem.[7] Silberstein uses here $\#$ instead of $+$.

216. *Composition of ratios.*—A strange misapplication of the $+$ sign is sometimes found in connection with the "composition" of ratios. If the ratios $\dfrac{NP}{CN}$ and $\dfrac{AN}{CN}$ are multiplied together, the product

---

[1] G. E. Fisher and I. J. Schwatt, *Text-Book of Algebra* (Philadelphia, 1898), p. 23.

[2] H. E. Slaught and U. J. Lennes, *High School Algebra* (Boston, 1907), p. 48.

[3] Otto Stolz und J. A. Gmeiner, *Theoretische Arithmetik* (2d ed.; Leipzig, 1911), Vol. I, p. 116.

[4] G. Peano, *Arithmetices principia nova methodo exposita* (Turin, 1889); "Sul concetto di numero," *Rivista di matem.*, Vol. I, p. 91; *Formulaire de mathématiques*, Vol. II, § 2 (Turin, 1898), p. 1.

[5] E. V. Huntington, *Transactions of the American Mathematical Society*, Vol. VI (1905), p. 27.

[6] Op. cit., Vol. I, p. 14. In the first edition Peano's notation was used.

[7] C. E. Weatherburn, *Advanced Vector Analysis* (London, 1924), p. xvi.

$\frac{NP}{CN} \cdot \frac{AN}{CN}$, according to an old phraseology, was "compounded" of the first two ratios.[1] Using the term "proportion" as synonymous with "ratio," the expression "composition of proportions" was also used. As the word "composition" suggests addition, a curious notation, using $+$, was at one time employed. For example, Isaac Barrow[2] denoted the "compounded ratio" $\frac{NP}{CN} \cdot \frac{AN}{CN}$ in this manner, "$NP \cdot CN +$ $AN \cdot CN$." That is, the sign of addition was used in place of a sign of multiplication, and the dot signified ratio as in Oughtred.

In another book[3] Barrow again multiplies equal ratios by equal ratios. In modern notation, the two equalities are

$$(PL+QO):QO=2BC:(BC-CP) \text{ and } QO:BC=BC:(BC+CP) \ .$$

Barrow writes the result of the multiplication thus:

$$PL+QO \cdot QO+QO \cdot BC=2BC \cdot BC-CP+BC \cdot BC+CP \ .$$

Here the $+$ sign occurs four times, the first and fourth times as a symbol of ordinary addition, while the second and third times it occurs in the "addition of equal ratios" which really means the multiplication of equal ratios. Barrow's final relation means, in modern notation,

$$\frac{PL+QO}{QO} \cdot \frac{QO}{BC}=\frac{2BC}{BC-CP} \cdot \frac{BC}{BC+CP} \ .$$

Wallis, in his *Treatise of Algebra* (London, 1685), page 84, comments on this subject as follows: "But now because *Euclide* gives to this the name of *Composition*, which word is known many times to impart an *Addition;* (as when we say the Line *ABC* is compounded of *AB* and *BC;*) some of our more ancient Writers have chanced to call it *Addition of Proportions;* and others, following them, have continued that form of speech, which abides in (in divers Writers) even to this day: And the Dissolution of this composition they call *Subduction of Proportion*. (Whereas that should rather have been called *Multiplication*, and this Division.)"

A similar procedure is found as late as 1824 in J. F. Lorenz' trans-

---

[1] See Euclid, *Elements*, Book VI, Definition 5. Consult also T. L. Heath, *The Thirteen Books of Euclid's "Elements,"* Vol. II (Cambridge, 1908), p. 132–35, 189, 190.

[2] *Lectiones opticae* (1669), Lect. VIII, § V, and other places.

[3] *Lectiones geometricae* (1674), Lect. XI, Appendix I, § V.

lation from the Greek of Euclid's *Elements* (ed. C. B. Mollweide; Halle, 1824), where on page 104 the Definition 5 of Book VI is given thus: "Of three or more magnitudes, A, B, C, D, which are so related to one another that the ratios of any two consecutive magnitudes $A:B$, $B:C$, $C:D$, are equal to one another, then the ratio of the first magnitude to the last is said to be *composed* of all these ratios so that

$$A:D = (A:B) + (B:C) + (C:D)" \left(\text{in modern notation, } \frac{A}{D} = \frac{A}{B} \cdot \frac{B}{C} \cdot \frac{C}{D} \right).$$

SIGNS OF MULTIPLICATION

217. Early symbols.—In the early Babylonian tablets there is, according to Hilprecht,[1] an ideogram *A-DU* signifying "times" or multiplication. The process of multiplication or division was known to the Egyptians[2] as *wshtp*, "to incline the head"; it can hardly be regarded as being a mathematical symbol. Diophantus used no symbol for multiplication (§ 102). In the Bakhshālī manuscript multiplication is usually indicated by placing the numbers side by side (§ 109). In some manuscripts of Bhāskara and his commentators a dot is placed between factors, but without any explanation (§ 112). The more regular mark for product in Bhāskara is the abbreviation *bha*, from *bhavita*, placed after the factors (§ 112).

Stifel in his *Deutsche Arithmetica* (Nürnberg, 1545) used the capital letter M to designate multiplication, and D to designate division. These letters were again used for this purpose by S. Stevin[3] who expresses our $3xyz^2$ thus: 3 ① M *sec* ① M *ter* ②, where *sec* and *ter* mean the "second" and "third" unknown quantities.

The M appears again in an anonymous manuscript of 1638 explaining Descartes' *Géométrie* of 1637, which was first printed in 1896;[4] also once in the Introduction to a book by Bartholinus.[5]

Vieta indicated the product of A and B by writing "*A in B*" (Fig. 84). Mere juxtaposition signified multiplication in the Bakhshālī tract, in some fifteenth-century manuscripts, and in printed algebras designating $6x$ or $5x^2$; but $5\frac{1}{3}$ meant $5+\frac{1}{3}$, not $5 \times \frac{1}{3}$.

---

[1] H. V. Hilprecht, *Babylonian Expedition*, Vol. XX, Part 1, *Mathematical etc. Tablets* (Philadelphia, 1906), p. 16, 23.

[2] T. Eric Peet, *The Rhind Mathematical Papyrus* (London, 1923), p. 13.

[3] *Œuvres mathematiques* (ed. Albert Girard; Leyden, 1634), Vol. I, p. 7.

[4] Printed in *Œuvres de Descartes* (éd. Adam et Tannery), Vol. X (Paris, 1908), p. 669, 670.

[5] Er. Bartholinus, *Renati des Cartes Principia matheseos universalis* (Leyden, 1651), p. 11. See J. Tropfke, *op. cit.*, Vol. II (2d ed., 1921), p. 21, 22.

218. *Early uses of the St. Andrew's cross, but not as a symbol of multiplication of two numbers.*—It is well known that the St. Andrew's cross (×) occurs as the symbol for multiplication in W. Oughtred's *Clavis mathematicae* (1631), and also (in the form of the letter *X*) in an anonymous Appendix which appeared in E. Wright's 1618 edition of John Napier's *Descriptio*. This Appendix is very probably from the pen of Oughtred. The question has arisen, Is this the earliest use of × to designate multiplication? It has been answered in the negative—incorrectly so, we think, as we shall endeavor to show.

In the *Encyclopédie des sciences mathématiques*, Tome I, Volume I (1904), page 40, note 158, we read concerning ×, "One finds it between factors of a product, placed one beneath the other, in the Commentary added by Oswald Schreckenfuchs to Ptolemy's *Almagest*, 1551."[1] As will be shown more fully later, this is not a correct interpretation of the symbolism. Not two, but four numbers are involved, two in a line and two others immediately beneath, thus:

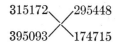

$$
\begin{array}{cc}
315172 & 295448 \\
395093 & 174715
\end{array}
$$

The cross does not indicate the product of any two of these numbers, but each bar of the cross connects two numbers which are multiplied. One bar indicates the product of 315172 and 174715, the other bar the product of 395093 and 295448. Each bar is used as a symbol singly; the two bars are not considered here as one symbol.

Another reference to the use of × before the time of Oughtred is made by E. Zirkel,[2] of Heidelberg, in a brief note in which he protests against attributing the "invention" of × to Oughtred; he states that it had a period of development of over one hundred years. Zirkel does

---

[1] *Clavdii Ptolemaei Pelusiensis Alexandrini Omnia quae extant Opera* (Basileae, 1551), Lib. ii, "Annotationes."

[2] Emil Zirkel, *Zeitschr. f. math. u. naturw. Unterricht*, Vol. LII (1921), p. 96. An article on the sign ×, which we had not seen before the time of proofreading, when R. C. Archibald courteously sent it to us, is written by N. L. W. A. Gravelaar in *Wiskundig Tijdschrift*, Vol. VI (1909–10), p. 1–25. Gravelaar cites a few writers whom we do not mention. His claim that, before Oughtred, the sign × occurred as a sign of multiplication, must be rejected as not borne out by the facts. It is one thing to look upon × as two symbols, each indicating a separate operation, and quite another thing to look upon × as only one symbol indicating only one operation. This remark applies even to the case in § 229, where the four numbers involved are conveniently placed at the four ends of the cross, and each stroke connects two numbers to be subtracted one from the other.

not make his position clear, but if he does not mean that $\times$ was used before Oughtred as a sign of multiplication, his protest is pointless.

Our own studies have failed to bring to light a clear and conclusive case where, before Oughtred, $\times$ was used as a symbol of multiplication. In medieval manuscripts and early printed books $\times$ was used as a mathematical sign, or a combination of signs, in eleven or more different ways, as follows: (1) in solutions of problems by the process of two false positions, (2) in solving problems in compound proportion involving integers, (3) in solving problems in simple proportion involving fractions, (4) in the addition and subtraction of fractions, (5) in the division of fractions, (6) in checking results of computation by the processes of casting out the 9's, 7's, or 11's, (7) as part of a group of lines drawn as guides in the multiplication of one integer by another, (8) in reducing radicals of different orders to radicals of the same order, (9) in computing on lines, to mark the line indicating "thousands," (10) to take the place of the multiplication table above 5 times 5, and (11) in dealing with amicable numbers. We shall briefly discuss each of these in order.

219. *The process of two false positions.*—The use of $\times$ in this process is found in the *Liber abbaci* of Leonardo[1] of Pisa, written in 1202. We must begin by explaining Leonardo's use of a single line or bar. A line connecting two numbers indicates that the two numbers are to be multiplied together. In one place he solves the problem: If 100 *rotuli* are worth 40 *libras*, how many *libras* are 5 *rotuli* worth? On the margin of the sheet stands the following:

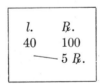

The line connecting 40 and 5 indicates that the two numbers are to be multiplied together. Their product is divided by 100, but no symbolism is used to indicate the division. Leonardo uses single lines over a hundred times in the manner here indicated. In more complicated problems he uses two or more lines, but they do not necessarily

---

[1] Leonardo of Pisa, *Liber abbaci* (1202) (ed. B. Boncompagni; Roma, 1857), Vol. I, p. 84.

form crosses. In a problem involving five different denominations of money he gives the following diagram:[1]

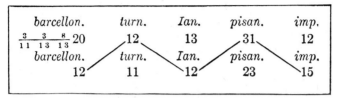

Here the answer 20+ is obtained by taking the product of the connected numbers and dividing it by the product of the unconnected numbers.

Leonardo uses a cross in solving, by double false position, the problem: If 100 *rotuli* cost 13 *libras*, find the cost of 1 *rotulus*. The answer is given in *solidi* and *denarii*, where 1 *libra* = 20 *solidi*, 1 *solidus* = 12 *denarii*. Leonardo assumes at random the tentative answers (the two false positions) of 3 *solidi* and 2 *solidi*. But 3 *solidi* would make this cost of 100 *rotuli* 15 *libra*, an error of +2 *libras*; 2 *solidi* would make the cost 10, an error of −3. By the underlying theory of two false positions, the errors in the answers (i.e., the errors $x-3$ and $x-2$ *solidi*) are proportional to the errors in the cost of 100 *rotuli* (i.e., +2 and −3 *libras*); this proportion yields $x=2$ *solidi* and $7\frac{1}{5}$ *denarii*. If the reader will follow out the numerical operations for determining our x he will understand the following arrangement of the work given by Leonardo (p. 319):

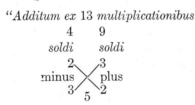

"*Additum ex* 13 *multiplicationibus*

Additum ex erroribus."

Observe that Leonardo very skilfully obtains the answer by multiplying each pair of numbers connected by lines, thereby obtaining the products 4 and 9, which are added in this case, and then dividing 13 by 5 (the sum of the errors). The cross occurring here is not one symbol, but two symbols. Each line singly indicates a multiplication. It would be a mistake to conclude that the cross is used here as a symbol expressing multiplication.

[1] *Ibid.*, Vol. I, p. 127.

The use of two lines crossing each other, in double or single false position, is found in many authors of later centuries. For example, it occurs in MS 14908 in the Munich Library,[1] written in the interval 1455–64; it is used by the German Widman,[2] the Italian Pacioli,[3] the Englishman Tonstall,[4] the Italian Sfortunati,[5] the Englishman Recorde,[6] the German Splenlin,[7] the Italians Ghaligai[8] and Benedetti,[9] the Spaniard Hortega,[10] the Frenchman Trenchant,[11] the Dutchman Gemma Frisius,[12] the German Clavius,[13] the Italian Tartaglia,[14] the Dutchman Snell,[15] the Spaniard Zaragoza,[16] the Britishers Jeake[17] and

[1] See M. Curtze, *Zeitschrift f. Math. u. Physik*, Vol. XL (Leipzig, 1895). Supplement, *Abhandlungen z. Geschichte d. Mathematik*, p. 41.

[2] Johann Widman, *Behēde vnd hubsche Rechenung* (Leipzig, 1489). We have used J. W. L. Glaisher's article in *Messenger of Mathematics*, Vol. LI (1922), p. 16.

[3] L. Pacioli, *Summa de arithmetica, geometria, etc.* (1494). We have used the 1523 edition, printed at Toscolano, fol. 99^b, 100^a, 182.

[4] C. Tonstall, *De arte supputandi* (1522). We have used the Strassburg edition of 1544, p. 393.

[5] Giovanni Sfortunati da Siena, *Nvovo Lvme. Libro di Arithmetica* (1534), fol. 89–100.

[6] R. Recorde, *Grovnd of Artes* (1543[?]). We have used an edition issued between 1636 and 1646 (title-page missing), p. 374.

[7] Gall Splenlin, *Arithmetica künstlicher Rechnung* (1645). We have used J. W. L. Glaisher's article in *op. cit.*, Vol. LI (1922), p. 62.

[8] Francesco Ghaligai, *Pratica d'arithmetica* (Nuovamente Rivista ... ; Firenze, 1552), fol. 76.

[9] *Io. Baptistae Benedicti Diversarvm specvlationvm mathematicarum, et physicarum Liber* (Turin, 1585), p. 105.

[10] Juan de Hortega, *Tractado subtilissimo de arismetica y de geometria* (emendado por Lonçalo Busto, 1552), fol. 138, 215b.

[11] Jan Trenchant, *L'arithmetiqve* (4th ed.; Lyon, 1578), p. 216.

[12] Gemma Frisius, *Arithmeticae Practicae methodvs facilis* (iam recens ab ipso authore emendata .... Parisiis, 1569), fol. 33.

[13] Christophori Clavii Bambergensis, *Opera mathematica* (Mogvntiae, 1612), Tomus secundus; "Numeratio," p. 58.

[14] *L'arithmetique de Nicolas Tartaglia Brescian* (traduit par Gvillavmo Gosselin de Caen ... Premier Partie; Paris, 1613), p. 105.

[15] *Willebrordi Snelli Doctrinae Triangvlorvm Canonicae liber qvatvor* (Leyden, 1627), p. 36.

[16] *Arithmetica Vniversal ... avthor* El M. R. P. Joseph Zaragoza (Valencia, 1669), p. 111.

[17] Samuel Jeake, ΛΟΓΙΣΤΙΚΗΛΟΓΊΑ or *Arithmetick* (London, 1696; Preface 1674), p. 501.

Wingate,[1] the Italian Guido Grandi,[2] the Frenchman Chalosse,[3] the Austrian Steinmeyer,[4] the Americans Adams[5] and Preston.[6] As a sample of a seventeenth-century procedure, we give Schott's solution[7] of $\frac{x}{2}-\frac{x}{6}-\frac{x}{8}=30$. He tries $x=24$ and $x=48$. He obtains errors -25 and -20. The work is arranged as follows:

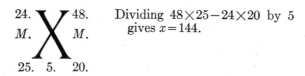

24.　　　48.　　　Dividing $48\times25-24\times20$ by 5
M.　　　M.　　　gives $x=144$.

25.　5.　20.

220. *Compound proportion with integers.*—We begin again with Leonardo of Pisa (1202)[8] who gives the problem: If 5 horses eat 6 quarts of barley in 9 days, for how many days will 16 quarts feed 10 horses? His numbers are arranged thus:

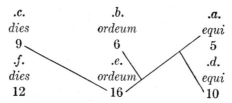

.c.　　　　.b.　　　　.a.
dies　　　ordeum　　　equi
9　　　　6　　　　5
.f.　　　.e.　　　.d.
dies　　　ordeum　　　equi
12　　　　16　　　　10

The answer is obtained by dividing $9\times16\times5$ by the product of the remaining known numbers. Answer 12.

Somewhat different applications of lines crossing each other are given by Nicolas Chuquet[9] and Luca Pacioli[10] in dealing with numbers in continued proportion.

[1] *Mr. Wingate's Arithmetick*, enlarged by John Kersey (11th ed.), with supplement by George Shelley (London, 1704), p. 128.

[2] Guido Grandi, *Instituzioni di arithmetia pratica* (Firenze, 1740), p. 104.

[3] *L'arithmetique par les fractions* ... par M. Chalosse (Paris, 1747), p. 158.

[4] *Tirocinium Arithmeticum a P. Philippo Steinmeyer* (Vienna and Freiburg, 1763), p. 475.

[5] Daniel Adams, *Scholar's Arithmetic* (10th ed.; Keene, N.H., 1816), p. 199.

[6] John Preston, *Lancaster's Theory of Education* (Albany, N.Y., 1817), p. 349.

[7] G. Schott, *Cursus mathematicus* (Würzburg, 1661), p. 36.

[8] *Op. cit.*, p. 132.

[9] Nicolas Chuquet, *Le Triparty en la Science des Nombres* (1484), edited by A. Marre, in *Bullettino Boncompagni*, Vol. XIII (1880), p. 700; reprint (Roma, 1881), p. 115.

[10] Luca Pacioli, *op. cit.*, fol. 93a.

Chuquet finds two mean proportionals between 8 and 27 by the scheme

where 12 and 18 are the two mean proportionals sought; i.e., 8, 12, 18, 27 are in continued proportion.

221. *Proportions involving fractions.*—Lines forming a cross (×), together with two horizontal parallel lines, were extensively applied to the solution of proportions involving fractions, and constituted a most clever device for obtaining the required answer mechanically. If it is the purpose of mathematics to resolve complicated problems by a minimum mental effort, then this device takes high rank.

The very earliest arithmetic ever printed, namely, the anonymous booklet gotten out in 1478 at Treviso,[1] in Northern Italy, contains an interesting problem of two couriers starting from Rome and Venice, respectively, the Roman reaching Venice in 7 days, the Venetian arriving at Rome in 9 days. If Rome and Venice are 250 miles apart, in how many days did they meet, and how far did each travel before they met? They met in $3\frac{15}{16}$ days. The computation of the distance traveled by the courier from Rome calls for the solution of the proportion which we write $7 : 250 = \frac{63}{16} : x$.

The Treviso arithmetic gives the following arrangement:

$$112$$

$$\frac{7}{1} \diagdown \diagup \frac{250\text{------}63}{1\text{------}16}$$

The connecting lines indicate what numbers shall be multiplied together; namely, 1, 250, and 63, also 7, 1, and 16. The product of the latter—namely, 112—is written above on the left. The author then finds $250 \times 63 = 15{,}750$ and divides this by 112, obtaining $140\frac{5}{8}$ miles.

These guiding lines served as Ariadne threads through the maze of a proportion involving fractions.

We proceed to show that this magical device was used again by Chuquet (1484), Widman (1489), and Pacioli (1494). Thus Chuquet[2]

---

[1] The Treviso arithmetic of 1478 is described and partly given in facsimile by Boncompagni in *Atti dell'Accademia Pontificia de' nuovi Lincei*, Tome XVI (1862–63; Roma, 1863), see p. 568.

[2] Chuquet, in Boncompagni, *Bullettino*, Vol. XIII, p. 636; reprint, p. (84).

uses the cross in the problem to find two numbers in the ratio of $\frac{2}{3}$
to $\frac{3}{4}$ and whose sum is 100. He writes $-\underset{4}{\overset{3}{\diagdown}}\underset{3}{\overset{2}{\diagup}}-$; multiplying 3 by 3,
and 2 by 4, he obtains two numbers in the proper ratio. As their
sum is only 17, he multiplies each by $\frac{100}{17}$ and obtains $47\frac{1}{17}$ and $52\frac{16}{17}$.
Johann Widman[1] solves the proportion $9 : \frac{53}{8} = \frac{89}{8} : x$ in this man-
ner: "Secz also $-\underset{1}{\overset{9}{\diagdown}}\underset{8}{\overset{53}{\diagup}} \underset{}{\overset{89}{}} \underset{}{\overset{8}{}}$ machss nach der Regel vnd küpt 8 fl.
35s 9 helr $\frac{5}{12}$." It will be observed that the computer simply took the
products of the numbers connected by lines. Thus $1 \times 53 \times 89 = 4{,}717$
gives the numerator of the fourth term; $9 \times 8 \times 8 = 576$ gives the
denominator. The answer is 8 florins and a fraction.

Such settings of numbers are found in Luca Pacioli,[2] Ch. Rudolph,[3]
G. Sfortunati,[4] O. Schreckenfuchs,[5] Hortega,[6] Tartaglia,[7] M. Stein-
metz,[8] J. Trenchant,[9] Hermann Follinus,[10] J. Alsted,[11] P. Hérigone,[12]
Chalosse,[13] J. Perez de Moya.[14] It is remarkable that in England neither
Tonstall nor Recorde used this device. Recorde[15] and Leonard Digges[16]

[1] Johann Widman, op. cit.; see J. W. L. Glaisher, op. cit., p. 6.

[2] Luca Pacioli, op. cit. (1523), fol. 18, 27, 54, 58, 59, 64.

[3] Christoph Rudolph, Kunstliche Rechnung (1526). We have used one of the
Augsburg editions, 1574 or 1588 (title-page missing), CVII.

[4] Giovanni Sfortunati da Siena, Nvovo Lvme. Libro di Arithmetica (1534),
fol. 37.

[5] O. Schreckenfuchs, op. cit. (1551).

[6] Juan de Hortega, op. cit. (1552), fol. 92a.

[7] N. Tartaglia, General Trattato di Nvmeri (la prima parte, 1556), fol. 111b,
117a.

[8] Arithmeticae Praecepta .... M. Mavricio Steinmetz Gersbachio (Leipzig,
1568) (no paging).

[9] J. Trenchant, op. cit., p. 142.

[10] Hermannvs Follinvs, Algebra sive liber de rebvs occvltis (Cologne, 1622), p. 72.

[11] Johannis-Henrici Alstedii Encyclopaedia (Hernborn, 1630), Lib. XIV,
Cossae libri III, p. 822.

[12] Pierre Herigone, Cvrsvs mathematici, Tomus VI (Paris, 1644), p. 320.

[13] L'Arithmetique par les fractions ... par M. Chalosse (Paris, 1747), p. 71.

[14] Juan Perez de Moya, Arithmetica (Madrid, 1784), p. 141. This text reads
the same as the edition that appeared in Salamanca in 1562.

[15] Robert Recorde, op. cit., p. 175.

[16] (Leonard Digges), A Geometrical Practical Treatise named Pantometria
(London, 1591).

use a slightly different and less suggestive scheme, namely, the capital letter Z for proportions involving either integers or fractions. Thus, $3:8=16:x$ is given by Recorde in the form $\underset{8}{\overset{3}{Z}}{\overset{16}{}}$. This rather unusual notation is found much later in the *American Accomptant* of Chauncey Lee (Lansinburgh, 1797, p. 223) who writes,

$$
\begin{array}{cc}
\text{``Cause} & \text{Effect''} \\
\text{4.5 yds.}\text{---}\overline{}18 \\
90\text{---}Q
\end{array}
$$

and finds $Q=90\times18\div4.5=360$ dollars.

222. *Addition and subtraction of fractions.*—Perhaps even more popular than in the solution of proportion involving fractions was the use of guiding lines crossing each other in the addition and subtraction of fractions. Chuquet[1] represents the addition of $\frac{2}{3}$ and $\frac{4}{5}$ by the following scheme:

$$
\begin{array}{cc}
\text{`` }10 & 12\text{ ''} \\
\hline
2 & 4 \\
\hline
3 & 5 \\
\end{array}
$$
$$\cdot 15 \cdot$$

The lower horizontal line gives $3\times5=15$; we have also $2\times5=10$, $3\times4=12$; hence the sum $\frac{22}{15}=1\frac{7}{15}$.

The same line-process is found in Pacioli,[2] Rudolph,[3] Apianus.[4] In England, Tonstall and Recorde do not employ this intersecting line-system, but Edmund Wingate[5] avails himself of it, with only slight variations in the mode of using it. We find it also in Oronce Fine,[6] Feliciano,[7] Schreckenfuchs,[8] Hortega,[9] Baëza,[10] the Italian

[1] Nicolas Chuquet, *op. cit.*, Vol. XIII, p. 606; reprint p. (54).

[2] Luca Pacioli, *op. cit.* (1523), fol. 51, 52, 53.

[3] Christoph Rudolph, *op. cit.*, under addition and subtraction of fractions.

[4] Petrus Apianus, *Kauffmansz Rechnung* (Ingolstadt, 1527).

[5] E. Wingate, *op. cit.* (1704), p. 152.

[6] *Orontii Finei Delphinatis, liberalivm Disciplinarvm professoris Regii Protomathesis: Opus varium* (Paris, 1532), fol. 4b.

[7] Francesco Feliciano, *Libro de arithmetica e geometria* (1550).

[8] O. Schreckenfuchs, *op. cit.*, "Annot.," fol. 25b.

[9] Hortega, *op. cit.* (1552), fol. 55a, 63b.

[10] *Nvmerandi doctrina*, authore Lodoico Baëza (Paris, 1556), fol. 38b.

translation of Fine's works,[1] Gemma Frisius,[2] Eyçaguirre,[3] Clavius,[4] the French translation of Tartaglia,[5] Follinus,[6] Girard,[7] Hainlin,[8] Caramuel,[9] Jeake,[10] Corachan,[11] Chalosse,[12] De Moya,[13] and in slightly modified form in Crusoe.[14]

223. *Division of fractions.*—Less frequent than in the preceding processes is the use of lines in the multiplication or division of fractions, which called for only one of the two steps taken in solving a proportion involving fractions. Pietro Borgi (1488)[15] divides $\frac{3}{4}$ by $\frac{4}{5}$

thus: $\;$ "$\dfrac{4}{5}\!\diagdown\!\!\!\diagup\dfrac{3}{4}\;\;\dfrac{15}{16}$" $\,$. $\;$ In dividing $\frac{1}{2}$ by $\frac{1}{3}$, Pacioli[16] writes

$$\text{``2}\qquad 3\text{''}$$
$$\frac{1}{3}\diagdown\!\!\!\!\diagup\frac{1}{2}$$

and obtains $\frac{3}{2}$ or $1\frac{1}{2}$.

Petrus Apianus (1527) uses the $\times$ in division. Juan de Hortega (1552)[17] divides $\frac{3}{8}$ by $\frac{5}{6}$, according to the following scheme:

$$\text{``}\;\;\dfrac{3}{8}\overset{18}{\underset{40}{\diagdown\!\!\!\!\diagup}}\dfrac{5}{6}\qquad val\hat{e}\;\;\dfrac{9}{20}\;\text{''}\,.$$

[1] *Opere di Orontio Fineo del Definato.* ... Tradotte da Cosimo Bartoli (Venice, 1587), fol. 31.

[2] *Arithmeticae Practicae methodvs facilis,* per Gemmam Frisium ... iam recèns ab ipso authore emendata ... (Paris, 1569), fol. 20.

[3] Sebastian Fernandez Eyçaguirre, *Libro de Arithmetica* (Brussels, 1608), p. 38.

[4] Chr. Clavius, *Opera omnia,* Tom. I (1611), Euclid, p. 383.

[5] *L'Arithmetique de Nicolas Tartaglia Brescian,* traduit ... par Gvillavmo Gosselin de Caen (Paris, 1613), p. 37.

[6] *Algebra sive Liber de Rebvs Occvltis,* ... Hermannvs Follinvs (Cologne, 1622), p. 40.

[7] Albert Girard, *Invention Nouvelle en L'Algebre* (Amsterdam, 1629).

[8] Johan. Jacob Hainlin, *Synopsis mathematica* (Tübingen, 1653), p. 32.

[9] *Joannis Caramvelis Mathesis Biceps Vetus et Nova* (Companiae, 1670), p. 20.

[10] Samuel Jeake, *op. cit.,* p. 51.

[11] Juan Bautista Corachan, *Arithmetica demonstrada* (Barcelona, 1719), p. 87.

[12] *L'Arithmetique par les fractions* ... par M. Chalosse (Paris, 1747), p. 8.

[13] J. P. de Moya, *op. cit.* (1784), p. 103.

[14] George E. Crusoe, *Y Mathematics?* ("Why Mathematics?") (Pittsburgh, Pa., 1921), p. 21.

[15] Pietro Borgi, *Arithmetica* (Venice, 1488), fol. 33*B.*

[16] L. Pacioli, *op. cit.* (1523), fol. 54*a.*

[17] Juan de Hortega, *op. cit.* (1552), fol. 66*a.*

We find this use of $\times$ in division in Sfortunati,[1] Blundeville,[2] Steinmetz,[3] Ludolf van Ceulen,[4] De Graaf,[5] Samuel Jeake,[6] and J. Perez de Moya.[7] De la Chapelle, in his list of symbols,[8] introduces $\times$ as a regular sign of division, *divisé par*, and $\times$ as a regular sign of multiplication, *multiplié par*. He employs the latter regularly in multiplication, but he uses the former only in the division of fractions, and he explains that in $\frac{6}{7}\times\frac{3}{4}=\frac{24}{21}$, "le sautoir $\times$ montre que 4 doit multiplier 6 & que 3 doit multiplier 7," thus really looking upon $\times$ as two symbols, one placed upon the other.

224. In the multiplication of fractions Apianus[9] in 1527 uses the parallel horizontal lines, thus, $\frac{1-3}{2-5}$. Likewise, Michael Stifel[10] uses two horizontal lines to indicate the steps. He says: "Multiplica numeratores inter se, et proveniet numerator productae summae. Multiplica etiam denominatores inter se, et proveniet denominator productae summae."

225. *Casting out the 9's, 7's, or 11's.*—Checking results by casting out the 9's was far more common in old arithmetics than by casting out the 7's or 11's. Two intersecting lines afforded a convenient grouping of the four results of an operation. Sometimes the lines appear in the form $\times$, at other times in the form $+$. Luca Pacioli[11] divides 97535399 by 9876, and obtains the quotient 9876 and remainder 23. Casting out the 7's (i.e., dividing a number by 7 and noting the residue), he obtains for 9876 the residue 6, for 97535399 the residue 3, for 23 the residue 2. He arranges these residues thus: $\dfrac{"6|2}{6|3}\,"$.

Observe that multiplying the residues of the divisor and quotient, 6 times $6=36$, one obtains 1 as the residue of 36. Moreover, $3-2$ is also 1. This completes the check.

[1] Giovanni Sfortvnati da Siena, *Nvovo Lvme. Libro di Arithmetica* (1534), fol. 26.

[2] *Mr. Blundevil. His Exercises contayning eight Treatises* (London, 1636), p. 29.

[3] M. Mavricio Steinmetz Gersbachio, *Arithmeticae praecepta* (1568) (no paging).

[4] Ludolf van Ceulen, *De arithm.* (title-page gone) (1615), p. 13.

[5] Abraham de Graaf, *De Geheele Mathesis of Wiskonst* (Amsterdam, 1694), p. 14.

[6] Samuel Jeake, *op. cit.*, p. 58.   [7] Juan Perez de Moya, *op. cit.*, p. 117.

[8] De la Chapelle, *Institutions de géométrie* (4th éd.; Paris, 1765), Vol. I, p. 44, 118, 185.

[9] Petrus Apianus, *op. cit.* (1527).

[10] M. Stifel, *Arithmetica integra* (Nuremberg, 1544), fol. 6.

[11] Luca Pacioli, *op. cit.* (1523), fol. 35.

Nicolas Tartaglia[1] checks, by casting out the 7's, the division $912345 \div 1987 = 459$ and remainder 312.

Casting the 7's out of 912345 gives 0, out of 1987 gives 6, out of 459 gives 4, out of 312 gives 4. Tartaglia writes down "$\frac{4|4}{6|0}$".

Here 4 times $6 = 24$ yields the residue 3; 0 minus 4, or better 7 minus 4, yields 3 also. The result "checks."

Would it be reasonable to infer that the two perpendicular lines + signified multiplication? We answer "No," for, in the first place, the authors do not state that they attached this meaning to the symbols and, in the second place, such a specialized interpretation does not apply to the other two residues in each example, which are to be *subtracted* one from the other. The more general interpretation, that the lines are used merely for the convenient grouping of the four residues, fits the case exactly.

Rudolph[2] checks the multiplication 5678 times $65 = 369070$ by casting out the 9's (i.e., dividing the sum of the digits by 9 and noting the residue); he finds the residue for the product to be 7, for the factors to be 2 and 8. He writes down

$$" \ \overset{7}{\underset{7}{8 \times 2}} \ " \ .$$

Here 8 times $2 = 16$, yielding the residue 7, written above. This residue is the same as the residue of the product; hence the check is complete. It has been argued that in cases like this Rudolph used $\times$ to indicate multiplication. This interpretation does not apply to other cases found in Rudolph's book (like the one which follows) and is wholly indefensible. We have previously seen that Rudolph used $\times$ in the addition and subtraction of fractions. Rudolph checks the proportion $9:11 = 48:x$, where $x = 58\frac{6}{9}$, by casting out the 7's, 9's, and 11's as follows:

$$"(7) \qquad\qquad (9) \qquad\qquad (11)"$$
$$\overset{6}{\underset{6}{2 \times 6}} \qquad\qquad \overset{0}{\underset{0}{0 \times 3}} \qquad\qquad \overset{0}{\underset{0}{9 \times 4}}$$

Take the check by 11's (i.e., division of a number by 11 and noting the residue). It is to be established that $9x = 48$ times 11, or that 9

---

[1] N. Tartaglia, *op. cit.* (1556), fol. 34*B*.

[2] Chr. Rudolph, *Kunstliche Rechnung* (Augsburg, 1574 or 1588 ed.) A VIII.

times 528=48 times 99. Begin by casting out the 11's of the factors 9 and 48; write down the residues 9 and 4. But the residues of 528 and 99 are both 0. Multiplying the residues 9 and 0, 4 and 0, we obtain in each case the product 0. This is shown in the figure. Note that here we do not take the product 9 times 4; hence $\times$ could not possibly indicate 9 times 4.

The use of $\times$ in casting out the 9's is found also in Recorde's *Grovnd of Artes* and in Clavius[1] who casts out the 9's and also the 7's.

Hortega[2] follows the Italian practice of using lines $+$, instead of $\times$, for the assignment of resting places for the four residues considered. Hunt[3] uses the Latin cross $-\!\!+$. The regular $\times$ is used by Regius (who also casts out the 7's),[4] Lucas,[5] Metius,[6] Alsted,[7] York,[8] Dechales,[9] Ayres,[10] and Workman.[11]

In the more recent centuries the use of a cross in the process of casting out the 9's has been abandoned almost universally; we have found it given, however, in an English mathematical dictionary[12] of 1814 and in a twentieth-century Portuguese cyclopedia.[13]

226. *Multiplication of integers.*—In Pacioli the square of 37 is found mentally with the aid of lines indicating the digits to be multiplied together, thus:

$$\text{``3} \underset{3}{\overset{7}{\times}} \text{7''}$$

1369

[1] Chr. Clavius, *Opera omnia* (1612), Tom. I (1611), "Numeratio," p. 11.

[2] Juan de Hortega, *op. cit.*, fol. 42b.

[3] Nicolas Hunt, *Hand-Maid to Arithmetick* (London 1633).

[4] Hudalrich Regius, *Vtrivsqve Arithmetices Epitome* (Strasburg, 1536), fol. 57; *ibid.* (Freiburg-in-Breisgau, 1543), fol. 56.

[5] Lossius Lucas, *Arithmetices Erotemata Pverilia* (Lüneburg, 1569), fol. 8.

[6] *Adriani Metii Alcmariani Arithmeticae libri dvo:* Leyden, Arith. Liber I, p. 11.

[7] Johann Heinrich Alsted, *Methodus Admirandorum mathematicorum novem Libris* (Tertia editio; Herbon, 1641), p. 32.

[8] Tho. York, *Practical Treatise of Arithmetick* (London, 1687), p. 38.

[9] R. P. Claudii Francisci Milliet Dechales Camberiensis, *Mundus Mathematicus. Tomus Primus, Editio altera* (Leyden, 1690), p. 369.

[10] John Ayres, *Arithmetick made Easie*, by E. Hatton (London, 1730), p. 53.

[11] Benjamin Workman, *American Accountant* (Philadelphia, 1789), p. 25.

[12] Peter Barlow, *Math. & Phil. Dictionary* (London, 1814), art. "Multiplication."

[13] *Encyclopedia Portugueza* (Porto), art. "Nove."

From the lower 7 two lines radiate, indicating 7 times 7, and 7 times 3. Similarly for the lower 3. We have here a cross as part of the line-complex. In squaring 456 a similar scheme is followed; from each digit there radiate in this case three lines. The line-complex involves three vertical lines and three well-formed crosses ×. The multiplication of 54 by 23 is explained in the manner of Pacioli by Mario Bettini[1] in 1642.

There are cases on record where the vertical lines are omitted, either as deemed superfluous or as the result of an imperfection in the typesetting. Thus an Italian writer, Unicorno,[2] writes:

$$\text{``7} \qquad \text{8''}$$
$$5 \qquad 6$$
$$\overline{4368}$$

It would be a rash procedure to claim that we have here a use of × to indicate the product of two numbers; these *lines* indicate the product of 6 and 70, and of 50 and 8; the lines are not to be taken as *one* symbol; they do not mean 78 times 56. The capital letter X is used by F. Ghaligai in a similar manner in his *Algebra*. The same remarks apply to J. H. Alsted[3] who uses the X, but omits the vertical lines, in finding the square of 32.

A procedure resembling that of Pacioli, but with the lines marked as arrows, is found in a recent text by G. E. Crusoe.[4]

227. *Reducing radicals to radicals of the same order.*—Michael Stifel[5] in 1544 writes: "Vt volo reducere $\sqrt{z}$ 5 et $\sqrt{c\!\!/}$ 4 ad idem signum, sic stabit exemplum ad regulam

$$5 \qquad 4$$
$$\sqrt{z} \qquad \sqrt{c\!\!/}$$

[1] Mario Bettino, *Apiaria Vniversae philosophiae mathematicae* (Bologna, 1642), "Apiarivm vndecimvm," p. 37.

[2] S. Joseppo Vnicorno, *De l'arithmetica universale* (Venetia, 1598), fol. 20. Quoted from C. le Paige, "Sur l'origine de certains signes d'opération," *Annales de la société scientifique de Bruxelles* (16th year, 1891–92), Part II, p. 82.

[3] J. H. Alsted, *Methodus Admirandorum Mathematicorum Novem libris exhibens universam mathesin* (tertiam editio; Herbon, 1641), p. 70.

[4] George E. Crusoe, *op. cit.*, p. 6.

[5] Michael Stifel, *Arithmetica integra* (1544), fol. 114.

$\sqrt{zcl}125$ et $\sqrt{zcl}16$." Here $\sqrt{5}$ and $\sqrt[3]{4}$ are reduced to radicals of the same order by the use of the cross $\times$. The orders of the given radicals are two and three, respectively; these orders suggest the cube of 5 or 125 and the square of 4, or 16. The answer is $\sqrt[6]{125}$ and $\sqrt[6]{16}$.

Similar examples are given by Stifel in his edition of Rudolff's *Coss*,[1] Peletier,[2] and by De Billy.[3]

228. *To mark the place for "thousands."*—In old arithmetics explaining the computation upon lines (a modified abacus mode of computation), the line on which a dot signified "one thousand" was marked with a $\times$. The plan is as follows:

$$\times\!\!-\!\!-\!\!-\!\!-\!\!-\!\!-\!\!1000$$
$$500$$
$$-\!\!-\!\!-\!\!-\!\!-\!\!-\!\!100$$
$$50$$
$$-\!\!-\!\!-\!\!-\!\!-\!\!-\!\!50$$
$$5$$
$$-\!\!-\!\!-\!\!-\!\!-\!\!-\!\!1$$

This notation was widely used in Continental and English texts.

229. *In place of multiplication table above 5×5.*—This old procedure is graphically given in Recorde's *Grovnd of Artes* (1543?). Required to multiply 7 by 8. Write the 7 and 8 at the cross as shown here; next, $10-8=2$, $10-7,=3$; write the 2 and 3 as shown:

Then, $2\times3=6$, write the 6; $7-2=5$, write the 5. The required product is 56. We find this process again in Oronce Fine,[4] Regius,[5]

---

[1] Michael Stifel, *Die Coss Christoffs Rudolffs* (Amsterdam, 1615), p. 136. (First edition, 1553.)

[2] *Jacobi Peletarii Cenomani, de occvlta parte nvmerorvm, qvam Algebram vocant, Libri duo* (Paris, 1560), fol. 52.

[3] Jacqves de Billy, *Abregé des Preceptes d'Algebre* (Reims, 1637), p. 22. See also the *Nova Geometriae Clavis*, authore P. Jacobo de Billy (Paris, 1643), p. 465.

[4] *Orontii Finei Delphinatis, liberalivm Disciplinarvm prefossoris* Regii Protomathesis: Opus uarium (Paris, 1532), fol. 4b.

[5] Hudalrich Regius, *Vtrivsqve arithmetices Epitome* (Strasburg, 1536), fol. 53; *ibid.* (Freiburg-in-Breisgau, 1543), fol. 56.

Stifel,[1] Boissiere,[2] Lucas,[3] the Italian translation of Oronce Fine,[4] the French translation of Tartaglia,[5] Alsted,[6] Bettini.[7] The French edition of Tartaglia gives an interesting extension of this process, which is exhibited in the product of 996 and 998, as follows:

$$994 \ 0 \ 0 \ 8$$

230. *Amicable numbers.*—N. Chuquet[8] shows graphically that 220 and 284 are amicable numbers (each the sum of the factors of the other) thus:

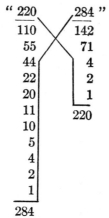

The old graphic aids to computation which we have described are interesting as indicating the emphasis that was placed by early arithmeticians upon devices that appealed to the eye and thereby contributed to economy of mental effort.

231. *The St. Andrew's cross used as a symbol of multiplication.*— As already pointed out, Oughtred was the first (§ 181) to use × as the

[1] Michael Stifel, *Arithmetica integra* (Nuremberg, 1544), fol. 3.

[2] Claude de Boissiere, Daulphinois, *L'Art d'Arythmetique* (Paris, 1554), fol. 15b.

[3] Lossius Lucas, *Arithmetices Erotemata Pverilia* (Lüneburg, 1569), fol. 8.

[4] *Opere di Orontio Fineo.* ... Tradotte da Cosimo Bartoli (Bologna, 1587), "Della arismetica," libro primo, fol. 6, 7.

[5] *L'arithmetique de Nicolas Tartaglia* ... traduit ... par Gvillavmo Gosselin de Caen. (Paris, 1613), p. 14.

[6] *Johannis-Henrici Alstedii Encyclopaedia* (Herbon, 1630), Lib. XIV, p. 810.

[7] Mario Bettino, *Apiaria* (Bologna, 1642), p. 30, 31.

[8] N. Chuquet, *op. cit.*, Vol. XIII, p. 621; reprint, p. (69).

sign of multiplication of two numbers, as $a \times b$ (see also §§ 186, 288). The cross appears in Oughtred's *Clavis mathematicae* of 1631 and, in the form of the letter X, in E. Wright's edition of Napier's *Descriptio* (1618). Oughtred used a small symbol $\times$ for multiplication (much smaller than the signs $+$ and $-$). In this practice he was followed by some writers, for instance, by Joseph Moxon in his *Mathematical Dictionary* (London, 1701), p. 190. It seems that some objection had been made to the use of this sign $\times$, for Wallis writes in a letter of September 8, 1668: "I do not understand why the sign of multiplication $\times$ should more trouble the convenient placing of the fractions than the older signs $+ - = > :: .$"[1] It may be noted that Oughtred wrote the $\times$ small and placed it high, between the factors. This practice was followed strictly by Edward Wells.[2]

On the other hand, in A. M. Legendre's famous textbook *Géométrie* (1794) one finds (p. 121) a conspicuously large-sized symbol $\times$, for multiplication. The following combination of signs was suggested by Stringham:[3] Since $\times$ means "multiplied by," and $/$ "divided by," the union of the two, viz., $\times/$, means "multiplied or divided by."

232. *Unsuccessful symbols for multiplication.*—In the seventeenth century a number of other designations of multiplication were proposed. Hérigone[4] used a rectangle to designate the product of two factors that were separated by a comma. Thus, "$\square 5+4+3, 7 \sim 3 : \sim 10, est\, 38$" meant $(5+4+3) \cdot (7-3) - 10 = 38$. Jones, in his *Synopsis palmariorum* (1706), page 252, uses the $\sqsubset$, the Hebrew letter *mem*, to denote a rectangular area. A six-pointed star was used by Rahn and, after him, by Brancker, in his translation of Rahn's *Teutsche Algebra* (1659). "The Sign of Multiplication is [✶] i.e., multiplied with." We encounter this use of ✶ in the *Philosophical Transactions*.[5]

Abraham de Graaf followed a practice, quite common among Dutch writers of the seventeenth and eighteenth centuries, of placing symbols on the right of an expression to signify direct operations (multiplication, involution), and placing the same symbols on the

---

[1] S. P. Rigaud, *Correspondence of Scientific Men of the Seventeenth Century* (Oxford, 1841), Vol. II, p. 494.

[2] Edward Wells, *The Young Gentleman's Arithmetic and Geometry* (2d ed.; London, 1723); "Arithmetic," p. 16, 41; "Geometry," p. 283, 291.

[3] Irving Stringham, *Uniplanar Algebra* (San Francisco, 1893), p. xiii.

[4] P. Herigone, *Cursus mathematici* (1644), Vol. VI, *explicatio notarum.* (First edition, 1642.)

[5] *Philosophical Transactions*, Vol. XVII, (1692–94), p. 680. See also §§ 194, 547.

left of an expression to signify inverse operations. Thus, Graaf[1] multiplies x^2+4 by $2\frac{1}{4}$ by using the following symbolism:

$$" \ \frac{x \ x \ tot \ 4}{als \ \frac{9}{4} \ xx \ tot \ 9} \ 2\frac{1}{4} \ "$$

In another place he uses this same device along with double commas, thus

$$" \ \frac{\overline{a+b} \ , \ , \ -cc}{\overline{a+b} \ , \ , \ -ccd} \ d \ "$$

to represent $(a+b)(-cc)(d) = (a+b)(-ccd)$.

Occasionally the comma was employed to mark multiplication, as in Hérigone (§ 189), F. Van Schooten,[2] who in 1657 gives $\dfrac{23,23,11,2}{3,3,3\sqrt{113,5}}$, where all the commas signify "times," as in Leibniz (§§ 197, 198, 547), in De Gua[3] who writes "3, 4, 5 . . . . &c. $\overline{n-m-2}$," in Petrus Horrebowius[4] who lets "A,B" stand for A times B, in Abraham de Graaf[5] who uses one or two commas, as in "$\overline{p-b},a$" for $(p-b)a$. The German Hübsch[6] designated multiplication by $\int$, as in $\frac{2}{5}\int\frac{1}{3}$.

233. *The dot for multiplication.*—The dot was introduced as a symbol for multiplication by G. W. Leibniz. On July 29, 1698, he wrote in a letter to John Bernoulli: "I do not like $\times$ as a symbol for multiplication, as it is easily confounded with x; . . . . often I simply relate two quantities by an interposed dot and indicate multiplication by $ZC \cdot LM$. Hence, in designating ratio I use not one point but two points, which I use at the same time for division." It has been stated that the dot was used as a symbol for multiplication before Leibniz, that Thomas Harriot, in his *Artis analyticae praxis* (1631), used the dot in the expressions "$aaa-3\cdot bba = +2\cdot ccc$." Similarly, in explaining cube root, Thomas Gibson[7] writes, in 1655, "$3\cdot bb$," "$3\cdot bcc$," but it

[1] Abraham de Graaf, *Algebra of Stelkonst* (Amsterdam, 1672), p. 8.

[2] *Francisci à Schooten. ... Exercitationum mathematicarum liber primus* (Leyden, 1657), p. 89.

[3] L'Abbe' de Gua, *Histoire de l'academie r. d. sciences, année 1741* (Paris, 1744), p. 81.

[4] *Petri Horrebowii. . . . . Operum mathematico-physicorum tomus primus* (Havniae, 1740), p. 4.

[5] Abraham de Graaf, *op. cit.* (1672), p. 87.

[6] J. G. G. Hübsch, *Arithmetica Portensis* (Leipzig, 1748). Taken from Wildermuth's article, "Rechnen," in K. A. Schmid's *Encyklopaedie des gesammten Erziehungs- und Unterrichtswesens* (1885).

[7] Tho. Gibson, *Syntaxis mathematica* (London, 1655), p. 36.

is doubtful whether either Harriot or Gibson meant these dots for multiplication. They are introduced without explanation. It is much more probable that these dots, which were placed after numerical coefficients, are survivals of the dots habitually used in old manuscripts and in early printed books to separate or mark off numbers appearing in the running text. Leibniz proposed the dot after he had used other symbols for over thirty years. In his first mathematical publication, the *De arte combinatoria*[1] of 1666, he used a capital letter C placed in the position ○ for multiplication, and placed in the position ○ for division. We have seen that in 1698 he advocated the point. In 1710 the Leibnizian symbols[2] were explained in the publication of the Berlin Academy (§ 198); multiplication is designated by apposition, and by a dot or comma (*punctum vel comma*), as in 3,2 or $a,b+c$ or $AB,CD+EF$. If at any time some additional symbol is desired, ○ is declared to be preferable to ×.

The general adoption of the dot for multiplication in Europe in the eighteenth century is due largely to Christian Wolf. It was thus used by L. Euler; it was used by James Stirling in Great Britain, where the Oughtredian × was very popular.[3] Whitworth[4] stipulates, "The full point is used for the sign of multiplication."

234. *The St. Andrew's cross in notation for transfinite ordinal numbers.*—The notation $\omega \times 2$, with the multiplicand on the left, was chosen by G. Cantor in the place of 2ω (where ω is the first transfinite ordinal number), because in the case of three ordinal transfinite numbers, a, β, γ, the product $a^{\beta} \cdot a^{\gamma}$ is equal to $a^{\beta+\gamma}$ when a^{β} is the multiplicand, but when a^{γ} is the multiplicand the product is $a^{\gamma+\beta}$. In transfinite ordinals, $\beta+\gamma$ is not equal to $\gamma+\beta$.

SIGNS FOR DIVISION AND RATIO

235. *Early symbols.*—Hilprecht[5] states that the Babylonians had an ideogram *IGI-GAL* for the expression of division. Aside from their fractional notation (§ 104), the Greeks had no sign for division. Diophantus[6] separates the dividend from the divisor by the words ἐν

---

[1] G. W. Leibniz, *Opera omnia*, Vol. II (Geneva, 1768), p. 347.

[2] *Miscellanea Berolinensia* (Berlin), Vol. I (1710), p. 156.

[3] See also §§ 188, 287, 288; Vol. II, §§ 541, 547.

[4] W. A. Whitworth, *Choice and Chance* (Cambridge, 1886), p. 19.

[5] H. V. Hilprecht, *The Babylonian Expedition Mathematical, etc., Tablets from the Temple Library of Nippur* (Philadelphia, 1906), p. 22.

[6] Diophantus, *Arithmetica* (ed. P. Tannery; Leipzig, 1893), p. 286. See also G. H. F. Nesselmann, *Algebra der Griechen* (Berlin, 1842), p. 299.

μορίῳ or μορίου, as in the expression $\deltaῦ\bar{\varsigma}$ λείψει $\bar{\varsigma\varsigma}$ $\bar{\kappa\delta}$ μορίου $\deltaῦ\bar{α}μ^{\delta}$ $\bar{ι\beta}$ λείψεις$\bar{\varsigma\varsigma}$, which means $(7x^2 - 24x) \div (x^2 + 12 - 7x)$. In the Bakhshālī arithmetic (§ 109) division is marked by the abbreviation *bhâ* from *bhâga*, "part." The Hindus often simply wrote the divisor beneath the dividend. Similarly, they designated fractions by writing the denominator beneath the numerator (§§ 106, 109, 113). The Arabic author[1] al-Ḥaṣṣâr, who belongs to the twelfth century, mentions the use of a fractional line in giving the direction: "Write the denominators below a [horizontal] line and over each of them the parts belonging to it; for example, if you are told to write three-fifths and a third of a fifth, write thus, $\dfrac{3}{5}\dfrac{1}{3}$." In a second example, four-thirteenths and three-elevenths of a thirteenth is written $\dfrac{4}{13}\dfrac{3}{11}$. This is the first appearance of the fractional line, known to us, unless indeed Leonardo of Pisa antedates al-Ḥaṣṣâr. That the latter was influenced in this matter by Arabic authors is highly probable. In his *Liber abbaci* (1202) he uses the fractional line (§ 122). Under the caption[2] "De diuisionibus integrorum numerorum" he says: "Cum super quemlibet numerum quedam uirgula protracta fuerit, et super ipsam quilibet alius numerus descriptus fuerit, superior numerus partem uel partes inferioris numeri affirmat; nam inferior denominatus, et superior denominans appellatur. Vt si super binarium protracta fuerit uirgula, et super ipsam unitas descripta sit ipsa unitas unam partem de duabus partibus unius integri affirmat, hoc est medietatem sic $\frac{1}{2}$." ("When above any number a line is drawn, and above that is written any other number, the superior number stands for the part or parts of the inferior number; the inferior is called the denominator, the superior the numerator. Thus, if above the two a line is drawn, and above that unity is written, this unity stands for one part of two parts of an integer, i.e., for a half, thus $\frac{1}{2}$.") With Leonardo, an indicated division and a fraction stand in close relation. Leonardo writes also $\cdot\dfrac{1}{2}\dfrac{5}{6}\dfrac{7}{10}$, which means, as he explains, seven-tenths, and five-sixths of one-tenth, and one-half of one-sixth of one-tenth.

236. One or two lunar signs, as in 8)24 or 8)24(, which are often employed in performing long and short division, may be looked upon as symbolisms for division. The arrangement 8)24 is found in Stifel's

---

[1] H. Suter, *Bibliotheca mathematica* (3d ser.), Vol. II (1901), p. 24.

[2] *Il Liber abbaci di Leonardo Pisano* (ed. B. Boncompagni; Roma, 1857), p. 23, 24.

Arithmetica integra (1544)[1], and in W. Oughtred's different editions of his *Clavis mathematicae.* In Oughtred's *Opuscula posthuma* one finds also $\frac{4}{3}|\frac{3}{2}[\frac{9}{8}$, (§ 182). Joseph Moxon[2] lets $D)A+B-C$ signify our $(A+B-C)\div D$.

Perhaps the earliest to suggest a special symbol for division other than the fractional line, and the arrangement 5)15 in the process of dividing, was Michael Stifel[3] in his *Deutsche Arithmetica* (1545). By the side of the symbols + and − he places the German capitals $\mathfrak{M}$ and $\mathfrak{D}$, to signify multiplication and division, respectively. Strange to say, he did not carry out his own suggestion; neither he nor seemingly any of his German followers used the $\mathfrak{M}$ and $\mathfrak{D}$ in arithmetic or algebraic manipulation. The letters M and D are found again in S. Stevin, who expressed our $\dfrac{5x^2}{y} \cdot z^2$ in this manner:[4]

$$5②D \ sec \ ①M \ ter \ ②\,,$$

where *sec* and *ter* signify the "second" and "third" unknown quantity.

The inverted letter Ɔ is used to indicate division by Gallimard,[5] as in

$$\text{``12 Ɔ 4}=3\text{'' and ``}a^2b^2 \ Ɔ \ a^2 \ .\text{''}$$

In 1790 Da Cunha[6] uses the horizontal letter ⌐ as a mark for division.

237. *Rahn's notation.*—In 1659 the Swiss Johann Heinrich Rahn published an algebra[7] in which he introduced ÷ as a sign of division (§ 194). Many writers before him had used ÷ as a minus sign (§§ 164, 208). Rahn's book was translated into English by Thomas Brancker (a graduate of Exeter College, Oxford) and published, with additions from the pen of Joh. Pell, at London in 1668. Rahn's *Teutsche Algebra* was praised by Leibniz[8] as an "elegant algebra," nevertheless it did not enjoy popularity in Switzerland and the symbol ÷ for division

---

[1] Michael Stifel, *Arithmetica integra* (Nürnberg, 1544), fol. 317$V°$, 318$r°$. This reference is taken from J. Tropfke, *op. cit.*, Vol. II (2d ed., 1921), p. 28, n. 114.

[2] Joseph Moxon, *Mathematical Dictionary* (3d ed.; London, 1701), p. 190, 191.

[3] Michael Stifel, *Deutsche Arithmetica* (Nürnberg, 1545), fol. 74$v°$. We draw this information from J. Tropfke, *op. cit.*, Vol. II (2d ed., 1921), p. 21.

[4] S. Stevin, *Œuvres* (ed. A. Girard, 1634), Vol. I, p. 7, def. 28.

[5] J. E. Gallimard, *La Science du calcul numerique*, Vol. I (Paris, 1751), p. 4; *Methode ... d'arithmetique, d'algèbre et de géométrie* (Paris, 1753), p. 32.

[6] J. A. da Cunha, *Principios mathematicos* (1790), p. 214.

[7] J. H. Rahn, *Teutsche Algebra* (Zürich, 1659).

[8] *Leibnizens mathematische Schriften* (ed. C. I. Gerhardt), Vol. VII, p. 214.

was not adopted by his countrymen. In England, the course of events was different. The translation met with a favorable reception; Rahn's ÷ and some other symbols were adopted by later English writers, and came to be attributed, not to Rahn, but to John Pell. It so happened that Rahn had met Pell in Switzerland, and had received from him (as Rahn informs us) the device in the solution of equations of dividing the page into three columns and registering the successive steps in the solution. Pell and Brancker never claimed for themselves the introduction of the ÷ and the other symbols occurring in Rahn's book of 1569. But John Collins got the impression that not only the three-column arrangement of the page, but all the new algebraic symbols were due to Pell. In his extensive correspondence with John Wallis, Isaac Barrow, and others, Collins repeatedly spoke of ÷ as "Pell's symbol." There is no evidence to support this claim (§ 194).[1]

The sign ÷ as a symbol for division was adopted by John Wallis and other English writers. It came to be adopted regularly in Great Britain and the United States, but not on the European Continent. The only text not in the English language, known to us as using it, is one published in Buenos Aires;[2] where it is given also in the modified form $\cdot/\cdot$, as in $\frac{3}{4}\cdot/\cdot\frac{5}{8}=\frac{3 4}{2 0}$. In an American arithmetic,[3] the abbreviation ÷rs was introduced for "divisors," and ÷nds for "dividends," but this suggestion met with no favor on the part of other writers.

238. *Leibniz' notations.*—In the *Dissertatio de arte combinatoria* (1668)[4] G. W. Leibniz proposed for division the letter C, placed horizontally, thus ᗡ, but he himself abandoned this notation and introduced the colon. His article of 1684 in the *Acta eruditorum* contains for the first time in print the colon (:) as the symbol for division.[5] Leibniz says: ". . . . notetur, me divisionem hic designare hoc modo: $x:y$, quod idem est ac x divis. per y seu $\frac{x}{y}$." In a publication of the year 1710[6] we read: "According to common practice, the division

---

[1] F. Cajori, "Rahn's Algebraic Symbols," *Amer. Math. Monthly*, Vol. XXXI (1924), p. 65–71.

[2] Florentino Garcia, *El aritmético Argentino* (5th ed.; Buenos Aires, 1871), p. 102. The symbol ÷ and its modified form are found in the first edition of this book, which appeared in 1833.

[3] *The Columbian Arithmetician*, "by an American" (Haverhill [Mass.], 1811), p. 41.

[4] Leibniz, *Opera omnia*, Tom. II (Geneva, 1768), p. 347.

[5] See *Leibnizens mathematische Schriften* (ed. C. I. Gerhardt), Vol. V (1858), p. 223. See also M. Cantor, *Gesch. d. Mathematik*, Vol. III (2d ed.; Leipzig), p. 194.

[6] *Miscellanea Berolinensia* (Berlin, 1710), p. 156. See our § 198.

is sometimes indicated by writing the divisor beneath the dividend, with a line between them; thus a divided by b is commonly indicated by $\dfrac{a}{b}$; very often however it is desirable to avoid this and to continue on the same line, but with the interposition of two points; so that $a:b$ means a divided by b. But if, in the next place $a:b$ is to be divided by c, one may write $a:b,:c$, or $(a:b):c$. Frankly, however, in this case the relation can be easily expressed in a different manner, namely $a:(bc)$ or $a:bc$, for the division cannot always be actually carried out but often can only be indicated and then it becomes necessary to mark the course of the deferred operation by commas or parentheses."

In Germany, Christian Wolf was influential through his textbooks in spreading the use of the colon (:) for division and the dot (·) for multiplication. His influence extended outside Germany. A French translation of his text[1] uses the colon for division, as in "$(a-b):b$." He writes: "$a:mac=b:mbc$."

239. In Continental Europe the Leibnizian colon has been used for division and also for ratio. This symbolism has been adopted in the Latin countries with only few exceptions. In 1878 Balbontin[2] used in place of it the sign ÷ preferred by the English-speaking countries. Another Latin-American writer[3] used a slanting line in this manner, $\left(\dfrac{6}{7}\backslash 3\right)=\dfrac{6:3}{7}=\dfrac{2}{7}$ and also $12\backslash 3=4$. An author in Peru[4] indicates division by writing the dividend and divisor on the same line, but inclosing the former in a parenthesis. Accordingly, "$(20)5$" meant $20\div 5$. Sometimes he uses brackets and writes the proportion $2:1\frac{1}{2}=20:15$ in this manner: "$2:1[1]2::20:15$."

240. There are perhaps no symbols which are as completely observant of political boundaries as are ÷ and : as symbols for division. The former belongs to Great Britain, the British dominions, and the United States. The latter belongs to Continental Europe and the Latin-American countries. There are occasional authors whose prac-

[1] C. Wolf, *Cours de mathématique*, Tom. I (Paris, 1747), p. 110, 118.

[2] Juan Maria Balbontin, *Tratado elemental de aritmetica* (Mexico, 1878), p. 13.

[3] Felipe Senillosa, *Tratado elemental de arismética* (neuva ed.; Buenos Aires, 1844), p. 16. We quote from p. 47: "Este signo deque hemos hecho uso en la particion (\) no es usado generalmente; siendo el que se usa los dos punctos (:) ó la forma de quebrado. Pero un quebrado denota mas bien un cociente ó particion ejecutada que la operacion ó acto del partir; así hemos empleado este signo \ con analogia al del multiplicar que es éste: ×."

[4] Juan de Dios Salazar, *Lecciones de aritmetica* (Arequipa, 1827), p. v, 74, 89.

tices present exceptions to this general statement of boundaries, but their number is surprisingly small. Such statements would not apply to the symbolisms for the differential and integral calculus, not even for the eighteenth century. Such statements would not apply to trigonometric notations, or to the use of parentheses or to the designation of ratio and proportion, or to the signs used in geometry.

Many mathematical symbols approach somewhat to the position of world-symbols, and approximate to the rank of a mathematical world-language. To this general tendency the two signs of division ÷ and : mark a striking exception. The only appearance of ÷ signifying division that we have seen on the European Continent is in an occasional translation of an English text, such as Colin Maclaurin's *Treatise of Algebra* which was brought out in French at Paris in 1753. Similarly, the only appearance of : as a sign for division that we have seen in Great Britain is in a book of 1852 by T. P. Kirkman.[1] Saverien[2] argues against the use of more than one symbol to mark a given operation. "What is more useless and better calculated to disgust a beginner and embarrass even a geometer than the three expressions ·, :, ÷, to mark division?"

241. *Relative position of divisor and dividend.*—In performing the operation of division, the divisor and quotient have been assigned various positions relative to the dividend. When the "scratch method" of division was practiced, the divisor was placed beneath the dividend and moved one step to the right every time a new figure of the quotient was to be obtained. In such cases the quotient was usually placed immediately to the right of the dividend, but sometimes, in early writers, it was placed above the dividend. In short division, the divisor was often placed to the left of the dividend, so that $a)b(c$ came to signify division.

A curious practice was followed in the Dutch journal, the *Maandelykse Mathematische Liefhebberye* (Vol. I [1759], p. 7), where $a)$—— signifies division by a, and ——$(a$ means multiplication by a. Thus:

$$\text{``}\quad x)\dfrac{xy=b-a+x}{\ \ \ \ \ \ \ }\quad\text{''}$$
$$ergo\ y=\dfrac{b-a+x}{x}\ \cdot$$

James Thomson called attention to the French practice of writing the divisor on the right. He remarks: "The French place the divisor

[1] T. P. Kirkman, *First Mnemonial Lessons in Geometry, Algebra and Trigonometry* (London, 1852).

[2] Alexandre Saverien, *Dictionnaire universel de mathematique et de physique* (Paris, 1753), "Caractere."

to the right of the dividend, and the quotient below it. . . . . This mode gives the work a more compact and neat appearance, and possesses the advantage of having the figures of the quotient near the divisor, by which means the practical difficulty of multiplying the divisor by a figure placed at a distance from it is removed. . . . . This method might, with much propriety, be adopted in preference to that which is employed in this country."[1]

The arrangement just described is given in Bézout's arithmetic,[2] in the division of 14464 by 8, as follows:

$$\text{``}14464\underline{8}_{\displaystyle 1808}\text{ ,,}$$

242. *Order of operations in terms containing both $\div$ and $\times$.*—If an arithmetical or algebraical term contains $\div$ and $\times$, there is at present no agreement as to which sign shall be used first. "It is best to avoid such expressions."[3] For instance, if in $24 \div 4 \times 2$ the signs are used as they occur in the order from left to right, the answer is 12; if the sign $\times$ is used first, the answer is 3.

Some authors follow the rule that the multiplications and divisions shall be taken in the order in which they occur.[4] Other textbook writers direct that multiplications in any order be performed first, then divisions as they occur from left to right.[5] The term $a \div b \times b$ is interpreted by Fisher and Schwatt[6] as $(a \div b) \times b$. An English committee[7] recommends the use of brackets to avoid ambiguity in such cases.

243. *Critical estimates of : and $\div$ as symbols.*—D. André[8] expresses himself as follows: "The sign : is a survival of old mathematical notations; it is short and neat, but it has the fault of being symmetrical toward the right and toward the left, that is, of being a symmetrical sign of an operation that is asymmetrical. It is used less and less.

---

[1] James Thomson, *Treatise on Arithmetic* (18th ed.; Belfast, 1837).

[2] *Arithmétique de Bézout* ... par F. Peyrard (13th ed.; Paris, 1833).

[3] M. A. Bailey, *American Mental Arithmetic* (New York, 1892), p. 41.

[4] Hawkes, Luby, and Touton, *First Course of Algebra* (New York, 1910), p. 10.

[5] Slaught and Lennes, *High School Algebra, Elementary Course* (Boston, 1907), p. 212.

[6] G. E. Fisher and I. J. Schwatt, *Text-Book of Algebra* (Philadelphia, 1898), p. 85.

[7] "The Report of the Committee on the Teaching of Arithmetic in Public Schools," *Mathematical Gazette*, Vol. VIII (1917), p. 238. See also p. 296.

[8] Désiré André, *Des Notations mathématiques* (Paris, 1909), p. 58, 59.

.... When it is required to write the quotient of a divided by b, in the body of a statement in ordinary language, the expression $a:b$ really offers the typographical advantage of not requiring, as does $\frac{a}{b}$, a wider separation of the line in which the sign occurs from the two lines which comprehend it."

In 1923 the National Committee on Mathematical Requirements[1] voiced the following opinion: "Since neither ÷ nor :, as signs of division, plays any part in business life, it seems proper to consider only the needs of algebra, and to make more use of the fractional form and (where the meaning is clear) of the symbol /, and to drop the symbol ÷ in writing algebraic expressions."

244. *Notations for geometrical ratio.*—William Oughtred introduced in his *Clavis mathematicae* the dot as the symbol for ratio (§ 181). He wrote (§ 186) geometrical proportion thus, $a.b::c.d$. This notation for ratio and proportion was widely adopted not only in England, but also on the European Continent. Nevertheless, a new sign, the colon (:), made its appearance in England in 1651, only twenty years after the first publication of Oughtred's text. This colon is due to the astronomer Vincent Wing. In 1649 he published in London his *Urania practica*, which, however, exhibits no special symbolism for ratio. But his *Harmonicon coeleste* (London, 1651) contains many times Oughtred's notation $A.B::C.D$, and many times also the new notation $A:B::C:D$, the two notations being used interchangeably. Later there appeared from his pen, in London, three books in one volume, *Logistica astronomica* (1656), *Doctrina spherica* (1655), and *Doctrina theorica* (1655), each of which uses the notation $A:B::C:D$.

A second author who used the colon nearly as early as Wing was a schoolmaster who hid himself behind the initials "R.B." In his book entitled *An Idea of Arithmetik*, at first designed for the use of "the Free Schoole at Thurlow in Suffolk .... by R.B., Schoolmaster there" (London, 1655), one finds $1.6::4.24$ and also $A:a::C:c$.

W. W. Beman pointed out in *L'Intermédiaire des mathématiciens*, Volume IX (1902), page 229, that Oughtred's Latin edition of his *Trigonometria* (1657) contains in the explanation of the use of the tables, near the end, the use of : for ratio. It is highly improbable that the colon occurring in those tables was inserted by Oughtred himself. In the *Trigonometria* proper, the colon does not occur, and Ought-

---

[1] *Report of the National Committee on Mathematical Requirements under the Auspices of the Mathematical Association of America, Inc.* (1923), p. 81.

red's regular notation for ratio and proportion $A.B::C.D$ is followed throughout. Moreover, in the English edition of Oughtred's trigonometry, printed in the same year (1657), but subsequent to the Latin edition, the passage of the Latin edition containing the : is recast, the new notation for ratio is abandoned, and Oughtred's notation is introduced. The : used to designate ratio (§ 181) in Oughtred's *Opuscula mathematica hactenus inedita* (1677) may have been introduced by the editor of the book.

It is worthy of note, also, that in a text entitled *Johnsons Arithmetik; In two Bookes* (2d ed.: London, 1633), the colon (:) is used to designate a fraction. Thus $\frac{3}{4}$ is written 3:4. If a fraction be considered as an indicated division, then we have here the use of : for division at a period fifty-one years before Leibniz first employed it for that purpose in print. However, dissociated from the idea of a fraction, division is not designated by any symbol in Johnson's text. In dividing 8976 by 15 he writes the quotient "598 6:15."

As shown more fully elsewhere (§ 258), the colon won its way as the regular symbol for geometrical ratio, both in England and the European Continent.

245. Oughtred's dot and Wing's colon did not prevent experimentation with other characters for geometric ratio, at a later date. But none of the new characters proposed became serious rivals of the colon. Richard Balam,[1] in 1653, used the colon as a decimal separatrix, and proceeded to express ratio by turning the colon around so that the two dots became horizontal; thus "3 . . 1" meant the geometrical ratio 1 to 3. This designation was used by John Kirkby[2] in 1735 for arithmetical ratio; he wrote arithmetical proportion "9 . . 6 = 6 . . 3." In the algebra of John Alexander,[3] of Bern, geometrical ratio is expressed by a dot, $a.b$, and also by $a \dot{-} b$. Thomas York[4] in 1687 wrote a geometrical proportion "33600   7::153600   32," using no sign at all between the terms of a ratio.

In the minds of some writers, a geometrical ratio was something more than an indicated division. The operation of division was associated with rational numbers. But a ratio may involve incommensu-

[1] Richard Balam, *Algebra: or The Doctrine of Composing, Inferring, and Resolving an Equation* (London, 1653), p. 4.

[2] John Kirkby, *Arithmetical Institutions* (London, 1735), p. 28.

[3] *Synopsis algebraica, opus posthumum Iohannis Alexandri, Bernatis-Helvetii. In usum scholae mathematicae apud Hospitium-Christi Londinense* (London, 1693), p. 16, 55. An English translation by Sam. Cobb appeared at London in 1709.

[4] Thomas York, *Practical Treatise of Arithmetik* (London, 1687), p. 146.

rable magnitudes which are expressible by two numbers, one or both of which are irrational. Hence ratio and division could not be marked by the same symbol. Oughtred's ratio *a.b* was not regarded by him as an indicated division, nor was it a fraction. In 1696 this matter was taken up by Samuel Jeake[1] in the following manner: "And so by some, to distinguish them [ratios] from Fractions, instead of the intervening Line, two Pricks are set; and so the Ratio Sesquialtera is thus expressed $\frac{3}{2}$." Jeake writes the geometrical proportion, "$\frac{7}{1} \cdot \frac{9}{1} :: 7 \cdot 9$."

Emanuel Swedenborg starts out, in his *Daedalus Hyperboreus* (Upsala, 1716), to designate geometric proportion by : :: :, but on page 126 he introduces ∹ as a *signum analogicum* which is really used as a symbol for the ratio of quantities. On the European Continent one finds Hérigone[2] using the letter π to stand for "proportional" or ratio; he writes π where we write : . On the other hand, there are isolated cases where : was assigned a different usage; the Italian L. Perini[3] employs it as separatrix between the number of feet and of inches; his "11:4" means 11 feet 4 inches.

246. Discriminating between ratio and division, F. Schmeisser[4] in 1817 suggested for geometric ratio the symbol . . , which (as previously pointed out) had been used by Richard Balam, and which was employed by Thomas Dilworth[5] in London, and in 1799 by Zachariah Jess,[6] of Wilmington, Delaware. Schmeisser comments as follows: "At one time ratio was indicated by a point, as in *a.b*, but as this signifies multiplication, Leibniz introduced two points, as in *a:b*, a designation indicating division and therefore equally inconvenient, and current only in Germany. For that reason have Mönnich, v. Winterfeld, Krause and other thoughtful mathematicians in more recent time adopted the more appropriate designation *a..b*." Schmeisser writes (p. 233) the geometric progression: "÷3..6..12 ..24..48..96 . . . . ."

[1] Samuel Jeake, ΛΟΓΙΣΤΙΚΗΛΟΓΙΑ, *or Arithmetick* (London, 1696), p. **410**.

[2] Peter Herigone, *Cursus mathematicus*, Vol. I (Paris, 1834), p. 8.

[3] Lodovico Perini, *Geometria pratica* (Venezia, 1750), p. 109.

[4] Friedrich Schmeisser, *Lehrbuch der reinen Mathesis*, Erster Theil, "**Die** Arithmetik" (Berlin, 1817), Vorrede, p. 58.

[5] Thomas Dilworth, *The Schoolmaster's Assistant* (2d ed.; London, 1784). (First edition, about 1744.)

[6] Zachariah Jess, *System of Practical Surveying* (Wilmington, 1799), p. 173.

Similarly, A. E. Layng,[1] of the Stafford Grammar School in England, states: "The Algebraic method of expressing a ratio $\dfrac{A}{B}$ being a very convenient one, will also be found in the Examples, where it should be regarded as a symbol for the words *the ratio of A to B*, and not as implying the operation of division; it should not be used for *book-work*."

247. *Division in the algebra of complex numbers.*—As, in the algebra of complex numbers, multiplication is in general not commutative, one has two cases in division, one requiring the solution of $a=bx$, the other the solution of $a=yb$. The solution of $a=bx$ is designated by Peirce[2] $\dfrac{a}{b\times}$, by Schröder[3] $\dfrac{a}{b}$, by Study[4] and Cartan $\dfrac{a}{b}$. The solution of $a=yb$ is designated by Peirce $\dfrac{a}{\times b}$ and by Schröder $a:b$, by Study and Cartan $\dfrac{a}{.b}$. The $\times$ and the . indicate in this notation the place of the unknown factor. Study and Cartan use also the notations of Peirce and Schröder.

SIGNS OF PROPORTION

248. *Arithmetical and geometrical progression.*—The notation ∺ was used by W. Oughtred (§ 181) to indicate that the numbers following were in continued geometrical proportion. Thus, ∺ 2, 6, 18, 54, 162 are in continued geometric proportion. During the seventeenth and eighteenth centuries this symbol found extensive application; beginning with the nineteenth century the need of it gradually passed away, except among the Spanish-American writers. Among the many English writers using ∺ are John Wallis,[5] Richard Sault[6], Edward Cocker,[7] John Kersey,[8] William Whiston,[9] Alexander Mal-

---

[1] A. E. Layng, *Euclid's Elements of Geometry* (London, 1891), p. 219.

[2] B. Peirce, *Linear Associative Algebra* (1870), p. 17; *Amer. Jour. of Math.*, Vol. IV (1881), p. 104.

[3] E. Schröder, *Formale Elemente der absoluten Algebra* (Progr. Bade, 1874).

[4] E. Study and E. Cartan, *Encyclopédie des scien. math.*, Tom. I, Vol. I (1908), p. 373.

[5] *Phil. Trans.*, Vol. V (London, 1670), p. 2203.

[6] Richard Sault, *A New Treatise of Algebra* (London [no date]).

[7] *Cocker's Artificial Arithmetick*, by Edward Cocker, perused and published by John Hawkes (London, 1684), p. 278.

[8] John Kersey, *Elements of Algebra* (London, 1674), Book IV, p. 177.

[9] A. Tacquet's edition of *W. Whiston's Elementa Euclidea geometriae* (Amsterdam, 1725), p. 124.

colm,[1] Sir Jonas Moore,[2] and John Wilson.[3] Colin Maclaurin indicates in his *Algebra* (1748) a geometric progression thus: "$\div\div1:q:q^2$: $q^3:q^4:q^5$: etc." E. Bézout[4] and L. Despiau[5] write for arithmetical progression "$\div1.3.5.7.9$," and "$\div\div3:6:12$" for geometrical progression.

Symbols for arithmetic progression were less common than for geometric progression, and they were more varied. Oughtred had no symbol. Wallis[6] denotes an arithmetic progression $A, B, C, D \rightleftharpoons$, or by $a, b, c, d, e, f \div\div$. The sign $\div$, which we cited as occurring in Bézout and Despiau, is listed by Saverien[7] who writes "$\div1.2.3.4.5$, etc." But Saverien gives also the six dots $:::$, which occur in Stone[8] and Wilson.[9] A still different designation, $\div$, for arithmetical progression is due to Kirkby[10] and Emerson,[11] another $\div\div\div$ to Clark,[12] again another $\div\div$ is found in Blassière.[13] Among French writers using $\div$ for arithmetic progression and $\div\div$ for geometric progression are Lamy,[14] De Belidor,[15] Suzanne,[16] and Fournier;[17] among Spanish-American

---

[1] Alexander Malcolm, *A New System of Arithmetick* (London, 1730), p. 115.

[2] Sir Jonas Moore, *Arithmetick in Four Books* (3d ed.; London, 1688), beginning of the Book IV.

[3] John Wilson, *Trigonometry* (Edinburgh, 1714), p. 24.

[4] E. Bézout, *Cours de mathématiques*, Tome I (2. éd.; Paris, 1797), "Arithmétique," p. 130, 165.

[5] *Select Amusements in Philosophy of Mathematics* . . . . translated from the French of M. L. Despiau, Formerly Professor of Mathematics and Philosophy at Paris. . . . . Recommended . . . . by Dr. Hutton (London, 1801), p. 19, 37, 43.

[6] John Wallis, *Operum mathematicorvm Pars Prima* (Oxford, 1657), p. 230, 236.

[7] A. Saverien, *Dictionnaire universel de mathematique et de physique* (Paris, 1753), art. "Caractere."

[8] E. Stone, *New Mathematical Dictionary* (London, 1726), art. "Characters."

[9] John Wilson, *Trigonometry* (Edinburgh, 1714).

[10] John Kirkby, *Arithmetical Institutions containing a compleat System of Arithmetic* (London, 1735), p. 36.

[11] W. Emerson, *Doctrine of Proportion* (1763), p. 27.

[12] Gilbert Clark, *Oughtredus explicatus* (London, 1682), p. 114.

[13] J. J. Blassière, *Institution du calcul numerique et litteral* (a La Haye, 1770), end of Part II.

[14] Bernard Lamy, *Elemens des mathematiques* (3d ed.; Amsterdam, 1692), p. 156.

[15] B. F. de Belidor, *Nouveau Cours de mathématique* (Paris, 1725), p. 71, 139.

[16] H. Suzanne, *De la Manière d' étudier. es Mathématiques* (2. éd.; Paris, 1810), p. 208.

[17] C. F. Fournier, *Éléments d'Arithmétique et d'Algèbre*, Vol. II (Nantes, 1822).

writers using these two symbols are Senillosa,[1] Izquierdo,[2] Liévano,[3] and Porfirio da Motta Pegado.[4] In German publications ÷ for arithmetical progression and ∷ for geometric progression occur less frequently than among the French. In the 1710 publication in the *Miscellanea Berolinensia*[5] ÷ is mentioned in a discourse on symbols (§ 198). The ÷ was used in 1716 by Emanuel Swedenborg.[6]

Emerson[7] designated harmonic progression by the symbol ∷ and harmonic proportion by ∴ .

249. *Arithmetical proportion* finds crude symbolic representation in the *Arithmetic* of Boethius as printed at Augsburg in 1488 (see Figure 103). Being, in importance, subordinate to geometrical proportion, the need of a symbolism was less apparent. But in the seventeenth century definite notations came into vogue. William Oughtred appears to have designed a symbolism. Oughtred's language (*Clavis* [1652], p. 21) is "Ut 7.4:12.9 vel 7.7−3:12.12−3. Arithmeticè proportionales sunt." As later in his work he does not use arithmetical proportion in symbolic analysis, it is not easy to decide whether the symbols just quoted were intended by Oughtred as part of his algebraic symbolism or merely as punctuation marks in ordinary writing. John Newton[8] says: "As 8,5:6,3. Here 8 exceeds 5, as much as 6 exceeds 3."

Wallis[9] says: "Et pariter 5,3; 11,9; 17,15; 19,17. sunt in eadem progressione arithmetica." In P. Chelucci's[10] *Institutiones analyticae*, arithmetical proportion is indicated thus: 6.8 ∴ 10.12. Oughtred's notation is followed in the article "Caractère" of the *Encyclopédie*

[1] Felipe Senillosa, *Tratado elemental de Arismetica* (Neuva ed.; Buenos Aires, 1844), p. 46.

[2] Gabriel Izquierdo, *Tratado de Aritmética* (Santiago [Chile], 1859), p. 167.

[3] Indalecio Liévano, *Tratado de Aritmetica* (2. éd.; Bogota, 1872), p. 147.

[4] Luiz Porfirio da Motta Pegado, *Tratade elementar de arithmetica* (2. éd.; Lisboa, 1875), p. 253.

[5] *Miscellanea Berolinensia* (Berolini, 1710), p. 159.

[6] Emanuel Swedberg, *Daedalus hyperboreus* (Upsala, 1716), p. 126. Facsimile reproduction in *Kungliga Vetenskaps Societetens i Upsala Tvåhundraårsminne* (Upsala, 1910).

[7] W. Emerson, *Doctrine of Proportion* (London, 1763), p. 2.

[8] John Newton, *Institutio mathematica or mathematical Institution* (London, 1654), p. 125.

[9] John Wallis, *op. cit.* (Oxford, 1657), p. 229.

[10] Paolino Chelucci, *Institutiones analyticae* (editio post tertiam Romanam prima in Germania; Vienna, 1761), p. 3. See also the first edition (Rome, 1738), p. 1–15.

méthodique (*Mathématiques*) (Paris: Liège, 1784). Lamy[1] says: "Proportion arithmétique, 5,7 ∵ 10,12.c'est à dire qu'il y a même différence entre 5 et 7, qu'entre 10 et 12."

In Arnauld's geometry[2] the same symbols are used for arithmetical progression as for geometrical progression, as in 7.3 :: 13.9 and 6.2 :: 12.4.

Samuel Jeake (1696)[3] speaks of "∶ Three Pricks or Points, sometimes in disjunct proportion for the words *is as*."

A notation for arithmetical proportion, noticed in two English seventeenth-century texts, consists of five dots, thus ∶∵; Richard Balam[4] speaks of "arithmetical disjunct proportionals" and writes "2.4 ∶∵ 3.5"; Sir Jonas Moore[5] uses ∶∙ and speaks of "disjunct proportionals." Balam adds, "They may also be noted thus, 2...4 = 3...5." Similarly, John Kirkby[6] designated arithmetrical proportion in this manner, 9..6 = 6..3, the symbolism for arithmetical ratio being 8..2. L'Abbé Deidier (1739)[7] adopts 20.2∴78.60. Before that Weigel[8] wrote "(o) 3| ∵ 4.7" and "(o) 2.| ∵ 3.5." Wolff (1710),[9] Panchaud,[10] Saverien,[11] L'Abbé Foucher,[12] Emerson,[13] place

[1] B. Lamy, *Elemens des mathematiques* (3. éd.; Amsterdam, 1692), p. 155.

[2] Antoine Arnauld, *Nouveaux elemens de geometrie* (Paris, 1667); also in the edition issued at The Hague in 1690.

[3] Samuel Jeake, ΛΟΓΙΣΤΙΚΗΛΟΓΙΑ or *Arithmetick* (London, 1696; Preface, 1674), p. 10–12.

[4] Richard Balam, *Algebra: or the Doctrine of Composing, Inferring, and Resolving an Equation* (London, 1653), p. 5.

[5] Sir Jonas Moore, *Moore's Arithmetick: In Four Books* (3d ed.; London, 1688), the beginning of Book IV.

[6] Rev. Mr. John Kirkby, *Arithmetical Institutions containing a compleat System of Arithmetic* (London, 1735), p. 27, 28.

[7] L'Abbé Deidier, *L'Arithmétiques des géomètres, ou nouveau élémens de mathématiques* (Paris, 1739), p. 219.

[8] *Erhardi Weigelii Specimina novarum inventionum* (Jenae, 1693), p. 9.

[9] Chr. v. Wolff, *Anfangsgründe aller math. Wissenschaften* (1710), Vol. I, p. 65. See J. Tropfke, *op. cit.*, Vol. III (2d ed., 1922), p. 12.

[10] Benjamin Panchaud, *Entretiens ou leçons mathématiques*, Premier Parti (Lausanne et Genève, 1743), p. vii.

[11] A. Saverien, *Dictionnaire universel* (Paris, 1753), art. "Proportion arithmetique."

[12] L'Abbé Foucher, *Géométrie métaphysique ou essai d'analyse* (Paris, 1758), p. 257

[13] W. Emerson, *The Doctrine of Proportion* (London, 1763), p. 27.

the three dots as did Chelucci and Deidier, viz., $a.b \because c.d$. Cosalli[1] writes the arithmetical proportion $a:b \because c:d$. Later Wolff[2] wrote $a-b =c-d$.

Blassière[3] prefers $2:7 \div 10:15$. Juan Gerard[4] transfers Oughtred's signs for geometrical proportion to arithmetical proportion and writes accordingly, $9.7::5.3$. In French, Spanish, and Latin-American texts Oughtred's notation, $8.6:5.3$, for arithmetical proportion has persisted. Thus one finds it in Benito Bails,[5] in a French text for the military,[6] in Fournier,[7] in Gabriel Izquierdo,[8] in Indalecio Liévano.[9]

250. *Geometrical proportion.*—A presentation of geometrical proportion that is not essentially rhetorical is found in the Hindu Bakhshālī arithmetic, where the proportion $10:\frac{163}{60}=4:\frac{163}{150}$ is written in the form[10]

10	163	4	*pha* 163
1	60	1	150

It was shown previously (§ 124) that the Arab al-Qalasâdî (fifteenth century) expresses the proportion $7.12=84:144$ in this manner: $144 \therefore 84 \therefore 12 \therefore 7$. Regiomontanus in a letter writes our modern $a:b:c$ in the form $a.b.c$, the dots being simply signs of separation.[11] In the edition of the *Arithmetica* of Boethius, published at Augsburg in 1488, a crude representation of geometrical and arithmetical propor-

---

[1] *Scritti inediti del P. D. Pietro Cossali* . . . . pubblicati da B. Boncompagni (Rome, 1857), p. 75.

[2] Chr. v. Wolff., *op. cit.* (1750), Vol. I, p. 73.

[3] J. J. Blassière, *Institution du calcul numerique et litteral* (a La Haye 1770), the end of Part II.

[4] Juan Gerard, *Tratado completo de aritmética* (Madrid, 1798), p. 69.

[5] Benito Bails, *Principios de matematica de la real academia de San Fernando* (2. ed.), Vol. I (Madrid, 1788), p. 135.

[6] *Cours de mathématiques, à l'usage des écoles impériales militaires* ... rédigé par ordre de M. le Général de Division Bellavène ... (Paris, 1809), p. 52. Dedication signed by "Allaize, Billy, Puissant, Boudrot, Professeurs de mathématiques à l'Ecole de Saint-Cyr."

[7] C. F. Fournier, *Eléments d'arithmétique et d'algèbre*, Tome II (Nantes, 1842), p. 87.

[8] Gabriel Izquierdo, *op. cit.* (Santiago [Chile], 1859), p. 155.

[9] Indalecio Liévano, *Tratado aritmetica* (2d ed.; Bogota, 1872), p. 147.

[10] G. R. Kaye, *The Bakhshālī Manuscript*, Parts I and II (Calcutta, 1927), p. 119.

[11] M. Curtze, *Abhandlungen z. Geschichte d. Mathematik*, Vol. XII (1902), p. 253.

tion is given, as shown in Figure 103. The upper proportion on the left is geometrical, the lower one on the left is arithmetical. In the latter, the figure 8 plays no part; the 6, 9, and 12 are in arithmetical proportion. The two exhibitions on the right relate to harmonical and musical proportion.

Proportion as found in the earliest printed arithmetic (in Treviso,

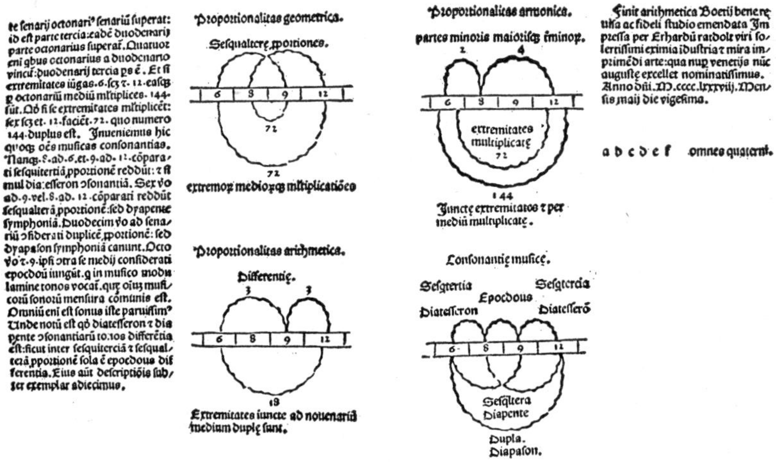

FIG. 103.—From the *Arithmetica* of Boethius, as printed in 1488, the last two pages. (Taken from D. E. Smith's *Rara arithmetica* [Boston, 1898], p. 28.)

1487) is shown in Figure 39. Stifel, in his edition of Rudolff's *Coss* (1553), uses vertical lines of separation, as in

$$\text{"100} \mid \tfrac{1}{6} z \mid 100 \ z \mid \text{Facit } \tfrac{1}{6} \ zz \text{ ."}$$

Tartaglia[1] indicates a proportion thus:

$$\text{"Se } £ \ 3 // \text{ val } \beta \ 4 \ // \text{ che valeranno } £ \ 28.\text{"}$$

Chr. Clavius[2] writes:

$$\text{"9 . 126 . 5 . ? fiunt 70 ."}$$

---

[1] N. Tartaglia, *La prima parte del General Tratato di Nvmeri, etc.* (Venice 1556), fol. 129*B*.

[2] Chr. Clavius, *Epitome arithmeticae practicae* (Rome, 1583), p. 137.

This notation is found as late as 1699 in Corachan's arithmetic[1] in such statements as

$$\text{``}A \ . \ B \ . \ C \ . \ D \ .$$
$$5 \ . \ 7 \ . \ 15 \ . \ 21 \ . \text{''}$$

Schwenter[2] marks the geometric proportion 68——51——85, then finds the product of the means $51 \times 85 = 4335$ and divides this by 68. In a work of Galileo,[3] in 1635, one finds:

"Regula aurea
58——95996. ——21600.
     21600

<hr>

57597600
95996
191992

<hr>

58	$\overset{3\,5\,7}{20735}$	13600
	3339	
	42 "	

In other places in Galileo's book the three terms in the proportion are not separated by horizontal lines, but by dots or simply by spacing. Johan Stampioen,[4] in 1639, indicates our $a:b=b:c$ by the symbolism:

$$\text{``}a_{,,} \ b \ \text{gel} : b \ _{,,} \ c \ .\text{''}$$

Further illustrations are given in § 221.

These examples show that some mode of presenting to the eye the numbers involved in a geometric proportion, or in the application of the rule of three, had made itself felt soon after books on mathematics came to be manufactured. Sometimes the exposition was rhetorical, short words being available for the writing of proportion. As late as 1601 Philip Lansberg[5] wrote "ut 5 ad 10; ita 10 ad 20," meaning

[1] Ivan Bavtista Corachan, *Arithmetica demonstrada* (Valencia, 1699), p. 199.

[2] Daniel Schwenter, *Geometriae practicae novae et auctae tractatus* (Nürnberg, 1623), p. 89.

[3] *Systema Cosmicvm, aucthore Galilaeo Galilaei. . . . . Ex Italica lingua latine conversum* (Florence, 1635), p. 294.

[4] Johan Stampioen, *Algebra ofte nieuwe Stel-Regel* (The Hague, 1639), p. 343.

[5] Philip Lansberg, *Triangulorum geometriae libri quatuor* (Middelburg [Zeeland], 1663), p. 5.

5:10=10:20. Even later the Italian Cardinal Michelangelo Ricci[1] wrote "esto *AC* ad *CB*, ut 9 ad 6." If the fourth term was not given, but was to be computed from the first three, the place for the fourth term was frequently left vacant, or it was designated by a question mark.

251. *Oughtred's notation.*—As the symbolism of algebra was being developed and the science came to be used more extensively, the need for more precise symbolism became apparent. It has been shown (§ 181) that the earliest noted symbolism was introduced by Oughtred. In his *Clavis mathematicae* (London, 1631) he introduced the notation 5.10::6.12 which he retained in the later editions of this text, as well as in his *Circles of Proportion* (1632, 1633, 1660), and in his *Trigonometria* (1657).

As previously stated (§ 169) the suggestion for this symbolism may have come to Oughtred from the reading of John Dee's Introduction to Billingley's *Euclid* (1570). Probably no mathematical symbol has been in such great demand in mathematics as the dot. It could be used, conveniently, in a dozen or more different meanings. But the avoidance of confusion necessitates the restriction of its use. Where then shall it be used, and where must other symbols be chosen? Oughtred used the dot to designate ratio. That made it impossible for him to follow John Napier in using the dot as the separatrix in decimal fractions. Oughtred could not employ two dots (:) for ratio, because the two dots were already pre-empted by him for the designation of aggregation, :*A*+*B*: signifying (*A*+*B*). Oughtred reserved the dot for the writing of ratio, and used four dots to separate the two equal ratios. The four dots were an unfortunate selection. The sign of equality (=) would have been far superior. But Oughtred adhered to his notation. Editions of his books containing it appeared repeatedly in the seventeenth century. Few symbols have met with more prompt adoption than those of Oughtred for proportion. Evidently the time was ripe for the introduction of a definite unambiguous symbolism. To be sure the adoption was not immediate. Nineteen years elapsed before another author used the notation *A*.*B*:: *C*.*D*. In 1650 John Kersey brought out in London an edition of Edmund Wingate's *Arithmetique made easie*, in which this notation is used. After this date, the publications employing it became frequent, some of them being the productions of pupils of Oughtred. We have

---

[1] *Michaelis Angeli Riccii exercitatio geometrica de maximis et minimis* (London, 1668), p. 3.

seen it in Vincent Wing,[1] Seth Ward,[2] John Wallis,[3] in "R.B.," a schoolmaster in Suffolk,[4] Samuel Foster,[5] Sir Jonas Moore,[6] and Isaac Barrow.[7] John Wallis[8] sometimes uses a peculiar combination of processes, involving the simplification of terms, during the very act of writing proportion, as in "$\frac{8}{2}A = 4A \cdot \frac{6}{2}A = 3A :: \frac{4}{2}A = 2A \cdot \frac{3}{2}A :: 8.6 ::$ $4.3.$" Here the dot signifies ratio.

The use of the dot, as introduced by Oughtred, did not become universal even in England. As early as 1651 the astronomer, Vincent Wing (§ 244), in his *Harmonicon Coeleste* (London), introduced the colon (:) as the symbol for ratio. This book uses, in fact, both notations for ratio. Many times one finds $A.B::C.D$ and many times $A:B::C:D$. It may be that the typesetter used whichever notation happened at the moment to strike his fancy. Later, Wing published three books (§ 244) in which the colon (:) is used regularly in writing ratios. In 1655 another writer, "R.B.," whom we have cited as using the symbols $A.B::C.D$, employed in the same publication also $A:B::C:D$. The colon was adopted in 1661 by Thomas Streete.[9]

That Oughtred himself at any time voluntarily used the colon as the sign for ratio does not appear. In the editions of his *Clavis* of 1648 and 1694, the use of : to signify ratio has been found to occur only once in each copy (§ 186); hence one is inclined to look upon this notation in these copies as printer's errors.

252. *Struggle in England between Oughtred's and Wing's notations, before 1700.*—During the second half of the seventeenth century there was in England competition between (.) and (:) as the symbols for the designation of the ratio (§§ 181, 251). At that time the dot maintained its ascendancy. Not only was it used by the two most influ-

---

[1] Vincent Wing, *Harmonicon coeleste* (London, 1651), p. 5.

[2] Seth Ward, *In Ismaelis Bullialdi astronomiae philolaicae fundamenta inquisitio brevis* (Oxford, 1653), p. 7.

[3] John Wallis, *Elenchus geometriae Hobbianae* (Oxford, 1655), p. 48; *Operum mathematicorum pars altera* (Oxford, 1656), the part on *Arithmetica infinitorum*, p. 181.

[4] *An Idea of Arithmetick, at first designed for the use of the Free Schoole at Thurlow in Suffolk. . . . .* By R. B., Schoolmaster there (London, 1655), p. 6.

[5] *Miscellanies: or mathematical Lucrubations of Mr. Samuel Foster . . . .* by John Twyden (London, 1659), p. 1.

[6] Jonas Moore, *Arithmetick in two Books* (London, 1660), p. 89; Moore's *Arithmetique in Four Books* (3d ed.; London, 1688), Book IV, p. 461.

[7] Isaac Barrow's edition of *Euclid's Data* (Cambridge, 1657), p. 2.

[8] John Wallis, *Adversus Marci Meibomii de Proportionibus Dialogum* (Oxford, 1657), "Dialogum," p. 54.

[9] Thomas Streete, *Astronomia Carolina* (1661). See J. Tropfke, *Geschichte der Elementar-Mathematik*, 3. Bd., 2. Aufl. (Berlin und Leipzig, 1922), p. 12.

ential English mathematicians before Newton, namely, John Wallis and Isaac Barrow, but also by David Gregory,[1] John Craig,[2] N. Mercator,[3] and Thomas Brancker.[4] I. Newton, in his letter to Oldenburg of October 24, 1676,[5] used the notation . :: . , but in Newton's *De analysi per aequationes terminorum infinitas*, the colon is employed to designate ratio, also in his *Quadratura curvarum*.

Among seventeenth-century English writers using the colon to mark ratio are James Gregory,[6] John Collins,[7] Christopher Wren,[8] William Leybourn,[9] William Sanders,[10] John Hawkins,[11] Joseph Raphson,[12] E. Wells,[13] and John Ward.[14]

253. *Struggle in England between Oughtred's and Wing's notations during 1700–1750.*—In the early part of the eighteenth century, the dot still held its place in many English books, but the colon gained in ascendancy, and in the latter part of the century won out. The single dot was used in John Alexander's *Algebra* (in which proportion is written in the form $a.b::c.X$ and also in the form $a\,\overline{..}\,b:c\,\overline{..}\,X$)[15] and, in John Colson's translation of Agnesi (before 1760).[16] It was used

[1] David Gregory in *Phil. Trans.*, Vol. XIX (1695–97), p. 645.

[2] John Craig, *Methodus figurarum lineis rectis et curvis* (London, 1685). Also his *Tractatus mathematicus* (London, 1693), but in 1718 he often used : :: . in his *De Calculo Fluentium Libri Duo*, brought out in London.

[3] N. Mercator, *Logarithmotechnia* (London, 1668), p. 29.

[4] Th. Brancker, *Introduction to Algebra* (trans. of Rhonius; London, 1668), p. 37.

[5] John Collins, *Commercium epistolicum* (London, 1712), p. 182.

[6] James Gregory, *Vera circuli et hyperbolae quadratura* (Patavia, 1668), p. 33.

[7] J. Collins, *Mariners Plain Scale New Plain'd* (London, 1659).

[8] *Phil. Trans.*, Vol. III (London), p. 868.

[9] W. Leybourn, *The Line of Proportion* (London, 1673), p. 14.

[10] William Sanders, *Elementa geometriae* (Glasgow, 1686), p. 3.

[11] *Cocker's Decimal Arithmetick* . . . . perused by John Hawkins (London, 1695) (Preface dated 1684), p. 41.

[12] J. Raphson, *Analysis aequationum universalis* (London, 1697), p. 26.

[13] E. Wells, *Elementa arithmeticae numerosae et speciosae* (Oxford, 1698), p. 107.

[14] John Ward, *A Compendium of Algebra* (2d ed.; London, 1698), p. 62.

[15] *A Synopsis of Algebra*. Being the Posthumous Work of John Alexander, of Bern in Swisserland. To which is added an Appendix by Humfrey Ditton. . . . . Done from the Latin by Sam. Cobb. M.A. (London, 1709), p. 16. The Latin edition appeared at London in 1693.

[16] Maria Gaetana Agnesi, *Analytical Institutions*, translated into English by the late Rev. John Colson. . . . . Now first printed . . . . under the inspection of Rev. John Hellins (London, 1801).

by John Wilson[1] and by the editors of Newton's *Universal arithmetick*.[2] In John Harris' *Lexicon technicum* (1704) the dot is used in some articles, the colon in others, but in Harris' translation[3] of G. Pardies' geometry the dot only is used. George Shelley[4] and Hatton[5] used the dot.

254. *Sporadic notations.*—Before the English notations . :: . and : :: : were introduced on the European Continent, a symbolism consisting of vertical lines, a modification of Tartaglia's mode of writing, was used by a few continental writers. It never attained popularity, yet maintained itself for about a century. René Descartes (1619–21)[6] appears to have been the first to introduce such a notation $a|b||c|d$. In a letter[7] of 1638 he replaces the middle double stroke by a single one. Slusius[8] uses single vertical lines in designating four numbers in geometrical proportion, $p|a|e|d-a$. With Slusius, two vertical strokes $||$ signify equality. Jaques de Billy[9] marks five quantities in continued proportion, thus $3-R5|R5-1|2|R5+1|3+R5$, where R means "square root." In reviewing publications of Huygens and others, the original notation of Descartes is used in the *Journal des Sçavans* (Amsterdam)[10] for the years 1701, 1713, 1716. Likewise, Picard,[11] De la Hire,[12] Abraham de Graaf,[13] and Parent[14] use the notation $a|b||xx|ab$.

---

[1] John Wilson, *Trigonometry* (Edinburgh, 1714), p. 24.

[2] I. Newton, *Arithmetica universalis* (ed. W. Whiston; Cambridge, 1707), p. 9; *Universal Arithmetick*, by Sir Isaac Newton, translated by Mr. Ralphson .... revised .... by Mr. Cunn (London, 1769), p. 17.

[3] *Plain Elements of Geometry and Plain Trigonometry* (London, 1701), p. 63.

[4] G. Shelley, *Wingate's Arithmetick* (London, 1704), p. 343.

[5] Edward Hatton, *An Intire System of Arithmetic* (London, 1721), p. 93.

[6] *Œuvres des Descartes* (éd. Adam et Tannery), Vol. X, p. 240.

[7] *Op. cit.*, Vol. II, p. 171.

[8] *Renati Francisci Slusii mesolabum seu duae mediae proportionales, etc.* (1668), p. 74. See also Slusius' reply to Huygens in *Philosophical Transactions* (London), Vols. III–IV (1668–69), p. 6123.

[9] Jaques de Billy, *Nova geometriae clavis* (Paris, 1643), p. 317.

[10] *Journal des Sçavans* (Amsterdam, année 1701), p. 376; *ibid.* (année 1713), p. 140, 387; *ibid.* (année 1716), p. 537.

[11] J. Picard in *Mémoires de l'Académie r. des sciences* (depuis 1666 jusqu'à 1699), Tome VI (Paris, 1730), p. 573.

[12] De la Hire, *Nouveaux elemens des sections coniques* (Paris, 1701), p. 184. J. Tropfke refers to the edition of 1679, p. 184.

[13] Abraham de Graaf, *De vervulling van der geometria en algebra* (Amsterdam, 1708), p. 97.

[14] A. Parent, *Essais et recherches de mathematique et de physique* (Paris, 1713), p. 224.

It is mentioned in the article "Caractere" in Diderot's *Encyclopédie* (1754). La Hire writes also "$aa\|xx\|ab$" for $a^2:x^2=x^2:ab$.

On a subject of such universal application in commercial as well as scientific publications as that of ratio and proportion, one may expect to encounter occasional sporadic attempts to alter the symbolism. Thus Hérigone[1] writes "$hg\ \pi\ ga\ 2|2\ hb\ \pi\ bd,\ signifi.$ *HG* est ad *GA*, *vt HB* ad *BD*," or, in modern notation, $hg:ga=hb:bd;$ here $2|2$ signifies equality, π signifies ratio. Again Peter Mengol,[2] of Bologna, writes "$a;r:a2;ar$" for $a:r=a^2:ar$. The London edition of the algebra of the Swiss J. Alexander[3] gives the signs . :: . but uses more often designations like $b\overline{}a:d\ \left|\dfrac{ad}{b}\right.$. Ade Mercastel,[4] of Rouen, writes $2,,3;;8,,12$. A close approach to the marginal symbolism of John Dee is that of the Spaniard Zaragoza[5] $4.3:12.9$. More profuse in the use of dots is J. Kresa[6] who writes $x\ldots r::r\ldots\dfrac{rr}{x}$, also $AE..EF::AD..DG$. The latter form is adopted by the Spaniard Cassany[7] who writes $128..119 ::3876;$ it is found in two American texts,[8] of 1797.

In greater conformity with pre-Oughtredian notations is van Schooten's notation[9] of 1657 when he simply separates the three given numbers by two horizontal dashes and leaves the place for the unknown number blank. Using Stevin's designation for decimal fractions, he writes "65——95,753③−1." Abraham de Graaf[10] is

[1] Pierre Herigone, *Cvrsvs mathematici* (Paris, 1644), Vol. VI, "Explicatio notarum." The first edition appeared in 1642.

[2] Pietro Mengoli, *Geometriae speciosae elementa* (Bologna, 1659), p. 8.

[3] *Synopsis algebraica*, Opus posthumum Johannis Alexandri, Bernatis-Helvetii (London, 1693), p. 135.

[4] Jean Baptiste Adrien de Mercastel, *Arithmétique démontrée* (Rouen, 1733), p. 99.

[5] Joseph Zaragoza, *Arithmetica vniversal* (Valencia, 1669), p. 48.

[6] Jacob Kresa, *Analysis speciosa trigonometriae sphericae* (Prague, 1720), p. 120, 121.

[7] Francisco Cassany, *Arithmetica Deseada* (Madrid, 1763), p. 102.

[8] *American Tutor's Assistant*. By sundry teachers in and near Philadelphia (3d ed.; Philadelphia, 1797), p. 57, 58, 62, 91–186. In the "explanation of characters," : :: : is given. The second text is Chauncey Lee's *American Accomptant* (Lansingburgh, 1797), where one finds (p. 63) $3..5::6..10$.

[9] Francis à Schooten, Leydensis, *Exercitationum mathematicarum liber primus* (Leyden, 1657), p. 19.

[10] Abraham de Graaf, *De Geheele mathesis of wiskonst* (Amsterdam, 1694), p. 16.

partial to the form $2-4=6-12$. Thomas York[1] uses three dashes 125—429—10—?, but later in his book writes "33600 7 :: 153600 32," the ratio being here indicated by a blank space. To distinguish ratios from fractions, Samuel Jeake[2] states that by some authors "instead of the intervening Line, two Pricks are set; and so the *Ratio* *sesquialtera* is thus expressed $\overset{.}{\underset{.}{\cdot}}\frac{3}{2}$." Accordingly, Jeake writes " $\frac{1}{7}\overset{.}{\cdot}\cdot\frac{1}{9}$ :: 9.7."

In practical works on computation with logarithms, and in some arithmetics a rhetorical and vertical arrangement of the terms of a proportion is found. Mark Forster[3] writes:

> "As Sine of 40 deg.   9,8080675
> To 1286   3,1092401
> So is Radius   10,0000000
> To the greatest Random 2000   3,3011726
>   Or, For Random at 36 deg."

As late as 1789 Benjamin Workman[4] writes " $\dfrac{\text{lb. d. lb.}}{1-7-112}$."

255. *Oughtred's notation on the European Continent.*—On the European Continent the dot as a symbol of geometrical ratio, and the four dots of proportion, . :: ., were, of course, introduced later than in England. They were used by Dulaurens,[5] Prestet,[6] Varignon,[7] Pardies,[8] De l'Hospital,[9] Jakob Bernoulli,[10] Johann Bernoulli,[11] Carré,[12] Her-

---

[1] Thomas York, *Practical Treatise of Arithmetick* (London, 1687), p. 132, 146.

[2] Samuel Jeake, ΛΟΓΙΣΤΙΚΗΛΟΓΊΑ, *or Arithmetick* (London, 1696 [Preface, 1674]), p. 411.

[3] Mark Forster, *Arithmetical Trigonometry* (London, 1690), p. 212.

[4] Benjamin Workman, *American Accountant* (Philadelphia, 1789), p. 62.

[5] Francisci Dulaurens, *Specimina mathematica* (Paris, 1667), p. 1.

[6] Jean Prestet, *Elemens des mathematiques* (Preface signed "J.P.") (Paris, 1675), p. 240. Also *Nouveaux elemens des mathematiques*, Vol. I (Paris, 1689), p. 355.

[7] P. Varignon in *Journal des Sçavans*, année 1687 (Amsterdam, 1688), p. 644. Also Varignon, *Eclaircissemens sur l'analyse des infiniment petits* (Paris, 1725), p. 16.

[8] *Œuvres du R. P. Ignace-Gaston Pardies* (Lyon, 1695), p. 121.

[9] De l'Hospital, *Analyse des infiniment petits* (Paris, 1696), p. 11.

[10] Jakob Bernoulli in *Acta eruditorum* (1687), p. 619 and many other places.

[11] Johann Bernoulli in *Histoire de l'académie r. des sciences*, année 1732 (Paris, 1735), p. 237.

[12] L. Carré, *Methode pour la Mesure des Surfaces* (Paris, 1700), p. 5.

mann,[1] and Rolle;[2] also by De Reaumur,[3] Saurin,[4] Parent,[5] Nicole,[6] Pitot,[7] Poleni,[8] De Mairan,[9] and Maupertuis.[10] By the middle of the eighteenth century, Oughtred's notation $A.B::C.D$ had disappeared from the volumes of the Paris Academy of Sciences, but we still find it in textbooks of Belidor,[11] Guido Grandi,[12] Diderot,[13] Gallimard,[14] De la Chapelle,[15] Fortunato,[16] L'Abbé Foucher,[17] and of Abbé Girault de Koudou.[18] This notation is rarely found in the writings of German authors. Erhard Weigel[19] used it in a philosophical work of 1693. Christian Wolf[20] used the notation "$DC.AD::EC.ME$" in 1707, and in 1710 "$3.12::5.20$" and also "$3:12=5:20$." Beguelin[21] used the dot for ratio in 1773. From our data it is evident that $A.B::C.D$ began

[1] J. Hermann in *Acta eruditorum* (1702), p. 502.

[2] M. Rolle in *Journal des Sçavans*, année 1702 (Amsterdam, 1703), p. 399.

[3] R. A. F. de Reaumur, *Histoire de l'académie r. des sciences*, année 1708 (Paris, 1730), "Mémoires," p. 209, but on p. 199 he used also the notation : :: :.

[4] J. Saurin, *op. cit.*, année 1708, "Mémoires," p. 26.

[5] Antoine Parent, *op. cit.*, année 1708, "Mémoires," p. 118.

[6] F. Nicole, *op. cit.*, année 1715 (Paris, 1741), p. 50.

[7] H. Pitot, *op. cit.*, année 1724 (Paris, 1726), "Mémoires," p. 109.

[8] Joannis Poleni, *Epistolarvm mathematicarvm Fascicvlvs* (Patavii, 1729).

[9] J. J. de Mairan, *Histoire de l'académie r. des sciences*, année 1740 (Paris, 1742), p. 7.

[10] P. L. Maupertuis, *op. cit.*, année 1731 (Paris, 1733), "Mémoires," p. 465.

[11] B. F. de Belidor, *Nouveau Cours de mathématique* (Paris, 1725), p. 481.

[12] Guido Grandi, *Elementi geometrici piani e solide de Euclide* (Florence, 1740).

[13] Denys Diderot, *Mémoires sur différens sujets de Mathématiques* (Paris, 1748), p. 16.

[14] J. E. Gallimard, *Géométrie élémentaire d'Euclide* (nouvelle éd.; Paris, 1749), p. 37.

[15] De la Chapelle, *Traité des sections coniques* (Paris, 1750), p. 150.

[16] F. Fortunato, *Elementa matheseos* (Brescia, 1750), p. 35.

[17] L'Abbé Foucher, *Géometrie métaphysique ou Essai d'analyse* (Paris, 1758), p. 257.

[18] L'Abbé Girault de Koudou, *Leçons analytiques du calcul des fluxions et des fluentes* (Paris, 1767), p. 35.

[19] *Erhardi Weigelii Philosophia mathematica* (Jenae, 1693), "Specimina novarum inventionum," p. 6, 181.

[20] C. Wolf in *Acta eruditorum* (1707), p. 313; Wolf, *Anfangsgründe aller mathematischen Wissenschaften* (1710), Band I, p. 65, but later Wolf adopted the notation of Leibniz, viz., $A:B=C:D$. See J. Tropfke, *Geschichte der Elementar-Mathematik*, Vol. III (2d ed.; Berlin und Leipzig, 1922), p. 13, 14.

[21] Nicolas de Beguelin in *Nouveaux mémoires de l'académie r. des sciences et belles-lettres*, année 1773 (Berlin, 1775), p. 211.

to be used in the Continent later than in England, and it was also later to disappear on the Continent.

256. An unusual departure in the notation for geometric proportion which involved an excellent idea was suggested by a Dutch author, Johan Stampioen,[1] as early as the year 1639. This was only eight years after Oughtred had proposed his . :: . Stampioen uses the designation $A,,B=C,,D$. We have noticed, nearly a century later, the use of two commas to represent ratio, in a French writer, Mercastel. But the striking feature with Stampioen is the use of Recorde's sign of equality in writing proportion. Stampioen anticipates Leibniz over half a century in using = to express the equality of two ratios. He is also the earliest writer that we have seen on the European Continent to adopt Recorde's symbol in writing ordinary equations. He was the earliest writer after Descartes to use the exponential form a^3. But his use of = did not find early followers. He was an opponent of Descartes whose influence in Holland at that time was great. The employment of = in writing proportion appears again with James Gregory[2] in 1668, but he found no followers in this practice in Great Britain.

257. *Slight modifications of Oughtred's notation.*—A slight modification of Oughtred's notation, in which commas took the place of the dots in designating geometrical ratios, thus $A,B::C,D$, is occasionally encountered both in England and on the Continent. Thus Sturm[3] writes "$3b,2b::2b,\dfrac{4bb}{3b}$ sive $\dfrac{4b}{3}$," Lamy[4] "$3,6::4,8$," as did also Ozanam,[5] De Moivre,[6] David Gregory,[7] L'Abbé Deidier,[8] Belidor,[9] who also uses the regular Oughtredian signs, Maria G. Agnesi,[10]

[1] Johan Stampioen d'Jonghe, *Algebra ofte Nieuwe Stel-Regel* ('s Graven-Haye, 1639).

[2] James Gregory, *Geometriae Pars Vniversalis* (Padua, 1668), p. 101.

[3] Christopher Sturm in *Acta eruditorum* (Leipzig, 1685), p. 260.

[4] R. P. Bernard Lamy, *Elemens des mathematiques*, troisième edition revue et augmentée sur l'imprismé à Paris (Amsterdam, 1692), p. 156.

[5] J. Ozanam, *Traité des lignes du premier genre* (Paris, 1687), p. 8; Ozanam, *Cours de mathématique*, Tome III (Paris, 1693), p. 139.

[6] A. de Moivre in *Philosophical Transactions*, Vol. XIX (London, 1698), p. 52; De Moivre, *Miscellanea analytica de seriebus* (London, 1730), p. 235.

[7] David Gregory, *Acta eruditorum* (1703), p. 456.

[8] L'Abbé Deidier, *La Mesure des Surfaces et des Solides* (Paris, 1740), p. 181.

[9] B. F. de Belidor, *op. cit.* (Paris, 1725), p. 70.

[10] Maria G. Agnesi, *Instituzioni analitiche*, Tome I (Milano, 1748), p. 76.

Nicolaas Ypey,[1] and Manfredi.[2] This use of the comma for ratio, rather than the Oughtredian dot, does not seem to be due to any special cause, other than the general tendency observable also in the notation for decimal fractions, for writers to use the dot and comma more or less interchangeably.

An odd designation occurs in an English edition of Ozanam,[3] namely, "$A.2.B.3::C.4.D.6$," where A,B,C,D are quantities in geometrical proportion and the numbers are thrown in by way of concrete illustration.

258. *The notation* : :: : *in Europe and America.*—The colon which replaced the dot as the symbol for ratio was slow in making its appearance on the Continent. It took this symbol about half a century to cross the British Channel. Introduced in England by Vincent Wing in 1651, its invasion of the Continent hardly began before the beginning of the eighteenth century. We find the notation $A:B::C:D$ used by Leibniz,[4] Johann Bernoulli,[5] De la Hire,[6] Parent,[7] Bomie,[8] Saulmon,[9] Swedenborg,[10] Lagny,[11] Senès,[12] Chevalier de Louville,[13] Clairaut,[14] Bouguer,[15] Nicole (1737, who in 1715 had used . :: .),[16] La

[1] Nicolaas Ypey, *Grondbeginselen der Keegelsneeden* (Amsterdam, 1769), p. 3.

[2] Gabriello Manfredi, *De Constructione Aequationum differentialium primi gradus* (1707), p. 123.

[3] J. Ozanam, *Cursus mathematicus*, translated "by several Hands" (London, 1712), Vol. I, p. 199.

[4] *Acta eruditorum* (1684), p. 472.

[5] Johanne (I) Bernoulli in *Journal des Sçavans*, année 1698 (Amsterdam, 1709), p. 576. See this notation used also in l'année 1791 (Amsterdam, 1702), p. 371.

[6] De la Hire in *Histoire de l'académie r. des sciences*, année 1708 (Paris, 1730), "Mémoires," p. 57.

[7] A. Parent in *op. cit.*, année 1712 (Paris, 1731), "Mémoires," p. 98.

[8] Bomie in *op. cit.*, p. 213.

[9] Saulmon in *op. cit.*, p. 283.

[10] Emanuel Swedberg, *Daedalus Hyperboreus* (Upsala, 1716).

[11] T. F. Lagny in *Histoire de l'académie r. des sciences*, année 1719 (Paris, 1721), "Mémoires," p. 139.

[12] Dominique de Senès in *op. cit.*, p. 363.

[13] De Louville in *op. cit.*, année 1724 (Paris, 1726), p. 67.

[14] Clairaut in *op. cit.*, année 1731 (Paris, 1733), "Mémoires," p. 484.

[15] Pierre Bougver in *op. cit.*, année 1733 (Paris, 1735), "Mémoires," p. 89.

[16] F. Nicole in *op. cit.*, année 1737 (Paris, 1740), "Mémoires," p. 64.

Caille,[1] D'Alembert,[2] Vicenti Riccati,[3] and Jean Bernoulli.[4] In the Latin edition of De la Caille's[5] *Lectiones* four notations are explained, namely, $3.12::2.8$, $3:12::2:8$, $3:12=2:8$, $3|12||2|8$, but the notation $3:12::2:8$ is the one actually adopted.

The notation : :: : was commonly used in England and the United States until the beginning of the twentieth century, and even now in those countries has not fully surrendered its place to : = : . As late as 1921 : :: : retains its place in Edwards' *Trigonometry*,[6] and it occurs in even later publications. The : :: : gained full ascendancy in Spain and Portugal, and in the Latin-American countries. Thus it was used in Madrid by Juan Gerard,[7] in Lisbon by Joao Felix Pereira[8] and Luiz Porfirio da Motta Pegado,[9] in Rio de Janeiro in Brazil by Francisco Miguel Pires[10] and C. B. Ottoni,[11] at Lima in Peru by Maximo Vazquez[12] and Luis Monsante,[13] at Buenos Ayres by Florentino Garcia,[14] at Santiage de Chile by Gabriel Izquierdo,[15] at Bogota in Colombia by Indalecio Liévano,[16] at Mexico by Juan Maria Balbontin.[17]

[1] La Caille in *op. cit.*, année 1741 (Paris, 1744), p. 256.

[2] D'Alembert in *op. cit.*, année 1745 (Paris, 1749), p. 367.

[3] Vincenti Riccati, *Opusculorum ad res physicas et mathematicas pertinentium. Tomus primus* (Bologna, 1757), p. 5.

[4] Jean Bernoulli in *Nouveaux mémoires de l'académie r. des sciences et belles-lettres*, année 1771 (Berlin, 1773), p. 286.

[5] N. L. de la Caille, *Lectiones elementares mathematicae* .... in Latinum traductae et ad editionem Parisinam anni MDCCLIX denuo exactae a C [arolo] S [cherffer] e S. J. (Vienna, 1762), p. 76.

[6] R. W. K. Edwards, *An Elementary Text-Book of Trigonometry* (new ed.; London, 1921), p. 152.

[7] Juan Gerard, Presbitero, *Tratado completo de aritmética* (Madrid, 1798), p. 69.

[8] J. F. Pereira, *Rudimentos de arithmetica* (Quarta Edição; Lisbon, 1863), p. 129.

[9] Luiz Porfirio da Motta Pegado, *Tratado elementar de arithmetica* (Secunda edição; Lisbon, 1875), p. 235.

[10] Francisco Miguel Pires, *Tratado de Trigonometria Espherica* (Rio de Janeiro, 1866), p. 8.

[11] C. B. Ottoni, *Elementos de geometria e trigonometria rectilinea* (4th ed.; Rio de Janeiro, 1874), "Trigon.," p. 36.

[12] Maximo Vazquez, *Aritmetica practica* (7th ed.; Lima, 1875), p. 130.

[13] Luis Monsante, *Lecciones de aritmetica demostrada* (7th ed.; Lima, 1872), p. 171.

[14] Florentino Garcia, *El aritmética Argentino* (5th ed.; Buenos Aires, 1871), p. 41; first edition, 1833.

[15] Gabriel Izquierdo, *Tratado de aritmética* (Santiago, 1859), p. 157.

[16] Indalecio Liévano, *Tratado de aritmetica* (2d ed.; Bogota, 1872), p. 148.

[17] Juan Maria Balbontin, *Tratado elemental de aritmetica* (Mexico, 1878), p. 96.

259. *The notation of Leibniz.*—In the second half of the eighteenth century this notation, $A:B::C:D$, had gained complete ascendancy over $A.B::C.D$ in nearly all parts of Continental Europe, but at that very time it itself encountered a serious rival in the superior Leibnizian notation, $A:B=C:D$. If a proportion expresses the equality of ratios, why should the regular accepted equality sign not be thus extended in its application? This extension of the sign of equality $=$ to writing proportions had already been made by Stampioen (§ 256). Leibniz introduced the colon (:) for ratio and for division in the *Acta eruditorum* of 1684, page 470 (§ 537). In 1693 Leibniz expressed his disapproval of the use of special symbols for ratio and proportion, for the simple reason that the signs for division and equality are quite sufficient. He[1] says: "Many indicate by $a \div b \dotdiv c \div d$ that the ratios a to b and c to d are alike. But I have always disapproved of the fact that special signs are used in ratio and proportion, on the ground that for ratio the sign of division suffices and likewise for proportion the sign of equality suffices. Accordingly, I write the ratio a to b thus: $a:b$ or $\frac{a}{b}$ just as is done in dividing a by b. I designate proportion, or the equality of two ratios by the equality of the two divisions or fractions. Thus when I express that the ratio a to b is the same as that of c to d, it is sufficient to write $a:b=c:d$ or $\frac{a}{b}=\frac{c}{d}$."

Cogent as these reasons are, more than a century passed before his symbolism for ratio and proportion triumphed over its rivals.

Leibniz's notation, $a:b=c:d$, is used in the *Acta eruditorum* of 1708, page 271. In that volume (p. 271) is laid the editorial policy that in algebra the Leibnizian symbols shall be used in the *Acta*. We quote the following relating to division and proportion (§ 197): "We shall designate division by two dots, unless circumstance should prompt adherence to the common practice. Thus, we shall have $a:b=\frac{a}{b}$. Hence with us there will be no need of special symbols for denoting proportion. For instance, if a is to b as c is to d, we have $a:b=c:d$."

The earliest influential textbook writer who adopted Leibniz' notation was Christian Wolf. As previously seen (§ 255) he sometimes

[1] G. W. Leibniz, *Matheseos universalis pars prior*, de Terminis incomplexis, No. 16; reprinted in *Gesammelte Werke* (C. I. Gerhardt), 3. Folge, II³, Band VII (Halle, 1863), p. 56.

wrote $a.b=c.d$. In 1710[1] he used both $3.12::5.20$ and $3:12=5:20$, but from 1713[2] on, the Leibnizian notation is used exclusively.

One of the early appearances of $a:b=c:d$ in France is in Clairaut's algebra[3] and in Saverien's dictionary,[4] where Saverien argues that the equality of ratios is best indicated by $=$ and that $::$ is superfluous. It is found in the publications of the Paris Academy for the year 1765,[5] in connection with Euler who as early as 1727 had used it in the commentaries of the Petrograd Academy.

Benjamin Panchaud brought out a text in Switzerland in 1743,[6] using $:=:$. In the Netherlands[7] it appeared in 1763 and again in 1775.[8] A mixture of Oughtred's symbol for ratio and the $=$ is seen in Pieter Venema[9] who writes $.=$.

In Vienna, Paulus Mako[10] used Leibniz' notation both for geometric and arithmetic proportion. The Italian Petro Giannini[11] used $:=:$ for geometric proportion, as does also Paul Frisi.[12] The first volume of *Acta Helvetia*[13] gives this symbolism. In Ireland, Joseph Fenn[14] used it about 1770. A French edition of Thomas Simpson's geometry[15] uses $:=:$. Nicolas Fuss[16] employed it in St. Petersburgh. In England,

[1] Chr. Wolf, *Anfangsgründe aller mathematischen Wissenschaften* (Magdeburg, 1710), Vol. I, p. 65. See J. Tropfke, *Geschichte der Elementar-Mathematik*, Vol. III (2d ed.; Berlin and Leipzig, 1922), p. 14.

[2] Chr. Wolf, *Elementa matheseos universae*, Vol. I (Halle, 1713), p. 31.

[3] A. C. Clairaut, *Elemens d'algebre* (Paris, 1746), p. 21.

[4] A. Saverien, *Dictionnaire universel de mathematique et physique* (Paris, 1753), arts. "Raisons semblables," "Caractere."

[5] *Histoire de l'académie r. des sciences*, année 1765 (Paris, 1768), p. 563; *Commentarii academiae scientiarum . . . .* ad annum 1727 (Petropoli, 1728), p. 14.

[6] Benjamin Panchaud, *Entretiens ou leçons mathématiques* (Lausanne, Genève, 1743), p. 226.

[7] A. R. Maudvit, *Inleiding tot de Keegel-Sneeden* (Shaage, 1763).

[8] J. A. Fas, *Inleiding tot de Kennisse en het Gebruyk der Oneindig Kleinen* (Leyden, 1775), p. 80.

[9] Pieter Venema, *Algebra ofte Stel-Konst*, Vierde Druk (Amsterdam, 1768), p. 118.

[10] Pavlvs Mako, *Compendiaria matheseos institutio* (editio altera; Vindobonae, 1766), p. 169, 170.

[11] Petro Giannini, *Opuscola mathematica* (Parma, 1773), p. 74.

[12] *Paulli Frisii Operum*, Tomus Secundus (Milan, 1783), p. 284.

[13] *Acta Helvetica, physico-mathematico-Botanico-Medica*, Vol. I (Basel, 1751), p. 87.

[14] Joseph Fenn, *The Complete Accountant* (Dublin, [n.d.]), p. 105, 128.

[15] Thomas Simpson, *Elémens de géométrie* (Paris, 1766).

[16] Nicolas Fuss, *Leçons de géométrie* (St. Pétersbourg, 1798), p. 112.

John Cole[1] adopted it in 1812, but a century passed after this date before it became popular there.

The Leibnizian notation was generally adopted in Europe during the nineteenth century.

In the United States the notation : :: : was the prevailing one during the nineteenth century. The Leibnizian signs appeared only in a few works, such as the geometries of William Chauvenet[2] and Benjamin Peirce.[3] It is in the twentieth century that the notation : = : came to be generally adopted in the United States.

A special symbol for variation sometimes encountered in English and American texts is $\propto$, introduced by Emerson.[4] "To the Common Algebraic Characters already receiv'd I add this $\propto$, which signifies a general Proportion; thus, $A \propto \dfrac{BC}{D}$, signifies that A is in a constant ratio to $\dfrac{BC}{D}$." The sign was adopted by Chrystal,[5] Castle,[6] and others.

SIGNS OF EQUALITY

260. *Early symbols.*—A symbol signifying "it gives" and ranking practically as a mark for equality is found in the linear equation of the Egyptian Ahmes papyrus (§ 23, Fig. 7). We have seen (§ 103) that Diophantus had a regular sign for equality, that the contraction *pha* answered that purpose in the Bakhshālī arithmetic (§ 109), that the Arab al-Qalasâdî used a sign (§ 124), that the dash was used for the expression of equality by Regiomontanus (§ 126), Pacioli (§ 138), and that sometimes Cardan (§ 140) left a blank space where we would place a sign of equality.

261. *Recorde's sign of equality.*—In the printed books before Recorde, equality was usually expressed rhetorically by such words as *aequales, aequantur, esgale, faciunt, ghelijck,* or *gleich,* and sometimes by the abbreviated form *aeq.* Prominent among the authors expressing equality in some such manner are Kepler, Galileo, Torricelli, Cavalieri, Pascal, Napier, Briggs, Gregory St. Vincent, Tacquet, and Fermat. Thus, about one hundred years after Recorde, some of

---

[1] John Cole, *Stereogoniometry* (London, 1812), p. 44, 265.

[2] William Chauvenet, *Treatise on Elementary Geometry* (Philadelphia, 1872), p. 69.

[3] Benjamin Peirce, *Elementary Treatise on Plane and Solid Geometry* (Boston, 1873), p. xvi.

[4] W. Emerson, *Doctrine of Fluxions* (3d ed.; London, 1768), p. **4**.

[5] G. Chrystal, *Algebra*, Part I, p. 275.

[6] Frank Castle, *Practical Mathematics for Beginners* (London, 1905), p. 317.

the most noted mathematicians used no symbol whatever for the
expression of equality. This is the more surprising if we remember
that about a century before Recorde, Regiomontanus (§ 126) in his
correspondence had sometimes used for equality a horizontal dash —,
that the dash had been employed also by Pacioli (§ 138) and Ghaligai
(§ 139). Equally surprising is the fact that apparently about the time
of Recorde a mathematician at Bologna should independently originate the same symbol (Fig. 53) and use it in his manuscripts.

Recorde's =, after its début in 1557, did not again appear in
print until 1618, or sixty-one years later. That some writers used
symbols in their private manuscripts which they did not exhibit in
their printed books is evident, not only from the practice of Regiomontanus, but also from that of John Napier who used Recorde's =
in an algebraic manuscript which he did not publish and which was
first printed in 1839.[1] In 1618 we find the = in an anonymous Appendix (very probably due to Oughtred) printed in Edward Wright's
English translation of Napier's famous *Descriptio*. But it was in
1631 that it received more than general recognition in England by
being adopted as the symbol for equality in three influential works,
Thomas Harriot's *Artis analyticae praxis*, William Oughtred's *Clavis
mathematicae*, and Richard Norwood's *Trigonometria*.

262. *Different meanings of* =.—As a source of real danger to
Recorde's sign was the confusion of symbols which was threatened on
the European Continent by the use of = to designate relations other
than that of equality. In 1591 Francis Vieta in his *In artem analyticen
isagoge* used = to designate arithmetical difference (§ 177). This
designation was adopted by Girard (§ 164), by Sieur de Var-Lezard[2]
in a translation of Vieta's *Isagoge* from the Latin into French, De
Graaf,[3] and by Franciscus à Schooten[4] in his édition of Descartes'
Géométrie. Descartes[5] in 1638 used = to designate *plus ou moins*,
i.e., ±.

Another complication arose from the employment of = by Johann

[1] Johannis Napier, *De Arte Logistica* (Edinburgh, 1839), p. 160.

[2] I. L. Sieur de Var-Lezard, *Introduction en l'art analytic ov nouvelle algèbre de
François Viète* (Paris, 1630), p. 36.

[3] Abraham de Graaf, *De beginselen van de Algebra of Stelkonst* (Amsterdam,
1672), p. 26.

[4] Renati Descartes, *Geometria* (ed. Franc. à Schooten; Francofvrti al Moenvm,
1695), p. 395.

[5] *Œuvres de Descartes* (éd. Adam et Tannery), Vol. II (Paris, 1898), p. 314,
426.

Caramuel[1] as the separatrix in decimal fractions; with him $102 = 857$ meant our 102.857. As late as 1706 G. H. Paricius[2] used the signs $=$, $:$, and $-$ as general signs to separate numbers occurring in the process of solving arithmetical problems. The confusion of algebraic language was further increased when Dulaurens[3] and Reyher[4] designated parallel lines by $=$. Thus the symbol $=$ acquired five different meanings among different continental writers. For this reason it was in danger of being discarded altogether in favor of some symbol which did not labor under such a handicap.

263. *Competing symbols.*—A still greater source of danger to our $=$ arose from competing symbols. Pretenders sprang up early on both the Continent and in England. In 1559, or two years after the appearance of Recorde's algebra, the French monk, J. Buteo,[5] published his *Logistica* in which there appear equations like "$1A$, $\frac{1}{3}B$, $\frac{1}{3}C$ [14" and "$3A.3B.15C[120$," which in modern notation are $x+\frac{1}{3}y+\frac{1}{3}z = 14$ and $3x+3y+15z=120$. Buteo's [functions as a sign of equality. In 1571, a German writer, Wilhelm Holzmann, better known under the name of Xylander, brought out an edition of Diophantus' *Arithmetica*[6] in which two parallel vertical lines $\parallel$ were used for equality. He gives no clue to the origin of the symbol. Moritz Cantor[7] suggests that perhaps the Greek word $\iota\sigma\omega$ ("equal") was abbreviated in the manuscript used by Xylander, by the writing of only the two letters $\iota\iota$. Weight is given to this suggestion in a Parisian manuscript on Diophantus where a single ι denoted equality.[8] In 1613, the Italian writer Giovanni Camillo Glorioso used Xylander's two vertical lines for equality.[9] It was used again by the Cardinal Michaelangelo Ricci.[10] This character was adopted by a few Dutch and French

[1] Joannis Caramuelis, *Mathesis Biceps vetus et nova* (1670), p. 7.

[2] Georg Heinrich Paricius, *Praxis arithmetices* (Regensburg, 1706). Quoted by M. Sterner, *Geschichte der Rechenkunst* (München und Leipzig, 1891), p. 348.

[3] François Dulaurens, *Specimina mathematica* (Paris, 1667).

[4] Samuel Reyher, *Euclides* (Kiel, 1698).

[5] J. Buteo, *Logistica* (Leyden, 1559), p. 190, 191. See J. Tropfke, *op. cit.*, Vol. III (2d ed.; Leipzig, 1922), p. 136.

[6] See Nesselmann, *Algebra der Griechen* (1842), p. 279.

[7] M. Cantor, *Vorlesungen über Geschichte der Mathematik*, Vol. II (2d ed.; Leipzig, 1913), p. 552.

[8] M. Cantor, *op. cit.*, Vol. I (3d ed.; 1907), p. 472.

[9] *Joannis Camillo Gloriosi, Ad theorema geometricvm* (Venetiis, 1613), p. 26.

[10] Michaelis Angeli Riccii, *Exercitatio geometrica de maximis et minimis* (Londini, 1668), p. 9.

mathematicians during the hundred years that followed, especially in the writing of proportion. Thus, R. Descartes,[1] in his *Opuscules de 1619–1621*, made the statement, "ex progressione $1|2||4|8||16|32||$ habentur numeri perfecti 6, 28, 496." Pierre de Carcavi, of Lyons, in a letter to Descartes (Sept. 24, 1649), writes the equation "$+1296 - 3060a + 2664a^2 - 1115a^3 + 239a^4 - 25a^5 + a^6 || 0$," where "la lettre a est l'inconnuë en la maniere de Monsieur Vieta" and $||$ is the sign of equality.[2] De Monconys[3] used it in 1666; De Sluse[4] in 1668 writes our $be = a^2$ in this manner "$be\ ||\ aa$." De la Hire (§ 254) in 1701 wrote the proportion $a:b = x^2:ab$ thus: "$a|b||xx|ab$." This symbolism is adopted by the Dutch Abraham de Graaf[5] in 1703, by the Frenchman Parent[6] in 1713, and by certain other writers in the *Journal des Sçavans*.[7] Though used by occasional writers for more than a century, this mark $||$ never gave promise of becoming a universal symbol for equality. A single vertical line was used for equality by S. Reyher in 1698. With him, "$A|B$" meant $A = B$. He attributes[8] this notation to the Dutch orientalist and astronomer Jacob Golius, saying: "Especially indebted am I to Mr. Golio for the clear algebraic mode of demonstration with the sign of equality, namely the rectilinear stroke standing vertically between two magnitudes of equal measure."

In England it was Leonard and Thomas Digges, father and son, who introduced new symbols, including a line complex ⧗ for equality (Fig. 78).[9]

The greatest oddity was produced by Hérigone in his *Cursus mathematicus* (Paris, 1644; 1st ed., 1634). It was the symbol "$2|2$." Based on the same idea is his "$3|2$" for "greater than," and his "$2|3$" for "less than." Thus, $a^2 + ab = b^2$ is indicated in his symbolism by

---

[1] *Œuvres de Descartes*, Vol. X (1908), p. 241.

[2] *Op. cit.*, Vol. V (1903), p. 418.

[3] *Journal des voyages de Monsieur de Monconys* (Troisième partie; Lyon, 1666), p. 2. Quoted by Henry in *Revue archéologique* (N.S.), Vol. XXXVII (1879), p. 333.

[4] *Renati Francisci Slusii Mesolabum*, Leodii Eburonum (1668), p. 51.

[5] Abraham de Graaf, *De Vervulling van de Geometria en Algebra* (Amsterdam, 1708), p. 97.

[6] A. Parent, *Essais et recherches de mathématique et de physique* (Paris, 1713), p. 224.

[7] *Journal des Sçavans* (Amsterdam, for 1713), p. 140; *ibid.* (for 1715), p. 537; and other years.

[8] Samuel Reyher, *op. cit.*, Vorrede.

[9] Thomas Digges, *Stratioticos* (1590), p. 35.

"$a2+ba2|2b2$." Though clever and curious, this notation did not appeal. In some cases Hérigone used also ⊔ to express equality. If this sign is turned over, from top to bottom, we have the one used by F. Dulaurens[1] in 1667, namely, ⊓; with Dulaurens ⊓ signifies "majus," ⊓ signifies "minus"; Leibniz, in some of his correspondence and unpublished papers, used[2] ⊓ and also[3] = ; on one occasion he used the Cartesian[4] ∞ for identity. But in papers which he printed, only the sign = occurs for equality.

Different yet was the equality sign .3 used by J. V. Andrea[5] in 1614.

The substitutes advanced by Xylander, Andrea, the two Digges, Dulaurens, and Hérigone at no time seriously threatened to bring about the rejection of Recorde's symbol. The real competitor was the mark ∞, prominently introduced by René Descartes in his *Géométrie* (Leyden, 1637), though first used by him at an earlier date.[6]

264. *Descartes' sign of equality.*—It has been stated that the sign was suggested by the appearance of the combined *ae* in the word *aequalis*, meaning "equal." The symbol has been described by Cantor[7] as the union of the two letters *ae*. Better, perhaps, is the description given by Wieleitner[8] who calls it a union of *oe* reversed; his minute examination of the symbol as it occurs in the 1637 edition of the *Géométrie* revealed that not all of the parts of the letter *e* in the combination *oe* are retained, that a more accurate way of describing that symbol is to say that it is made up of two letters *o*, that is, *oo* pressed against each other and the left part of the first excised. In some of the later appearances of the symbol, as given, for example, by van Schooten in 1659, the letter *e* in *oe*, reversed, remains intact. We incline to the opinion that Descartes' symbol for equality, as it appears in his *Géométrie* of 1637, is simply the astronomical symbol

[1] F. Dulaurens, *Specimina mathematica* (Paris, 1667).

[2] C. I. Gerhardt, *Leibnizens mathematische Schriften*, Vol. I, p. 100, 101, 155, 163, etc.

[3] *Op. cit.*, Vol. I, p. 29, 49, 115, etc.

[4] *Op. cit.*, Vol. V, p. 150.

[5] *Joannis Valentini Andreae, Collectaneorum Mathematicorum decades XI* (Tubingae, 1614). Taken from P. Treutlein, "Die deutsche Coss," *Abhandlungen zur Geschichte der Mathematik*, Vol. II (1879), p. 60.

[6] *Œuvres de Descartes* (éd. Ch. Adam et P. Tannery), Vol. X (Paris, 1908), p. 292, 299.

[7] M. Cantor, *op. cit.*, Vol. II (2d ed., 1913), p. 794.

[8] H. Wieleitner in *Zeitschr. für math. u. naturwiss. Unterricht*, Vol. XLVII (1916), p. 414.

for Taurus, placed sideways, with the opening turned to the left.
This symbol occurs regularly in astronomical works and was therefore available in some of the printing offices.

Descartes does not mention Recorde's notation; his *Géométrie* is
void of all bibliographical and historical references. But we know that
he had seen Harriot's *Praxis*, where the symbol is employed regularly.
In fact, Descartes himself[1] used the sign = for equality in a letter of
1640, where he wrote "$1C-6N=40$" for $x^3-6x=40$. Descartes does
not give any reason for advancing his new symbol ∞. We surmise that
Vieta's, Girard's, and De Var-Lezard's use of = to denote arithmetical "difference" operated against his adoption of Recorde's sign.
Several forces conspired to add momentum to Descartes' symbol ∞.
In the first place, the *Géométrie*, in which it first appeared in print,
came to be recognized as a work of genius, giving to the world analytic
geometry, and therefore challenging the attention of mathematicians.
In the second place, in this book Descartes had perfected the exponential notation, a^n (n, a positive integer), which in itself marked a
tremendous advance in symbolic algebra; Descartes' ∞ was likely to
follow in the wake of the exponential notation. The ∞ was used by
F. Debeaune[2] as early as October 10, 1638, in a letter to Roberval.

As Descartes had lived in Holland several years before the appearance of his *Géométrie*, it is not surprising that Dutch writers should be
the first to adopt widely the new notation. Van Schooten used the
Cartesian sign of equality in 1646.[3] He used it again in his translation
of Descartes' *Géométrie* into Latin (1649), and also in the editions of
1659 and 1695. In 1657 van Schooten employed it in a third publication.[4] Still more influential was Christiaan Huygens[5] who used ∞ as
early as 1646 and in his subsequent writings. He persisted in this
usage, notwithstanding his familiarity with Recorde's symbol through
the letters he received from Wallis and Brouncker, in which it occurs
many times.[6] The Descartian sign occurs in the writings of Hudde
and De Witt, printed in van Schooten's 1659 and later editions of
Descartes' *Géométrie*. Thus, in Holland, the symbol was adopted by

---

[1] *Œuvres de Descartes*, Vol. III (1899), p. 190.

[2] *Ibid.*, Vol. V (1903), p. 519.

[3] Francisci à Schooten, *De organica conicarum sectionum* (Leyden, 1646), p. 91.

[4] Francisci à Schooten, *Exercitationvm mathematicarum liber primus* (Leyden,
1657), p. 251.

[5] *Œuvres complètes de Christiaan Huygens*, Tome I (La Haye, 1888), p. 26, 526.

[6] *Op. cit.*, Tome II, p. 296, 519; Tome IV, p. 47, 88.

the most influential mathematicians of the seventeenth century. It worked its way into more elementary textbooks. Jean Prestet[1] adopted it in his *Nouveaux Élémens*, published at Paris in 1689. This fact is the more remarkable, as in 1675 he[2] had used the sign =. It seems to indicate that soon after 1675 the sign ∞ was gaining over = in France. Ozanam used ∞ in his *Dictionaire mathematique* (Amsterdam, 1691), though in other books of about the same period he used ∼, as we see later. The Cartesian sign occurs in a French text by Bernard Lamy.[3]

In 1659 Descartes' equality symbol invaded England, appearing in the Latin passages of Samuel Foster's *Miscellanies*. Many of the Latin passages in that volume are given also in English translation. In the English version the sign = is used. Another London publication employing Descartes' sign of equality was the Latin translation of the algebra of the Swiss Johann Alexander.[4] Michael Rolle uses ∞ in his *Traité d'algèbre* of 1690, but changes to = in 1709.[5] In Holland, Descartes' equality sign was adopted in 1660 by Kinckhvysen,[6] in 1694 by De Graaf,[7] except in writing proportions, when he uses =. Bernard Nieuwentiit uses Descartes' symbol in his *Considerationes* of 1694 and 1696, but preferred = in his *Analysis infinitorum* of 1695. De la Hire[8] in 1701 used the Descartian character, as did also Jacob Bernoulli in his *Ars Conjectandi* (Basel, 1713). Descartes' sign of equality was widely used in France and Holland during the latter part of the seventeenth and the early part of the eighteenth centuries, but it never attained a substantial foothold in other countries.

265. *Variations in the form of Descartes' symbol.*—Certain variations of Descartes' symbol of equality, which appeared in a few texts, are probably due to the particular kind of symbols available or improvisable in certain printing establishments. Thus Johaan Cara-

[1] Jean Prestet, *Nouveaux Élémens dés mathématiques*, Vol. I (Paris, 1689), p. 261.

[2] J. P. [restet] *Élémens des mathématiques* (Paris, 1675), p. 10.

[3] Bernard Lamy, *Elemens des mathematiques* (3d ed.; Amsterdam, 1692), p. 93.

[4] *Synopsis Algebraica, Opus posthumum Johannis Alexandri, Bernatis-Helvetii. In usum scholae mathematicae apud Hospitium-Christi Londinense* (Londini, 1693), p. 2.

[5] *Mem. de l'académie royale des sciences*, année 1709 (Paris), p. 321.

[6] Gerard Kinckhvysen, *De Grondt der Meet-Konst* (Te Haerlem, 1660), p. 4.

[7] Abraham de Graaf, *De Geheele Mathesis of Wiskonst* (Amsterdam, 1694), p. 45.

[8] De la Hire, *Nouveaux élémens des sections coniques* (Paris, 1701), p. 184.

muel[1] in 1670 employed the symbol Æ; the 1679 edition of Fermat's[2] works gives ∞ in the treatise *Ad locos planos et solidos isagoge*, but in Fermat's original manuscripts this character is not found.[3] On the margins of the pages of the 1679 edition occur also expressions of which "*DA{BE*" is an example, where $DA = BE$. J. Ozanam[4] employs ⌣ in 1682 and again in 1693; he refers to ⊐ as used to mark equality, "mais nous le changerons en celuy-cy, ∽ ; que nous semble plus propre, et plus naturel." Andreas Spole[5] said in 1692: "∿ vel = est nota aequalitates." Wolff[6] gives the Cartesian symbol inverted, thus ∝.

266. *Struggle for supremacy.*—In the seventeenth century, Recorde's = gained complete ascendancy in England. We have seen its great rival ∞ in only two books printed in England. After Harriot and Oughtred, Recorde's symbol was used by John Wallis, Isaac Barrow, and Isaac Newton. No doubt these great names helped the symbol on its way into Europe.

On the European Continent the sign = made no substantial headway until 1650 or 1660, or about a hundred years after the appearance of Recorde's algebra. When it did acquire a foothold there, it experienced sharp competition with other symbols for half a century before it fully established itself. The beginning of the eighteenth century may be designated roughly as the time when all competition of other symbols practically ceased. Descartes himself used = in a letter of September 30, 1640, to Mersenne. A Dutch algebra of 1639 and a tract of 1640, both by J. Stampioen,[7] and the *Teutsche Algebra* of the Swiss Johann Heinrich Rahn (1659), are the first continental textbooks that we have seen which use the symbol. Rahn says, p. 18: "Bey disem anlaasz habe ich das namhafte gleichzeichen = zum ersten gebraucht, bedeutet ist gleich, $2a = 4$ heisset $2a$ ist gleich 4." It was used by Bernhard Frenicle de Bessy, of magic-squares fame, in a

---

[1] J. Caramuel, *op. cit.*, p. 122.

[2] *Varia opera mathematica D. Petri de Fermat* (Tolosae, 1679), p. 3, 4, 5.

[3] *Œuvres de Fermat* (ed. P. Tannery et C. Henry), Vol. I (Paris, 1891), p. 91.

[4] *Journal des Sçavans* (de l'an 1682), p. 160; Jacques Ozanam, *Cours de Mathematiques*, Tome I (Paris, 1692), p. 27; also Tome III (Paris, 1693), p. 241.

[5] Andreas Spole, *Arithmetica vulgaris et specioza* (Upsaliae, 1692), p. 16. See G. Eneström in *L'Intermédiaire des mathématiciens*, Tome IV (1897), p. 60.

[6] Christian Wolff, *Mathematisches Lexicon* (Leipzig, 1716), "Signa," p. 1264.

[7] Johan Stampioen d'Jonghe, *Algebra ofte Nieuwe Stel-Regel* ('s Graven-Hage, 1639); *J. Stampioenii Wisk-Konstich ende Reden-maetich Bewijs* (s'Graven-Hage, 1640).

letter[1] to John Wallis of December 20, 1661, and by Huips[2] in the same year. Leibniz, who had read Barrow's *Euclid* of 1655, adopted the Recordean symbol, in his *De arte combinatoria of 1666* (§ 545), but then abandoned it for nearly twenty years. The earliest textbook brought out in Paris that we have seen using this sign is that of Arnauld[3] in 1667; the earliest in Leyden is that of C. F. M. Dechales[4] in 1674.

The sign = was used by Prestet,[5] Abbé Catelan and Tschirnhaus,[6] Hoste,[7] Ozanam,[8] Nieuwentijt,[9] Weigel,[10] De Lagny,[11] Carré,[12] L'Hospital,[13] Polynier,[14] Guisnée,[15] and Reyneau.[16]

This list constitutes an imposing array of names, yet the majority of writers of the seventeenth century on the Continent either used Descartes' notation for equality or none at all.

267. With the opening of the eighteenth century the sign = gained rapidly; James Bernoulli's *Ars Conjectandi* (1713), a posthumous publication, stands alone among mathematical works of prominence of that late date, using ∞. The dominating mathematical advance of the time was the invention of the differential and integral calculus. The fact that both Newton and Leibniz used Recorde's symbol led to its general adoption. Had Leibniz favored Descartes'

[1] *Œuvres complètes des Christiaan-Huygens* (La Haye), Tome IV (1891), p. 45.

[2] Frans van der Huips, *Algebra ofte een Noodige* (Amsterdam, 1661), p. 178. Reference supplied by L. C. Karpinski.

[3] Antoine Arnauld, *Nouveaux Elemens de Geometrie* (Paris, 1667; 2d ed., 1683).

[4] C. F. Dechales, *Cvrsvs sev Mvndvs Mathematicvs*, Tomvs tertivs (Lvgdvni, 1674), p. 666; Editio altera, 1690.

[5] J. P[restet], *op. cit.* (Paris, 1675), p. 10.

[6] *Acta eruditorum* (anno 1682), p. 87, 393.

[7] P. Hoste, *Recueil des traites de mathematiques*, Tome III (Paris, 1692), p. 93.

[8] Jacques Ozanam, *op. cit.*, Tome I (nouvelle éd.; Paris, 1692), p. 27. In various publications between the dates 1682 and 1693 Ozanam used as equality signs ∼, ∞, and =.

[9] Bernard Nieuwentijt, *Analysis infinitorum.*

[10] *Erhardi Weigelii Philosophia mathematica* (Jenae, 1693), p. 135.

[11] Thomas F. de Lagny, *Nouveaux élémens d'arithmétique, et d'algèbre* (Paris, 1697), p. 232.

[12] Louis Carré, *Methode pour la mesure des surfaces* (Paris, 1700), p. 4.

[13] Marquis de l'Hospital, *Analyse des Infiniment Petits* (Paris, 1696, 1715).

[14] Pierre Polynier, *Élémens des Mathématiques* (Paris, 1704), p. 3.

[15] Guisnée, *Application de l'algèbre à la géométrie* (Paris, 1705).

[16] Charles Reyneau, *Analyse demontrée*, Tome I (1708).

∞, then Germany and the rest of Europe would probably have joined France and the Netherlands in the use of it, and Recorde's symbol would probably have been superseded in England by that of Descartes at the time when the calculus notation of Leibniz displaced that of Newton in England. The final victory of = over ∞ seems mainly due to the influence of Leibniz during the critical period at the close of the seventeenth century.

The sign of equality = ranks among the very few mathematical symbols that have met with universal adoption. Recorde proposed no other algebraic symbol; but this one was so admirably chosen that it survived all competitors. Such universality stands out the more prominently when we remember that at the present time there is still considerable diversity of usage in the group of symbols for the differential and integral calculus, for trigonometry, vector analysis, in fact, for every branch of mathematics.

The difficulty of securing uniformity of notation is further illustrated by the performance of Peter van Musschenbroek,[1] of Leyden, an eighteenth-century author of a two-volume text on physics, widely known in its day. In some places he uses = for equality and in others for ratio; letting $S. s.$ be distances, and $T. t$ .times, he says: "Erit $S. s.$ :: $T. t.$ exprimunt hoc Mathematici scribendo, est $S = T.$ sive Spatium est uti tempus, nam signum = non exprimit aequalitatem, sed rationem." In writing proportions, the ratio is indicated sometimes by a dot, and sometimes by a comma. In 1754, Musschenbroek had used ∞ for equality.[2]

268. *Variations in the form of Recorde's symbol.*—There has been considerable diversity in the form of the sign of equality. Recorde drew the two lines very long (Fig. 71) and close to each other, ═════. This form is found in Thomas Harriot's algebra (1631), and occasionally in later works, as, for instance, in a paper of De Lagny[3] and in Schwab's edition of Euclid's *Data*.[4] Other writers draw the two lines very short, as does Weigel[5] in 1693. At Upsala, Emanuel

[1] Petro van Musschenbroek, *Introductio ad philosophiam naturalem*, Vol. I (Leyden, 1762), p. 75, 126.

[2] Petri van Musschenbroek, *Dissertatio physica experimentalis de magnete* (Vienna), p. 239.

[3] De Lagny in *Mémoires de l'académie r. d. sciences* (depuis 1666 jusqu'à 1699), Vol. II (Paris, 1733), p. 4.

[4] Johann Christoph Schwab, *Euclids Data* (Stuttgart, 1780), p. 7.

[5] *Erhardi Weigeli Philosophia mathematica* (Jena, 1693), p. 181.

Swedenborg[1] makes them very short and slanting upward, thus //.
At times one encounters lines of moderate length, drawn far apart $\equiv$,
as in an article by Nicole[2] and in other articles, in the *Journal des
Sçavans*. Frequently the type used in printing the symbol is the figure
1, placed horizontally, thus[3] $\sqcup$ or[4] $\sqcap$.

In an American arithmetic[5] occurs, "$1+6, =7, \times 6=42, \div 2=21$."

Wolfgang Bolyai[6] in 1832 uses $\doteq$ to signify absolute equality; $\overline{\overline{}}$,
equality in content; $A(=B$ or $B=)A$, to signify that each value of A
is equal to some value of B; $A(=)B$, that each of the values of A is
equal to some value of B, and vice versa.

To mark the equality of vectors, Bellavitis[7] used in 1832 and later
the sign $\doteq$.

Some recent authors have found it expedient to assign $=$ a more
general meaning. For example, Stolz and Gmeiner[8] in their theoretical
arithmetic write $a \circ b = c$ and read it "*a* mit *b* ist *c*," the $=$ signifying
"is explained by" or "is associated with." The small circle placed
between a and b means, in general, any relation or *Verknüpfung*.

De Morgan[9] used in one of his articles on logarithmic theory a
double sign of equality $= =$ in expressions like $(be^{\beta \sqrt{-1}})^x = = ne^{\nu \sqrt{-1}}$,
where β and ν are angles made by b and n, respectively, with the initial
line. He uses this double sign to indicate "that every symbol shall
express not merely the length and direction of a line, but also the
quantity of revolution by which a line, setting out from the unit line,
is supposed to attain that direction."

[1] Emanuel Swedberg, *Daedalus Hyperboreus* (Upsala, 1716), p. 39. See facsimile reproduction in *Kungliga Vetenskaps Societetens i Upsala Tvåhundraårsminne*
(Upsala, 1910).

[2] François Nicole in *Journal des Sçavans*, Vol. LXXXIV (Amsterdam, 1728),
p. 293. See also année 1690 (Amsterdam, 1691), p. 468; année 1693 (Amsterdam,
1694), p. 632.

[3] James Gregory, *Geometria Pars Vniversalis* (Padua, 1668); Emanuel Swedberg, *op. cit.*, p. 43.

[4] H. Vitalis, *Lexicon mathematicum* (Rome, 1690), art. "Algebra."

[5] *The Columbian Arithmetician*, "by an American" (Haverhill [Mass.], 1811),
p. 149.

[6] Wolfgangi Bolyai de Bolya, *Tentamen* (2d ed.), Tom. I (Budapestini, 1897),
p. xi.

[7] Guisto Bellavitis in *Annali del R. Lomb.-Ven.* (1832), Tom. II, p. 250–53.

[8] O. Stolz und J. A. Gmeiner, *Theoretische Arithmetik* (Leipzig), Vol. I (2d ed.;
1911), p. 7.

[9] A. de Morgan, *Trans. Cambridge Philos. Society*, Vol. VII (1842), p. 186.

269. *Variations in the manner of using it.*—A rather unusual use of equality signs is found in a work of Deidier[1] in 1740, viz.,

$$\frac{0+1+2=3}{2+2+2=6}=\frac{1}{2}\ ; \qquad \frac{0.\ 1.\ 4.\ =5}{4,\ 4,\ 4,\ =12}=\frac{1}{3}+\frac{1}{12}\ .$$

H. Vitalis[2] uses a modified symbol: "Nota $\backsimeq$ significat repetitam aequationem .... *vt* $10\sqsupseteq 6.\ +4\backsimeq 8+2$." A discrimination between $=$ and ∞ is made by Gallimard[3] and a few other writers; " $=$, est égale à; ∞ qui signifie tout simplement, égal à , ou , qui est égal à."

A curious use, in the same expressions, of $=$, the comma, and the word *aequalis* is found in a Tacquet-Whiston[4] edition of Euclid, where one reads, for example, "erit $8\times 432=3456$ aequalis $8\times 400=3200$, $+8\times 30=240, +8\times 2=16$."

L. Gustave du Pasquier[5] in discussing general complex numbers employs the sign of double equality $\equiv$ to signify "equal by definition."

The relations between the coefficients of the powers of x in a series may be expressed by a formal equality involving the series as a whole, as in

$$1+n_{(1)}x+n_{(2)}x^2+\ \cdots\ \overline{=}_{f}(1+x)\{1+(n+1)_{(1)}x+(n-1)_{(2)}x^2+\ \cdots\ ,$$

where the symbol $\overline{=}_{f}$ indicates that the equality is only formal, not arithmetical.[6]

270. *Nearly equal.*—Among the many uses made in recent years of the sign $\sim$ is that of "nearly equal to," as in "$e\sim\frac{1}{4}$"; similarly, $e\cong\frac{1}{4}$ is allowed to stand for "equal or nearly equal to."[7] A. Eucken[8] lets $\simeq$ stand for the lower limit, as in "$J\simeq 45.10^{-40}$ (untere Grenze)," where J means a mean moment of inertia. Greenhill[9] denotes approximate

---

[1] L'Abbé Deidier, *La mesure des surfaces et des solides* (Paris, 1740), p. 9.

[2] H. Vitalis, *loc. cit.*

[3] J. E. Gallimard, *La Science du calcul numerique*, Vol. I (Paris, 1751), p. 3.

[4] Andrea Tacquet, *Elementa Euclidea geometriae* [after] Gulielmus Whiston (Amsterdam, 1725), p. 47.

[5] *Comptes Rendus du Congrès International des Mathématicians* (Strasbourg, 22–30 Septembre 1920), p. 164.

[6] Art. "Algebra" in *Encyclopaedia Britannica* (11th ed., 1910).

[7] A. Kratzer in *Zeitschrift für Physik*, Vol. XVI (1923), p. 356, 357.

[8] A. Eucken in *Zeitschrift der physikalischen Chemie*, Band C, p. 159.

[9] A. G. Greenhill, *Applications of Elliptic Functions* (London, 1892), p. 303, 340, 341.

equality by $\underset{\displaystyle\sim\!\!\sim}{\displaystyle\sim\!\!\sim}$. An early suggestion due to Fischer[1] was the sign $\asymp$ for "approximately equal to." This and three other symbols were proposed by Boon[2] who designed also four symbols for "greater than but approximately equal to" and four symbols for "less than but approximately equal to."

SIGNS OF COMMON FRACTIONS

271. *Early forms.*—In the Egyptian Ahmes papyrus unit fractions were indicated by writing a special mark over the denominator (§§ 22, 23). Unit fractions are not infrequently encountered among the Greeks (§ 41), the Hindus and Arabs, in Leonardo of Pisa (§ 122), and in writers of the later Middle Ages in Europe.[3] In the text *Triśatika*, written by the Hindu Śrīdhara, one finds examples like the following: "How much money is there when half a *kākini*, one-third of this and one-fifth of this are added together?

Statement
$$\begin{array}{|cc|} \hline 1 & 1 \\ 1 & 2 \\ \hline \end{array} \quad \begin{array}{|ccc|} \hline 1 & 1 & 1 \\ 1 & 2 & 3 \\ \hline \end{array} \quad \begin{array}{|cccc|} \hline 1 & 1 & 1 & 1 \\ 1 & 2 & 3 & 5 \\ \hline \end{array}$$
Answer. *Varātikas* 14."

This means $1\times\frac{1}{2}+1\times\frac{1}{2}\times\frac{1}{3}+1\times\frac{1}{2}\times\frac{1}{3}\times\frac{1}{5}=\frac{7}{10}$, and since 20 *varātikas* $=1$ *kākini*, the answer is 14 *varātikas*.

John of Meurs (early fourteenth century)[4] gives $\frac{7}{9}$ as the sum of three unit fractions $\frac{1}{2}$, $\frac{1}{4}$, and $\frac{1}{36}$, but writes "$\frac{1}{2}\ \frac{1}{2}\ \frac{1}{9}$," which is an ascending continued fraction. He employs a slightly different notation for $\frac{1}{24}$, namely, "$\frac{1}{4}\ \frac{1}{3}\ 0\ \frac{1}{2}\ 0$."

Among Heron of Alexandria and some other Greek writers the numerator of any fraction was written with an accent attached, and was followed by the denominator marked with two accents (§ 41). In some old manuscripts of Diophantus the denominator is placed above the numerator (§ 104), and among the Byzantines the denominator is found in the position of a modern exponent;[5] $\vartheta^{\iota\alpha}$ signified accordingly $\frac{9}{11}$.

---

[1] Ernst Gottfried Fischer, *Lehrbuch der Elementar-Mathematik*, 4. Theil, *Anfangsgründe der Algebra* (Berlin und Leipzig, 1829), p. 147. Reference given by R. C. Archibald in *Mathematical Gazette*, Vol. VIII (London, 1917), p. 49.

[2] C. F. Boon, *Mathematical Gazette*, Vol. VII (London, 1914), p. 48.

[3] See G. Eneström in *Bibliotheca mathematica* (3d ser.), Vol. XIV (1913–14), p. 269, 270.

[4] Vienna Codex 4770, the *Quadripartitum numerorum*, described by L. C. Karpinski in *Bibliotheca mathematica* (3d ser.), Vol. XIII (1912–13), p. 109.

[5] F. Hultsch, *Metrologicorum scriptorum reliquiae*, Vol. I (Leipzig, 1864), p. 173–75.

The Hindus wrote the denominator beneath the numerator, but without a separating line (§§ 106, 109, 113, 235).

In the so-called arithmetic of John of Seville,[1] of the twelfth century (?), which is a Latin elaboration of the arithmetic of al-Khowârizmî, as also in a tract of Alnasavi (1030 A.D.),[2] the Indian mode of writing fractions is followed; in the case of a mixed number, the fractional part appears below the integral part. Alnasavi pursues this course consistently[3] by writing a zero when there is no integral part; for example, he writes $\frac{1}{11}$ thus: $\begin{smallmatrix}``0\ ''\\ \mathsf{I}\\ \mathsf{II}\end{smallmatrix}$.

272. *The fractional line* is referred to by the Arabic writer al-Ḥaṣṣâr (§§ 122, 235, Vol. II § 422), and was regularly used by Leonardo of Pisa (§§ 122, 235). The fractional line is absent in a twelfth-century Munich manuscript;[4] it was not used in the thirteenth-century writings of Jordanus Nemorarius,[5] nor in the *Gernardus algorithmus demonstratus*, edited by Joh. Schöner (Nürnberg, 1534), Part II, chapter i.[6] When numerator and denominator of a fraction are letters, Gernardus usually adopted the form *ab* (*a* numerator, *b* denominator), probably for graphic reasons. The fractional line is absent in the Bamberger arithmetic of 1483, but occurs in Widman (1489), and in a fifteenth-century manuscript at Vienna.[7] While the fractional line came into general use in the sixteenth century, instances of its omission occur as late as the seventeenth century.

273. Among the sixteenth- and seventeenth-century writers omitting the fractional line were Baëza[8] in an arithmetic published at Paris, Dibuadius[9] of Denmark, and Paolo Casati.[10] The line is

[1] Boncompagni, *Trattati d'aritmetica*, Vol. II, p. 16–72.

[2] H. Suter, *Bibliotheca mathematica* (3d ser.), Vol. VII (1906–7), p. 113–19.

[3] M. Cantor, *op. cit.*, Vol. I (3d ed.), p. 762.

[4] Munich MS Clm 13021. See *Abhandlungen über Geschichte der Mathematik*, Vol. VIII (1898), p. 12–13, 22–23, and the peculiar mode of operating with fractions.

[5] *Bibliotheca mathematica* (3d ser.), Vol. XIV, p. 47.

[6] *Ibid.*, p. 143.

[7] Codex Vindob. 3029, described by E. Rath in *Bibliotheca mathematica* (3d ser.), Vol. XIII (1912–13), p. 19. This manuscript, as well as Widman's arithmetic of 1489, and the anonymous arithmetic printed at Bamberg in 1483, had as their common source a manuscript known as Algorismus Ratisponensis.

[8] *Nvmerandi doctrina authore Lodoico Baeza* (Lvtetia, 1556), fol. 45.

[9] *C. Dibvadii in arithmeticam irrationalivm Evclidis* (Arnhemii, 1605).

[10] Paolo Casati, *Fabrica et Vso Del Compasso di Proportione* (Bologna, 1685) [Privilege, 1662], p. 33, 39, 43, 63, 125.

usually omitted in the writings of Marin Mersenne[1] of 1644 and
1647. It is frequently but not usually omitted by Tobias Beutel.[2]

In the middle of a fourteenth-century manuscript[3] one finds the
notation $3\,\overline{5}$ for $\frac{3}{5}$, $4\,\overline{7}$ for $\frac{4}{7}$. A Latin manuscript,[4] Paris 7377A,
which is a translation from the Arabic of Abu Kamil, contains the
fractional line, as in $\frac{1}{9}$, but $\frac{1}{9}\frac{1}{9}$ is a continued fraction and stands for $\frac{1}{9}$
plus $\frac{1}{81}$, whereas $\frac{6}{9}\frac{1}{2}$ as well as $\frac{91}{92}$ represent simply $\frac{1}{18}$. Similarly,
Leonardo of Pisa,[5] who drew extensively from the Arabic of Abu
Kamil, lets $\frac{5}{8}\frac{6}{8}$ stand for $\frac{5}{64}$, there being a difference in the order of
reading. Leonardo read from right to left, as did the Arabs, while
authors of Latin manuscripts of about the fourteenth century read
as we do from left to right. In the case of a mixed number, like $3\frac{1}{5}$,
Leonardo and the Arabs placed the integer to the right of the fraction.

274. *Special symbols for simple fractions* of frequent occurrence
are found. The Ahmes papyrus has special signs for $\frac{1}{2}$ and $\frac{2}{3}$ (§ 22);
there existed a hieratic symbol for $\frac{1}{4}$ (§ 18). Diophantus employed
special signs for $\frac{1}{2}$ and $\frac{2}{3}$ (§ 104). A notation to indicate one-half,
almost identical with one sometimes used during the Middle Ages in
connection with Roman numerals, is found in the fifteenth century
with the Arabic numerals. Says Cappelli: "I remark that for the designation of one-half there was used also in connection with the Arabic
numerals, in the XV. century, a line between two points, as $4 \div$ for
$4\frac{1}{2}$, or a small cross to the right of the number in place of an exponent,
as $4^{\dagger}$, presumably a degeneration of 1/1, for in that century this form
was used also, as 7 1/1 for $7\frac{1}{2}$. Toward the close of the XV. century
one finds also often the modern form $\frac{1}{2}$."[6] The Roman designation of
certain unit fractions are set forth in § 58. The peculiar designations
employed in the Austrian cask measures are found in § 89. In a fifteenth-century manuscript we find: "Whan pou hayst write pat, for
pat pat leues, write such a merke as is here ^{w} vpon his hede, pe quych

---

[1] Marin Mersenne, *Cogitata Physico-mathematica* (Paris, 1644), "Phaenomena
ballistica"; *Novarvm observationvm Physico-mathematicarvm*, Tomvs III (Paris,
1647), p. 194 ff.

[2] Tobias Beutel, *Geometrische Gallerie* (Leipzig, 1690), p. 222, 224, 236, 239,
240, 242, 243, 246.

[3] *Bibliotheca mathematica* (3d ser.), Vol. VII, p. 308-9.

[4] L. C. Karpinski in *ibid.*, Vol. XII (1911-12), p. 53, 54.

[5] Leonardo of Pisa, *Liber abbaci* (ed. B. Boncompagni, 1857), p. 447. Noteworthy here is the use of e to designate the absence of a number.

[6] A. Cappelli, *Lexicon Abbreviaturarum* (Leipzig, 1901), p. L.

merke schal betoken halfe of þe odde þat was take away";[1] for example, half of 241 is 120^w. In a mathematical roll written apparently in the south of England at the time of Recorde, or earlier, the character $\sim$ stands for one-half, a dot · for one-fourth, and $\sim$ for three-fourths.[2] In some English archives[3] of the sixteenth and seventeenth centuries one finds one-half written in the form $\frac{1}{\sim}$. In the earliest arithmetic printed in America, the *Arte para aprendar todo el menor del arithmetica* of Pedro Paz (Mexico, 1623), the symbol $\underline{\circ}$ is used for $\frac{1}{2}$ a few times in the early part of the book. This symbol is taken from the *Arithmetica practica* of the noted Spanish writer, Juan Perez de Moya, 1562 (14th ed., 1784, p. 13), who uses $\underline{\circ}$ and also $\underset{m}{\circ}$ for $\frac{1}{2}$ or *medio*.

This may be a convenient place to refer to the origin of the sign % for "per cent," which has been traced from the study of manuscripts by D. E. Smith.[4] He says that in an Italian manuscript an "unknown writer of about 1425 uses a symbol which, by natural stages, developed into our present %. Instead of writing ' per 100', 'ꝑ 100' or 'ꝑ cento,' as had commonly been done before him, he wrote 'ꝑc°' for 'ꝑ $\overset{\circ}{c}$,' just as the Italians wrote $\overset{\circ}{1}$, $\overset{\circ}{2}$, ... and 1°, 2°, ... for primo, secundo, etc. In the manuscripts which I have examined the evolution is easily traced, the c° becoming $\frac{0}{0}$ about 1650, the original meaning having even then been lost. Of late the 'per' has been dropped; leaving only $\frac{0}{0}$ or %." By analogy to %, which is now made up of two zeros, there has been introduced the sign %₀, having as many zeros as 1,000 and signifying *per mille*.[5] Cantor represents the fraction $(100+p)/100$ "by the sign 1, 0p, not to be justified mathematically but in practice extremely convenient."

275. *The solidus.*[6]—The ordinary mode of writing fractions $\dfrac{a}{b}$ is typographically objectionable as requiring three terraces of type. An effort to remove this objection was the introduction of the solidus, as in a/b, where all three fractional parts occur in the regular line of type. It was recommended by De Morgan in his article on "The Calculus

[1] R. Steele, *The Earliest Arithmetics in English* (London, 1922), p. 17, 19. The *p* in "pou," "pat," etc., appears to be our modern *th*.

[2] D. E. Smith in *American Mathematical Monthly*, Vol. XXIX (1922), p. 63.

[3] *Antiquaries Journal*, Vol. VI (London, 1926), p. 272.

[4] D. E. Smith, *Rara arithmetica* (1898), p. 439, 440.

[5] Moritz Cantor, *Politische Arithmetik* (2. Aufl.; Leipzig, 1903), p. 4.

[6] The word "solidus" in the time of the Roman emperors meant a gold coin (a "solid" piece of money); the sign / comes from the old form of the initial letter *s*, namely, ∫, just as £ is the initial of *libra* ("pound"), and *d* of *denarius* ("penny").

of Functions," published in the *Encyclopaedia Metropolitana* (1845). But practically that notation occurs earlier in Spanish America. In the *Gazetas de Mexico* (1784), page 1, Manuel Antonio Valdes used a curved line resembling the sign of integration, thus $1/4$, $3/4$; Henri Cambuston[1] brought out in 1843, at Monterey, California, a small arithmetic employing a curved line in writing fractions. The straight solidus is employed, in 1852, by the Spaniard Antonio Serra Y Oliveres.[2] In England, De Morgan's suggestion was adopted by Stokes[3] in 1880. Cayley wrote Stokes, "I think the 'solidus' looks very well indeed . . . . ; it would give you a strong claim to be President of a Society for the Prevention of Cruelty to Printers." The solidus is used frequently by Stolz and Gmeiner.[4]

While De Morgan recommended the solidus in 1843, he used $a:b$ in his subsequent works, and as Glaisher remarks, "answers the purpose completely and it is free from the objection to ÷ viz., that the pen must be twice removed from the paper in the course of writing it."[5] The colon was used frequently by Leibniz in writing fractions (§ 543, 552) and sometimes also by Karsten,[6] as in $1:3 = \frac{1}{3}$; the ÷ was used sometimes by Cayley.

G. Peano adopted the notation b/a whenever it seemed convenient.[7]

Alexander Macfarlane[8] adds that Stokes wished the solidus to take the place of the horizontal bar, and accordingly proposed that the terms immediately preceding and following be welded into one, the welding action to be arrested by a period. For example, $m^2 - n^2/m^2 + n^2$ was to mean $(m^2 - n^2)/(m^2 + n^2)$, and a/bcd to mean $\frac{a}{bcd}$, but $a/bc \cdot d$ to mean $\frac{a}{bc} d$. "This solidus notation for algebraic expressions oc-

[1] Henri Cambuston, *Definicion de las principales operaciones de arismetica* (1843), p. 26.

[2] Antonio Serra Y Oliveres, *Manuel de la Tipografia Española* (Madrid, 1852), p. 71.

[3] G. G. Stokes, *Math. and Phys. Papers*, Vol. I (Cambridge, 1880), p. vii. See also J. Larmor, *Memoirs and Scient. Corr. of G. G. Stokes*, Vol. I (1907), p. 397.

[4] O. Stolz and J. A. Gmeiner, *Theoretische Arithmetik* (2d ed.; Leipzig, 1911), p. 81.

[5] J. W. L. Glaisher, *Messenger of Mathematics*, Vol. II (1873), p. 109.

[6] W. J. G. Karsten, *Lehrbegrif der gesamten Mathematik*, Vol. I (Greifswald, 1767), p. 50, 51, 55.

[7] G. Peano, *Lezioni di analisi infinitesimale*, Vol. I (Torino, 1893), p. 2.

[8] Alexander Macfarlane, *Lectures on Ten British Physicists* (New York, 1919), p. 100, 101.

curring in the text has since been used in the *Encyclopaedia Britannica*, in Wiedemann's *Annalen* and quite generally in mathematical literature." It was recommended in 1915 by the Council of the London Mathematical Society to be used in the current text.

"The use of small fractions in the midst of letterpress," says Bryan,[1] "is often open to the objection that such fractions are difficult to read, and, moreover, very often do not come out clearly in printing. It is especially difficult to distinguish $\frac{1}{3}$ from $\frac{1}{8}$. .... For this reason it would be better to confine the use of these fractions to such common forms as $\frac{1}{4}, \frac{1}{2}, \frac{3}{4}, \frac{1}{3}$, and to use the notation 18/22 for other fractions."

SIGNS OF DECIMAL FRACTIONS

276. *Stevin's notation.*—The invention of decimal fractions is usually ascribed to the Belgian Simon Stevin, in his *La Disme*, published in 1585 (§ 162). But at an earlier date several other writers came so close to this invention, and at a later date other writers advanced the same ideas, more or less independently, that rival candidates for the honor of invention were bound to be advanced. The *La Disme* of Stevin marked a full grasp of the nature and importance of decimal fractions, but labored under the burden of a clumsy notation. The work did not produce any immediate effect. It was translated into English by R. Norton[2] in 1608, who slightly modified the notation by replacing the circles by round parentheses. The fraction .3759 is given by Norton in the form $3^{(1)}7^{(2)}5^{(3)}9^{(4)}$.

277. Among writers who adopted Stevin's decimal notation is Wilhelm von Kalcheim[3] who writes 693 ② for our 6.93. He applies it also to mark the decimal subdivisions of linear measure: "Die Zeichen sind diese: ◎ ist ein ganzes oder eine ruthe: ① ist ein erstes / prime oder schuh: ② ist ein zweites / secunde oder Zoll: ③ ein drittes / korn oder gran: ④ ist ein viertes stipflin oder minuten: und so forthan." Before this J. H. Beyer writes[4] 8 7̌98 for 8.00798; also

---

[1] G. H. Bryan, *Mathematical Gazette*, Vol. VIII (1917), p. 220.

[2] *Disme: the Art of Tenths, or Decimall Arithmetike,* .... *invented by the excellent mathematician, Simon Stevin*. Published in English with some additions by Robert Norton, Gent. (London, 1608). See also A. de Morgan in *Companion to the British Almanac* (1851), p. 11.

[3] *Zusammenfassung etlicher geometrischen Aufgaben.* .... Durch Wilhelm von Kalcheim, genant Lohausen Obristen (Bremen, 1629), p. 117.

[4] Johann Hartmann Beyer, *Logistica decimalis, das ist die Kunstrechnung mit den zehntheiligen Brüchen* (Frankfurt a/M., 1603). We have not seen Beyer's

$$14.3761 \overset{\text{viii}}{} \text{for } 14.00003761, \quad 123.\overset{0}{4}.\overset{i}{5}.\overset{ii}{9}.\overset{iii}{8}.\overset{iv}{7}.\overset{v}{2}. \quad \text{or } 123.\overset{i}{4}.\overset{ii}{5}.\overset{iii}{9}.\overset{iv}{8}.\overset{v}{7}.\overset{vi}{2}$$

or $123.\overset{0}{459}.\overset{iii}{872}$ for $123.\overset{vi}{459872}$, 643 for $0.\overset{iv}{0643}$.

That Stevin's notation was not readily abandoned for a simpler one is evident from Ozanam's use[1] of a slight modification of it as late as 1691, in passages like "$\frac{6667}{10000}$ ég. à $\overset{(1)}{6}\,\overset{(2)}{6}\,\overset{(3)}{6}\,\overset{(4)}{7}$," and $\overset{(0)}{3}\,\overset{(1)}{9}\,\overset{(2)}{8}$ for our 3.98.

278. *Other notations used before 1617.*—Early notations which one might be tempted to look upon as decimal notations appear in works whose authors had no real comprehension of decimal fractions and their importance. Thus Regiomontanus,[2] in dividing 85869387 by 60000, marks off the last four digits in the dividend and then divides by 6 as follows:

$$8\ 5\ 8\ 6\ |\ 9\ 3\ 8\ 7$$
$$1\ 4\ 3\ 1$$

In the same way, Pietro Borgi[3] in 1484 uses the stroke in dividing 123456 by 300, thus

$$\text{"}per\ 300$$
$$1\ 2\ 3\ 4\ |\ 5\ 6$$
$$4\ 1\ 1 \quad \text{------}$$
$$4\ 1\ 1\tfrac{156}{300}.\text{"}$$

Francesco Pellos (Pellizzati) in 1492, in an arithmetic published at Turin, used a point and came near the invention of decimal fractions.[4]

Christoff Rudolff[5] in his *Coss* of 1525 divides 652 by 10. His words are: "Zu exempel / ich teile 652 durch 10. stet also 65/2. ist 65 der quocient vnnd 2 das übrig. Kompt aber ein Zal durch 100 zů teilen / schneid ab die ersten zwo figuren / durch 1000 die ersten drey / also weiter für yede o ein figur." ("For example, I divide 652 by 10. It gives 65/2; thus, 65 is the quotient and 2 the remainder. If a number is to be divided by 100, cut off the first two figures, if by

---

book; our information is drawn from J. Tropfke, *Geschichte der Elementar-Mathematik*, Vol. I (2d ed.; Berlin and Leipzig, 1921), p. 143; S. Günther, *Geschichte der Mathematik*, Vol. I (Leipzig, 1908), p. 342.

[1] J. Ozanam, *L'Usage du Compas de Proportion* (a La Haye, 1691), p. 203, 211.

[2] *Abhandlungen zur Geschichte der Mathematik*, Vol. XII (1902), p. 202, 225.

[3] See G. Eneström in *Bibliotheca mathematica* (3d ser.), Vol. X (1909–10), p. 240.

[4] D. E. Smith, *Rara arithmetica* (1898), p. 50, 52.

[5] Quoted by J. Tropfke, *op. cit.*, Vol. I (2d ed., 1921), p. 140.

1,000 the first three, and so on for each 0 a figure.") This rule for division by 10,000, etc., is given also by P. Apian[1] in 1527.

In the *Exempel Büchlin* (Vienna, 1530), Rudolff performs a multiplication involving what we now would interpret as being decimal fractions.[2] Rudolff computes the values $375 \ (1+\frac{5}{100})^n$ for $n=1$, $2, \ldots, 10$. For $n=1$ he writes $393 \mid 75$, which really denotes 393.75; for $n=3$ he writes $434 \mid 109375$. The computation for $n=4$ is as follows:

$$4\ 3\ 4 \mid 1\ 0\ 9\ 3\ 7\ 5$$
$$2\ 1 \quad 7\ 0\ 5\ 4\ 6\ 8\ 7\ 5$$
$$4\ 5\ 5 \mid 8\ 1\ 4\ 8\ 4\ 3\ 7\ 5$$

Here Rudolff uses the vertical stroke as we use the comma and, in passing, uses decimals without appreciating the importance and generality of his procedure.

F. Vieta fully comprehends decimal fractions and speaks of the advantages which they afford;[3] he approaches close to the modern notations, for, after having used (p. 15) for the fractional part smaller type than for the integral part, he separated the decimal from the integral part by a vertical stroke (p. 64, 65); from the vertical stroke to the actual comma there is no great change.

In 1592 Thomas Masterson made a close approach to decimal fractions by using a vertical bar as separatrix when dividing £337652643 by a million and reducing the result to shillings and pence. He wrote:[4]

$$``\quad \text{facit} \begin{cases} l. & 3\ 3\ 7 \mid 6\ 5\ 2\ 6\ 4\ 3\ " \\ s. & 1\ 3 \mid 0\ 5\ 2\ 8\ 6\ 0 \\ d. & \text{———} \mid 6\ 3\ 4\ 3\ 2\ 0 \end{cases}$$

John Kepler in his *Oesterreichisches Wein-Visier-Büchlein* (Lintz, MDCXVI), reprinted in Kepler's *Opera omnia* (ed. Ch. Frisch), Volume V (1864), page 547, says: "Fürs ander, weil ich kurtze Zahlen brauche, derohalben es offt Brüche geben wirdt, so mercke, dass alle Ziffer, welche nach dem Zeichen (◖) folgen, die gehören zu

[1] P. Apian, *Kauffmannsz Rechnung* (Ingolstadt, 1527), fol. ciijr°. Taken from J. Tropfke, *op. cit.*, Vol. I (2d ed., 1921), p. 141.

[2] See D. E. Smith, "Invention of the Decimal Fraction," *Teachers College Bulletin* (New York, 1910–11), p. 18; G. Eneström, *Bibliotheca mathematica* (3d ser.), Vol. X (1909–10), p. 243.

[3] F. Vieta, *Universalium inspectionum*, p. 7; Appendix to the *Canon mathematicus* (1st ed.; Paris, 1579). We copy this reference from the *Encyclopédie des scienc. math.*, Tome I, Vol. I (1904), p. 53, n. 180.

[4] A. de Morgan, *Companion to the British Almanac* (1851), p. 8.

dem Bruch, als der Zehler, der Nenner darzu wird nicht gesetzt, ist aber allezeit eine runde Zehnerzahl von so vil Nullen, als vil Ziffer nach dem Zeichen kommen. Wann kein Zeichen nicht ist, das ist eine gantze Zahl ohne Bruch, vnd wann also alle Ziffern nach dem Zeichen gehen, da heben sie bissweilen an von einer Nullen. Dise Art der Bruch-rechnung ist von Jost Bürgen zu der sinusrechnung erdacht, vnd ist darzu gut, dass ich den Bruch abkürtzen kan, wa er vnnötig lang werden wil, ohne sonderen Schaden der vberigen Zahlen; kan ihne auch etwa auff Erhaischung der Notdurfft erlengern. Item lesset sich also die gantze Zahl vnd der Bruch mit einander durch alle species Arithmeticae handlen wie nur eine Zahl. Als wann ich rechne 365 Gulden mit 6 per cento, wievil bringt es dess Jars Interesse? dass stehet nun also:

$$3(65$$
$$6 \text{ mal}$$

$$\text{facit } 21(90$$

vnd bringt 21 Gulden vnd 90 hundertheil, oder 9 zehentheil, das ist 54 kr."

Joost Bürgi[1] wrote 1414 for 141.4 and 001414 for 0.01414; on the title-page of his *Progress-Tabulen* (Prag, 1620) he wrote 23027̊0022 for our 230270.022. This small circle is referred to often in his *Gründlicher Unterricht*, first published in 1856.[2]

279. *Did Pitiscus use the decimal point?*—If Bartholomaeus Pitiscus of Heidelberg made use of the decimal point, he was probably the first to do so. Recent writers[3] on the history of mathematics are

---

[1] See R. Wolf, *Viertelj. Naturf. Ges.* (Zürich), Vol. XXXIII (1888), p. 226.

[2] *Grunert's Archiv der Mathematik und Physik*, Vol. XXVI (1856), p. 316–34.

[3] A. von Braunmühl, *Geschichte der Trigonometrie*, Vol. I (Leipzig, 1900), p. 225.

M. Cantor, *Vorlesungen über Geschichte der Mathematik*, Vol. II (2d ed.; Leipzig, 1913), p. 604, 619.

G. Eneström in *Bibliotheca mathematica* (3d ser.), Vol. VI (Leipzig, 1905), p. 108, 109.

J. W. L. Glaisher in *Napier Tercentenary Memorial Volume* (London, 1913), p. 77.

N. L. W. A. Gravelaar in *Nieuw Archief voor Wiskunde* (2d ser.; Amsterdam), Vol. IV (1900), p. 73.

S. Günther, *Geschichte der Mathematik*, 1. Teil (Leipzig, 1908), p. 342.

L. C. Karpinski in *Science* (2d ser.), Vol. XLV (New York, 1917), p. 663–65.

D. E. Smith in *Teachers College Bulletin, Department of Mathematics* (New York, 1910–11), p. 19.

J. Tropfke, *Geschichte der Elementar-Mathematik*, Vol. I (2d ed.; Leipzig, 1921), p. 143.

divided on the question as to whether or not Pitiscus used the decimal point, the majority of them stating that he did use it. This disagreement arises from the fact that some writers, apparently not having access to the 1608 or 1612 edition of the *Trigonometria*[1] of Pitiscus, reason from insufficient data drawn from indirect sources, while others fail to carry conviction by stating their conclusions without citing the underlying data.

Two queries are involved in this discussion: (1) Did Pitiscus employ decimal fractions in his writings? (2) If he did employ them, did he use the dot as the separatrix between units and tenths?

Did Pitiscus employ decimal fractions? As we have seen, the need of considering this question arises from the fact that some early writers used a symbol of separation which we could interpret as separating units from tenths, but which they themselves did not so interpret. For instance,[2] Christoff Rudolff in his *Coss* of 1525 divides 652 by 10, "stet also 65|2. ist 65 der quocient vnnd 2 das übrig." The figure 2 looks like two-tenths, but in Rudolff's mind it is only a remainder. With him the vertical bar served to separate the 65 from this remainder; it was not a decimal separatrix, and he did not have the full concept of decimal fractions. Pitiscus, on the other hand, did have this concept, as we proceed to show. In computing the chord of an arc of 30° (the circle having 10^7 for its radius), Pitiscus makes the statement (p. 44): "All these chords are less than the radius and as it were certain parts of the radius, which parts are commonly written $\frac{5176381}{100000000}$. But much more brief and necessary for the work, is this writing of it .05176381. For those numbers are altogether of the same value, as these two numbers 09. and $\frac{9}{10}$ are." In the original Latin the last part reads as follows: " .... quae partes vulgo sic scriberentur $\frac{5176381}{100000000}$. Sed multò compendiosior et ad calculum accommodatior est ista scriptio .05176381. Omnino autem idem isti numeri valent, sicut hi duo numeri 09. et $\frac{9}{10}$ idem valent."

One has here two decimals. The first is written .05176381. The dot on the left is not separating units from tenths; it is only a rhetorical mark. The second decimal fraction he writes 09., and he omits the dot on the left. The zero plays here the rôle of decimal separatrix.

[1] I have used the edition of 1612 which bears the following title: *Bartholamei* | *Pitisci Grunbergensis* | *Silesij* | *Trigonometriae* | *Sioe. De dimensione Triangulor* [um] *Libri Qvinqve. Jtem* | *Problematvm variorv.* [m] *nempe* | *Geodaeticorum,* | *Altimetricorum,* | *Geographicorum,* | *Gnomonicorum, et* | *Astronomicorum:* | *Libri Decem.* | *Editio Tertia.* | *Cui recens accessit Pro* | *blematum Arckhitectonicorum Liber* | *unus* | *Francofurti.* | *Typis Nicolai Hofmanni:* | *Sumptibus Ionae Rosae.*| *M.DCXII.*

[2] Quoted from J. Tropfke, *op. cit.*, Vol. I (1921), p. 140.

The dots appearing here are simply the punctuation marks written after (sometimes also before) a number which appears in the running text of most medieval manuscripts and many early printed books on mathematics. For example, Clavius[1] wrote in 1606: "Deinde quia minor est $\frac{4}{7}$. quam $\frac{3}{5}$. erit per propos .8. minutarium libri 9. Euclid. minor proportio 4. ad 7. quam 3. ad 5."

Pitiscus makes extensive use of decimal fractions. In the first five books of his *Trigonometria* the decimal fractions are not preceded by integral values. The fractional numerals are preceded by a zero; thus on page 44 he writes 02679492 (our 0.2679492) and finds its square root which he writes 05176381 (our 0.5176381). Given an arc and its chord, he finds (p. 54) the chord of one-third that arc. This leads to the equation (in modern symbols) $3x - x^3 = .5176381$, the radius being unity. In the solution of this equation by approximation he obtains successively 01, 017, 0174 . . . . and finally 01743114. In computing, he squares and cubes each of these numbers. Of 017, the square is given as 00289, the cube as 0004913. This proves that Pitiscus understood operations with decimals. In squaring 017 appears the following:

$$
\begin{array}{r}
"\ 001.7 \\
2\ 7 \\
1\ 89 \\
\hline
002\ 89.4\ "
\end{array}
$$

What rôle do these dots play? If we put $a = \frac{1}{10}$, $b = \frac{7}{100}$, then $(a+b)^2 = a^2 + (2a+b)b$; $001 = a^2$, $027 = (2a+b)$, $00189 = (2a+b)b$, $00289 \equiv (a+b)^2$. The dot in 001.7 serves simply as a separator between the 001 and the digit 7, found in the second step of the approximation. Similarly, in 00289.4, the dot separates 00289 and the digit 4, found in the third step of the approximation. It is clear that the dots used by Pitiscus in the foreging approximation are not decimal points.

The part of Pitiscus' *Trigonometria* (1612) which bears the title "Problematvm variorvm . . . . libri vndecim" begins a new pagination. Decimal fractions are used extensively, but integral parts appear and a vertical bar is used as decimal separatrix, as (p. 12) where he says, "pro . . . . 13|00024. assumo 13. fractione scilicet $\frac{24}{100000}$ neglecta." ("For 13.00024 I assume 13, the fraction, namely, $\frac{24}{100,000}$ being neglected.") Here again he displays his understanding of decimals, and he uses the dot for other purposes than a decimal separatrix. The writer has carefully examined every appearance of

[1] *Christophori Clavius* . . . . *Geometria practica* (Mogvntiae, 1606), p. 343.

dots in the processes of arithmetical calculation, but has failed to find the dot used as a decimal separatrix. There are in the Pitiscus of 1612 three notations for decimal fractions, the three exhibited in 0522 (our .522), 5|269 (our 5.269), and the form (p. 9) of common fractions, $121\frac{418}{1000}$. In one case (p. 11) there occurs the tautological notation $29|\frac{95}{100}$ (our 29.95).

280. But it has been affirmed that Pitiscus used the decimal point in his trigonometric Table. Indeed, the dot does appear in the Table of 1612 hundreds of times. Is it used as a decimal point? Let us quote from Pitiscus (p. 34): "Therefore the radius for the making of these Tables is to be taken so much the more, that there may be no error in so many of the figures towards the left hand, as you will have placed in the Tables: And as for the superfluous numbers they are to be cut off from the right hand toward the left, after the ending of the calculation. So did Regiomontanus, when he would calculate the tables of sines to the radius of 6000000; he took the radius 60000000000. and after the computation was ended, he cut off from every sine so found, from the right hand toward the left four figures, so Rhaeticus when he would calculate a table of sines to the radius of 10000000000 took for the radius 1000000000000000 and after the calculation was done, he cut off from every sine found from the right hand toward the left five figures: But I, to find out the numbers in the beginning of the Table, took the radius of 100000 00000 00000 00000 00000. But in the Canon itself have taken the radius divers numbers for necessity sake: As hereafter in his place shall be declared."

On page 83 Pitiscus states that the radius assumed is unity followed by 5, 7, 8, 9, 10, 11, or 12 ciphers, according to need. In solving problems he takes, on page 134, the radius 10^7 and writes sin 61°46′ = 8810284 (the number in the table is 88102.838); on page 7 ("Probl. var.") he takes the radius 10^5 and writes sin 41°10′ = 65825 (the number in the Table is 66825.16). Many examples are worked, but in no operation are the trigonometric values taken from the Table written down as decimal fractions. In further illustration we copy the following numerical values from the Table of 1612 (which contains sines, tangents, and secants):

" sin 2″ = 97   sec   3″ = 100000.00001.06
  sin 3″ = 1.45   sec 2°30′ = 100095.2685.
  tan 3″ = 1.45   sec 3°30′ = 100186.869
          sin 89°59′59″ = 99999.99999.88
          tan 89°59′59″ = 20626480624.
sin 30°31′ = 50778.90   sec 30°31′ = 116079.10"

To explain all these numbers the radius must be taken 10^{12}. The $100000.00001.06$ is an integer. The dot on the right is placed between tens and hundreds. The dot on the left is placed between millions and tens of millions.

When a number in the Table contains two dots, the left one is always between millions and tens of millions. The right-hand dot is between tens and hundreds, except in the case of the secants of angles between $0°19'$ and $2°31'$ and in the case of sines of angles between $87°59'$ and $89°40'$; in these cases the right-hand dot is placed (probably through a printer's error) between hundreds and thousands (see sec. $2°30'$). The tangent of $89°59'59''$ (given above) is really 20626480624-0000000, when the radius is 10^{12}. All the figures below ten millions are omitted from the Table in this and similar cases of large functional values.

If a sine or tangent has one dot in the Table and the secant for the same angle has two dots, then the one dot for the sine or tangent lies between millions and tens of millions (see sin $3''$, sec $3''$).

If both the sine and secant of an angle have only one dot in the Table and $r = 10^{12}$, that dot lies between millions and tens of millions (see sin $30°31'$ and sec $30°31'$). If the sine or tangent of an angle has no dots whatever (like sin $2''$), then the figures are located immediately below the place for tens of millions. For all angles above $2°30'$ and below $88°$ the numbers in the Table contain each one and only one dot. If that dot were looked upon as a decimal point, correct results could be secured by the use of that part of the Table. It would imply that the radius is always to be taken 10^5. But this interpretation is invalid for any one of the following reasons: (1) Pitiscus does not always take the $r = 10^5$ (in his early examples he takes $r = 10^7$), and he explicitly says that the radius may be taken 10^5, 10^7, 10^8, 10^9, 10^{10}, 10^{11}, or 10^{12}, to suit the degrees of accuracy demanded in the solution. (2) In the numerous illustrative solutions of problems the numbers taken from the Table are always in integral form. (3) The two dots appearing in some numbers in the Table could not both be decimal points. (4) The numbers in the Table containing no dots could not be integers.

The dots were inserted to facilitate the selection of the trigonometric values for any given radius. For $r = 10^5$, only the figures lying to the left of the dot between millions and tens of millions were copied. For $r = 10^{10}$, the figures to the left of the dot between tens and hundreds were chosen, zeroes being supplied in cases like sin $30°31'$, where there was only one dot, so as to yield sin $30°31' = 5077890000$.

For $r = 10^7$, the figures for 10^5 and the two following figures were copied from the Table, yielding, for example, sin $30°31' = 5077890$. Similarly for other cases.

In a Table[1] which Pitiscus brought out in 1613 one finds the sine of $2°52'30''$ given as 5015.71617.47294, thus indicating a different place assignment of the dots from that of 1612. In our modern tables the natural sine of $2°52'30''$ is given as .05015. This is in harmony with the statement of Pitiscus on the title-page that the Tables are computed "ad radium 1.00000.00000.00000." The observation to be stressed is that these numbers in the Table of Pitiscus (1613) are not decimal fractions, but integers.

Our conclusions, therefore, are that Pitiscus made extended use of decimal fractions, but that the honor of introducing the dot as the separatrix between units and tenths must be assigned to others.

J. Ginsburg has made a discovery of the occurrence of the dot in the position of a decimal separatrix, which he courteously permits to be noted here previous to the publication of his own account of it. He has found the dot in Clavius' *Astrolabe*, published in Rome in 1593, where it occurs in a table of sines and in the explanation of that table (p. 228). The table gives sin $16°12' = 2789911$ and sin $16°13' = 2792704$. Clavius places in a separate column 46.5 as a correction to be made for every second of arc between $16°12'$ and $16°13'$. He obtained this 46.5 by finding the difference 2793 "between the two sines 2789911.2792704," and dividing that difference by 60. He identifies 46.5 as signifying $46\frac{5}{10}$. This dot separates units and tenths. In his works, Clavius uses the dot regularly to separate any two successive numbers. The very sentence which contains 46.5 contains also the integers "2789911.2792704." The question arises, did Clavius in that sentence use both dots as general separators of two pairs of numbers, of which one pair happened to be the integers 46 and the five-tenths, or did Clavius consciously use the dot in 46.5 in a more restricted sense as a decimal separatrix? His use of the plural "duo hi numeri 46.5" goes rather against the latter interpretation. If a more general and more complete statement can be found in Clavius, these doubts may be removed. In his *Algebra* of 1608, Clavius writes all decimal fractions in the form of common fractions. Nevertheless, Clavius unquestionably deserves a place in the history of the introduction of the dot as a decimal separatrix.

More explicit in statement was John Napier who, in his *Rabdologia*

---

[1] B. Pitiscus, *Thesavrvs mathematicvs, sive Canon sinum* (Francofurti, 1613), p. 19.

of 1617, recommended the use of a "period or comma" and uses the comma in his division. Napier's *Constructio* (first printed in 1619) was written before 1617 (the year of his death). In section 5 he says: "Whatever is written after the period is a fraction," and he actually uses the period. In the Leyden edition of the *Constructio* (1620) one finds (p. 6) "25.803. idem quod $25\frac{803}{1000}$."

281. The point occurs in E. Wright's 1616 edition of Napier's *Descriptio*, but no evidence has been advanced, thus far, to show that the sign was intended as a separator of units and tenths, and not as a more general separator as in Pitiscus.

282. *The decimal comma and point of Napier.*—That John Napier in his *Rabdologia* of 1617 introduced the comma and point as separators of units and tenths, and demonstrated that the comma was intended to be used in this manner by performing a division, and properly placing the comma in the quotient, is admitted by all historians. But there are still historians inclined to the belief that he was not the first to use the point or comma as a separatrix between units and tenths. We copy from Napier the following: "Since there is the same facility in working with these fractions as with whole numbers, you will be able after completing the ordinary division, and adding a period or comma, as in the margin, to add to the dividend or to the remainder one cypher to obtain

```
              6 4
            1 3 6
          3 1 6
        1 1 8,0 0 0
      1 4 1
    4 0 2
  4 2 9
  8 6 1 0 9 4,0 0 0(1 9 9 3,2 7 3
  4 3 2
  3 8 8 8
    3 8 8 8
      1 2 9 6
    _____

        8 6 4
        3 0 2 4
          1 2 9 6
```

tenths, two for hundredths, three for thousandths, or more afterwards as required: And with these you will be able to proceed with the working as above. For instance, in the preceding example, here repeated, to which we have added three cyphers, the quotient will become 1 9 9 3,2 7 3, which signifies 1 9 9 3 units and 2 7 3 thousandth parts or $\frac{273}{1000}$."[1]

Napier gives in the *Rabdologia* only three examples in which decimals occur, and even here he uses in the text the sexagesimal exponents for the decimals in the statement of the results.[2] Thus he writes 1994.9160 as 1994,9 $\overset{/}{1}$ $\overset{//}{6}$ $\overset{///}{0}$; in the edition brought out at Leyden in 1626, the circles used by S. Stevin in his notation of decimals are used in place of Napier's sexagesimal exponents.

Before 1617, Napier used the decimal point in his *Constructio*, where he explains the notation in sections 4, 5, and 47, but the *Constructio* was not published until 1619, as already stated above. In section 5 he says: "Whatever is written after the period is a fraction," and he actually uses the period. But in the passage we quoted from *Rabdologia* he speaks of a "period or comma" and actually uses a comma in his illustration. Thus, Napier vacillated between the period and the comma; mathematicians have been vacillating in this matter ever since.

In the 1620 edition[3] of the *Constructio*, brought out in Leyden, one reads: "Vt 10000000.04, valet idem, quod $10000000\frac{4}{100}$. Item 25.803. idem quod $25\frac{803}{1000}$. Item 9999998.0005021, idem valet quod $9999998\frac{5021}{10000000}$. & sic de caeteris."

283. *Seventeenth-century notations after 1617.*—The dot or comma attained no ascendancy over other notations during the seventeenth century.

In 1623 John Johnson (the *survaighour*)[4] published an *Arithmatick* which stresses decimal fractions and modifies the notation of Stevin by omitting the circles. Thus, £ 3. 2 2 9 1 6 is written

$$£\,3 \left|\overset{\text{1. 2. 3. 4. 5.}}{2\,2\,9\,1\,6}\right.,$$

while later in the text there occurs the symbolism 31 | 2500 and 54 | 2625, and also the more cautious "358 | 49411 fifths" for our 358.49411.

---

[1] John Napier, *Rabdologia* (Edinburgh, 1617), Book I, chap. iv. This passage is copied by W. R. Macdonald, in his translation of John Napier's *Constructio* (Edinburgh, 1889), p. 89.

[2] J. W. L. Glashier, "Logarithms and Computation," *Napier Tercentenary Memorial Volume* (ed. Cargill Gilston Knott; London, 1915), p. 78.

[3] *Mirifici logarithmorvm Canonis Constructio . . . . authore & Inventore Ioanne Nepero, Barone Merchistonii, etc.* (Scoto. Lvgdvni, M.DC.XX.), p. 6.

[4] From A. de Morgan in *Companion to the British Almanac* (1851), p. 12.

Henry Briggs[1] drew a horizontal line under the numerals in the decimal part which appeared in smaller type and in an elevated position; Briggs wrote 5_{9321} for our 5.9321. But in his Tables of 1624 he employs commas, not exclusively as a decimal separatrix, although one of the commas used for separation falls in the right place between units and tenths. He gives $-0,22724,3780$ as the logarithm of $\frac{16}{27}$.

A. Girard[2] in his *Invention nouvelle* of 1629 uses the comma on one occasion; he finds one root of a cubic equation to be $1\frac{532}{1000}$ and then explains that the three roots expressed in decimals are 1,532 and 347 and $-1,879$. The 347 is .347; did Girard consider the comma unnecessary when there was no integral part?

Bürgi's and Kepler's notation is found again in a work which appeared in Poland from the pen of Joach. Stegman;[3] he writes 39(063. It occurs again in a geometry written by the Swiss Joh. Ardüser.[4]

William Oughtred adopted the sign $2|5$ in his *Clavis mathematicae* of 1631 and in his later publications.

In the second edition of Wingate's *Arithmetic* (1650; ed. John Kersey) the decimal point is used, thus: .25, .0025.

In 1651 Robert Jager[5] says that the common way of natural arithmetic being tedious and prolix, God in his mercy directed him to what he published; he writes upon decimals, in which $16|7249$ is our 16.7249.

Richard Balam[6] used the colon and wrote 3:04 for our 3.04. This same symbolism was employed by Richard Rawlyns,[7] of Great Yarmouth, in England, and by H. Meissner[8] in Germany.

---

[1] Henry Briggs, *Arithmetica logarithmica* (London, 1624), Lectori. S.

[2] De Morgan, *Companion to the British Almanac* (1851), p. 12; *Invention nouvelle*, fol. *E*2.

[3] Joach. Stegman, *Institutionum mathematicarum libri II* (Rakow, 1630), Vol. I, cap. xxiv, "De logistica decimali." We take this reference from J. Tropfke, *op. cit.*, Vol. I (2d ed., 1921), p. 144.

[4] Joh. Ardüser, *Geometriae theoricae et practicae XII libri* (Zürich, 1646), fol. 306, 180*b*, 270*a*.

[5] Robert Jager, *Artificial Arithmetick in Decimals* (London, 1651). Our information is drawn from A. de Morgan in *Companion to the British Almanac* (1851), p. 13.

[6] Rich. Balam, *Algebra* (London, 1653), p. 4.

[7] Richard Rawlyns, *Practical Arithmetick* (London, 1656), p. 262.

[8] H. Meissner, *Geometria tyronica* (1696[?]). This reference is taken from J. Tropfke, *op. cit.*, Vol. I (2d ed., 1921), p. 144.

Sometimes one encounters a superposition of one notation upon another, as if one notation alone might not be understood. Thus F. van Schooten[1] writes 58,5 ① for 58.5, and 638,82 ② for 638.82. Tobias Beutel[2] writes 645.$\frac{8\ 7\ 9}{1\ 0\ 0\ 0}$. A. Tacquet[3] sometimes writes 25.8 $\overset{\text{i}}{0}\,\overset{\text{ii}}{0}\,\overset{\text{iii}}{0}\,\overset{\text{iv}}{7}\,\overset{\text{v}}{9}$, at other times omits the dot, or the Roman superscripts.

Samuel Foster[4] of Gresham College, London, writes 31.$\underline{008}$; he does not rely upon the dot alone, but adds the horizontal line found in Briggs.

Johann Caramuel[5] of Lobkowitz in Bohemia used two horizontal parallel lines, like our sign of equality, as 22=3 for 22.3, also 92= 123,345 for 92.123345. In a Parisian text by Jean Prestet[6] 272097792$\overset{\text{vi}}{}$ is given for 272.097792; this mode of writing had been sometimes used by Stevin about a century before Prestet, and in 1603 by Beyer.

William Molyneux[7] of Dublin had three notations; he frequently used the comma bent toward the right, as in 30,24. N. Mercator[8] in his *Logarithmotechnia* and Dechales[9] in his course of mathematics used the notation as in 12[345.

284. The great variety of forms for separatrix is commented on by Samuel Jeake in 1696 as follows: "For distinguishing of the Decimal Fraction from Integers, it may truly be said, *Quot Homines, tot Sententiae;* every one fancying severally. For some call the Tenth Parts, the *Primes;* the Hundredth Parts, *Seconds;* the Thousandth Parts, *Thirds*, etc. and mark them with *Indices* equivalent over their heads. As to express 34 integers and $\frac{1\ 4\ 2\ 6}{1\ 0\ 0\ 0\ 0}$ Parts of an Unit, they do it thus, 34.$\overset{/}{1}$. $\overset{//}{4}$. $\overset{///}{2}$. $\overset{////}{6}$. Or thus, 34.$\overset{(1)}{1}$. $\overset{(2)}{4}$. $\overset{(3)}{2}$. $\overset{(4)}{6}$. Others thus, 34,1426''''; or thus, 34,1426$^{(4)}$. And some thus, 34.1 . 4 . 2 . 6 . setting the Decimal Parts

[1] Francisci à Schooten, *Exercitationvm mathematicarum liber primus* (Leyden, 1657), p. 33, 48, 49.

[2] Tobias Beutel, *Geometrischer Lust-Garten* (Leipzig, 1690), p. 173.

[3] *Arithmeticae theoria et praxis, autore Andrea Tacqvet* (2d ed.; Antwerp, 1665), p. 181–88.

[4] Samuel Foster, *Miscellanies: or Mathematical Lvcvbrations* (London, 1659), p. 13.

[5] *Joannis Caramvels Mathesis Biceps. Vetus, et Nova* (Companiae, 1670), "Arithmetica," p. 191.

[6] Jean Prestet, *Nouveaux elemens des mathematiques*, Premier volume (Paris, 1689), p. 293.

[7] William Molyneux, *Treatise of Dioptricks* (London, 1692), p. 165.

[8] N. Mercator, *Logarithmotechnia* (1668), p. 19.

[9] A. de Morgan, *Companion to the British Almanac* (1851), p. 13.

at little more than ordinary distance one from the other. . . . . Others distinguish the Integers from the Decimal Parts only by placing a Cöma before the Decimal Parts thus, 34,1426; a good way, and very useful. Others draw a Line under the Decimals thus, 34$\underline{_{1426}}$, writing them in smaller Figures than the Integers. And others, though they use the Cöma in the work for the best way of distinguishing them, yet after the work is done, they use a Rectangular Line after the place of the Units, called *Separatrix*, a separating Line, because it separates the Decimal Parts from the Integers, thus 34|1426. And sometimes the Cöma is inverted thus, 34'1426, contrary to the true Cöma, and set at top. I sometimes use the one, and sometimes the other, as cometh to hand." The author generally uses the comma. This detailed statement from this seventeenth-century writer is remarkable for the omission of the point as a decimal separatrix.

285. *Eighteenth-century discard of clumsy notations.*—The chaos in notations for decimal fractions gradually gave way to a semblance of order. The situation reduced itself to trials of strength between the comma and the dot as separatrices. To be sure, one finds that over a century after the introduction of the decimal point there were authors who used besides the dot or comma the strokes or Roman numerals to indicate primes, seconds, thirds, etc. Thus, Chelucci[1] in 1738 writes
$$\overset{0}{5}.\overset{\text{I}}{8}\ \overset{\text{II}}{6}\ \overset{\text{III}}{4}\ \overset{\text{IV}}{2}, \text{ also } \overset{\text{I}}{4}.\overset{}{2}\ \overset{\text{IV}}{5} \text{ for } 4.2005,\ \overset{\text{II}}{3}.\overset{}{5}\ \overset{\text{V}}{7} \text{ for } 3.05007.$$

W. Whiston[2] of Cambridge used the semicolon a few times, as in 0;9985, though ordinarily he preferred the comma. O. Gherli[3] in Modena, Italy, states that some use the sign 35|345, but he himself uses the point. E. Wells[4] in 1713 begins with 75.25, but later in his arithmetic introduces Oughtred's |75. Joseph Raphson's translation into English of I. Newton's *Universal Arithmetick* (1728),[5] contains 732,|569 for our 732.569. L'Abbé Deidier[6] of Paris writes the

[1] Paolino Chelucci, *Institutiones analyticae* . . . . *auctore Paulino A. S. Josepho Lucensi* (Rome), p. 35, 37, 41, 283.

[2] Isaac Newton, *Arithmetica Vniversalis* (Cambridge, 1707), edited by G. W[histon], p. 34.

[3] O. Gherli, *Gli elementi* . . . . *delle mathematiche pure*, Vol. I (Modena, 1770), p. 60.

[4] Edward Wells, *Young gentleman's arithmetick* (London, 1713), p. 59, 105, 157.

[5] *Universal Arithmetick, or Treatise of Arithmetical Composition and Resolution* . . . . transl. by the late Mr. Joseph Ralphson, & revised and corrected by Mr. Cunn (2d ed.; London, 1728), p. 2.

[6] L'Abbé Deidier, *L'Arithmétique des Géomètres, ou nouveaux élémens de mathématiques* (Paris, 1739), p. 413.

decimal point and also the strokes for tenths, hundredths, etc. He says: "Pour ajouter ensemble 32.6$'$ 3$''$ 4$'''$ et 8.5$'$ 4$''$.3$'''$—

$$32\ 6\ 3\ 4^{\text{III}}$$
$$8\ 5\ 4\ 3^{\text{III}}$$
$$\overline{}$$
$$41\ 1\ 7\ 7^{\text{III}}\ "$$

A somewhat unusual procedure is found in Sherwin's *Tables*[1] of 1741, where a number placed inside a parenthesis is used to designate the number of zeroes that precede the first significant figure in a decimal; thus, (4) 2677 means .00002677.

In the eighteenth century, trials of strength between the comma and the dot as the separatrix were complicated by the fact that Leibniz had proposed the dot as the symbol of multiplication, a proposal which was championed by the German textbook writer Christian Wolf and which met with favorable reception throughout the Continent. And yet Wolf[2] himself in 1713 used the dot also as separatrix, as "loco $5\frac{47}{10000}$ scribimus 5.0047." As a symbol for multiplication the dot was seldom used in England during the eighteenth century, Oughtred's $\times$ being generally preferred. For this reason, the dot as a separatrix enjoyed an advantage in England during the eighteenth century which it did not enjoy on the Continent. Of fifteen British books of that period, which we chose at random, nine used the dot and six the comma. In the nineteenth century hardly any British authors employed the comma as separatrix.

In Germany, France, and Spain the comma, during the eighteenth century, had the lead over the dot, as a separatrix. During that century the most determined continental stand in favor of the dot was made in Belgium[3] and Italy.[4] But in recent years the comma has finally won out in both countries.

[1] H. Sherwin, *Mathematical Tables* (3d ed.; rev. William Gardiner, London, 1741), p. 48.

[2] Christian Wolf, *Elementa matheseos universae*, Tomus I (Halle, 1713), p. 77.

[3] Désiré André, *Des Notations Mathématiques* (Paris, 1909), p. 19, 20.

[4] Among eighteenth-century writers in Italy using the dot are Paulino A. S. Josepho Lucensi who in his *Institutiones analyticae* (Rome, 1738) uses it in connection with an older symbolism, "3.05007"; G. M. della Torre, *Istituzioni arimmetiche* (Padua, 1768); Odoardo Gherli, *Elementi delle matematiche pure*, Modena, Tomo I (1770); Peter Ferroni, *Magnitudinum exponentialium logarithmorum et trigonometriae sublimis theoria* (Florence, 1782); F. A. Tortorella, *Arithmetica degl'idioti* (Naples, 1794).

286. *Nineteenth century: different positions for dot and comma.*—
In the nineteenth century the dot became, in England, the favorite
separatrix symbol. When the brilliant but erratic Randolph Churchill
critically spoke of the "damned little dots," he paid scant respect to
what was dear to British mathematicians. In that century the dot
came to serve in England in a double capacity, as the decimal symbol
and as a symbol for multiplication.

Nor did these two dots introduce confusion, because (if we may
use a situation suggested by Shakespeare) the symbols were placed in
Romeo and Juliet positions, the Juliet dot stood on high, above
Romeo's reach, her joy reduced to a decimal over his departure, while
Romeo below had his griefs multiplied and was "a thousand times the
worse" for want of her light. Thus, $2\cdot 5$ means $2\frac{5}{10}$, while 2.5 equals
10. It is difficult to bring about a general agreement of this kind,
but it was achieved in Great Britain in the course of a little over half
a century. Charles Hutton[1] said in 1795: "I place the point near the
upper part of the figures, as was done also by Newton, a method which
prevents the separatrix from being confounded with mere marks of
punctuation." In the Latin edition[2] of Newton's *Arithmetica uni-
versalis* (1707) one finds, "Sic numerus 732'|569. denotat septingentas
triginta duas unitates, ... qui et sic 732,|569, vel sic $732\cdot 569$. vel
etiam sic 732|569, nunnunquam scribitur . . . . 57104'2083 . . . .
0'064." The use of the comma prevails; it is usually placed high, but
not always. In Horsely's and Castillon's editions of Newton's *Arithmetica universalis* (1799) one finds in a few places the decimal notation 35'72; it is here not the point but the comma that is placed on
high. Probably as early as the time of Hutton the expression "decimal point" had come to be the synonym for "separatrix" and was
used even when the symbol was not a point. In most places in Horsley's and Castillon's editions of Newton's works, the comma 2,5 is
used, and only in rare instances the point 2.5. The sign $2\cdot 5$ was used
in England by H. Clarke[3] as early as 1777, and by William Dickson[4]
in 1800. After the time of Hutton the $2\cdot 5$ symbolism was adopted by
Peter Barlow (1814) and James Mitchell (1823) in their mathematical
dictionaries. Augustus de Morgan states in his *Arithmetic:* "The

---

[1] Ch. Hutton, *Mathematical and Philosophical Dictionary* (London, 1795),
art. "Decimal Fractions."

[2] I. Newton, *Arithmetica universalis* (ed. W. Whiston; Cambridge, 1707), p. 2.
See also p. 15, 16.

[3] H. Clarke, *Rationale of Circulating Numbers* (London, 1777).

[4] W. Dickson in *Philosophical Transactions*, Vol. VIII (London, 1800), p. 231.

student is recommended always to write the decimal point in a line with the top of the figures, or in the middle, as is done here, and never at the bottom. The reason is that it is usual in the higher branches of mathematics to use a point placed between two numbers or letters which are multiplied together."[1] A similar statement is made in 1852 by T. P. Kirkman.[2] Finally, the use of this notation in Todhunter's texts secured its general adoption in Great Britain.

The extension of the usefulness of the comma or point by assigning it different vertical positions was made in the arithmetic of Sir Jonas Moore[3] who used an elevated and inverted comma, 116'64. This notation never became popular, yet has maintained itself to the present time. Daniel Adams,[4] in New Hampshire, used it, also Juan de Dios Salazar[5] in Peru, Don Gabriel Ciscar[6] of Mexico, A. de la Rosa Toro[7] of Lima in Peru, and Federico Villareal[8] of Lima. The elevated and inverted comma occurs in many, but not all, the articles using decimal fractions in the *Enciclopedia-vniversal ilvstrada Evropeo-Americana* (Barcelona, 1924).

Somewhat wider distribution was enjoyed by the elevated but not inverted comma, as in 2'5. Attention has already been called to the occurrence of this symbolism, a few times, in Horsley's edition of Newton's *Arithmetica universalis*. It appeared also in W. Whiston's edition of the same work in 1707 (p. 15). Juan de Dios Salazar of Peru, who used the elevated inverted comma, also uses this. It is Spain and the Spanish-American countries which lead in the use of this notation. De La-Rosa Toro, who used the inverted comma, also used this. The 2'5 is found in Luis Monsante[9] of Lima; in Maximo

[1] A. de Morgan, *Elements of Arithmetic* (4th ed.; London, 1840), p. 72.

[2] T. P. Kirkman, *First Mnemonical Lessons in Geometry, Algebra and Trigonometry* (London, 1852), p. 5.

[3] *Moore's Arithmetick: In Four Books* (3d ed.; London, 1688), p. 369, 370, 465.

[4] Daniel Adams, *Arithmetic* (Keene, N.H., 1827), p. 132.

[5] Juan de Dios Salazar, *Lecciones de Aritmetica*, Teniente del Cosmògrafo major de esta Republica del Perŭ (Arequipa, 1827), p. 5, 74, 126, 131. This book has three different notations: 2,5; 2'5; 2'5.

[6] Don Gabriel Ciscar, *Curso de estudios elementales de Marina* (Mexico, 1825).

[7] Agustin de La-Rosa Toro, *Aritmetica Teorico-Practica* (tercera ed.; Lima, 1872), p. 157.

[8] D. Federico Villareal, *Calculo Binomial* (P. I. Lima [Peru], 1898), p. 416.

[9] Luis Monsante, *Lecciones de Aritmetica Demostrada* (7th ed.; Lima, 1872), p. 89.

Vazquez[1] of Lima; in Manuel Torres Torija[2] of Mexico; in D. J. Cortazár[3] of Madrid. And yet, the Spanish-speaking countries did not enjoy the monopoly of this symbolism. One finds the decimal comma placed in an elevated position, 2'5, by Louis Bertrand[4] of Geneva, Switzerland.

Other writers use an inverted wedge-shaped comma,[5] in a lower position, thus: 2ᴧ5. In Scandinavia and Denmark the dot and the comma have had a very close race, the comma being now in the lead. The practice is also widely prevalent, in those countries, of printing the decimal part of a number in smaller type than the integral part.[6] Thus one frequently finds there the notations 2,₅ and 2.₅. To sum up, in books printed within thirty-five years we have found the decimal notations[7] 2·5, 2'5, 2,5, 2'5, 2'5, 2ᴧ5, 2,₅, 2.₅.

287. The earliest arithmetic printed on the American continent which described decimal fractions came from the pen of Greenwood,[8] professor at Harvard College. He gives as the mark of separation "a Comma, a Period, or the like," but actually uses a comma. The arithmetic of "George Fisher" (Mrs. Slack), brought out in England, and also her *The American Instructor* (Philadelphia, 1748) contain both the comma and the period. Dilworth's *The Schoolmaster's Assistant*, an English book republished in America (Philadelphia, 1733), used the period. In the United States the decimal point[9] has always had the

---

[1] Maximo Vazquez, *Aritmetica practica* (septiema ed.; Lima, 1875), p. 57.

[2] Manuel Torres Torija, *Nociones de Algebra Superior y elementos fundamentales de cálculo differencial é Integral* (México, 1894), p. 137.

[3] D. J. Cortazár, *Tratado de Aritmética* (42d ed.; Madrid, 1904).

[4] L. Bertrand, *Developpment nouveaux de la partie elementaire des mathematiques*, Vol. I (Geneva, 1778), p. 7.

[5] As in A. F. Vallin, *Aritmética para los niños* (41st ed.; Madrid, 1889), p. 66.

[6] Gustaf Haglund, *Samlying of Öfningsexempel till Lärabok i Algebra*, Fjerde Upplagan (Stockholm, 1884), p. 19; *Öfversigt af Kongl. Vetenskaps-Akademiens Förhandlingar*, Vol. LIX (1902; Stockholm, 1902, 1903), p. 183, 329; *Oversigt over det Kongelige Danske Videnskabernes Selskabs, Fordhandlinger* (1915; Kobenhavn, 1915), p. 33, 35, 481, 493, 545.

[7] An unusual use of the elevated comma is found in F. G. Gausz's *Fünfstellige vollständige Logar. u. Trig. Tafeln* (Halle a. S., 1906), p. 125; a table of squares of numbers proceeds from $N = 0'00$ to $N = 10'00$. If the square of 63 is wanted, take the form 6'3; its square is 39'6900. Hence $63^2 = 3969$.

[8] Isaac Greenwood, *Arithmetick Vulgar and Decimal* (Boston, 1729), p. 49. See facsimile of a page showing decimal notation in L. C. Karpinski, *History of Arithmetic* (Chicago, New York, 1925), p. 134.

[9] Of interest is Chauncey Lee's explanation in his *American Accomptan'* (Lasingburgh, 1797), p. 54, that, in writing denominate numbers, he separates

lead over the comma, but during the latter part of the eighteenth and the first half of the nineteenth century the comma in the position of 2,5 was used quite extensively. During 1825–50 it was the influence of French texts which favored the comma. We have seen that Daniel Adams used 2‘5 in 1827, but in 1807 he[1] had employed the ordinary 25,17 and ,375. Since about 1850 the dot has been used almost exclusively. Several times the English elevated dot was used in books printed in the United States. The notation 2·5 is found in Thomas Sarjeant's *Arithmetic*,[2] in F. Nichols' *Trigonometry*,[3] in American editions of Hutton's *Course of Mathematics* that appeared in the interval 1812–31, in Samuel Webber's *Mathematics*,[4] in William Griev's *Mechanics Calculator*, from the fifth Glasgow edition (Philadelphia, 1842), in *The Mathematical Diary* of R. Adrain[5] about 1825, in Thomas Sherwin's *Common School Algebra* (Boston, 1867; 1st ed., 1845), in George R. Perkins' *Practical Arithmetic* (New York, 1852). Sherwin writes: "To distinguish the sign of Multiplication from the period used as a decimal point, the latter is elevated by inverting the type, while the former is larger and placed down even with the lower extremities of the figures or letters between which it stands." In 1881 George Bruce Halsted[6] placed the decimal point halfway up and the multiplication point low.

It is difficult to assign definitely the reason why the notation 2·5 failed of general adoption in the United States. Perhaps it was due to mere chance. Men of influence, such as Benjamin Peirce, Elias Loomis, Charles Davies, and Edward Olney, did not happen to become interested in this detail. America had no one of the influence of De Morgan and Todhunter in England, to force the issue in favor of 2·5. As a result, 2.5 had for a while in America a double meaning, namely, 2 5/10 and 2 times 5. As long as the dot was seldom used to

---

the denominations "in a vulgar table" by two commas, but "in a decimal table" by the decimal point; he writes £ 175,, 15,, 9, and 1.41.

[1] Daniel Adams, *Scholar's Arithmetic* (4th ed.; Keene, N.H., 1807).

[2] Thomas Sarjeant, *Elementary Principles of Arithmetic* (Philadelphia, 1788), p. 80.

[3] F. Nichols, *Plane and Spherical Trigonometry* (Philadelphia, 1811), p. 33.

[4] Samuel Webber, *Mathematics*, Vol. I (Cambridge, 1801; also 1808, 2d ed.), p. 227.

[5] R. Adrain, *The Mathematical Diary*, No. 5, p. 101.

[6] George Bruce Halsted, *Elementary Treatise on Mensuration* (Boston, 1881).

express multiplication, no great inconvenience resulted, but about 1880 the need of a distinction arose. The decimal notation was at that time thoroughly established in this country, as 2.5, and the dot for multiplication was elevated to a central position. Thus with us $2 \cdot 5$ means 2 times 5.

Comparing our present practice with the British the situation is this: We write the decimal point low, they write it high; we place the multiplication dot halfway up, they place it low. Occasionally one finds the dot placed high to mark multiplication also in German books, as, for example, in Friedrich Meyer[1] who writes $2 \cdot 3 = 6$.

288. It is a notable circumstance that at the present time the modern British decimal notation is also the notation in use in Austria where one finds the decimal point placed high, but the custom does not seem to prevail through any influence emanating from England. In the eighteenth century P. Mako[2] everywhere used the comma, as in 3,784. F. S. Mozhnik[3] in 1839 uses the comma for decimal fractions, as in 3,1344, and writes the product "2 . 3..n." The *Sitzungsberichte der philosophisch-historischen Classe d. K. Akademie der Wissenschaften*, Erster Band (Wien, 1848), contains decimal fractions in many articles and tables, but always with the low dot or low comma as decimal separatrix; the low dot is used also for multiplication, as in "1.2.3. . .r."

But the latter part of the nineteenth century brought a change. The decimal point is placed high, as in $1 \cdot 63$, by I. Lemoch[4] of Lemberg. N. Fialkowski of Vienna in 1863 uses the elevated dot[5] and also in 1892.[6] The same practice is followed by A. Steinhauser of Vienna,[7] by Johann Spielmann[8] and Richard Supplantschitsch,[9] and by Karl

---

[1] Friedrich Meyer, *Dritter Cursus der Planimetrie* (Halle a/S., 1885), p. 5.

[2] P. Mako e S.I., *De . . . . aequationvm resolvtionibvs libri dvo* (Vienna, 1770), p. 135; *Compendiaria Matheseos Institvtio. . . . .* Pavlvs Mako e S.I. in Coll. Reg. Theres Prof. Math. et Phys. Experim. (editio tertia; Vienna, 1771).

[3] Franz Seraphin Mozhnik, *Theorie der numerischen Gleichungen* (Wien, 1839), p. 27, 33.

[4] Ignaz Lemoch, *Lehrbuch der praktischen Geometrie*, 2. Theil, 2. Aufl. (Wien, 1857), p. 163.

[5] Nikolaus Fialkowski, *Das Decimalrechnen mit Rangziffern* (Wien, 1863), p. 2.

[6] N. Fialkowski, *Praktische Geometrie* (Wien, 1892), p. 48.

[7] Anton Steinhauser, *Lehrbuch der Mathematik. Algebra* (Wien, 1875), p. 111, 138.

[8] Johann Spielmann, *Močniks Lehrbuch der Geometrie* (Wien, 1910), p. 66.

[9] Richard Supplantschitsch, *Mathematisches Unterrichtswerk, Lehrbuch der Geometrie* (Wien, 1910), p. 91.

Rosenberg.[1] Karl Zahradníček[2] writes 0˙35679.1˙0765.1˙9223.0˙3358, where the lower dots signify multiplication and the upper dots are decimal points. In the same way K. Wolletz[3] writes $(-0˙0462)$. 0˙0056.

An isolated instance of the use of the elevated dot as decimal separatrix in Italy is found in G. Peano.[4]

In France the comma placed low is the ordinary decimal separatrix in mathematical texts. But the dot and also the comma are used in marking off digits of large numbers into periods. Thus, in a political and literary journal of Paris (1908)[5] one finds "2,251,000 drachmes," "Fr. 2.638.370 75," the francs and centimes being separated by a vacant place. One finds also "601,659 francs 05" for Fr. 601659. 05. It does not seem customary to separate the francs from centimes by a comma or dot.

That no general agreement in the notation for decimal fractions exists at the present time is evident from the publication of the International Mathematical Congress in Strasbourg (1920), where decimals are expressed by commas[6] as in 2,5 and also by dots[7] as in 2.5. In that volume a dot, placed at the lower border of a line, is used also to indicate multiplication.[8]

The opinion of an American committee of mathematicians is expressed in the following: "Owing to the frequent use of the letter x, it is preferable to use the dot (a raised period) for multiplication in the few cases in which any symbol is necessary. For example, in a case like $1·2·3 \ldots (x-1)·x$, the center dot is preferable to the symbol $\times$; but in cases like $2a(x-a)$ no symbol is necessary. The committee recognizes that the period (as in $a.b$) is more nearly international than the center dot (as in $a·b$); but inasmuch as the period will continue to be used in this country as a decimal point,

---

[1] Karl Rosenberg, *Lehrbuch der Physik* (Wien, 1913), p. 125.

[2] Karl Zahradníček, *Močniks Lehrbuch der Arithmetik und Algebra* (Wien, 1911), p. 141.

[3] K. Wolletz, *Arithmetik und Algebra* (Wien, 1917), p. 163.

[4] Giuseppe Peano, *Risoluzione graduale delle equazioni numeriche* (Torino, 1919), p. 8. Reprint from *Atti della r. Accad. delle Scienze di Torino*, Vol. LIV (1918–19).

[5] *Les Annales*, Vol. XXVI, No. 1309 (1908), p. 22, 94.

[6] *Comptes rendus du congrès international des mathématiques* (Strasbourg, 22–30 Septembre 1920; Toulouse, 1921), p. 253, 543, 575, 581.

[7] *Op. cit.*, p. 251.

[8] *Op. cit.*, p. 153, 252, 545.

it is likely to cause confusion, to elementary pupils at least, to attempt to use it as a symbol for multiplication."[1]

289. *Signs for repeating decimals.*—In the case of repeating decimals, perhaps the earliest writer to use a special notation for their designation was John Marsh,[2] who, "to avoid the Trouble for the future of writing down the Given Repetend or Circulate, whether Single or Compound, more than once," distinguishes each "by placing a Period over the first Figure, or over the first and last Figures of the given Repetend." Likewise, John Robertson[3] wrote 0,3̇ for 0,33 . . . . , 0,2̇3̇ for 0,2323 . . . . , 0,7̇85̇ for 0,785785. . . . . H. Clarke[4] adopted the signs .6́ for .666 . . . . , .6́42́ for .642642. . . . . A choice favoring the dot is shown by Nicolas Pike[5] who writes, 37̇9̇, and by Robert Pott[6] and James Pryde[7] who write ·3̇, ·4̇5̇, ·34̇56̇7̇. A return to accents is seen in the *Dictionary* of Davies and Peck[8] who place accents over the first, or over the first and last figure, of the repetend, thus: .'2, .'5723', 2.4'18'.

SIGNS OF POWERS

290. *General remarks.*—An ancient symbol for squaring a number occurs in a hieratic Egyptian papyrus of the late Middle Empire, now in the Museum of Fine Arts in Moscow.[9] In the part containing the computation of the volume of a frustrated pyramid of square base there occurs a hieratic term, containing a pair of walking legs ⋀ and signifying "make in going," that is, squaring the number. The Diophantine notation for powers is explained in § 101, the Hindu notation in §§ 106, 110, 112, the Arabic in § 116, that of Michael Psellus in § 117. The additive principle in marking powers is referred

[1] *The Reorganization of Mathematics in Secondary Schools*, by the National Committee on Mathematical Requirements, under the auspices of the Mathematical Association of America (1923), p. 81.

[2] John Marsh, *Decimal Arithmetic Made Perfect* (London, 1742), p. 5.

[3] John Robertson, *Philosophical Transactions* (London, 1768), No. 32, p. 207–13. See Tropfke, *op. cit.*, Vol. I (1921), p. 147.

[4] H. Clarke, *The Rationale of Circulating Numbers* (London, 1777), p. 15, 16.

[5] Nicolas Pike, *A New and Complete System of Arithmetic* (Newbury-port, 1788), p. 323.

[6] Robert Pott, *Elementary Arithmetic, etc.* (Cambridge, 1876), Sec. X, p. 8.

[7] James Pryde, *Algebra Theoretical and Practical* (Edinburgh, 1852), p. 278.

[8] C. Davies and W. G. Peck, *Mathematical Dictionary* (1855), art. "Circulating Decimal."

[9] See *Ancient Egypt* (1917), p. 100–102.

to in §§ 101, 111, 112, 124. The multiplicative principle in marking powers is elucidated in §§ 101, 111, 116, 135, 142.

Before proceeding further, it seems desirable to direct attention to certain Arabic words used in algebra and their translations into Latin. There arose a curious discrepancy in the choice of the principal unknown quantity; should it be what we call x, or should it be x^2? al-Khowârizmî and the older Arabs looked upon x^2 as the principal unknown, and called it *māl* ("assets," "sum of money").[1] This viewpoint may have come to them from India. Accordingly, x (the Arabic *jidr*, "plant-root," "basis," "lowest part") must be the square root of *māl* and is found from the equation to which the problem gives rise. By squaring x the sum of money could be ascertained.

Al-Khowârizmî also had a general term for the unknown, *shai* ("thing"); it was interpreted broadly and could stand for either *māl* or *jidr* (x^2 or x). Later, John of Seville, Gerard of Cremona, Leonardo of Pisa, translated the Arabic *jidr* into the Latin *radix*, our x; the Arabic *shai* into *res*. John of Seville says in his arithmetic:[2] "Quaeritur ergo, quae res cum. X. radicibus suis idem decies accepta radice sua efficiat 39." ("It is asked, therefore, what thing together with 10 of its roots or what is the same, ten times the root obtained from it, yields 39.") This statement yields the equation $x^2+10x=39$. Later *shai* was also translated as *causa*, a word which Leonardo of Pisa used occasionally for the designation of a second unknown quantity. The Latin *res* was translated into the Italian word *cosa*, and from that evolved the German word *coss* and the English adjective "cossic." We have seen that the abbreviations of the words *cosa* and *cubus*, viz., *co.* and *cu.*, came to be used as algebraic symbols. The words *numerus, dragma, denarius*, which were often used in connection with a given absolute number, experienced contractions sometimes employed as symbols. Plato of Tivoli,[3] in his translation from the Hebrew of the *Liber embadorum* of 1145, used a new term, *latus* ("side"), for the first power of the unknown, x, and the name *embadum* ("content") for the second power, x^2. The term *latus* was found mainly in early Latin writers drawing from Greek sources and was used later by Ramus (§ 322), Vieta (§ 327), and others.

291. *Double significance of "R" and "l."*—There came to exist considerable confusion on the meaning of terms and symbols, not only

---

[1] J. Ruska, *Sitzungsberichte Heidelberger Akad., Phil.-hist. Klasse* (1917), Vol. II, p. 61 f.; J. Tropfke, *op. cit.*, Vol. II (2d ed., 1921), p. 106.

[2] Tropfke, *op. cit.*, Vol. II (2d ed., 1921), p. 107.

[3] M. Curtze, *Bibliotheca mathematica* (3d ser.), Vol. I (1900), p. 322, n. 1.

because *res* (*x*) occasionally was used for x^2, but more particularly because both *radix* and *latus* had two distinct meanings, namely, x and $\sqrt{x}$. The determination whether x or $\sqrt{x}$ was meant in any particular case depended on certain niceties of designation which the unwary was in danger of overlooking (§ 137).

The letter l (*latus*) was used by Ramus and Vieta for the designation of roots. In some rare instances it also represented the first power of the unknown x. Thus, in Schoner's edition of Ramus[1] $5l$ meant $5x$, while $l5$ meant $\sqrt{5}$. Schoner marks the successive powers "*l.*, *q.*, *c.*, *bq.*, *ſ.*, *qc.*, *b.ſ.*, *tq.*, *cc.*" and named them *latus, quadratus, cubus, biquadratus*, and so on. Ramus, in his *Scholarvm mathematicorvm libri unus et triginti* (1569), uses the letter l only for square root, not for x or in the designation of powers of x; but he uses (p. 253) the words *latus, quadratus, latus cubi* for x, x^2, x^3.

This double use of l is explained by another pupil of Ramus, Bernardus Salignacus,[2] by the statement that if a number precedes the given sign it is the coefficient of the sign which stands for a power of the unknown, but if the number comes immediately after the l the root of that number is to be extracted. Accordingly, $2q$, $3c$, $5l$ stand respectively for $2x^2$, $3x^3$, $5x$; on the other hand, $l5$, $lc8$, $lbq16$ stand respectively for $\sqrt{5}$, $\sqrt[3]{8}$, $\sqrt[4]{16}$. The double use of the capital L is found in G. Gosselin (§§ 174, 175).

B. Pitiscus[3] writes our $3x - x^3$ thus, $3l - 1c$, and its square $9q - 6.bq + 1qc$, while Willebrord Snellius[4] writes our $5x - 5x^3 + x^5$ in the form $5l - -5c + 1\beta$. W. Oughtred[5] writes $x^5 - 15x^4 + 160x^3 - 1250x^2 + 6480x = 170304782$ in the form $1qc - 15qq + 160c - 1250q + 6480l = 170304782$.

Both $l.$ and $R.$ appear as characters designating the first power

---

[1] *Petri Rami Veromandui Philosophi* .... *arithmetica libri duo et geometriae septem et viginti. Dudum quidem a Lazaro Schonero* .... (Francofvrto ad moenvm, MDCXXVII). P. 139 begins: "De Nvmeris figvratis Lazari Schoneri liber." See p. 177.

[2] *Bernardi Salignaci Burdegalensis Algebrae libri duo* (Francofurti, 1580). See P. Treutlein in *Abhandl. zur Geschichte der Mathematik*, Vol. II (1879), p. 36.

[3] *Batholomaei Pitisci* .... *Trigonometriae editio tertia* (Francofurti, 1612), p. 60.

[4] Willebrord Snellius, *Doctrinae triangvlorvm cononicae liber qvatvor* (Leyden, 1627), p. 37.

[5] William Oughtred, *Clavis mathematicae*, under "De aequationum affectarum resolutione in numeris" (1647 and later editions).

in a work of J. J. Heinlin[1] at Tübingen in 1679. He lets N stand for *unitas, numerus absolutus,* 4, $l.$, $R.$ for *latus vel radix; z., q.* for *quadratus, zensus; ce, c* for *cubus; zz, qq, bq* for *biquadratus.* But he utilizes the three signs 4, l, R also for indicating roots. He speaks[2] of "Latus cubicum, vel Radix cubica, cujus nota est *Lc. R.c.* 4. *ce.*"

John Wallis[3] in 1655 says "Est autem lateris l, numerus pyramidalis $l^3 + 3l^2 + 2l$" and in 1685 writes[4] "$ll - 2laa + a^4 : \div bb$," where the l takes the place of the modern x and the colon is a sign of aggregation, indicating that all three terms are divided by b^2.

292. The use of $R.$ (*Radix*) to signify root and also power is seen in Leonardo of Pisa (§ 122) and in Luca Pacioli (§§ 136, 137). The sign R was allowed to stand for the first power of the unknown x by Peletier in his algebra, by K. Schott[5] in 1661, who proceeds to let $Q.$ stand for x^2, $C.$ for x^3, $Biqq$ or $qq.$ for x^4, $Ss.$ for x^5, $Cq.$ for x^6, $SsB.$ for x^7, $Triq.$ or qqq for x^8, $Cc.$ for x^9. One finds R in W. Leybourn's publication of J. Billy's[6] *Algebra*, where powers are designated by the capital letters N, R, Q, QQ, S, QC, $S2$, $QQQ.$, and where $x^2 = 20 - x$ is written "$1Q = 20 - 1R.$"

Years later the use of $R.$ for x and of $?$. (an inverted capital letter E, rounded) for x^2 is given by Tobias Beutel[7] who writes "21 $?$, gleich 2100, 1 $?$. gleigh 100, $1R.$ gleich 10."

293. *Facsimilis of symbols in manuscripts.*—Some of the forms for radical signs and for x, x^2, x^3, x^4, and x^5, as found in early German manuscripts and in Widman's book, are tabulated by J. Tropfke, and we reproduce his table in Figure 104.

In the Munich manuscript *cosa* is translated *ding;* the symbols in Figure 104, C2a, seem to be modified d's. The symbols in C3 are signs for *res*. The manuscripts C3b, C6, C7, C9, H6 bear on the evolution of the German symbol for x. Paleographers incline to the view that it is a modification of the Italian *co*, the *o* being highly disfigured. In B are given the signs for dragma or *numerus*.

[1] *Joh. Jacobi Heinlini Synopsis mathematica universalis* (3d ed.; Tübingen, 1679), p. 66.

[2] *Ibid.*, p. 65.   [3] John Wallis, *Arithmetica infinitorum* (Oxford, 1655), p. 144.

[4] John Wallis, *Treatise of Algebra* (London, 1685), p. 227.

[5] P. Gasparis Schotti . . . . *Cursus mathematicus* (Herbipoli [Würtzburg], 1661), p. 530.

[6] *Abridgement of the Precepts of Algebra.* The Fourth Part. Written in French by James Billy and now translated into English. . . . . Published by Will. Leybourn (London, 1678), p. 194.

[7] Tobias Beutel, *Geometrische Galleri* (Leipzig, 1690), p. 165.

294. *Two general plans for marking powers.*—In the early development of algebraic symbolism, no signs were used for the powers of given numbers in an equation. As given numbers and coefficients were not represented by letters in equations before the time of Vieta, but were specifically given in numerals, their powers could be computed on the spot and no symbolism for powers of such numbers was needed. It was different with the unknown numbers, the determination of which constituted the purpose of establishing an equation. In consequence,

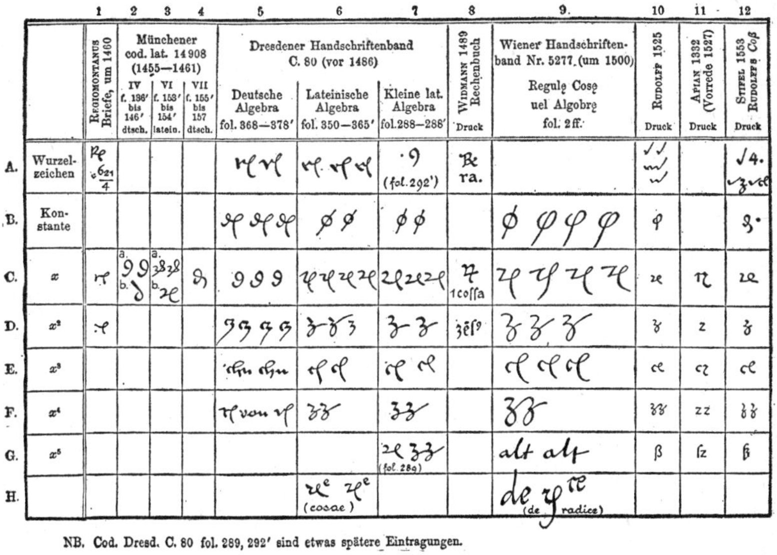

		2 — Münchener cod. lat. 14908 (1455—1461)			5 — Dresdener Handschriftenband C. 80 (vor 1486)			8	9 — Wiener Handschriftenband Nr. 5277. (um 1500)	10	11	12
	1 Regiomontanus Briefe, um 1460	IV f.196' bis 146' dtsch.	VI f.153' bis 154' latein.	VII f.155' bis 157 dtsch.	Deutsche Algebra fol.368—376'	Lateinische Algebra fol.350—365'	Kleine lat. Algebra fol.288—288'	Widmann 1489 Rechenbuch Druck	Regule Cose uel Algobre fol. 2 ff.	Rudolff 1525 Druck	Apian 1532 (Vorrede 1527) Druck	Stifel 1553 Rudolff's Coß Druck
A. Wurzelzeichen							(fol. 292')	ra.				
B. Konstante												
C. x		a. b.					1 coſſa					
D. x^2												
E. x^3												
F. x^4												
G. x^5					(fol. 28a)							
H.				(cosae)			(de radice)					

NB. Cod. Dresd. C. 80 fol. 289, 292' sind etwas spätere Eintragungen.

Fig. 104.—Signs found in German manuscripts and early German books. (Taken from J. Tropfke, *op. cit.*, Vol. II [2d ed., 1921], p. 112.)

one finds the occurrence of symbolic representation of the unknown and its powers during a period extending over a thousand years before the introduction of the literal coefficient and its powers.

For the representation of the unknown there existed two general plans. The first plan was to use some abbreviation of a name signifying unknown quantity and to use also abbreviations of the names signifying the square and the cube of the unknown. Often special symbols were used also for the fifth and higher powers whose orders were prime numbers. Other powers of the unknown, such as the fourth, sixth, eighth powers, were represented by combinations of those symbols. A good illustration is a symbolism of Luca Paciola, in which *co.* (*cosa*) represented x, *ce.* (*censo*) x^2, *cu.* (*cubo*) x^3, *p.r.* (*primo relato*) x^5; combinations of these yielded *ce.ce.* for x^4, *ce.cu.* for

x^6, etc. We have seen these symbols also in Tartaglia and Cardan, in the Portuguese Nuñez (§ 166), the Spanish Perez de Moya in 1652, and Antich Rocha[1] in 1564. We may add that outside of Italy Pacioli's symbols enjoyed their greatest popularity in Spain. To be sure, the German Marco Aurel wrote in 1552 a Spanish algebra (§ 165) which contained the symbols of Rudolff, but it was Perez de Moya and Antich Rocha who set the fashion, for the sixteenth century in Spain; the Italian symbols commanded some attention there even late in the eighteenth century, as is evident from the fourteenth unrevised impression of Perez de Moya's text which appeared at Madrid in 1784. The 1784 impression gives the symbols as shown in Figure 105, and also the explanation, first given in 1562, that the printing office does not have these symbols, for which reason the ordinary letters of the alphabet will be used.[2] Figure 105 is interesting, for it purports to show the handwritten forms used by De Moya. The symbols are not the German, but are probably derived from them. In a later book, the *Tratado de Mathematicas* (Alcala, 1573), De Moya gives on page 432 the German symbols for the powers of the unknown, all except the first power, for which he gives the crude imitation *Ze*. Antich Rocha, in his *Arithmetica*, folio 253, is partial to capital letters and gives the successive powers thus: *N*, *Co*, *Ce*, *Cu*, *Cce*, *R*, *CeCu*, *RR*, *Ccce*, *Ccu*, etc. The same fondness for capitals is shown in his *Mas* for "more" (§ 320).

We digress further to state that the earliest mathematical work published in America, the *Sumario compendioso* of Juan Diez Freyle[3]

[1] *Arithmetica por Antich Rocha de Gerona compuesta, y de varios Auctores recopilada* (Barcelona, 1564, also 1565).

[2] Juan Perez de Moya, *Aritmetica practica, y especulativa* (14th ed.; Madrid, 1784), p. 263: "Por los diez caractéres, que en el precedente capítulo se pusieron, uso estos. Por el qual dicen numero *n.* por la cosa, *co.* por el censo, *ce.* por cubo, *cu.* por censo, de censo, *cce.* por el primero relato, *R.* por el censo, y cubo, *ce.cu.* por segundo relato, *RR.* por censo de censo de censo, *cce.* por cubo de cubo, *ccu.* Esta figura *r.* quiere decir raíz quadrada. Esta figura *rr.* denota raíz quadrada de raíz quadrada. Estas *rrr.* denota raíz cúbica. De estos dos caractéres, *p. m.* notarás, que la *p.* quiere decir mas, y la *m.* menos, el uno es copulativo, el otro disyuntivo, sirven para sumar, y restar cantidades diferentes, como adelante mejor entenderás. Quando despues de *r.* se pone *u.* denota raíz quadrada universal: y asi *rru.* raíz de raíz quadrada universal: y de esta suerte *rrru.* raíz cúbica universal. Esta figura *ig.* quiere decir igual. Esta *q.* denota cantidad, y asi *qs.* cantidades: estos caractéres me ha parecido poner, porque no habia otros en la Imprenta; tú podrás usar, quando hagas demandas, de los que se pusieron en el segundo capítulo, porque son mas breves, en lo demás todos son de una condicion."

[3] Edition by D. E. Smith (Boston and London, 1921).

(City of Mexico, 1556) gives six pages to algebra. It contains the words *cosa*, *zenso*, or *censo*, but no abbreviations for them. The work does not use the signs $+$ or $-$, nor the $\bar{p}$ and $\bar{m}$. It is almost purely rhetorical.

The data which we have presented make it evident that in Perez de Moya, Antich Rocha, and P. Nuñez the symbols of Pacioli are used and that the higher powers are indicated by the combinations of symbols of the lower powers. This *general principle* underlies the notations of Diophantus, the Hindus, the Arabs, and most of the Germans and Italians before the seventeenth century. For convenience we shall call this the "Abbreviate Plan."

Cap. II. *En el qual se ponen algunos caractéres, que sirven por cantidades proporcionales.*

En este capítulo se ponen algunos caractéres, dando à cada uno el nombre y valor que le conviene. Los quales son inventados por causa de brevedad; y es de saber, que no es de necesidad, que estos, y no otros hayan de ser, porque cada uno puede usar de lo que quisiere, è inventar mucho mas, procediendo con la proporcion que le pareciere. Los caractéres son estos.

Fig. 105.—The written algebraic symbols for powers, as given in Perez de Moya's *Arithmetica* (Madrid, 1784), p. 260 (1st ed., 1562). The successive symbols are called *cosa es raíz, censo, cubo, censo de censo, primero relato, censo y cubo, segundo relato, censo de censo de censo, cubo de cubo.*

The second plan was not to use a symbol for the unknown quantity itself, but to limit one's self in some way to simply indicating by a numeral the power of the unknown quantity. As long as powers of only one unknown quantity appeared in an equation, the writing of the index of its power was sufficient. In marking the first, second, third, etc., powers, only the numerals for "one," "two," "three," etc., were written down. A good illustration of this procedure is Chuquet's 10^2 for $10x^2$, 10^1 for $10x$, and 10^0 for 10. We shall call this the "Index Plan." It was stressed by Chuquet, and passed through several stages of development in Bombelli, Stevin, and Girard. Then, after the introduction of special letters to designate one or more unknown quantities, and the use of literal coefficients, this notation was perfected by Hérigone and Hume; it finally culminated in the present-day form in the writings of Descartes, Wallis, and Newton.

295. *Early symbolisms.*—In elaborating the notations of powers according to the "Abbreviate Plan" cited in § 294, one or the other of two distinct principles was brought into play in combining the symbols of the lower powers to mark the higher powers. One was the additive principle of the Greeks in combining powers; the other was the multiplicative principle of the Hindus. Diophantus expressed the fifth power of the unknown by writing the symbols for x^2 and for x^3, one following the other; the indices 2 and 3 were *added*. Now, Bhāskara writes his symbols for x^2 and x^3 in the same way, but lets the two designate, not x^5, but x^6; the indices 2 and 3 are *multiplied*. This difference in designation prevailed through the Arabic period, the later Middle Ages in Europe down into the seventeenth century. It disappeared only when the notations of powers according to the "Abbreviate Plan" passed into disuse. References to the early symbolisms, mainly as exhibited in our accounts of individual authors, are as follows:

ABBREVIATE PLAN

ADDITIVE PRINCIPLE

Diophantus, and his editors Xylander, Bachet, Fermat (§ 101)
al-Karkhî, eleventh century (§ 116)
Leonardo of Pisa (§ 122)
Anonymous Arab (§ 124)
Dresden Codex C. 80 (§ 305, Fig. 104)
M. Stifel, (1545), *sum; sum; sum: x^3* (§ 154)
F. Vieta (1591), and in later publications (§ 177)
C. Glorioso, 1527 (§ 196)
W. Oughtred, 1631 (§ 182)
Samuel Foster, 1659 (§ 306)

MULTIPLICATIVE PRINCIPLE

Bhāskara, twelfth century (§ 110–12)
Arabic writers, except al-Karkhî (§ 116)
L. Pacioli, 1494, *ce. cu.* for x^6 (§ 136)
H. Cardano, 1539, 1545 (§ 140)
N. Tartaglia, 1556–60 (§ 142)
Ch. Rudolff, 1525 (§ 148)
M. Stifel, 1544 (§ 151)
J. Scheubel, 1551, follows Stifel (§ 159)
A. Rocha, 1565, follows Pacioli (§ 294)
C. Clavius, 1608, follows Stifel (§ 161)
P. Nuñez, 1567, follows Pacioli and Cardan (§ 166)
R. Recorde, 1557, follows Stifel (§ 168)

L. and T. Digges, 1579 (§ 170)
A. M. Visconti, 1581 (§ 145)
Th. Masterson, 1592 (§ 171)
J. Peletier, 1554 (§ 172)
G. Gosselin, 1577 (§ 174)
L. Schoner, 1627 (§ 291)

NEW NOTATIONS ADOPTED

Ghaligai and G. del Sodo, 1521 (§ 139)
M. Stifel, 1553, repeating factors (§ 156)
J. Buteon, 1559 (§ 173)
J. Scheubel, N, Ra, Pri, Se (§ 159)
Th. Harriot, repeating factors (§ 188)
Johann Geysius, repeating factors (§ 196, 305)
John Newton, 1654 (§ 305)
Nathaniel Torporley (§ 305)
Joseph Raphson, 1702 (§ 305)
Samuel Foster, 1659, use of lines (§ 306)

INDEX PLAN

Psellus, nomenclature without signs (§ 117)
Neophytos, scholia (§§ 87, 88)
Nicole Oresme, notation for fractional powers (§ 123)
N. Chuquet, 1484, 12^3 for $12x^3$ (§ 131)
E. de la Roche, 1520 (§ 132)
R. Bombelli, 1572 (§ 144)
Grammateus, 1518, pri, $se.$, $ter.$ $quart.$ (§ 147)
G. van der Hoecke, 1537, pri, se, 3^a (§ 150)
S. Stevin (§ 162)
A. Girard, 1629 (§ 164)
L. & T. Digges, 1579 (§ 170, Fig. 76)
P. Hérigone, 1634 (§ 189)
J. Hume, 1635, 1636 (§ 190)

296. *Notations applied only to an unknown quantity, the base being omitted.*—As early as the fourteenth century, Oresme had the exponential concept, but his notation stands in historical isolation and does not constitute a part of the course of evolution of our modern exponential symbolism. We have seen that the earliest important steps toward the modern notation were taken by the Frenchman Nicolas Chuquet, the Italian Rafael Bombelli, the Belgian Simon Stevin, the Englishmen L. and T. Digges. Attention remains to be called to a symbolism very similar to that of the Digges, which was contrived by Pietro Antonio Cataldi of Bologna, in an algebra of 1610

and a book on roots of 1613. Cataldi wrote the numeral exponents in their natural upright position,[1] and distinguished them by crossing them out. His "5 $\mathfrak{Z}$ via 8 $\mathcal{4}$ fa 40 $\mathcal{7}$" means $5x^3 \cdot 8x^4 = 40x^7$. His sign for x is $\mathcal{1}$. He made only very limited use of this notation.

The drawback of Stevin's symbolism lay in the difficulty of writing and printing numerals and fractions within the circle. Apparently as a relief from this cumbrousness, we find that the Dutch writer, Adrianus Romanus, in his *Ideae Mathematicae pars prima* (Antwerp, 1593), uses in place of the circle two rounded parentheses and vinculums above and below; thus, with him $1\overline{(45)}$ stands for x^{45}. He uses this notation in writing his famous equation of the forty-fifth degree. Franciscus van Schooten[2] in his early publications and when he quotes from Girard uses the notation of Stevin.

A notation more in line with Chuquet's was that of the Swiss Joost Bürgi who, in a manuscript now kept in the library of the observatory at Pulkowa, used Roman numerals for exponents and wrote[3]

$$\overset{\text{vi}}{8} + \overset{\text{v}}{12} - \overset{\text{iv}}{9} + \overset{\text{iii}}{10} + \overset{\text{ii}}{3} + \overset{\text{i}}{7} - \overset{0}{4} \text{ for } 8x^6 + 12x^5 - 9x^4 + 10x^3 + 3x^2 + 7x - 4 .$$

In this notation Bürgi was followed by Nicolaus Reymers (1601) and J. Kepler.[4] Reymers[5] used also the cossic symbols, but chose R in place of $\mathcal{R}$; occasionally he used a symbolism as in $25\text{IIII} + 20\text{II} - 10\text{III} - 8\text{I}$ for the modern $25x^4 + 20x^2 - 10x^3 - 8x$. We see that Cataldi, Romanus, Fr. van Schooten, Bürgi, Reymers, and Kepler belong in the list of those who followed the "Index Plan."

297. *Notations applied to any quantity, the base being designated.*— As long as literal coefficients were not used and numbers were not generally represented by letters, the notations of Chuquet, Bombelli,

[1] G. Wertheim, *Zeitschr. f. Math. u. Physik*, Vol. XLIV (1899), Hist.-Lit. Abteilung, p. 48.

[2] Francisci à Schooten, *De Organica conicarum sectionum . . . . Tractatus* (Leyden, 1646), p. 96; Schooten, *Renati Descartes Geometria* (Frankfurt a./M., 1695), p. 359.

[3] P. Treutlein in *Abhandlungen zur Geschichte der Mathematik*, Vol. II (Leipzig, 1879), p. 36, 104.

[4] In his "De Figurarum regularium" in *Opera omnia* (ed. Ch. Frisch), Vol. V (1864), p. 104, Kepler lets the radius AB of a circle be 1 and the side BC of a regular inscribed heptagon be R. He says: "In hac proportione continuitatem fingit, ut sicut est $AB1$ ad BC $1R$, sic sit $1R$ ad $1z$, et $1z$ as 1 $\mathcal{cl}$, et 1 $\mathcal{cl}$ ad $1zz$, et $1zz$ ad $1z$ $\mathcal{cl}$ et sic perpetuo, quod nos commodius signabimus per apices six, 1, 1^{I}, 1^{II}, 1^{III}, 1^{IV}, 1^{V}, 1^{VI}, 1^{VII}, etc."

[5] N. Raimarus Ursus, *Arithmetica analytica* (Frankfurt a. O., 1601), Bl. $C3v°$. See J. Tropfke, *op. cit.*, Vol. II (2d ed., 1921), p. 122.

Stevin, and others were quite adequate. There was no pressing need of indicating the powers of a given number, say the cube of twelve; they could be computed at once. Moreover, as only the unknown quantity was raised to powers which could not be computed on the spot, why should one go to the trouble of writing down the base? Was it not sufficient to put down the exponent and omit the base? Was it not easier to write $\overset{v}{16}$ than $16x^5$? But when through the innovations of Vieta and others, literal coefficients came to be employed, and when several unknowns or variables came to be used as in analytic geometry, then the omission of the base became a serious defect in the symbolism. It will not do to write $15x^2 - 16y^2$ as $\overset{ii}{15} - \overset{ii}{16}$. In watching the coming changes in notation, the reader will bear this problem in mind. Vieta's own notation of 1591 was clumsy: *D quadratum* or *D. quad.* stood for D^2, *D cubum* for D^3; *A quadr.* for x^2, *A* representing the unknown number.

In this connection perhaps the first writer to be mentioned is Luca Pacioli who in 1494 explained, as an alternative notation of powers, the use of R as a base, but in place of the exponent he employs an ordinal that is too large by unity (§ 136). Thus $R.\ 30^a$ stood for x^{29}. Evidently Pacioli did not have a grasp of the exponential concept.

An important step was taken by Romanus[1] who uses letters and writes bases as well as the exponents in expressions like

$$A\,(4)+B\,(4)+4A\,(3)\ in\ B+6A\,(2)\ in\ B(2)+4A\ in\ B(3)$$

which signifies

$$A^4+B^4+4A^3B+6A^2B^2+4AB^3\ .$$

A similar suggestion came from the Frenchman, Pierre Hérigone, a mathematician who had a passion for new notations. He wrote our a^3 as $a3$, our $2b^4$ as $2b4$, and our $2ba^2$ as $2ba2$. The coefficient was placed *before* the letter, the exponent *after*.

In 1636 James Hume[2] brought out an edition of the algebra of Vieta, in which he introduced a superior notation, writing down the base and elevating the exponent to a position above the regular line and a little to the right. The exponent was expressed in Roman

[1] See H. Bosmans in *Annales Société scient. de Bruxelles*, Vol. XXX, Part II (1906), p. 15.

[2] James Hume, *L'Algèbre de Viète, d'une methode nouvelle claire et facile* (Paris 1636). See *Œuvres de Descartes* (ed. Charles Adam et P. Tannery), Vol. V, p. 504, 506–12.

numerals. Thus, he wrote A^{iii} for A^3. Except for the use of the Roman numerals, one has here our modern notation. Thus, this Scotsman, residing in Paris, had almost hit upon the exponential symbolism which has become universal through the writings of Descartes.

298. *Descartes' notation of 1637.*—Thus far had the notation advanced before Descartes published his *Géométrie* (1637) (§ 191). Hérigone and Hume almost hit upon the scheme of Descartes. The only difference was, in one case, the position of the exponent, and, in the other, the exponent written in Roman numerals. Descartes expressed the exponent in Arabic numerals and assigned it an elevated position. Where Hume would write $5a^{iv}$ and Hérigone would write $5a4$, Descartes wrote $5a^4$. From the standpoint of the printer, Hérigone's notation was the simplest. But Descartes' elevated exponent offered certain advantages in interpretation which the judgment of subsequent centuries has sustained. Descartes used positive integral exponents only.

299. *Did Stampioen arrive at Descartes' notation independently?*— Was Descartes alone in adopting the notation $5a^4$ or did others hit upon this particular form independently? In 1639 this special form was suggested by a young Dutch writer, Johan Stampioen.[1] He makes no acknowledgment of indebtedness to Descartes. He makes it appear that he had been considering the two forms ^3a and a^3, and had found the latter preferable.[2] Evidently, the symbolism a^3 was adopted by Stampioen after the book had been written; in the body of his book[3] one finds aaa, $bbbb$, $fffff$, $gggggg$, but the exponential notation above noted, as described in his passage following the Preface, is not used. Stampioen uses the notation a^4 in some but not all parts of a controversial publication[4] of 1640, on the solution of cubic equations, and directed against Waessenaer, a personal friend of Descartes. In view of the fact that Stampioen does not state the originators of any of the notations which he uses, it is not improbable that his a^3 was taken from Descartes, even though Stampioen stands out as an opponent of Descartes.[5]

[1] Johan Stampioen d'Jonghe, *Algebra ofte Nieuwe Stel-Regel* (The Hague, 1639). See his statement following the Preface.

[2] Stampioen's own words are: "*aaa·* dit is *a* drievoudich in hem selfs gemennichvuldicht. men soude oock daer voor konnen stellen ^3a ofte better a^3."

[3] J. Stampioen, *op. cit.*, p. 343, 344, 348.

[4] *I. I. Stampioenii Wis-Konstigh ende Reden-Maetigh Bewijs* ('s Graven-Hage, 1640), unpaged Introduction and p. 52–55.

[5] *Œuvres de Descartes*, Vol. XII (1910), p. 32, 272–74.

300. *Notations used by Descartes before 1637.*—Descartes' indebtedness to his predecessors for the exponential notation has been noted. The new features in Descartes' notation, $5a^3$, $6ab^4$, were indeed very slight. What notations did Descartes himself employ before 1637?

In his *Opuscules de 1619–1621* he regularly uses German symbols as they are found in the algebra of Clavius; Descartes writes[1]

$$\text{``} 36 - 3z - 6 \; \mathfrak{X} \; aequ. \; 1 \; \curlyvee \text{,''}$$

which means $36 - 3x^2 - 6x = x^3$. These *Opuscules* were printed by Foucher de Careil (Paris, 1859–60), but this printed edition contains corruptions in notation, due to the want of proper type. Thus the numeral 4 is made to stand for the German symbol $\mathfrak{X}$; the small letter γ is made to stand for the radical sign $\sqrt{\ }$. The various deviations from the regular forms of the symbols are set forth in the standard edition of Descartes' works. Elsewhere (§ 264) we call attention that Descartes[2] in a letter of 1640 used the Recordian sign of equality and the symbols N and C of Xylander, in the expression "$1C - 6N = 40$." Writing to Mersenne, on May 3, 1638, Descartes[3] employed the notation of Vieta, "$Aq + Bq + A \; in \; B \; bis$" for our $a^2 + b^2 + 2ab$. In a posthumous document,[4] of which the date of composition is not known, Descartes used the sign of equality found in his *Géométrie* of 1637, and P. Hérigone's notation for powers of given letters, as $b3x$ for b^3x, $a3z$ for a^3z. Probably this document was written before 1637. Descartes[5] used once also the notation of Dounot (or Deidier, or Bar-le-Duc, as he signs himself in his books) in writing the equation $1C - 9Q + 13N \; eq. \; \sqrt{288} - 15$, but Descartes translates it into $y^3 - 9y^2 + 13y - 12\sqrt{2} + 15 \infty 0$.

301. *Use of Hérigone's notation after 1637.*—After 1637 there was during the seventeenth century still very great diversity in the exponential notation. Hérigone's symbolism found favor with some writers. It occurs in Florimond Debeaune's letter[6] of September 25, 1638, to Mersenne in terms like $2y4$, $y3$, $2l2$ for $2y^4$, y^3, and $2l^2$, re-

---

[1] *Ibid.*, Vol. X (1908), p. 249–51. See also E. de Jonquières in *Bibliotheca mathematica* (2d ser.), Vol. IV (1890), p. 52, also G. Eneström, *Bibliotheca mathematica* (3d ser.), Vol. VI (1905), p. 406.

[2] *Œuvres de Descartes*, Vol. III (1899), p. 190.

[3] *Ibid.*, Vol. II (1898), p. 125; also Vol. XII, p. 279.

[4] *Ibid.*, Vol. X (1908), p. 299.

[5] *Ibid.*, Vol. XII, p. 278.   [6] *Ibid.*, Vol. V (1903), p. 516.

spectively. G. Schott[1] gives it along with older notations. Pietro Mengoli[2] uses it in expressions like $a4+4a3r+6a2r2+4ar3+r4$ for our $a^4+4a^3r+6a^2r^2+4ar^3+r^4$. The Italian Cardinal Michelangelo Ricci[3] writes "AC_2 in CB_3" for $\overline{AC}^2 . \overline{CB}^3$. In a letter[4] addressed to Ozanam one finds $b4+c4\frown a4$ for $b^4+c^4=a^4$. Chr. Huygens[5] in a letter of June 8, 1684, wrote $a3+aab$ for a^3+a^2b. In the same year an article by John Craig[6] in the *Philosophical Transactions* contains $a3y+a4$ for a^3y+a^4, but a note to the "Benevole Lector" appears at the end apologizing for this notation. Dechales[7] used in 1674 and again in 1690 (along with older notations) the form $A4+4A3B+6A2B2+4AB3+B4$. A Swedish author, Andreas Spole,[8] who in 1664–66 sojourned in Paris, wrote in 1692 an arithmetic containing expressions $3a3+3a2-2a-2$ for $3a^3+3a^2-2a-2$. Joseph Moxon[9] lets "$A-B.(2)$" stand for our $(A-B)^2$, also "$A-B.(3)$" for our $(A-B)^3$. With the eighteenth century this notation disappeared.

302. *Later use of Hume's notation of 1636.*—Hume's notation of 1636 was followed in 1638 by Jean de Beaugrand[10] who in an anonymous letter to Mersenne criticized Descartes and states that the equation $x^{IV}+4x^{III}-19x^{II}-106x-120$ has the roots $+5$, -2, -3, -4. Beaugrand also refers to Vieta and used vowels for the unknowns, as in "$A'''+3AAB+ADP$ *esgale a* ZSS." Again Beaugrand writes "$E'''\circlearrowleft-13E-12$" for $x^3-13x-12$, where the $\circlearrowleft$ apparently designates the omission of the second term, as does $\divideontimes$ with Descartes.

303. *Other exponential notations suggested after 1637.*—At the time of Descartes and the century following several other exponential notations were suggested which seem odd to us and which serve to

---

[1] G. Schott, *Cursus mathematicus* (Würzburg, 1661), p. 576.

[2] *Ad Maiorem Dei Gloriam Geometriae speciosae Elementa*, ... Petri Mengoli (Bologna, 1659), p. 20.

[3] *Michaelis Angeli Riccii Exercitatio geometrica* (Londini, 1668), p. 2. [Preface, 1666.]

[4] *Journal des Sçavans*, l'année 1680 (Amsterdam, 1682), p. 160.

[5] *Ibid.*, l'année 1684, Vol. II (2d ed.; Amsterdam, 1709), p. 254.

[6] *Philosophical Transactions*, Vol. XV–XVI (London, 1684–91), p. 189.

R. P. *Claudii Francisci Milliet Dechales Camberiensis Mundus mathematicus*, Tomus tertius (Leyden, 1674), p. 664; Tomus primus (editio altera; Leyden, 1690), p. 635.

[8] Andreas Spole, *Arithmetica vulgaris et specioza* (Upsala, 1692). See G. Eneström in *L'Intermédiaire des mathématiciens*, Vol. IV (1897), p. 60.

[9] Joseph Moxon, *Mathematical Dictionary* (London, 1701), p. 190, 191.

[10] *Œuvres de Descartes*, Vol. V (1903), p. 506, 507.

indicate how the science might have been retarded in its progress under the handicap of cumbrous notations, had such wise leadership as that of Descartes, Wallis, and Newton not been available. Rich. Balam[1] in 1653 explains a device of his own, as follows: "(2) $:3:$, the Duplicat, or Square of 3, that is, 3×3; (4) $:2:$, the Quadruplicat of 2, that is, $2 \times 2 \times 2 \times 2 = 16$." The Dutch J. Stampioen[2] in 1639 wrote $\square A$ for A^2; as early as 1575 F. Maurolycus[3] used $\square$ to designate the square of a line. Similarly, an Austrian, Johannes Caramuel,[4] in 1670 gives "$\square 25$. est Quadratum Numeri 25. hoc est, 625." Huygens[5] wrote "1000(3)10" for $1,000 = 10^3$, and "1024(10)2" for $1024 = 2^{10}$. A Leibnizian symbolism[6] explained in 1710 indicates the cube of $AB+BC$ thus: $\boxed{3}$ $(AB+BC)$; in fact, before this time, in 1695 Leibniz[7] wrote $\boxed{m}\,\overline{y+a}$ for $(y+a)^m$.

304. Descartes preferred the notation aa to a^2. Fr. van Schooten,[8] in 1646, followed Descartes even in writing qq, xx rather than q^2, x^2, but in his 1649 Latin edition of Descartes' geometry he wrote preferably x^2. The symbolism xx was used not only by Descartes, but also by Huygens, Rahn, Kersey, Wallis, Newton, Halley, Rolle, Euler— in fact, by most writers of the second half of the seventeenth and of the eighteenth centuries. Later, Gauss[9] was in the habit of writing xx, and he defended his practice by the statement that x^2 did not take up less space than xx, hence did not fulfil the main object of a symbol. The x^2 was preferred by Leibniz, Ozanam, David Gregory, and Pascal.

305. The reader should be reminded at this time that the representation of positive integral powers by the repetition of the factors was suggested very early (about 1480) in the Dresden Codex C. 80 under the heading *Algorithmus de additis et minutis* where $x^2 = z$ and $x^{10} = zzzzz$; it was elaborated more systematically in 1553 by M.

[1] Rich. Balam, *Algebra, or The Doctrine of Composing, Inferring, and Resolving an Equation* (London, 1653), p. 9.

[2] Johan Stampioen, *Algebra* (The Hague, 1639), p. 38.

[3] *D. Francisc' Mavrolyci Abbatis messanensis Opuscula Mathematica* (Venice, 1575) (Euclid, Book XIII), p. 107.

[4] *Joannis Caramvelis Mathesis Biceps. Vetus et Nova* (Companiae, 1670), p. 131, 132.

[5] *Christiani Hugenii Opera* .... quae collegit .... Guilielmus Jacobus's Gravesande (Leyden, 1751), p. 456.

[6] *Miscellanea Berolinensia* (Berlin, 1710), p. 157.

[7] *Acta eruditorum* (1695), p. 312.

[8] *Francisci à Schooten Leydensis de Organica conicarum sectionum* .... *Tractatus* (Leyden, 1646), p. 91 ff.

[9] M. Cantor, *op. cit.*, Vol. II (2d ed.), p. 794 n.

Stifel (§ 156). One sees in Stifel the exponential notation applied, not to the unknown but to several different quantities, all of them known. Stifel understood that a quantity with the exponent zero had the value 1. But this notation was merely a suggestion which Stifel himself did not use further. Later, in Alsted's *Encyclopaedia*,[1] published at Herborn in Prussia, there is given an explanation of the German symbols for *radix, zensus, cubus*, etc.; then the symbols from Stifel, just referred to, are reproduced, with the remark that they are preferred by some writers. The algebra proper in the *Encyclopaedia* is from the pen of Johann Geysius[2] who describes a similar notation $2a$, $4aa$, $8aaa$, ... , $512aaaaaaaaa$ and suggests also the use of Roman numerals as indices, as in $2l \overset{\text{I}}{} 4q \overset{\text{II}}{} 8c \overset{\text{III}}{} . . . . 512cc \overset{\text{IX}}{}$. Forty years after, Caramvel[3] ascribes to Geysius the notation aaa for the cube of a, etc.

In England the repetition of factors for the designation of powers was employed regularly in Thomas Harriot. In a manuscript preserved in the library of Sion College, Nathaniel Torporley (1573–1632) makes strictures on Harriot's book, but he uses Harriot's notation.[4] John Newton[5] in 1654 writes $aaaaa$. John Collins writes in the *Philosophical Transactions* of 1668 $aaa-3aa+4a=N$ to signify $x^3-3x^2+4x=N$. Harriot's mode of representation is found again in the *Transactions*[6] for 1684. Joseph Raphson[7] uses powers of g up to g^{10}, but in every instance he writes out each of the factors, after the manner of Harriot.

306. The following curious symbolism was designed in 1659 by Samuel Foster[8] of London:

⌐	⌐	⌐	⌐	⌐	⌐	⌐	⌐
q	c	qq	qc	cc	qqc	qcc	ccc
2	3	4	5	6	7	8	9

[1] *Johannis-Henrici Alstedii Encyclopaedia* (Herborn, 1630), Book XIV, "Arithmetica," p. 844.

[2] *Ibid.*, p. 865–74.

[3] *Joannis Caramvelis Mathesis Biceps* (Campaniae, 1670), p. 121.

[4] J. O. Halliwell, *A Collection of Letters Illustrative of the Progress of Science in England* (London, 1841), p. 109–16.

[5] John Newton, *Institutio Mathematica or a Mathematical Institution* (London, 1654), p. 85.

[6] *Philosophical Transactions*, Vol. XV–XVI (London, 1684–91), p. 247, 340.

[7] Josepho Raphson, *Analysis Aequationum universalis* (London, 1702). [First edition, 1697.]

[8] Samuel Foster, *Miscellanies, or Mathematical Lucubrations* (London, 1659), p. 10.

Foster did not make much use of it in his book. He writes the proportion

$$\text{“ } At\ AC\ .\ AR::\overline{CD}|:\overline{RP}|\ \text{,”}$$

which means

$$AC:AR=\overline{CD}^2:\overline{RP}^2\ .$$

An altogether different and unique procedure is encountered in the *Maandelykse Mathematische Liefhebberye* (1754–69), where $\overset{m}{\sqrt{}}\text{----}$ signifies extracting the mth root, and $\text{----}\overset{}{\sqrt{}}$ signifies raising to the mth power. Thus,

$$\text{“ }\frac{y^n}{b}=\overset{m}{\sqrt{}}\,ay \qquad \text{”}$$
$$\frac{y^{mn}}{b^m}=ay\ \overset{m}{\sqrt{}}\ .$$

307. *Spread of Descartes' notation.*—Since Descartes' *Géométrie* appeared in Holland, it is not strange that the exponential notation met with prompter acceptance in Holland than elsewhere. We have already seen that J. Stampioen used this notation in 1639 and 1640. The great disciple of Descartes, Fr. van Schooten, used it in 1646, and in 1649 in his Latin edition of Descartes' geometry. In 1646 van Schooten indulges[1] in the unusual practice of raising some (but not all) of his coefficients to the height of exponents. He writes $x^3-{}^3aax-{}^2a^3 \infty$ 0 to designate $x^3-3a^2x-2a^3=0$. Van Schooten[2] does the same thing in 1657, when he writes ${}^2ax$ for $2ax$. Before this Marini Ghetaldi[3] in Italy wrote coefficients in a *low* position, as subscripts, as in the proportion,

$$\text{“ }ut\ AQ\ ad\ A_2\ in\ B\ ita\ \frac{x}{y}\ ad\ m_2{+}n\ \text{,”}$$

which stands for $A^2:2AB=\frac{x}{y}:(2m+n)$. Before this Albert Girard[4] placed the coefficients where we now write our exponents. I quote: "Soit un binome conjoint $B+C$. Son Cube era $B(B_q+C_q^3)+C(B_q^3+C_q)$." Here the cube of $B+C$ is given in the form corresponding

[1] Francisci à Schooten, *De organica conicarum sectionum . . . . tractatus* (Leyden, 1646), p. 105.

[2] Francisci à Schooten, *Exercitationum mathematicarum liber primus* (Leyden, 1657), p. 227, 274, 428, 467, 481, 483.

[3] Marini Ghetaldi, *De resolutione et compositione mathematica libri quinque. Opus posthumum* (Rome, 1630). Taken from E. Gelcich, *Abhandlungen zur Geschichte der Mathematik*, Vol. IV (1882), p. 198.

[4] A. Girard, *Invention nouvelle en l'algebre* (1629), "3 C."

to $B(B^2+3C^2)+C(3B^2+C^2)$. Much later, in 1679, we find in the collected works of P. Fermat[1] the coefficients in an elevated position: ^2D in A for $2DA$, ^2R in E for $2RE$.

The Cartesian notation was used by C. Huygens and P. Mersenne in 1646 in their correspondence with each other,[2] by J. Hudde[3] in 1658, and by other writers.

In England, J. Wallis[4] was one of the earliest writers to use Descartes' exponential symbolism. He used it in 1655, even though he himself had been trained in Oughtred's notation.

The Cartesian notation is found in the algebraic parts of Isaac Barrow's[5] geometric lectures of 1670 and in John Kersey's *Algebra*[6] of 1673. The adoption of Descartes' a^4 in strictly algebraic operations and the retention of the older A_q, A_c for A^2, A^3 in geometric analysis is of frequent occurrence in Barrow and in other writers. Seemingly, the impression prevailed that A^2 and A^3 suggest to the pupil the purely arithmetical process of multiplication, AA and AAA, but that the symbolisms A_q and A_c conveyed the idea of a geometric square and geometric cube. So we find in geometrical expositions the use of the latter notation long after it had disappeared from purely algebraic processes. We find it, for instance, in W. Whiston's edition of Tacquet's *Euclid*,[7] in Sir Isaac Newton's *Principia*[8] and *Opticks*,[9] in B. Robins' *Tracts*,[10] and in a text by K. F. Hauber.[11] In the *Philosophical Transactions* of London none of the pre-Cartesian notations for powers appear, except a few times in an article of 1714 from the pen of R. Cotes, and an occasional tendency to adhere to the primitive, but very

---

[1] P. Fermat, *Varia opera* (Toulouse, 1679), p. 5.

[2] C. Huygens, *Œuvres*, Vol. I (La Haye, 1888), p. 24.

[3] *Joh. Huddeni Epist. I de reductione aequationum* (Amsterdam, 1658); Matthiessen, *Grundzüge der Antiken u. Modernen Algebra* (Leipzig, 1878), p. 349.

[4] John Wallis, *Arithmetica infinitorum* (Oxford, 1655), p. 16 ff.

[5] Isaac Barrow, *Lectiones Geometriae* (London, 1670), Lecture XIII (W. Whewell's ed.), p. 309.

[6] John Kersey, *Algebra* (London, 1673), p. 11.

[7] See, for instance, *Elementa Euclidea geometriae auctore Andrea Tacquet, ...* Gulielmus Whiston (Amsterdam, 1725), p. 41.

[8] Sir Isaac Newton, *Principia* (1687), Book I, Lemma xi, Cas. 1, and in other places.

[9] Sir Isaac Newton, *Opticks* (3d ed.; London, 1721), p. 30.

[10] Benjamin Robins, *Mathematical Tracts* (ed. James Wilson, 1761), Vol. II, p. 65.

[11] Karl Friderich Hauber, *Archimeds zwey Bücher über Kugel und Cylinder* (Tübingen, 1798), p. 56 ff.

lucid method of repeating the factors, as *aaa* for a^3. The modern exponents did not appear in any of the numerous editions of William Oughtred's *Clavis mathematicae;* the last edition of that popular book was issued in 1694 and received a new impression in 1702. On February 5, 1666–67, J. Wallis[1] wrote to J. Collins, when a proposed new edition of Oughtred's *Clavis* was under discussion: "It is true, that as in other things so in mathematics, fashions will daily alter, and that which Mr. Oughtred designed by great letters may be now by others designed by small; but a mathematician will, with the same ease and advantage, understand A_c or *aaa*." As late as 1790 the Portuguese J. A. da Cunha[2] occasionally wrote A_q and A_c. J. Pell wrote r^2 and t^2 in a letter written in Amsterdam on August 7, 1645.[3] J. H. Rahn's *Teutsche Algebra*, printed in 1659 in Zurich, contains for positive integral powers two notations, one using the Cartesian exponents, a^3, x^4, the other consisting of writing an Archemidean spiral (Fig. 96) between the base and the exponent on the right. Thus $a\odot3$ signifies a^3. This symbol is used to signify involution, a process which Rahn calls *involviren*. In the English translation, made by T. Brancker and published in 1668 in London, the Archimedean spiral is displaced by the omicron-sigma (Fig. 97), a symbol found among several English writers of textbooks, as, for instance, J. Ward,[4] E. Hatton,[5] Hammond,[6] C. Mason,[7] and by P. Ronayne[8]—all of whom use also Rahn's and Brancker's ᴡᴡ/ to signify evolution. The omicron-sigma is found in Birks;[9] it is mentioned by Saverien,[10] who objects to it as being superfluous.

Of interest is the following passage in Newton's *Arithmetick*,[11] which consists of lectures delivered by him at Cambridge in the period 1669–85 and first printed in 1707: "Thus $\sqrt{}64$ denotes 8; and $\sqrt{}3:64$

---

[1] Rigaud, *Correspondence of Scientific Men of the Seventeenth Century*, Vol. I (Oxford, 1841), p. 63.

[2] J. A. da Cunha, *Principios mathematicos* (1790), p. 158.

[3] J. O. Halliwell, *Progress of Science in England* (London, 1841), p. 89.

[4] John Ward, *The Young Mathematician's Guide* (London, 1707), p. 144.

[5] Edward Hatton, *Intire System of Arithmetic* (London, 1721), p. 287.

[6] Nathaniel Hammond, *Elements of Algebra* (London, 1742).

[7] C. Mason in the *Diarian Repository* (London, 1774), p. 187.

[8] Philip Ronayne, *Treatise of Algebra* (London, 1727), p. 3.

[9] Anthony and John Birks, *Arithmetical Collections* (London, 1766), p. viii.

[10] A. Saverien, *Dictionnaire universel de mathematique et de physique* (Paris, 1753), "Caractere."

[11] Newton's *Universal Arithmetick* (London, 1728), p. 7.

denotes 4. . . . . There are some that to denote the Square of the first Power, make use of q, and of c for the Cube, qq for the Biquadrate, and qc for the Quadrato-Cube, etc. . . . . Others make use of other sorts of Notes, but they are now almost out of Fashion."

In the eighteenth century in England, when parentheses were seldom used and the vinculum was at the zenith of its popularity, bars were drawn horizontally and allowed to bend into a vertical stroke[1] (or else were connected with a vertical stroke), as in $A \times B^{\frac{1}{n}}\big|^n$ or in $\overline{a+b}\big|^{\frac{m}{n}}$.

In France the Cartesian exponential notation was not adopted as early as one might have expected. In J. de Billy's *Nova geometriae clavis* (Paris, 1643), there is no trace of that notation; the equation $x+x^2=20$ is written "$1R{+}1Q$ *aequatur* 20." In Fermat's edition[2] of Diophantus of 1670 one finds in the introduction $1QQ+4C+10Q+20N+1$ for $x^4+4x^3+10x^2+20x+1$. But in an edition of the works of Fermat, brought out in 1679, after his death, the algebraic notation of Vieta which he had followed was discarded in favor of the exponents of Descartes.[3] B. Pascal[4] made free use of positive integral exponents in several of his papers, particularly the *Potestatum numericarum summa* (1654).

In Italy, C. Renaldini[5] in 1665 uses both old and new exponential notations, with the latter predominating.

308. *Negative, fractional, and literal exponents.*—Negative and fractional exponential notations had been suggested by Oresme, Chuquet, Stevin, and others, but the modern symbolism for these is due to Wallis and Newton. Wallis[6] in 1656 used positive integral exponents and speaks of negative and fractional "indices," but he does not actually write a^{-1} for $\dfrac{1}{a}$, $a^{\frac{3}{2}}$ for $\sqrt{a^3}$. He speaks of the series

---

[1] See, for instance, A. Malcolm, *A New System of Arithmetick* (London, 1730), p. 143.

[2] *Diophanti Alexandrini arithmeticarum Libri Sex, cum commentariis G. B. Bacheti V. C. et observationibus D. P. de Fermat* (Tolosae, 1670), p. 27.

[3] See *Œuvres de Fermat* (éd. Paul Tannery et Charles Henry), Tome I (Paris, 1891), p. 91 n.

[4] *Œuvres de Pascal* (éd. Leon Brunschvicg et Pierre Boutroux), Vol. III (Paris, 1908), p. 349–58.

[5] *Caroli Renaldinii, Ars analytica* . . . . (Florence, 1665), p. 11, 80, 144.

[6] J. Wallis, *Arithmetica infinitorum* (1656), p. 80, Prop. CVI.

$\dfrac{1}{\sqrt{1}}, \dfrac{1}{\sqrt{2}}, \dfrac{1}{\sqrt{3}},$ etc., as having the "index $-\frac{1}{2}$." Our modern notation involving fractional and negative exponents was formally introduced a dozen years later by Newton[1] in a letter of June 13, 1676, to Oldenburg, then secretary of the Royal Society of London, which explains the use of negative and fractional exponents in the statement, "Since algebraists write a^2, a^3, a^4, etc., for aa, aaa, $aaaa$, etc., so I write $a^{\frac{1}{2}}$, $a^{\frac{3}{2}}$, $a^{\frac{5}{3}}$, for $\sqrt{a}$, $\sqrt{a^3}$, $\sqrt{c\,a^5}$; and I write a^{-1}, a^{-2}, a^{-3}, etc., for $\dfrac{1}{a}, \dfrac{1}{aa}, \dfrac{1}{aaa},$ etc." He exhibits the general exponents in his binomial formula first announced in that letter:

$$\overline{P+PQ}\Big|^{\frac{m}{n}} = P^{\frac{m}{n}} + \frac{m}{n}\,AQ + \frac{m-n}{2n}\,BQ + \frac{m-2n}{3n}\,CQ + \frac{m-3n}{4n}\,DQ+\ ,\ \text{etc.},$$

where $A = P^{\frac{m}{n}}$, $B = \dfrac{m}{n}P^{\frac{m}{n}}Q$, etc., and where $\dfrac{m}{n}$ may represent any real and rational number. It should be observed that Newton wrote here literal exponents such as had been used a few times by Wallis,[2] in 1657, in expressions like $\sqrt{^d R^d} = R$, $AR^m \times AR^n = A^2 R^{m+n}$, which arose in the treatment of geometric progression. Wallis gives also the division $AR^m)AR^{m+n}(R^n$. Newton[3] employs irrational exponents in his letter to Oldenburg of the date October 24, 1676, where he writes $\overline{x\sqrt{2}+x\sqrt{7}}\Big|^{\sqrt[3]{}\,\frac{2}{3}} = y$. Before Wallis and Newton, Vieta indicated general exponents a few times in a manner almost rhetorical;[4] his

$$A\ potestas + \frac{E\ potestate - A\ potesta}{E\ gradui + A\ gradu}\ in\ A\ gradum$$

is our

$$x^m + \frac{y^m - x^m}{y^n + x^n} \cdot x^n\ ,$$

the two distinct general powers being indicated by the words *potestas* and *gradus*. Johann Bernoulli[5] in 1691-92 still wrote $3\,\square\,\sqrt[3]{ax+xx}$ for

[1] *Isaaci Newtoni Opera* (ed. S. Horsely), Tom. IV (London, 1782), p. 215.

[2] J. Wallis, *Mathesis universalis* (Oxford, 1657), p. 292, 293, 294.

[3] See J. Collins, *Commercium epistolicum* (ed. J. B. Biot and F. Lefort; Paris, 1856), p. 145.

[4] Vieta, *Opera mathematica* (ed. Fr. van Schooten, 1634), p. 197.

[5] Iohannis I Bernoulli, *Lectiones de calculo differentialium* .... **von Paul Schafheitlin.** Separatabdruck aus den *Verhandlungen der Naturforschenden Gesellschaft in Basel,* Vol. XXXIV, 1922.

$3 \sqrt[3]{(ax+x^2)^2}$, $4C\sqrt[4]{yx+xx}$ for $4\sqrt[4]{(yx+x^2)^3}$, $5QQ\sqrt[5]{ayx+x^3+zyx}$ for $5\sqrt[5]{(ayx+x^3+zyx)^4}$. But fractional, negative, and general exponents were freely used by D. Gregory[1] and were fully explained by W. Jones[2] and by C. Reyneau.[3] Reyneau remarks that this theory is not explained in works on algebra.

309. *Imaginary exponents.*—The further step of introducing imaginary exponents is taken by L. Euler in a letter to Johann Bernoulli,[4] of October 18, 1740, in which he announces the discovery of the formula $e^{+x\sqrt{-1}}+e^{-x\sqrt{-1}}=2 \cos x$, and in a letter to C. Goldbach,[5] of December 9, 1741, in which he points out as a curiosity that the fraction $\dfrac{2+\sqrt{-1}+2^{-\sqrt{-1}}}{2}$ is nearly equal to $\frac{10}{13}$. The first appearance of imaginary exponents in print is in an article by Euler in the *Miscellanea Berolinensia* of 1743 and in Euler's *Introductio in analysin* (Lausannae, 1747), Volume I, page 104, where he gives the all-important formula $e^{+v\sqrt{-1}}=\cos v+\sqrt{-1} \sin v$.

310. At an earlier date occurred the introduction of variable exponents. In a letter of 1679, addressed to C. Huygens, G. W. Leibniz[6] discussed equations of the form $x^x-x=24, x^z+z^x=b, x^x+z^z=c$. On May 9, 1694, Johann Bernoulli[7] mentions expressions of this sort in a letter to Leibniz who, in 1695, again considered exponentials in the *Acta eruditorum*, as did also Johann Bernoulli in 1697.

311. Of interest is the following quotation from a discussion by T. P. Nunn, in the *Mathematical Gazette*, Volume VI (1912), page 255, from which, however, it must not be inferred that Wallis actually wrote down fractional and negative exponents: "Those who are acquainted with the work of John Wallis will remember that he invented negative and fractional indices in the course of an investigation into methods of evaluating areas, etc. He had

[1] David Gregory, *Exercitatio geometrica de dimensione figurarum* (Edinburgh, 1684), p. 4–6.

[2] William Jones, *Synopsis palmariorum matheseos* (London, 1706), p. 67, 115–19.

[3] Charles Reyneau, *Analyse demontrée* (Paris, 1708), Vol. I, Introduction.

[4] See G. Eneström, *Bibliotheca mathematica* (2d ser.), Vol. XI (1897), p. 49.

[5] P. H. Fuss, *Correspondance mathématique et physique* (Petersburg, 1843), Vol. I, p. 111.

[6] C. I. Gerhardt, *Briefwechsel von G. W. Leibniz mit Mathematikern* (2d ed.; Berlin, 1899), Vol. I, p. 568.

[7] Johann Bernoulli in *Leibnizens Mathematische Schriften* (ed. C. I. Gerhardt), Vol. III (1855), p. 140.

discovered that if the ordinates of a curve follow the law $y = kx^n$ its area follows the law $A = \dfrac{1}{n+1} \cdot kx^{n+1}$, n being (necessarily) a positive integer. This law is so remarkably simple and so powerful as a method that Wallis was prompted to inquire whether cases in which the ordinates follow such laws as $y = \dfrac{k}{x^n}$, $y = k \sqrt[n]{x}$ could not be brought within its scope. He found that this extension of the law would be possible if $\dfrac{k}{x^n}$ could be written kx^{-n}, and $k\sqrt[n]{x}$ as $kx^{\frac{1}{n}}$. From this, from numerous other historical instances, and from general psychological observations, I draw the conclusion that extensions of notation should be taught because and when they are needed for the attainment of some practical purpose, and that logical criticism should come after the suggestion of an extension to assure us of its validity."

312. *Notation for principal values.*—When in the early part of the nineteenth century the multiplicity of values of a^n came to be studied, where a and n may be negative or complex numbers, and when the need of defining the principal values became more insistent, new notations sprang into use in the exponential as well as the logarithmic theories. A. L. Cauchy[1] designated all the values that a^n may take, for given values of a and n $[a \neq 0]$, by the symbol $((a))^x$, so that $((a))^x = e^{xla} \cdot e^{2kx\pi i}$, where l means the tabular logarithm of $|a|$, $e = 2.718 \ldots$, $\pi = 3.141 \ldots$, $k = 0$, ± 1, ± 2, $\ldots$. This notation is adopted by O. Stolz and J. A. Gmeiner[2] in their *Theoretische Arithmetik*.

Other notations sprang up in the early part of the last century. Martin Ohm elaborated a general exponential theory as early as 1821 in a Latin thesis and later in his *System der Mathematik* (1822–33).[3] In a^x, when a and x may both be complex, log a has an infinite number of values. When, out of this infinite number some particular value of log a, say $\propto$, is selected, he indicates this by writing $(a \| \propto)$. With this understanding he can write $x \log a + y \log a = (x+y) \log a$, and consequently $a^x \cdot a^y = a^{x+y}$ is a *complete* equation, that is, an equation in which both sides have the same number of values, representing exactly the same expressions. Ohm did not introduce the particular value of a^x which is now called the "principal value."

[1] A. L. Cauchy, *Cours d'analyse* (Paris, 1821), chap. vii, § 1.

[2] O. Stolz and J. A. Gmeiner, *Theoretische Arithmetik* (Leipzig), Vol. II (1902), p. 371–77.

[3] Martin Ohm, *Versuch eines vollkommen consequenten Systems der Mathematik*, Vol. II (2d ed., 1829), p. 427. [First edition of Vol. II, 1823.]

Crelle[1] let $|u|^k$ indicate some fixed value of u^k, preferably a real value, if one exists, where k may be irrational or imaginary; the two vertical bars were used later by Weierstrass for the designation of absolute value (§ 492).

313. *Complicated exponents.*—When exponents themselves have exponents, and the latter exponents also have exponents of their own, then clumsy expressions occur, such as one finds in Johann I Bernoulli,[2] Goldbach,[3] Nikolaus II Bernoulli,[4] and Waring.

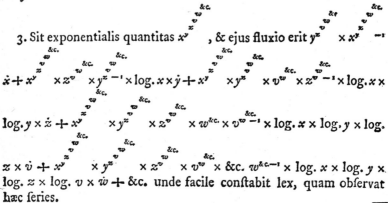

FIG. 106.—E. Waring's "repeated exponents." (From *Meditationes analyticae* [1785], p. 8.)

De Morgan[5] suggested a new notation for cases where exponents are complicated expressions. Using a solidus, he proposes $a \wedge \{(a+bx) /(c+ex)\}$, where the quantity within the braces is the exponent of a. He returned to this subject again in 1868 with the statement: "A convenient notation for repeated exponents is much wanted: not a working symbol, but a contrivance for preventing the symbol from wasting a line of text. The following would do perfectly well, $x|a|b|c|d$,

[1] A. L. Crelle in *Crelle's Journal*, Vol. VII (1831), p. 265, 266.

[2] Iohann I Bernoulli, *Acta eruditorum* (1697), p. 125–33.

[3] P. H. Fuss, *Correspondance math. et phys. ... du XVIIIe siècle*, Vol. II (1843), p. 128.

[4] *Op. cit.*, p. 133.

[5] A. de Morgan, "Calculus of Functions," *Encyclopaedia Metropolitana*, Vol. II (1845), p. 388.

in which each *post* means all which follows is to be placed on the top of it. Thus:[1]

$$x|a|b|c|d = x^{a|b|c|d} = x^{a^{b|c|d}} = x^{a^{b^{c|d}}} = x^{a^{b^{c^d}}} \text{ ''}.$$

When the base and the successive exponents are all alike, say a, Woepcke[2] used the symbol $\frac{m}{a}$ for $\left((a^a)^{a\cdots}\right)^a$, and $\frac{m}{a}$ for $_a\!\left(a^{\cdots(a^a)}\right)$, where m indicates the number of repetitions of a. He extended this notation to cases where a is real or imaginary, not zero, and m is a positive or negative integer, or zero. A few years later J. W. L. Glaisher suggested still another notation for complicated exponents, namely, $a\!\uparrow\!x^{\frac{1}{n}}+\dfrac{b}{x^n}\!\uparrow$, the arrows merely indicating that the quantity between them is to be raised so as to become the exponent of a. Glaisher prefers this to "a *Exp*. u" for a^u. Harkness and Morley[3] state, "It is usual to write $exp\ (z) = e^z$, when z is complex." The contraction "exp" was recommended by a British Committee (§ 725) in 1875, but was ignored in the suggestions of 1916, issued by the Council of the London Mathematical Society. G. H. Bryan stresses the usefulness of this symbol.[4]

Another notation was suggested by H. Schubert. If a^a is taken as an exponent of a, one obtains $a^{(a^a)}$ or a^{a^a}, and so on. Schubert designates the result by $(a; b)$, indicating that a has been thus written b times.[5] For the expression $(a; b)^{(a;\ c)}$ there has been adopted the sign $(a; b+c)$, so that $(a; b)^{(a;\ c)} = (a; c+1)^{(a;\ b-1)}$.

314. *D. F. Gregory*[6] in 1837 made use of the sign $(+)^r$, r an integer, to designate the repetition of the operation of multiplication. Also, $(+a^2)^{\frac{1}{2}} = +^{\frac{1}{2}}(a^2)^{\frac{1}{2}} = +^{\frac{1}{2}}a$, where the $+^{\frac{1}{2}}$ "will be different, according as we suppose the $+$ to be equivalent to the operation repeated an even or an odd number of times. In the former case it will be equal to $+$, in the latter to $-$. And generally, if we raise $+a$ to any power m, whether whole or fractional, we have $(+a)^m = +^m a^m$. . . . . So long as

[1] A. de Morgan, *Transactions of the Cambridge Philosophical Society*, Vol. XI, Part III (1869), p. 450.

[2] F. Woepcke in *Crelle's Journal*, Vol. XLII (1851), p. 83.

[3] J. Harkness and F. Morley, *Theory of Functions* (New York, 1893), p. 120.

[4] *Mathematical Gazette*, Vol. VIII (London, 1917), p. 172, 220.

[5] H. Schubert in *Encyclopédie d. scien. math.*, Tome I, Vol. I (1904), p. 61. L. Euler considered a^{a^a}, etc. See E. M. Lémeray, *Proc. Edinb. Math. Soc.*, Vol. XVI (1897), p. 13.

[6] *The Mathematical Writings of Duncan Farquharson Gregory* (ed. William Walton; Cambridge, 1865), p. 124–27, 145.

m is an integer, rm is an integer, and $+^{rm}a^m$ has only one value; but if m be a fraction of the form $\frac{p}{q}$, $+^{r\frac{p}{q}}$ will acquire different values, according as we assign different values to r .... $\sqrt{(-a)}\times\sqrt{(-a)}=\sqrt{(+a^2)}=\sqrt{(+)}\sqrt{(a^2)}=-a$; for in this case we know how the $+$ has been derived, namely from the product $--=+$, or $-^2=+$, which of course gives $+^{\frac{1}{2}}=-$, there being here nothing indeterminate about the $+$. It was in consequence of sometimes tacitly assuming the existence of $+$, and at another time neglecting it, that the errors in various trigonometrical expressions arose; and it was by the introduction of the factor $\cos 2r\pi+-^{\frac{1}{2}}\sin 2r\pi$ (which is equivalent to $+^r$) that Poinsot established the formulae in a more correct and general shape." Gregory finds "$\sin(+^{\frac{p}{q}}c)=+^{\frac{p}{q}}\sin c$."

A special notation for the positive integral powers of an imaginary root r of $x^{n-1}+x^{n-2}+ \dots +x+1=0$, n being an odd prime, is given by Gauss;[1] to simplify the typesetting he designates r, rr, r^3, etc., by the symbols [1], [2], [3], etc.

315. *Conclusions.*—There is perhaps no symbolism in ordinary algebra which has been as well chosen and is as elastic as the Cartesian exponents. Descartes wrote a^3, x^4; the extension of this to general exponents a^n was easy. Moreover, the introduction of fractional and negative numbers, as exponents, was readily accomplished. The irrational exponent, as in $a^{\sqrt{2}}$, found unchallenged admission. It was natural to try exponents in the form of pure imaginary or of complex numbers (L. Euler, 1740). In the nineteenth century valuable interpretations were found which constitute the general theory of b^n where b and n may both be complex. Our exponential notation has been an aid for the advancement of the science of algebra to a degree that could not have been possible under the old German or other early notations. Nowhere is the importance of a good notation for the rapid advancement of a mathematical science exhibited more forcibly than in the exponential symbolism of algebra.

SIGNS FOR ROOTS

316. *Early forms.*—Symbols for roots appear very early in the development of mathematics. The sign ⌐ for square root occurs in two Egyptian papyri, both found at Kahun. One was described by F. L.

---

[1] C. F. Gauss, *Disquisitiones arithmeticae* (Leipzig, 1801), Art. 342; *Werke*, Vol. I (1863), p. 420.

Griffith[1] and the other by H. Schack-Schackenburg.[2] For Hindu signs see §§ 107, 108, 112; for Arabic signs see § 124.

317. *General statement.*—The principal symbolisms for the designation of roots, which have been developed since the influx of Arabic learning into Europe in the twelfth century, fall under four groups having for their basic symbols, respectively, R (*radix*), l (*latus*), the sign $\sqrt{}$, and the fractional exponent.

318. *The sign ℞; first appearance.*—In a translation[3] from the Arabic into Latin of a commentary of the tenth book of the *Elements* of Euclid, the word *radix* is used for "square root." The sign ℞ came to be used very extensively for "root," but occasionally it stood also for the first power of the unknown quantity, x. The word *radix* was used for x in translations from Arabic into Latin by John of Seville and Gerard of Cremona (§ 290). This double use of the sign ℞ for x and also for square root is encountered in Leonardo of Pisa (§§ 122, 292)[4] and Luca Pacioli (§§ 135–37, 292).

Before Pacioli, the use of ℞ to designate square root is also met in a correspondence that the German astronomer Regiomontanus (§ 126)[5] carried on with Giovanni Bianchini, who was court astronomer at Ferrara in Italy, and with Jacob von Speier, a court astronomer at Urbino (§ 126).

In German manuscripts referred to as the Dresden MSS C. 80, written about the year 1480, and known to have been in the hands of J. Widman, H. Grammateus, and Adam Riese, there is a sign consisting of a small letter, with a florescent stroke attached (Fig. 104). It has been interpreted by some writers as a letter r with an additional stroke. Certain it is that in Johann Widman's arithmetic of 1489 occurs the crossed capital letter ℞, and also the abbreviation *ra* (§ 293).

Before Widman, the Frenchman Chuquet had used ℞ for "root"

[1] F. L. Griffith, *The Petrie Papyri, I. Kahun Papyri,* Plate VIII.

[2] H. Schack-Schackenburg in *Zeitschrift für aegyptische Sprache und Altertumskunde,* Vol. XXXVIII (1900), p. 136; also Plate IV. See also Vol. XL, p. 65.

[3] M. Curtze, *Anaritii in decem libros elementorum Euclidis commentarii* (Leipzig, 1899), p. 252–386.

[4] *Scritti di Leonardo Pisano* (ed. B. Boncompagni), Vol. II (Rome, 1862), "La practica geometriae," p. 209, 231. The word *radix,* meaning x, is found also in Vol. I, p. 407.

[5] M. Curtze, "Der Briefwechsel Riomantan's, etc.," *Abhandlungen zur Geschichte der mathematischen Wissenschaften,* Vol. XII (Leipzig, 1902), p. 234, 318.

in his manuscript, *Le Triparty* (§ 130). He[1] indicates R^2 16. as 4, "R^4. 16. *si est* . 2.," "R^5. 32. *si est* . 2."

319. *Sixteenth-century use of* R.—The different uses of R. made in Pacioli's *Summa* (1494, 1523) are fully set forth in §§ 134–38. In France, De la Roche followed Chuquet in the use of R. (§ 132). The symbol appears again in Italy in Ghaligai's algebra (1521), and in later editions (§ 139), while in Holland it appeared as early as 1537 in the arithmetic of Giel Van der Hoecke (§ 150) in expressions like "Item wildi aftrecken R $\frac{1}{5}$ van R $\frac{4}{5}$ resi R $\frac{1}{5}$"; i.e., $\sqrt{\frac{4}{5}}-\sqrt{\frac{1}{5}}=\sqrt{\frac{1}{5}}$. The employment of R in the calculus of radicals by Cardan is set forth in §§ 141, 199. A promiscuous adoption of different notations is found in the algebra of Johannes Scheubel (§§ 158, 159) of the University of Tübingen. He used Widman's abbreviation *ra*, also the sign $\sqrt{}$; he indicates cube root by *ra. cu.* or by $\wedge\!\wedge\!\!/$, fourth root by *ra. ra.* or by $\wedge\!\!/$. He suggests a notation of his own, of which he makes no further use, namely, *radix se.*, for cube root, which is the abbreviation of *radix secundae quantitatis*. As the sum "*ra*. 15 *ad ra*. 17" he gives "*ra. col.* 32+$\sqrt{1020}$," i.e., $\sqrt{15}+\sqrt{17}=\sqrt{32+\sqrt{1020}}$. The *col., collecti,* signifies here aggregation.

Nicolo Tartaglia in 1556 used R extensively and also parentheses (§§ 142, 143). Francis Maurolycus[2] of Messina in 1575 wrote "*r*. 18" for $\sqrt{18}$, "*r. v. 6 $\tilde{m}$. r* $7\frac{1}{5}$" for $\sqrt{6-\sqrt{7\frac{1}{5}}}$. Bombelli's radical notation is explained in § 144. It thus appears that in Italy the R had no rival during the sixteenth century in the calculus of radicals. The only variation in the symbolism arose in the marking of the order of the radical and in the modes of designation of the aggregation of terms that were affected by R.

320. In Spain[3] the work of Marco Aurel (1552) (§ 204) employs the signs of Stifel, but Antich Rocha, adopting the Italian abbreviations in adjustment to the Spanish language, lets, in his *Arithmetica* of 1564, "15 *Mas ra. q.* 50 *Mas ra. q.* 27 *Mas ra. q.* 6" stand for $15+\sqrt{50}+\sqrt{27}+\sqrt{6}$. A few years earlier, J. Perez de Moya, in his *Aritmetica practica y speculativa* (1562), indicates square root by *r*,

---

[1] *Le Triparty en la science des nombres par Maistre Nicolas Chuquet Parisien* ... par M. Aristide Marre in Boncompagni's *Bullettino*, Vol. XIII, p. 655; (reprint, Rome, 1881), p. 103.

[2] *D. Francisci Mavrolyci, Abbatis Messanensis, Opuscula mathematica* (Venice, 1575), p. 144.

[3] Our information on these Spanish authors is drawn partly from Julio Rey Pastor, *Los Matemáticos españoles de siglo XVI* (Oviedo, 1913), p. 42.

cube root by rrr, fourth root by rr, marks powers by $co.$, $ce.$, $cu.$, $c.$ $ce.$, and "plus" by p, "minus" by m, "equal" by $eq.$

In Holland, Adrianus Romanus[1] used a small r, but instead of v wrote a dot to mark a root of a binomial or polynomial; he wrote

r $bin.$ $2+r$ $bin.$ $2+r$ $bin.$ $2+r$ $2.$ to designate $\sqrt{2+\sqrt{2+\sqrt{2+\sqrt{2}}}}$.

In Tartaglia's arithmetic, as translated into French by Gosselin[2] of Caen, in 1613, one finds the familiar $\mathbb{R}$ cu to mark cube root. A modification was introduced by the Scotsman James Hume,[3] residing in Paris, who in his algebra of 1635 introduced Roman numerals to indicate the order of the root (§ 190). Two years later, the French text by Jacqves de Billy[4] used $\mathbb{R}Q$, $\mathbb{R}C$, $\mathbb{R}QC$ for $\sqrt{}$, $\sqrt[3]{}$, $\sqrt[6]{}$, respectively.

321. *Seventeenth-century use of* $\mathbb{R}$.—During the seventeenth century, the symbol $\mathbb{R}$ lost ground steadily but at the close of the century it still survived; it was used, for instance, by Michael Rolle[5] who employed the signs $2+\overline{R.-121.}$ to represent $2+\sqrt{-121}$, and $R.$ $trin.$ $\overline{6aabb-9a^4b-b^3}$ to represent $\sqrt{6a^2b^2-9a^4b-b^3}$. In 1690 H. Vitalis[6] takes R_2 to represent *secunda radix*, i.e., the radix next after the square root. Consequently, with him, as with Scheubel, 3. $R.$ 2^* 8, meant $3\sqrt[2]{8}$, or 6.

The sign R or $\mathbb{R}$, representing a radical, had its strongest foothold in Italy and Spain, and its weakest in England. With the close of the seventeenth century it practically passed away as a radical sign; the symbol $\sqrt{}$ gained general ascendancy. Elsewhere it will be pointed out in detail that some authors employed R to represent the unknown x. Perhaps its latest regular appearance as a radical sign is in the Spanish text of Perez de Moya (§ 320), the first edition of which appeared in 1562. The fourteenth edition was issued in 1784; it still gave rrr as signifying cube root, and rr as fourth root. Moya's book offers a most striking example of the persistence for centuries of old and clumsy notations, even when far superior notations are in general use.

[1] *Ideae Mathematicae Pars Prima*, ... *Adriano Romano Lovaniensi* (Antwerp, 1593), following the Preface.

[2] *L'Arithmetique de Nicolas Tartaglia Brescian*, traduit ... par Gvillavmo Gosselin de Caen, Premier Partie (Paris, 1613), p. 101.

[3] James Hume, *Traité de l'algebre* (Paris, 1635), p. 53.

[4] Jacqves de Billy, *Abregé des Preceptes d'Algebre* (Rheims, 1637), p. 21.

[5] *Journal des Sçavans* de l'An 1683 (Amsterdam, 1709), p. 97.

[6] *Lexicon mathematicum ... authore Hieronymo Vitali* (Rome, 1690), art. "Algebra."

322. *The sign l.*—The Latin word *latus* ("side of a square") was introduced into mathematics to signify root by the Roman surveyor Junius Nipsus,[1] of the second century A.D., and was used in that sense by Martianus Capella,[2] Gerbert,[3] and by Plato of Tivoli in 1145, in his translation from the Arabic of the *Liber embadorum* (§ 290). The symbol l (*latus*) to signify root was employed by Peter Ramus[4] with whom "l 27 *ad* l 12" gives "l 75," i.e., $\sqrt{27}+\sqrt{12}=\sqrt{75}$; "$ll$ 32 *de* ll 162" gives "ll 2," i.e., $\sqrt[4]{32}$ from $\sqrt[4]{162}=\sqrt[4]{2}$. Again,[5] "$8-l$ 20 *in* 2 *quotus est* $(4-l$ 5." means $8-\sqrt{20}$, divided by 2, gives the quotient $4-\sqrt{5}$. Similarly,[6] "$lr.$ $l112-l76$" meant $\sqrt{\sqrt{112}-\sqrt{76}}$; the r signifying here *residua*, or "remainder," and therefore $lr.$ signified the square root of the binomial difference.

In the 1592 edition[7] of Ramus' arithmetic and algebra, edited by Lazarus Schoner, "lc 4" stands for $\sqrt[3]{4}$, and "$l\,bq$ 5" for $\sqrt[4]{5}$, in place of Ramus' "ll 5." Also, $\sqrt{2}$. $\sqrt{3}=\sqrt{6}$, $\sqrt{6}\div\sqrt{2}=\sqrt{3}$ is expressed thus:[8]

"Esto multiplicandum l_2 per l_3 factus erit l 6.

$$l\,2.\qquad l\,6\!\left(l\,3\;.\text{"}\right.$$
$$l\,3.\qquad l\,2$$
$$\overline{}$$
$$l\,6.$$

It is to be noted that with Schoner the l received an extension of meaning, so that $5l$ and $l5$, respectively, represent $5x$ and $\sqrt{5}$, the l standing for the first power of the unknown quantity when it is not

---

[1] *Die Schriften der römischen Feldmesser* (ed. Blume, Lachmann, Rudorff; Berlin, 1848–52), Vol. I, p. 96.

[2] Martianus Capella, *De Nuptiis* (ed. Kopp; Frankfort, 1836), lib. VII, § 748.

[3] *Gerberti opera mathematica* (ed. Bubnow; Berlin, 1899), p. 83. See J. Tropfke, *op. cit.*, Vol. II (2d ed., 1921), p. 143.

[4] *P. Rami Scholarvm mathematicarvm libri unus et triginti* (Basel, 1569), Lib. XXIV, p. 276, 277.

[5] *Ibid.*, p. 179.

[6] *Ibid.*, p. 283.

[7] *Petri Rami ... Arithmetices libri duo, et algebrae totidem: à Lazaro Schonero* (Frankfurt, 1592), p. 272 ff.

[8] *Petri Rami ... Arithmeticae libri duo et geometriae septem et viginti, Dudum quidem, à Lazaro Schonero* (Frankfurt a/M., 1627), part entitled "De Nvmeri figvratis Lazari Schoneri liber," p. 178.

followed by a number (see also § 290). A similar change in meaning resulting from reversing the order of two symbols has been observed in Pacioli in connection with R (§§ 136, 137) and in A. Girard in connection with the circle of Stevin (§ 164). The double use of the sign l, as found in Schoner, is explained more fully by another pupil of Ramus, namely, Bernardus Salignacus (§ 291).

Ramus' l was sometimes used by the great French algebraist Francis Vieta who seemed disinclined to adopt either R or $\sqrt{}$ for indicating roots (§ 177).

This use of the letter l in the calculus of radicals never became popular. After the invention of logarithms, this letter was needed to mark logarithms. For that reason it is especially curious that Henry Briggs, who devoted the latter part of his life to the computation and the algorithm of logarithms, should have employed l in the sense assigned it by Ramus and Vieta. In 1624 Briggs used l, $l_{(3)}$, ll for square, cube, and fourth root, respectively. "Sic $l_{(3)}$ 8 [i.e., $\sqrt[3]{8}$], latus cubicum Octonarii, id est 2. sic l bin 2 $+$ l 3. [i.e., $\sqrt{2+\sqrt{3}}$] latus binomii 2$+l$3." Again, "ll 85$\frac{1}{3}$ [i.e., $\sqrt[4]{85\frac{1}{3}}$]. Latus 85$\frac{1}{3}$ est 9$\underline{237604307}$, et huius lateris latus est 3$\underline{03934274}$. cui numero aequatur ll 85$\frac{1}{3}$."[1]

323. *Napier's line symbolism.*—John Napier[2] prepared a manuscript on algebra which was not printed until 1839. He made use of Stifel's notation for radicals, but at the same time devised a new scheme of his own. "It is interesting to notice that although Napier invented an excellent notation of his own for expressing roots, he did not make use of it in his algebra, but retained the cumbrous, and in some cases ambiguous notation generally used in his day. His notation was derived from this figure

$$
\begin{array}{|c|c|c|}
\hline
1 & 2 & 3 \\
\hline
4 & 5 & 6 \\
\hline
7 & 8 & 9 \\
\hline
\end{array}
$$

in the following way: ⊔ prefixed to a number means its square root, ⊐ its fourth root, □ its fifth root, ⌐ its ninth root, and so on, with extensions of obvious kinds for higher roots."[3]

---

[1] Henry Briggs, *Arithmetica logarithmica* (London, 1624), Introduction.

[2] *De Arte Logistica Joannis Naperi Merchistonii Baronis Libri qui supersunt* (Edinburgh, 1839), p. 84.

[3] J. E. A. Steggall, "De arte logistica," *Napier Tercentenary Memorial Volume* (ed. Cargill Gilston Knott; London, 1915), p. 160.

324. *Origin of* $\sqrt{}$.—This symbol originated in Germany. L. Euler guessed that it was a deformed letter *r*, the first letter in *radix*.[1] This opinion was held generally until recently. The more careful study of German manuscript algebras and the first printed algebras has convinced Germans that the old explanation is hardly tenable; they have accepted the a priori much less probable explanation of the evolution of the symbol from a dot. Four manuscript algebras have been available for the study of this and other questions.

The oldest of these is in the Dresden Library, in a volume of manuscripts which contains different algebraic treatises in Latin and one in German.[2] In one of the Latin manuscripts (see Fig. 104, *A*7), probably written about 1480, dots are used to signify root extraction. In one place it says: "In extraccione radicis quadrati alicuius numeri preponatur numero vnus punctus. In extraccione radicis quadrati radicis quadrati prepone numero duo puncta. In extraccione cubici radicis alicuius numeri prepone tria puncta. In extraccione cubici radicis alicuius radicis cubici prepone 4 puncta."[3] That is, one dot (·) placed before the radicand signifies square root; two dots (..) signify the square root of the square root; three dots (...) signify cube root; four dots (....), the cube root of the cube root or the ninth root. Evidently this notation is not a happy choice. If one dot meant square root and two dots meant square root of square root (i.e., $\sqrt{\sqrt{\ }}$), then three dots should mean square root of square root of square root, or eighth root. But such was not actually the case; the three dots were made to mean cube root, and four dots the ninth root. What was the origin of this dot-system? No satisfactory explanation has been found. It is important to note that this Dresden manuscript was once in the possession of Joh. Widman, and that Adam Riese, who in 1524 prepared a manuscript algebra of his own, closely followed the Dresden algebra.

325. The second document is the Vienna MS[4] No. 5277, Regule-

---

[1] L. Euler, *Institutiones calculi differentialis* (1775), p. 103, art. 119; J. Tropfke, *op. cit.*, Vol. II (2d ed., 1921), p. 150.

[2] M. Cantor, *Vorles. über Geschichte der Mathematik*, Vol. II (2. Aufl., 1900), p. 241.

[3] E. Wappler, *Zur Geschichte der deutschen Algebra im XV. Jahrhundert*, Zwickauer Gymnasialprogramm von 1887, p. 13. Quoted by J. Tropfke, *op. cit.*, Vol. II (1921), p. 146, and by M. Cantor, *op. cit.*, Vol. II (2. Aufl., 1900), p. 243.

[4] C. J. Gerhardt, *Monatsberichte Akad.* (Berlin, 1867), p. 46; *ibid.* (1870), p. 143–47; Cantor. *op. cit.*, Vol. II (2d ed., 1913), p. 240, 424.

Cose–uel Algobre–. It contains the passage: "Quum ʒ assimiletur radici de radice punctus deleatur de radice, ʒ in se ducatur et remanet adhuc inter se aequalia"; that is, "When $x^2 = \sqrt{x}$, erase the point before the x and multiply x^2 by itself, then things equal to each other are obtained." In another place one finds the statement, *per punctum intellige radicem*—"by a point understand a root." But no dot is actually used in the manuscript for the designation of a root.

The third manuscript is at the University of Göttingen, Codex Gotting. Philos. 30. It is a letter written in Latin by Initius Algebras,[1] probably before 1524. An elaboration of this manuscript was made in German by Andreas Alexander.[2] In it the radical sign is a heavy point with a stroke of the pen up and bending to the right, thus $\int$. It is followed by a symbol indicating the index of the root; $\int$ʒ indicates square root; $\int c^e$, cube root; $\int cc^e$, the ninth root, etc. Moreover, $\int cs|8 + \int 22_ʒ$ stands for $\sqrt{8 + \sqrt{22}}$, where cs (i.e., *communis*) signifies the root of the binomial which is designated as one quantity, by lines, vertical and horizontal. Such lines are found earlier in Chuquet (§ 130). The ʒ, indicating the square root of the binomial, is placed as a subscript after the binomial. Calling these two lines a "gnomon," M. Curtze adds the following:

"This gnomon has here the signification, that what it embraces is not a length, but a power. Thus, the simple 8 is a length or simple number, while $|8_ʒ$ is a square consisting of eight areal units whose linear unit is $\int$ʒ8. In the same way $|8_{c^e}$ would be a cube, made up of 8 cubical units, of which $\int c^e|8$ is its side, etc. A double point, with the tail attached to the last, signifies always the root of the root. For example, $.\int c^e|88$ would mean the cube root of the cube root of 88. It is identical with $\int cc^e 88$, but is used only when the radicand is a so-called median [*Mediale*] in the Euclidean sense."[3]

326. The fourth manuscript is an algebra or *Coss* completed by Adam Riese[4] in 1524; it was not printed until 1892. Riese was familiar with the small Latin algebra in the Dresden collection, cited above;

---

[1] *Initius Algebras: Algebrae Arabis Arithmetici viri clarrisimi Liber ad Ylem geometram magistrum suum.* This was published by M. Curtze in *Abhandlungen zur Geschichte der mathematischen Wissenschaften*, Heft XIII (1902), p. 435–611. Matters of notation are explained by Curtze in his introduction, p. 443–48.

[2] G. Eneström, *Bibliotheca mathematica* (3d ser.), Vol. III (1902), p. 355–60.

[3] M. Curtze, *op. cit.*, p. 444.

[4] B. Berlet, *Adam Riese, sein Leben, seine Rechenbücher und seine Art zu rechnen; die Coss von Adam Riese* (Leipzig-Frankfurt a/M., 1892).

he refers also to Andreas Alexander.[1] For indicating a root, Riese does not use the dot, pure and simple, but the dot with a stroke attached to it, though the word *punct* ("point") occurs. Riese says: "Ist, so ʒ vergleicht wird √ vom radix, so mal den ʒ in sich multipliciren vnnd das punct vor.dem Radix aussleschn."[2] This passage has the same interpretation as the Latin passage which we quoted from the Vienna manuscript.

We have now presented the main facts found in the four manuscripts. They show conclusively that the dot was associated as a symbol with root extraction. In the first manuscript, the dot actually appears as a sign for roots. The dot does not appear as a sign in the second manuscript, but is mentioned in the text. In the third and fourth manuscripts, the dot, pure and simple, does not occur for the designation of roots; the symbol is described by recent writers as a dot with a stroke or tail attached to it. The question arises whether our algebraic sign √ took its origin in the dot. Recent German writers favor that view, but the evidence is far from conclusive. Johannes Widman, the author of the *Rechnung* of 1489, was familiar with the first manuscript which we cited. Nevertheless he does not employ the dot to designate root, easy as the symbol is for the printer. He writes down ℞ and *ra*. Christoff Rudolff was familiar with the Vienna manuscript which uses the dot with a tail. In his *Coss* of 1525 he speaks of the *Punkt* in connection with root symbolism, but uses a mark with a very short heavy downward stroke (almost a point), followed by a straight line or stroke, slanting upward (see Fig. 59). As late as 1551, Scheubel,[3] in his printed *Algebra*, speaks of points. He says: "Solent tamen multi, et bene etiam, has desideratas radices, suis punctis cū linea quadam a dextro latere ascendente, notare. . . . ." ("Many are accustomed, and quite appropriately, to designate the desired roots by points, from the right side of which there ascends a kind of stroke.") It is possible that this use of "point" was technical, signifying "sign for root," just as at a later period the expression "decimal point" was used even when the symbol actually written down to mark a decimal fraction was a comma. It should be added that if Rudolff looked upon his radical sign as really a dot, he would have been less likely to have used the dot again for a second purpose in his radical symbolism, namely, for the purpose of designating that

---

[1] B. Berlet, *op. cit.*, p. 29, 33.

[2] C. I. Gerhardt, *op. cit.* (1870) p. 151.

[3] J. Scheubel, *Algebra compendiosa* (Paris, 1551), fol. 25*B*. Quoted from J. Tropfke, *op. cit.*, Vol. II (2d ed., 1921), p. 149.

the introducing of typesetting machines and the great cost of type-setting by hand operate against a double or multiple line notation. The dots have not generally prevailed in the marking of aggregation for the reason, no doubt, that there was danger of confusion since dots are used in many other symbolisms—those for multiplication, division, ratio, decimal fractions, time-derivatives, marking a number into periods of two or three digits, etc.

343. *Aggregation expressed by letters.*—The expression of aggrega-tion by the use of letters serving as abbreviations of words expressing aggregation is not quite as old as the use of horizontal bars, but it is more common in works of the sixteenth century. The need of marking the aggregation of terms arose most frequently in the treatment of radicals. Thus Pacioli, in his *Summa* of 1494 and 1523, employs v (*vniversale*) in marking the root of a binomial or polynomial (§ 135). This and two additional abbreviations occur in Cardan (§ 141). The German manuscript of Andreas Alexander (1524) contains the letters cs for *communis* (§ 325); Chr. Rudolff sometimes used the word "collect," as in "$\sqrt{}$ des collects $17+\sqrt{208}$" to designate $\sqrt{17+\sqrt{208}}$.[1] J. Scheubel adopted *Ra. col.* (§ 159). S. Stevin, Fr. Vieta, and A. Romanus wrote *bin.*, or *bino.*, or *binomia, trinom.*, or similar abbrevia-tions (§ 320). The u or v is found again in Pedro Nuñez (who uses also L for "ligature"),[2] Leonard and Thomas Digges (§ 334), in J. R. Brasser[3] who in 1663 lets v signify "universal radix" and writes "$v\sqrt{}.8\div\sqrt{45}$" to represent $\sqrt{8-\sqrt{45}}$. W. Oughtred sometimes wrote $\sqrt{}u$ or $\sqrt{}b$ (§§ 183, 334). In 1685 John Wallis[4] explains the notations $\sqrt{}b:2+\sqrt{3}$, $\sqrt{}r:2-\sqrt{3}$, $\sqrt{}u:2\pm\sqrt{3}$, $\sqrt{2\pm\sqrt{3}}$, $\sqrt{}:2\pm\sqrt{3}$, where b means "binomial," u "universal," r "residual," and sometimes uses redundant forms like $\sqrt{}b:\sqrt{5}+1:$.

344. *Aggregation expressed by horizontal bars or vinculums.*—The use of the horizontal bar to express the aggregation of terms goes back to the time of Nicolas Chuquet who in his manuscript (1484) under-lines the parts affected (§ 130). We have seen that the same idea is followed by the German Andreas Alexander (§ 325) in a manuscript of 1545, and by the Italian Raffaele Bombelli in the manuscript edition

[1] J. Tropfke, *op. cit.*, Vol. II (1921), p. 150.

[2] Pedro Nuñez, *Libra de algebra en arithmetica y geometria* (Anvers, 1567), fol. 52.

[3] J. R. Brasser, *Regula of Algebra* (Amsterdam, 1663), p. 27.

[4] John Wallis, *Treatise of Algebra* (London, 1685), p. 109, 110. The use of letters for aggregation practically disappeared in the seventeenth century.

of his algebra (about 1550) where he wrote $\sqrt[3]{2+\sqrt{-121}}$ in this manner:[1] $R^3[2.p.R[0 \ \tilde{m} \ 121]]$; parentheses were used and, in addition, vinculums were drawn underneath to indicate the range of the parentheses. The employment of a long horizontal brace in connection with the radical sign was introduced by Thomas Harriot[2] in 1631; he expresses aggregation thus: $\sqrt{ccc+\sqrt{ccccc-bbbbb}}$. This notation may, perhaps, have suggested to Descartes his new radical symbolism of 1637. Before that date, Descartes had used dots in the manner of Stifel and Van Ceulen. He wrote[3] $\sqrt{.2-\sqrt{2}}$. for $\sqrt{2-\sqrt{2}}$. He attaches the vinculum to the radical sign $\sqrt{}$ and writes $\sqrt{a^2+b^2}$, $\sqrt{-\frac{1}{2}a+\sqrt{\frac{1}{4}aa+bb}}$, and in case of cube roots $\sqrt{C. \frac{1}{2}q+\sqrt{\frac{1}{4}gg-27p^3}}$. Descartes does not use parentheses in his *Géométrie*. Descartes uses the horizontal bar only in connection with the radical sign. Its general use for aggregation is due to Fr. van Schooten, who, in editing Vieta's collected works in 1646, discarded parentheses and placed a horizontal bar above the parts affected. Thus Van Schooten's "$B \ in \ \overline{D \ quad.+B \ in \ D}$" means $B(D^2+BD)$. Vieta[4] himself in 1593 had written this expression differently, namely, in this manner:

$$" B \ in \ \begin{cases} D. \ quadratum \ " \\ +B \ in \ D \end{cases}.$$

B. Cavalieri in his *Geometria indivisibilibae* and in his *Exercitationes geometriae sex* (1647) uses the vinculum in this manner, $\overline{AB}$, to indicate that the two letters A and B are not to be taken separately, but conjointly, so as to represent a straight line, drawn from the point A to the point B.

Descartes' and Van Schooten's stressing the use of the vinculum led to its adoption by J. Prestet in his popular text, *Elemens des Mathématiques* (Paris, 1675). In an account of Rolle[5] the cube root is to be taken of $2+\overline{R.-121}$, i.e., of $2+\sqrt{-121}$. G. W. Leibniz[6] in a

---

[1] See E. Bertolotti in *Scientia*, Vol. XXXIII (1923), p. 391 n.

[2] Thomas Harriot, *Artis analyticae praxis* (London, 1631), p. 100.

[3] R. Descartes, *Œuvres* (éd. Ch. Adam et P. Tannery), Vol. X (Paris, 1908), p. 286 f., also p. 247, 248.

[4] See J. Tropfke, *op. cit.*, Vol. II (1921), p. 30.

[5] *Journal des Sçavans* de l'an 1683 (Amsterdam, 1709), p. 97.

[6] G. W. Leibniz' letter to D. Oldenburgh, Feb. 3, 1672–73, printed in J. Collin's *Commercium epistolicum* (1712).

letter of 1672 uses expressions like $\overline{a \backsim b} \backsim \overline{b \backsim c} \backsim \overline{b \backsim c} \backsim \overline{c \backsim d}$, where $\backsim$ signifies "difference." Occasionally he uses the vinculum until about 1708, though usually he prefers round parentheses. In 1708 Leibniz' preference for round parentheses (§ 197) is indicated by a writer in the *Acta eruditorum*. Joh. (1) Bernoulli, in his *Lectiones de calculo differ-entialium*, uses vinculums but no parentheses.[1]

345. In England the notations of W. Oughtred, Thomas Harriot, John Wallis, and Isaac Barrow tended to retard the immediate intro-duction of the vinculum. But it was used freely by John Kersey (1673)[2] who wrote $\sqrt{(2)} : \frac{1}{2}r - \sqrt{\frac{1}{4}rr - s}$: and by Newton, as, for in-stance, in his letter to D. Oldenburgh of June 13, 1676, where he gives the binomial formula as the expansion of $\overline{P + PQ}\Big|^{\frac{m}{n}}$. In his *De Analysi per Aequationes numero terminorum Infinitas*, Newton writes[3]

$\overline{\overline{y - 4 \times y + 5 \times y - 12 \times y + 17}} = 0$ to represent $\{[(y - 4)y - 5]y - 12\}y + 17 = 0$. This notation was adopted by Edmund Halley,[4] David Gregory, and John Craig; it had a firm foothold in England at the close of the seventeenth century. During the eighteenth century it was the regular symbol of aggregation in England and France; it took the place very largely of the parentheses which are in vogue in our day. The vincu-lum appears to the exclusion of parentheses in the *Geometria organica* (1720) of Colin Maclaurin, in the *Elements of Algebra* of Nicholas Saunderson (Vol. I, 1741), in the *Treatise of Algebra* (2d ed.; London, 1756) of Maclaurin. Likewise, in Thomas Simpson's *Mathematical Dissertations* (1743) and in the 1769 London edition of Isaac Newton's *Universal Arithmetick* (translated by Ralphson and revised by Cunn), vinculums are used and parentheses do not occur. Some use of the vinculum was made nearly everywhere during the eighteenth century, especially in connection with the radical sign $\sqrt{}$, so as to produce $\sqrt{}$. This last form has maintained its place down to the present time. However, there are eighteenth-century writers who avoid the vincu-lum altogether even in connection with the radical sign, and use

[1] The Johannis (1) Bernoullii *Lectiones de calculo differentialium*, which re-mained in manuscript until 1922, when it was published by Paul Schafheitlin in *Verhandlungen der Naturforschenden Gesellschaft in Basel*, Vol. XXXIV (1922).

[2] John Kersey, *Algebra* (London, 1673), p. 55.

[3] *Commercium epistolicum* (éd. Biot et Lefort; Paris, 1856), p. 63.

[4] *Philosophical Transactions* (London), Vol. XV–XVI (1684–91), p. 393; Vol. XIX (1695–97), p. 60, 645, 709.

parentheses exclusively. Among these are Poleni (1729),[1] Cramer (1750),[2] and Cossali (1797).[3]

346. There was considerable vacillation on the use of the vinculum in designating the square root of minus unity. Some authors wrote $\sqrt{-1}$; others wrote $\sqrt{-1}$ or $\sqrt{(-1)}$. For example, $\sqrt{-1}$ was the designation adopted by J. Wallis,[4] J. d'Alembert,[5] I. A. Segner,[6] C. A. Vandermonde,[7] A. Fontaine.[8] Odd in appearance is an expression of Euler,[9] $\sqrt{(2\sqrt{-1}-4)}$. But $\sqrt{(-1)}$ was preferred by Du Séjour[10] in 1768 and by Waring[11] in 1782; $\sqrt{-1}$ by Laplace[12] in 1810.

347. It is not surprising that, in times when a notation was passing out and another one taking its place, cases should arise where both are used, causing redundancy. For example, J. Stampioen in Holland sometimes expresses aggregation of a set of terms by three notations, any one of which would have been sufficient; he writes[13] in 1640, $\sqrt{\cdot\,\overline{(aaa+6aab+9bba)}}$, where the dot, the parentheses, and the vinculum appear; John Craig[14] writes $\sqrt{2ay-y^2}$: and $\sqrt{:\sqrt{6a^4-\frac{3}{2}a^2}}$, where the colon is the old Oughtredian sign of aggregation, which is here superfluous, because of the vinculum. Tautology in notation is found in Edward Cocker[15] in expressions like $\sqrt{\overline{aa+bb}}$, $\sqrt{:c+\frac{1}{4}bb}-\frac{1}{2}b$, and

---

[1] Ioannis Poleni, *Epistolarvm mathematicarvm fascicvlvs* (Padua, 1729).

[2] Gabriel Cramer, *L'Analyse des lignes courbes algébriques* (Geneva, 1750).

[3] Pietro Cossali, *Origini ... dell'algebra*, Vol. I (Parma, 1797).

[4] John Wallis, *Treatise of Algebra* (London, 1685), p. 266.

[5] J. d'Alembert in *Histoire de l'académie r. des sciences*, année 1745 (Paris, 1749), p. 383.

[6] I. A. Segner, *Cursus mathematici*, Pars IV (Halle, 1763), p. 44.

[7] C. A. Vandermonde in *op. cit.*, année 1771 (Paris, 1774), p. 385.

[8] A. Fontaine, *ibid.*, année 1747 (Paris, 1752), p. 667.

[9] L. Euler in *Histoire de l'académie r. d. sciences et des belles lettres*, année 1749 (Berlin, 1751), p. 228.

[10] Du Séjour, *ibid.* (1768; Paris, 1770), p. 207.

[11] E. Waring, *Meditationes algebraicae* (Cambridge; 3d ed., 1782), p. xxxvl, etc.

[12] P. S. Laplace in *Mémoires d. l'académie r. d. sciences*, année 1817 (Paris, 1819), p. 153.

[13] *I. I. Stampionii Wis-Konstigh ende Reden-Maetigh Bewijs* (The Hague, 1640), p. 7.

[14] John Craig, *Philosophical Transactions*, Vol. XIX (London, 1695–97), p. 709.

[15] *Cockers Artificial Arithmetick. . . . . Composed by Edward Cocker. . . . . Perused, corrected and published by John Hawkins* (London, 1702) ["To the Reader," 1684], p. 368, 375.

a few times in John Wallis.[1] In the *Acta eruditorum* (1709), page 327, one finds $ny\sqrt{a}=\frac{2}{3}\sqrt{[(x-nna)^3]}$, where the [ ] makes, we believe, its first appearance in this journal, but does so as a redundant symbol.

348. *Aggregation expressed by dots.*—The denoting of aggregation by placing a dot before the expression affected is first encountered in Christoff Rudolff (§ 148). It is found next in the *Arithmetica integra* of M. Stifel, who sometimes places a dot also at the end. He writes[2]

$\sqrt{z}.12+\sqrt{z}\ 6+.\sqrt{z}.12-\sqrt{z}\ 6$ for our $\sqrt{12+\sqrt{6}}+\sqrt{12-\sqrt{6}}$; also $\sqrt{z}.144-6+\sqrt{z}.144-6$ for $\sqrt{144-6}+\sqrt{144-6}$. In 1605 C. Dibuadius[3] writes $\sqrt{.2-\sqrt{.2+\sqrt{.2+\sqrt{.2+\sqrt{.2+\sqrt{2}}}}}}$ as the side of a regular polygon of 128 sides inscribed in a circle of unit radius, i.e.,

$$\sqrt{2-\sqrt{2+\sqrt{2+\sqrt{2+\sqrt{2+\sqrt{2}}}}}}$$ (see also § 332). It must be admitted that this old notation is simpler than the modern. In Snell's translation[4] into Latin (1610) of Ludolph van Ceulen's work on the circle is given the same notation, $\sqrt{.2+\sqrt{.2-\sqrt{.2-\sqrt{.2+\sqrt{2\frac{1}{2}}}}}}-\sqrt{2\frac{1}{4}}$. In Snell's 1615 translation[5] into Latin of Ludolph's arithmetic and geometry is given the number $\sqrt{.2-\sqrt{.2\frac{1}{2}+\sqrt{1\frac{1}{4}}}}$ which, when divided by $\sqrt{.2+\sqrt{.2\frac{1}{2}+\sqrt{1\frac{1}{4}}}}$, gives the quotient $\sqrt{5}+1-\sqrt{.5+\sqrt{20}}$. The Swiss Joh. Ardüser[6] in 1646 writes $\sqrt{.2\div\sqrt{.2+\sqrt{.2+\sqrt{.2+\sqrt{2+\sqrt{.2+\sqrt{.2+\sqrt{3}}}}}}}}$, etc., as the side of an inscribed polygon of 768 sides, where ÷ means "minus."

The substitution of two dots (the colon) in the place of the single dot was effected by Oughtred in the 1631 and later editions of his *Clavis mathematicae*. With him this change became necessary when he adopted the single dot as the sign of ratio. He wrote ordinarily $\sqrt{q}:BC_q-BA_q:$ for $\sqrt{\overline{BC}^2-\overline{BA}^2}$, placing colons before and after the terms to be aggregated (§ 181).[7]

---

[1] John Wallis, *Treatise of Algebra* (London, 1685), p. 133.

[2] M. Stifel, *Arithmetica integra* (Nürnberg, 1544), fol. 135v°. See J. Tropfke, *op. cit.*, Vol. III (Leipzig, 1922), p. 131.

[3] *C. Dibvadii in arithmeticam irrationalivm Evclidis decimo elementorum libro* (Arnhem, 1605).

[4] Willebrordus Snellius, *Lvdolphi à Cevlen de Circvlo et adscriptis liber ... é vernaculo Latina fecit ...* (Leyden, 1610), p. 1, 5.

[5] *Fvndamenta arithmetica et geometrica. ... Lvdolpho a Cevlen, ... in Latinum translata a Wil. Sn.* (Leyden, 1615), p. 27.

[6] Joh. Ardüser, *Geometriae theoricae et practicae XII libri* (Zürich, 1646), fol. 181b.

[7] W Oughtred, *Clavis mathematicae* (1652), p. 104.

Sometimes, when all the terms to the end of an expression are to be aggregated, the closing colon is omitted. In rare instances the opening colon is missing. A few times in the 1694 English edition, dots take the place of the colon. Oughtred's colons were widely used in England. As late as 1670 and 1693 John Wallis[1] writes $\sqrt{}:5-2\sqrt{3}:$. It occurs in Edward Cocker's[2] arithmetic of 1684, Jonas Moore's arithmetic[3] of 1688, where $C:A+E$ means the cube of $(A+E)$. James Bernoulli[4] gives in 1689 $\sqrt{}:a+\sqrt{}:a+\sqrt{}:a+\sqrt{}:a+\sqrt{}:a+$, etc. These methods of denoting aggregation practically disappeared at the beginning of the eighteenth century, but in more recent time they have been reintroduced. Thus, R. Carmichael[5] writes in his *Calculus of Operations:* "$D.\ uv=u.\ Dv+Du.\ v$." G. Peano has made the proposal to employ points as well as parentheses.[6] He lets $a.bc$ be identical with $a(bc)$, $a:bc.d$ with $a[(bc)d]$, $ab.cd:e.fg \therefore hk.l$ with $\{[(ab)(cd)]\ [e(fg)]\}\ [(hk)l]$.

349. *Aggregation expressed by commas.*—An attempt on the part of Hérigone (§ 189) and Leibniz to give the comma the force of a symbol of aggregation, somewhat similar to Rudolff's, Stifel's, and van Ceulen's previous use of the dot and Oughtred's use of the colon, was not successful. In 1702 Leibniz[7] writes $c-b,\ l$ for $(c-b)l$, and $c-b$, $d-b,\ l$ for $(c-b)(d-b)l$. In 1709 a reviewer[8] in the *Acta eruditorum* represents $(m\div[m-1])x^{(m-1)\div m}$ by $(m,:m-1)x^{m-1,:m}$, a designation somewhat simpler than our modern form.

350. *Aggregation expressed by parenthesis* is found in rare instances as early as the sixteenth century. Parentheses present comparatively no special difficulties to the typesetter. Nevertheless, it took over two centuries before they met with general adoption as mathematical symbols. Perhaps the fact that they were used quite extensively as purely rhetorical symbols in ordinary writing helped to

---

[1] John Wallis in *Philosophical Transactions*, Vol. V (London, for the year 1670), p. 2203; *Treatise of Algebra* (London, 1685), p. 109; Latin ed. (1693), p. 120.

[2] Cocker's *Artificial Arithmetick* . . . . perused . . . . by John Hawkins (London, 1684), p. 405.

[3] Moore's *Arithmetick: in Four Books* (London, 1688; 3d ed.), Book IV, p. 425.

[4] *Positiones arithmeticae de seriebvs infinitis* . . . . *Jacobo Bernoulli* (Basel, 1689).

[5] R. Carmichael, *Der Operationscalcul*, deutsch von C. H. Schnuse (Braunschweig, 1857), p. 16.

[6] G. Peano, *Formulaire mathématique*, Édition de l'an 1902–3 (Turin, 1903), p. 4.

[7] G. W. Leibniz in *Acta eruditorum* (1702), p. 212.

[8] Reviewer in *ibid.* (1709), p. 230. See also p. 180.

retard their general adoption as mathematical symbols. John Wallis, for example, used parentheses very extensively as symbols containing parenthetical rhetorical statements, but made practically no use of them as symbols in algebra.

As a rhetorical sign to inclose an auxiliary or parenthetical statement parentheses are found in Newton's *De analysi per equationes numero terminorum infinitas*, as given by John Collins in the *Commercium epistolicum* (1712). In 1740 De Gua[1] wrote equations in the running text and inclosed them in parentheses; he wrote, for example, "... seroit $(\overline{7a-3x}\cdot dx = 3\sqrt{2ax-xx}\cdot dx)$ et où l'arc de cercle. ..."

English mathematicians adhered to the use of vinculums, and of colons placed before and after a polynomial, more tenaciously than did the French; while even the French were more disposed to stress their use than were Leibniz and Euler. It was Leibniz, the younger Bernoullis, and Euler who formed the habit of employing parentheses more freely and to resort to the vinculum less freely than did other mathematicians of their day. The straight line, as a sign of aggregation, is older than the parenthesis. We have seen that Chuquet, in his *Triparty* of 1484, underlined the terms that were to be taken together.

351. *Early occurrence of parentheses.*—Brackets[2] are found in the manuscript edition of R. Bombelli's *Algebra* (about 1550) in the expressions like $R^3[2\tilde{m}R[0\tilde{m}.121]]$ which stands for $\sqrt[3]{2-\sqrt{-121}}$. In the printed edition of 1572 an inverted capital letter L was employed to express *radix legata;* see the facsimile reproduction (Fig. 50). Michael Stifel does not use parentheses as signs of aggregation in his printed works, but in one of his handwritten marginal notes[3] occurs the following: "... *faciant aggregatum* $(12-\sqrt{44})$ *quod sumptum cum* $(\sqrt{44}-2)$ *faciat* 10" (i.e., "... One obtains the aggregate $(12-\sqrt{44})$, which added to $(\sqrt{44}-2)$ makes 10"). It is our opinion that these parentheses are punctuation marks, rather than mathematical symbols; signs of aggregation are not needed here. In the 1593 edition of F. Vieta's *Zetetica*, published in Turin, occur braces and brackets (§ 177) sometimes as open parentheses, at other times as closed ones. In Vieta's collected works, edited by Fr. van Schooten

---

[1] Jean Paul de Gua de Malves, *Usages de l'analyse de Descartes* (Paris, 1740), p. 302.

[2] See E. Bortolotti in *Scientia*, Vol. XXXIII (1923), p. 390.

[3] E. Hoppe, "Michael Stifels handschriftlicher Nachlass," *Mitteilungen Math. Gesellschaft Hamburg*, III (1900), p. 420. See J. Tropfke, *op. cit.*, Vol. II (2d ed., 1921), p. 28, n. 114.

in 1646, practically all parentheses are displaced by vinculums. However, in J. L. de Vaulezard's translation[1] into French of Vieta's *Zetetica* round parentheses are employed. Round parentheses are encountered in Tartaglia,[2] Cardan (but only once in his *Ars Magna*),[3] Clavius (see Fig. 66), Errard de Bar-le-Duc,[4] Follinus,[5] Girard,[6] Norwood,[7] Hume,[8] Stampioen, Henrion, Jacobo de Billy,[9] Renaldini[10] and Foster.[11] This is a fairly representative group of writers using parentheses, in a limited degree; there are in this group Italians, Germans, Dutch, French, English. And yet the mathematicians of none of the countries represented in this group adopted the general use of parentheses at that time. One reason for this failure lies in the fact that the vinculum, and some of the other devices for expressing aggregation, served their purpose very well. In those days when machine processes in printing were not in vogue, and when typesetting was done by hand, it was less essential than it is now that symbols should, in orderly fashion, follow each other in a line. If one or more vinculums were to be placed above a given polynomial, such a demand upon the printer was less serious in those days than it is at the present time.

[1] J. L. de Vaulezard's *Zététiques de F. Viète* (Paris, 1630), p. 218. Reference taken from the *Encyclopédie d. scien. math.*, Tom I, Vol. I, p. 28.

[2] N. Tartaglia, *General trattato di numeri e misure* (Venice), Vol. II (1556), fol. 167*b*, 169*b*, 170*b*, 174*b*, 177*a*, etc., in expressions like "$R v.(R$ 28 men R 10)" for $\sqrt{\sqrt{28}-\sqrt{10}}$; fol. 168*b*, "men (22 men R6" for $-(22-\sqrt{6})$, only the opening part being used. See G. Eneström in *Bibliotheca mathematica* (3d ser.), Vol. VII (1906-7), p. 296. Similarly, in *La Quarta Parte del general trattato* (1560), fol. 40*B*, he regularly omits the second part of the parenthesis when occurring on the margin, but in the running text both parts occur usually.

[3] H. Cardano, *Ars magna*, as printed in *Opera*, Vol. IV (1663), fol. 438.

[4] I. Errard de Bar-le-Duc, *La geometrie et practique generale d'icelle* (3d ed.; revué par D. H. P. E. M.; Paris, 1619), p. 216.

[5] Hermann Follinus, *Algebra sive liber de rebvs occvltis* (Cologne, 1622), p. 157.

[6] A. Girard, *Invention nouvelle en l'algebre* (Amsterdam, 1629), p. 17.

[7] R. Norwood, *Trigonometrie* (London, 1631), Book I, p. 30.

[8] Jac. Humius, *Traite de l'algebre* (Paris, 1635).

[9] Jacobo de Billy, *Novae geometriae clavis algebra* (Paris, 1643), p. 157; also in an *Abridgement of the Precepts of Algebra* (written in French by James de Billy; London, 1659), p. 346.

[10] Carlo Renaldini, *Opus algebricum* (1644; enlarged edition, 1665). Taken from Ch. Hutton, *Tracts on Mathematical and Philosophical Subjects*, Vol. II (1812), p. 297.

[11] Samuel Foster, *Miscellanies: or Mathematical Lucubrations* (London, 1659), p. 7.

And so it happened that in the second half of the seventeenth century, parentheses occur in algebra less frequently than during the first half of that century. However, voices in their favor are heard. The Dutch writer, J. J. Blassière,[1] explained in 1770 the three notations $(2a+5b)(3a-4b)$, $(2a+5b)\times(3a-4b)$, and $\overline{2a+5b}\times\overline{3a-4b}$, and remarked: "Mais comme la première manière de les enfermer entre des Parenthèses, est la moins sujette à erreur, nous nous en servirons dans la suite." E. Waring in 1762[2] uses the vinculum but no parentheses; in 1782[3] he employs parentheses and vinculums interchangeably. Before the eighteenth century parentheses hardly ever occur in the *Philosophical Transactions* of London, in the publications of the Paris Academy of Sciences, in the *Acta eruditorum* published in Leipzig. But with the beginning of the eighteenth century, parentheses do appear. In the *Acta eruditorum*, Carré[4] of Paris uses them in 1701, G. W. Leibniz[5] in 1702, a reviewer of Gabriele Manfredi[6] in 1708. Then comes in 1708 (§ 197) the statement of policy[7] in the *Acta eruditorum* in favor of the Leibnizian symbols, so that "in place of $\sqrt{aa+bb}$ we write $\sqrt{}(aa+bb)$ and for $\overline{aa+bb}\times c$ we write $aa+bb,c$ .... we shall designate $\overline{aa+bb}^m$ by $(aa+bb)^m$: whence $\overset{m}{V}\sqrt{aa+bb}$ will be $=(aa+bb)^{1:m}$ and $\overset{m}{V}\sqrt{\overline{aa+bb}^n}=(aa+bb)^{n:m}$. Indeed, we do not doubt that all mathematicians reading these *Acta* recognize the preeminence of Mr. Leibniz' symbolism and agree with us in regard to it."

From now on round parentheses appear frequently in the *Acta eruditorum*. In 1709 square brackets make their appearance.[8] In the *Philosophical Transactions* of London[9] one of the first appearances of parentheses was in an article by the Frenchman P. L. Maupertuis in 1731, while in the *Histoire de l'académie royale des sciences* in Paris,[10]

---

[1] J. J. Blassière, *Institution du calcul numerique et litteral.* (a la Haye, 1770), 2. Partie, p. 27.

[2] E. Waring, *Miscellanea analytica* (Cambridge, 1762).

[3] E. Waring, *Meditationes algebraicae* (Cambridge; 3d ed., 1782).

[4] L. Carré in *Acta eruditorum* (1701), p. 281.

[5] G. W. Leibniz, *ibid.* (1702), p. 219.

[6] Gabriel Manfredi, *ibid.* (1708), p. 268.

[7] *Ibid.* (1708), p. 271.

[8] *Ibid.* (1709), p. 327.

[9] P. L. Maupertuis in *Philosophical Transactions*, for 1731–32, Vol. XXXVII (London), p. 245.

[10] Johann II Bernoulli, *Histoire de l'académie royale des sciences*, année 1732 (Paris, 1735), p. 240 ff.

Johann (John) Bernoulli of Bale first used parentheses and brackets in the volume for the year 1732. In the volumes of the Petrograd Academy, J. Hermann[1] uses parentheses, in the first volume, for the year 1726; in the third volume, for the year 1728, L. Euler[2] and Daniel Bernoulli used round parentheses and brackets.

352. The constant use of parentheses in the stream of articles from the pen of Euler that appeared during the eighteenth century contributed vastly toward accustoming mathematicians to their use. Some of his articles present an odd appearance from the fact that the closing part of a round parenthesis is much larger than the opening part,[3] as in $(1-\frac{z}{\pi})(1-\frac{z}{\pi-s})$. Daniel Bernoulli[4] in 1753 uses round parentheses and brackets in the same expression while T. U. T. Aepinus[5] and later Euler use two types of round parentheses of this sort, $\mathsf{C}(\beta+\gamma)(M-1)+AM\mathsf{O}$. In the publications of the Paris Academy, parentheses are used by Johann Bernoulli (both round and square ones),[6] A. C. Clairaut,[7] P. L. Maupertuis,[8] F. Nicole,[9] Ch. de Montigny,[10] Le Marquis de Courtivron,[11] J. d'Alembert,[12] N. C. de Condorcet,[13] J. Lagrange.[14] These illustrations show that about the middle of the eighteenth century parentheses were making vigorous inroads upon the territory previously occupied in France by vinculums almost exclusively.

[1] J. Hermann, *Commentarii academiae scientiarum imperialis Petropolitanae*, Tomus I ad annum 1726 (Petropoli, 1728), p. 15.

[2] *Ibid.*, Tomus III (1728; Petropoli, 1732), p. 114, 221.

[3] L. Euler in *Miscellanea Berolinensia*, Vol. VII (Berlin, 1743), p. 93, 95, 97, 139, 177.

[4] D. Bernoulli in *Histoire de l'académie r. des sciences et belles lettres*, année 1753 (Berlin, 1755), p. 175.

[5] Aepinus in *ibid.*, année 1751 (Berlin, 1753), p. 375; année 1757 (Berlin, 1759), p. 308–21.

[6] *Histoire de l'académie r. des sciences*, année 1732 (Paris, 1735), p. 240, 257.

[7] *Ibid.*, année 1732, p. 385, 387.

[8] *Ibid.*, année 1732, p. 444.

[9] *Ibid.*, année 1737 (Paris, 1740), "Mémoires," p. 64; also année 1741 (Paris, 1744), p. 36.

[10] *Ibid.*, année 1741, p. 282.

[11] *Ibid.*, année 1744 (Paris, 1748), p. 406.

[12] *Ibid.*, année 1745 (Paris, 1749), p. 369, 380.

[13] *Ibid.*, année 1769 (Paris, 1772), p. 211.

[14] *Ibid.*, année 1774 (Paris, 1778), p. 103.

353. *Terms in an aggregate placed in a vertical column.*—The employment of a brace to indicate the sum of coefficients or factors placed in a column was in vogue with Vieta (§ 176), Descartes, and many other writers. Descartes in 1637 used a single brace,[1] as in

$$x^4 - 2ax^3 + 2aa \atop - cc \Big\} xx - 2a^3x + a^4 \approx 0 \ ,$$

or a vertical bar[2] as in

$$x^4 + \tfrac{1}{2}aa \Big| {-a^3 \atop zz \ -acc} \Big| {z +\tfrac{5}{16}a^4 \atop -\tfrac{1}{4}aacc} \approx 0 \ .$$

Wallis[3] in 1685 puts the equation $aaa + baa + cca = ddd$, where a is the unknown, also in the form

$$\frac{1}{aaa} \Big| {+ \ b \atop aa} \Big| {+cc \atop a} \Big| {= ddd \atop 1} \ .$$

Sometimes terms containing the same power of x were written in a column without indicating the common factor or the use of symbols of aggregation; thus, John Wallis[4] writes in 1685,

$$aaa + baa + bca = +bcd$$
$$+caa - bda$$
$$-daa - cda$$

Giovanni Poleni[5] writes in 1729,

$$y^6 + xxy^4 - 2ax^3yy + aax^4 = 0$$
$$- 2axy^4 + aaxxyy$$
$$- \ aay^4$$

The use of braces for the combination of terms arranged in columns has passed away, except perhaps in recording the most unusual algebraic expressions. The tendency has been, whenever possible, to discourage symbolism spreading out vertically as well as horizontally. Modern printing encourages progression line by line.

354. *Marking binomial coefficients.*—In the writing of the factors in binomial coefficients and in factorial expressions much diversity of practice prevailed during the eighteenth century, on the matter of

[1] Descartes, *Œuvres* (éd. Adam et Tannery), Vol. VI, p. 450.

[2] *Ibid.*

[3] John Wallis, *Treatise of Algebra* (London, 1685), p. 160.

[4] John Wallis, *op. cit.*, p. 153.

[5] Joannis Poleni, *Epistolarvm mathematicarvm fascicvlvs* (Padua, 1729) (no pagination).

the priority of operations indicated by $+$ and $-$, over the operations of multiplication marked by $\cdot$ and $\times$. In $n \cdot n - 1 \cdot n - 2$ or $n \times n - 1 \times n - 2$, or $n, n - 1, n - 2$, it was understood very generally that the subtractions are performed first, the multiplications later, a practice contrary to that ordinarily followed at that time. In other words, these expressions meant $n(n-1)(n-2)$. Other writers used parentheses or vinculums, which removed all inconsistency and ambiguity. Nothing was explicitly set forth by early writers which would attach different meanings to nn and $n \cdot n$ or $n \times n$. And yet, $n \cdot n - 1 \cdot n - 2$ was not the same as $nn - 1n - 2$. Consecutive dots or crosses tacitly conveyed the idea that what lies between two of them must be aggregated as if it were inclosed in a parenthesis. Some looseness in notation occurs even before general binomial coefficients were introduced. Isaac Barrow[1] wrote "$L - M \times : R + S$" for $(L-M)(R+S)$, where the colon designated aggregation, but it was not clear that $L-M$, as well as $R+S$, were to be aggregated. In a manuscript of Leibniz[2] one finds the number of combinations of n things, taken k at a time, given in the form

$$\frac{n \frown n - 1 \frown n - 2, \text{ etc., } n - k + 1}{1 \frown 2 \frown 3, \text{ etc., } \frown k} .$$

This diversity in notation continued from the seventeenth down into the nineteenth century. Thus, Major Edward Thornycroft (1704)[3] writes $m \times m - 1 \times m - 2 \times m - 3$, etc. A writer[4] in the *Acta eruditorum* gives the expression $n, n - 1$. Another writer[5] gives $\dfrac{(n, n-1, n-2)}{2, 3}$. Leibniz'[6] notation, as described in 1710 (§ 198), contains $e \cdot e - 1 \cdot e - 2$ for $e(e-1)(e-2)$. Johann Bernoulli[7] writes $n \cdot n - 1 \cdot n - 2$. This same notation is used by Jakob (James) Bernoulli[8] in a

[1] Isaac Barrow, *Lectiones mathematicae*, Lect. XXV, Probl. VII. See also Probl. VIII.

[2] D. Mahnke, *Bibliotheca mathematica* (3d ser.), Vol. XIII (1912–13), p. 35. See also *Leibnizens Mathematische Schriften*, Vol. VII (1863), p. 101.

[3] E. Thornycroft in *Philosophical Transactions*, Vol. XXIV (London, 1704–5), p. 1963.

[4] *Acta eruditorum* (Leipzig, 1708), p. 269.

[5] *Ibid.*, Suppl., Tome IV (1711), p. 160.

[6] *Miscellanea Berolinensia* (Berlin, 1710), p. 161.

[7] Johann Bernoulli in *Acta eruditorum* (1712), p. 276.

[8] Jakob Bernoulli, *Ars Conjectandi* (Basel, 1713), p. 99.

posthumous publication, by F. Nicole[1] who uses $x+n\cdot x+2n\cdot x+3n$, etc., by Stirling[2] in 1730, by Cramer[3] who writes in a letter to J. Stirling $a\cdot a+b\cdot a+2b$, by Nicolaus Bernoulli[4] in a letter to Stirling $r\cdot r+b\cdot r+2b\cdot$ ... by Daniel Bernoulli[5] $l-1\cdot l-2$, by Lambert[6] $4m-1\cdot$ $4m-2$, and by König[7] $n\cdot n-5\cdot n-6\cdot n-7$. Euler[8] in 1764 employs in the same article two notations: one, $n-5\cdot n-6\cdot n-7$; the other, $n(n-1)(n-2)$. Condorcet[9] has $n+2\times n+1$. Hindenburg[10] of Göttingen uses round parentheses and brackets, nevertheless he writes binomial factors thus, $m\cdot m-1\cdot m-2\ldots m-s+1$. Segner[11] and Ferroni[12] write $n\cdot n-1\cdot n-2$. Cossali[13] writes $4\times-2=-8$. As late as 1811 A. M. Legendre[14] has $n\cdot n-1\cdot n-2\ldots 1$.

On the other hand, F. Nicole,[15] who in 1717 avoided vinculums, writes in 1723, $x\cdot\overline{n+n}\cdot\overline{x+2n}$, etc. Stirling[16] in 1730 adopts $\overline{z-1}\cdot\overline{z-2}$. De Moivre[17] in 1730 likewise writes $\overline{m-p}\times\overline{m-q}\times\overline{m-s}$, etc. Similarly, Dodson,[18] $n\cdot\overline{n-1}\cdot\overline{n-2}$, and the Frenchman F. de Lalande,[19]

[1] Nicole in *Histoire de l'académie r. des sciences*, année 1717 (Paris, 1719), "Mémoires," p. 9.

[2] J. Stirling, *Methodus differentialis* (London, 1730), p. 9.

[3] Ch. Tweedie, *James Stirling* (Oxford, 1922), p. 121.   [4] *Op. cit.*, p. 144.

[5] Daniel I. Bernoulli, "Notationes de aequationibus," *Comment. Acad. Petrop.*, Tome V (1738), p. 72.

[6] J. H. Lambert, *Observationes in Acta Helvetica*, Vol. III.

[7] S. König, *Histoire de l'académie r. des sciences et des belles lettres*, année 1749 (Berlin, 1751), p. 189.

[8] L. Euler, *op. cit.*, année 1764 (Berlin, 1766), p. 195, 225.

[9] N. C. de Condorcet in *Histoire de l'académie r. des sciences*, année 1770 (Paris, 1773), p. 152.

[10] Carl Friedrich Hindenburg, *Infinitinomii dignatum leges . . . . ac formulae* (Göttingen, 1779), p. 30.

[11] J. A. de Segner, *Cursus mathematici*, pars II (Halle, 1768), p. 190.

[12] P. Ferroni, *Magnitudinum exponentialium . . . . theoria* (Florence, 1782), p. 29.

[13] Pietro Cossali, *Origine, trasporto in Italia ... dell'algebra*, Vol. I (Parma, 1797), p. 260.

[14] A. M. Legendre, *Exercices de calcul intégral*, Tome I (Paris, 1811), p. 277.

[15] *Histoire de l'académie r. des sciences*, année 1723 (Paris, 1753), "Mémoires," p. 21.

[16] James Stirling, *Methodus differentialis* (London, 1730), p. 6.

[17] Abraham de Moivre, *Miscellanea analytica de seriebus* (London, 1730), p. 4.

[18] James Dodson, *Mathematical Repository*, Vol. I (London, 1748), p. 238.

[19] F. de Lalande in *Histoire de l'académie r. des sciences*, année 1761 (Paris, 1763), p. 127.

$m \cdot (m+1) \cdot (m+2)$. In Lagrange[1] we encounter in 1772 the strictly modern form $(m+1)(m+2)(m+3), \ldots$, in Laplace[2] in 1778 the form $(i-1) \cdot (i-2) \ldots (i-r+1)$.

The omission of parentheses unnecessarily aggravates the interpretation of elementary algebraic expressions, such as are given by Kirkman,[3] viz., $-3 = 3 \times -1$ for $-3 = 3 \times (-1)$, $-m \times -n$ for $(-m)(-n)$.

355. *Special uses of parentheses.*—A use of round parentheses and brackets which is not strictly for the designation of aggregation is found in Cramer[4] and some of his followers. Cramer in 1750 writes two equations involving the variables x and y thus:

$$A \ldots x' - [-1]x^{n-1} + [1^2]x^{n-2} - [1^3]x^{n-3} + \&c. \ldots [1^n] = 0 \ ,$$
$$B \ldots (0)x^\circ + (1)x' + (2)x^2 + (3)x^3 + \&c. \ldots + (m)x^m = 0 \ ,$$

where $1, 1^2, 1^3, \ldots$, within the brackets of equation A do not mean powers of unity, but the coefficients of x, which are rational functions of y. The figures $0, 1, 2, 3$, in B are likewise coefficients of x and functions of y. In the further use of this notation, (02) is made to represent the product of (0) and (2); (30) the product of (3) and (0), etc. Cramer's notation is used in Italy by Cossali[5] in 1799.

Special uses of parentheses occur in more recent time. Thus W. F. Sheppard[6] in 1912 writes

$$(n,r) \text{ for } n(n-1) \ldots (n-r+1)/r!$$
$$[n,r] \text{ for } n(n+1) \ldots (n+r-1)/r!$$
$$(n,2s+1] \text{ for } (n-s)(n-s+1) \ldots (n+s)/(2s+1)!$$

356. *A star to mark the absence of terms.*—We find it convenient to discuss this topic at this time. René Descartes, in *La Géométrie* (1637), arranges the terms of an algebraic equation according to the descending order of the powers of the unknown quantity x, y, or z. If any power of the unknown below the highest in the equation is

---

[1] J. Lagrange in *ibid.*, année 1772, Part I (Paris, 1775), "Mémoires," p. 523.

[2] P. S. Laplace in *ibid.*, année 1778 (Paris, 1781), p. 237.

[3] T. P. Kirkman, *First Mnemonical Lessons in Geometry, Algebra and Trigonometry* (London, 1852), p. 8, 9.

Gabriel Cramer, *Analyse des Lignes courbes algébriques* (Geneva, 1750), p. 660.

[5] Pietro Cossali, *op. cit.*, Vol. II (Parma, 1799), p. 41.

[6] W. F. Sheppard in *Fifth International Mathematical Congress*, Vol. II, p. 355.

lacking, that fact is indicated by a *, placed where the term would have been. Thus, Descartes writes $x^6 - a^4bx = 0$ in this manner:[1]

$$x^6 \ * \ * \ * \ * \ - \ - \ a^4bx^* \infty 0 \ .$$

He does not explain why there was need of inserting these stars in the places of the missing terms. But such a need appears to have been felt by him and many other mathematicians of the seventeenth and eighteenth centuries. Not only were the stars retained in later editions of *La Géométrie*, but they were used by some but not all of the leading mathematicians, as well as by many compilers of textbooks. Kinckhuysen[2] writes "$x^5 \ * \ * \ * \ * \ -b \infty 0$." Prestet[3] in 1675 writes a^3**+b^3, and retains the * in 1689. The star is used by Baker,[4] Varignon,[5] John Bernoulli,[6] Alexander,[7] A. de Graaf,[8] E. Halley.[9] Fr. van Schooten used it not only in his various Latin editions of Descartes' *Geometry*, but also in 1646 in his *Conic Sections*,[10] where he writes $z^3 \infty *-pz+q$ for $z^3 = -pz+q$. In W. Whiston's[11] 1707 edition of I. Newton's *Universal Arithmetick* one reads aa^*-bb and the remark ". . . . locis vacuis substituitur nota * ." Raphson's English 1728 edition of the same work also uses the *. Jones[12] uses * in 1706, Reyneau[13] in 1708; Simpson[14] employs it in 1737 and Waring[15] in 1762. De Lagny[16]

---

[1] René Descartes, *La géométrie* (Leyden, 1637); *Œuvres de Descartes* (éd. Adam et Tannery), Vol. VI (1903), p. 483.

[2] Gerard Kinckhuysen, *Algebra ofte Stel-Konst* (Haarlem, 1661), p. 59.

[3] *Elemens des mathematiques* (Paris, 1675), Epître, by J. P.[restet], p. 23. *Nouveaux elemens des Mathematiques*, par Jean Prestet (Paris, 1689), Vol. II, p. 450.

[4] Thomas Baker, *Geometrical Key* (London, 1684), p. 13.

[5] *Journal des Sçavans*, année 1687, Vol. XV (Amsterdam, 1688), p. 459. The star appears in many other places of this *Journal*.

[6] John Bernoulli in *Acta eruditorum* (1688), p. 324. The symbol appears often in this journal.

[7] John Alexander, *Synopsis Algebraica* ... (Londini, 1693), p. 203.

[8] Abraham de Graaf, *De Geheele Mathesis* (Amsterdam, 1694), p. 259.

[9] E. Halley in *Philosophical Transactions*, Vol. XIX (London, 1695–97), p. 61.

[10] Francisci à Schooten, *De organica conicarum sectionum* (Leyden, 1646), p. 91.

[11] *Arithmetica universalis* (Cambridge, 1707), p. 29.

[12] W. Jones, *Synopsis palmariorum matheseos* (London, 1706), p. 178.

[13] Charles Reyneau, *Analyse demontrée*, Vol. I (Paris, 1708), p. 13, 89.

[14] Thomas Simpson, *New Treatise of Fluxions* (London, 1737), p. 208.

[15] Edward Waring, *Miscellanea Analytica* (1762), p. 37.

[16] *Mémoires de l'académie r. d. sciences. Depuis 1666 jusqu'à 1699*, Vol. XI (Paris, 1733), p. 241, 243, 250.

employs it in 1733, De Gua[1] in 1741, MacLaurin[2] in his *Algebra*, and Fenn[3] in his *Arithmetic*. But with the close of the eighteenth century the feeling that this notation was necessary for the quick understanding of elementary algebraic polynomials passed away. In more advanced fields the star is sometimes encountered in more recent authors. Thus, in the treatment of elliptic functions, Weierstrass[4] used it to mark the absence of a term in an infinite series, as do also Greenhill[5] and Fricke.[6]

[1] *Histoire de l'académie r. d. sciences*, année 1741(Paris, 1744), p. 476.

[2] Colin Maclaurin, *Treatise of Algebra* (2d éd.; London, 1756), p. 277.

[3] Joseph Fenn, *Universal Arithmetic* (Dublin, 1772), p. 33.

[4] H. A. Schwarz, *Formeln und Lehrsätze* . . . . *nach Vorlesungen des Weierstrass* (Göttingen, 1885), p. 10, 11.

[5] A. G. Greenhill, *Elliptic Functions* (1892), p. 202, 204.

[6] R. Fricke, *Encyklopädie d. Math. Wissenschaften*, Vol. II² (Leipzig, 1913), p. 269.

IV

SYMBOLS IN GEOMETRY
(ELEMENTARY PART)

A. ORDINARY ELEMENTARY GEOMETRY

357. The symbols sometimes used in geometry may be grouped roughly under three heads: (1) pictographs or pictures representing geometrical concepts, as △ representing a triangle; (2) ideographs designed especially for geometry, as ∼ for "similar"; (3) symbols of elementary algebra, like + and −.

Early use of pictographs.—The use of geometrical drawings goes back at least to the time of Ahmes, but the employment of pictographs in the place of words is first found in Heron's *Dioptra*. Heron (150 A.D.) wrote △ for triangle, $\underline{ov}$ for parallel and parallelogram, also $\underline{\rho'}$ for parallelogram, □' for rectangle, ☉ for circle.[1] Similarly, Pappus (fourth century A.D.) writes ○ and ⊙ for circle, ▽ and △ for triangle, ∟ for right angle, $\underline{ot}$ or = for parallel, □ for square.[2] But these were very exceptional uses not regularly adopted by the authors and occur in few manuscripts only. They were not generally known and are not encountered in other mathematical writers for about one thousand years. Paul Tannery calls attention to the use of the symbol □ in a medieval manuscript to represent, not a square foot, but a cubic foot; Tannery remarks that this is in accordance with the ancient practice of the Romans.[3] This use of the square is found in the *Triparty* of Chuquet (§ 132) and in the arithmetic of De la Roche.

358. Geometric figures were used in astrology to indicate roughly the relative positions of two heavenly bodies with respect to an observer. Thus ☌, ☍, □, △, ✳ designated,[4] respectively, conjunction,

---

[1] *Notices et extraits des manuscrits de la Bibliothèque impériale*, Vol. XIX, Part II (Paris, 1858), p. 173.

[2] *Pappi Alexandrini Collectionis quae supersunt* (ed. F. Hultsch), Vol. III, Tome I (Berlin, 1878), p. 126–31.

[3] Paul Tannery, *Mémoires scientifiques*, Vol. V (Toulouse and Paris, 1922), p. 73.

[4] Kepler says: "Quot sunt igitur aspectus? Vetus astrologia agnoscit tantum quinque: conjunctionem (☌), cum radii planetarum binorum in Terram descendentes in unam conjunguntur lineam; quod est veluti principium aspectuum omnium. 2) Oppositionem (☍), cum bini radii sunt ejusdem rectae partes, seu

opposition, at right angles, at 120°, at 60°. These signs are reproduced in Christian Wolff's *Mathematisches Lexicon* (Leipzig, 1716), page 188. The $*$, consisting of three bars crossing each other at 60°, was used by the Babylonians to indicate degrees. Many of their war carriages are pictured as possessing wheels with six spokes.[1]

359. In Plato of Tivoli's translation (middle of twelfth century) of the *Liber embadorum* by Savasorda who was a Hebrew scholar, at Barcelona, about 1100 A.D., one finds repeatedly the designations $\widehat{abc}$, $\widehat{ab}$ for arcs of circles.[2] In 1555 the Italian Fr. Maurolycus[3] employs $\triangle$, $\square$, also $*$ for hexagon and $\therefore$ for pentagon, while in 1575 he also used $\square$. About half a century later, in 1623, Metius in the Netherlands exhibits a fondness for pictographs and adopts not only $\triangle$, $\square$, but a circle with a horizontal diameter and small drawings representing a sphere, a cube, a tetrahedron, and an octohedron. The last four were never considered seriously for general adoption, for the obvious reason that they were too difficult to draw. In 1634, in France, Hérigone's *Cursus mathematicus* (§ 189) exhibited an eruption of symbols, both pictographs and arbitrary signs. Here is the sign $<$ for angle, the usual signs for triangle, square, rectangle, circle, also $\lrcorner$ for right angle, the Heronic $=$ for parallel, $\Diamond$ for parallelogram, $\frown$ for arc of circle, $\cap$ for segment, $-$ for straight line, $\perp$ for perpendicular, $5<$ for pentagon, $6<$ for hexagon.

In England, William Oughtred introduced a vast array of characters into mathematics (§§ 181–85); over forty of them were used in symbolizing the tenth book of Euclid's *Elements* (§§ 183, 184), first printed in the 1648 edition of his *Clavis mathematicae*. Of these symbols only three were pictographs, namely, $\square$ for rectangle, $\square$ for square, $\triangle$ for triangle (§ 184). In the first edition of the *Clavis* (1631), the $\square$ alone occurs. In the *Trigonometria* (1657), he employed $\angle$ for angle and $\measuredangle$ for angles (§ 182), $\|$ for parallel occurs in Oughtred's

---

cum duae quartae partes circuli a binis radiis interceptae sunt, id est unus semicirculus. 3) Tetragonum seu quadratum ($\square$), cum una quarta. 4) Trigonum seu trinum ($\triangle$), cum una tertia seu duae sextae. 5) Hexagonum seu sextilem ($*$), cum una sexta." See Kepler, *Opera omnia* (ed. Ch. Frisch), Vol. VI (1866), p. 490, quoted from "Epitomes astronomiae" (1618).

[1] C. Bezold, *Ninive und Babylon* (1903), p. 23, 54, 62, 124. See also J. Tropfke, *op. cit.*, Vol. I (2d ed., 1921), p. 38.

[2] See M. Curtze in *Bibliotheca mathematica* (3d ser.), Vol. I (1900), p. 327, 328.

[3] *Francisci Maurolyci Abbatis Messanensis Opuscula Mathematica* (Venice, 1575), p. 107, 134. See also Francisco Maurolyco in Boncompagni's *Bulletino*, Vol. IX, p. 67.

Opuscula mathematica hactenus inedita (1677), a posthumous work (§ 184).

Klügel[1] mentions a cube ▢ as a symbol attached to cubic measure, corresponding to the use of □ in square measure.

Euclid in his *Elements* uses lines as symbols for magnitudes, including numbers,[2] a symbolism which imposed great limitations upon arithmetic, for he does not add lines to squares, nor does he divide a line by another line.

360. *Signs for angles.*—We have already seen that Hérigone adopted < as the sign for angle in 1634. Unfortunately, in 1631, Harriot's *Artis analyticae praxis* utilized this very symbol for "less than." Harriot's > and < for "greater than" and "less than" were so well chosen, while the sign for "angle" could be easily modified so as to remove the ambiguity, that the change of the symbol for angle was eventually adopted. But < for angle persisted in its appearance, especially during the seventeenth and eighteenth centuries. We find it in W. Leybourn,[3] J. Kersey,[4] E. Hatton,[5] E. Stone,[6] J. Hodgson,[7] D'Alembert's *Encyclopédie*,[8] Hall and Steven's *Euclid*,[9] and Th. Reye.[10] John Caswell[11] used the sign ⋛ to express "equiangular."

A popular modified sign for angle was ∠, in which the lower stroke is horizontal and usually somewhat heavier. We have encountered this in Oughtred's *Trigonometria* (1657), Caswell,[12] Dulaurens,[13]

[1] G. S. Klügel, *Math. Wörterbuch*, 1. Theil (Leipzig, 1803), art. "Bruchzeichen."

[2] See, for instance, Euclid's *Elements*, Book V; see J. Gow, *History of Greek Mathematics* (1884), p. 106.

[3] William Leybourn, *Panorganon: or a Universal Instrument* (London, 1672), p. 75.

[4] John Kersey, *Algebra* (London, 1673), Book IV, p. 177.

[5] Edward Hatton, *An Intire System of Arithmetic* (London, 1721), p. 287.

[6] Edmund Stone, *New Mathematical Dictionary* (London, 1726; 2d ed., 1743), art. "Character."

[7] James Hodgson, *A System of Mathematics*, Vol. I (London, 1723), p. 10.

[8] *Encyclopédie ou Dictionnaire raissonné, etc.* (Diderot), Vol. VI (Lausanne et Berne, 1781), art. "Caractere."

[9] H. S. Hall and F. H. Stevens, *Euclid's Elements*, Parts I and II (London, 1889), p. 10.

[10] Theodor Reye, *Die Geometrie der Lage* (5th ed.; Leipzig, 1909), 1. Abteilung, p. 83.

[11] John Caswell, "Doctrine of Trigonometry," in Wallis' *Algebra* (1685).

[12] John Caswell, "Trigonometry," in *ibid.*

[13] Francisci Dulaurens, *Specimina mathematica duobus libris comprehensa* (Paris, 1667), "Symbols."

Jones,[1] Emerson,[2] Hutton,[3] Fuss,[4] Steenstra,[5] Klügel,[6] Playfair,[7] Kambly,[8] Wentworth,[9] Fiedler,[10] Casey,[11] Lieber and von Lühmann,[12] Byerly,[13] Müller,[14] Mehler,[15] C. Smith,[16] Beman and Smith,[17] Layng,[18] Hopkins,[19] Robbins,[20] the National Committee (in the U.S.A.).[21] The plural "angles" is designated by Caswell $\angle \angle$; by many others thus, $\measuredangle$. Caswell also writes $Z \angle \angle$ for the "sum of two angles," and $X \angle \angle$ for the "difference of two angles." From these quotations it is evident that the sign $\angle$ for angle enjoyed wide popularity in different countries. However, it had rivals.

361. Sometimes the same sign is inverted, thus $\diagup$ as in John Ward.[22] Sometimes it is placed so as to appear $\wedge$, as in the *Ladies*

---

[1] William Jones, *Synopsis palmariorum matheseos* (London, 1706), p. 221.

[2] [W. Emerson], *Elements of Geometry* (London, 1763), p. 4.

[3] Charles Hutton, *Mathematical and Philosophical Dictionary* (1695), art. "Characters."

[4] Nicolas Fuss, *Leçons de géométrie* (St. Petersbourg, 1798), p. 38.

[5] Pibo Steenstra, *Grondbeginsels der Meetkunst* (Leyden, 1779), p. 101.

[6] G. S. Klügel, *Math. Wörterbuch*, fortgesetzt von C. B. Mollweide und J. A. Grunert, 5. Theil (Leipzig, 1831), art. "Zeichen."

[7] John Playfair, *Elements of Geometry* (Philadelphia, 1855), p. 114.

[8] L. Kambly, *Die Elementar-Mathematik*, 2. Theil: *Planimetrie*, 43. Aufl. (Breslau, 1876).

[9] G. A. Wentworth, *Elements of Geometry* (Boston, 1881; Preface, 1878).

[10] W. Fiedler, *Darstellende Geometrie*, 1. Theil (Leipzig, 1883), p. 7.

[11] John Casey, *Sequel to the First Six Books of the Elements of Euclid* (Dublin, 1886).

[12] H. Lieber und F. von Lühmann, *Geometrische Konstruktions-Aufgaben*, 8. Aufl. (Berlin, 1887), p. 1.

[13] W. E. Byerly's edition of Chauvenet's *Geometry* (Philadelphia, 1905), p. 44.

[14] G. Müller, *Zeichnende Geometrie* (Esslingen, 1889), p. 12.

[15] F. G. Mehler, *Hauptsätze der Elementar Mathematik*, 8. Aufl. (Berlin, 1894), p. 4.

[16] Charles Smith, *Geometrical Conics* (London, 1894).

[17] W. W. Beman and D. E. Smith, *Plane and Solid Geometry* (Boston, 1896), p. 10.

[18] A. E. Layng, *Euclid's Elements of Geometry* (London, 1890), p. 4.

[19] G. Irving Hopkins, *Inductive Plane Geometry* (Boston, 1902), p. 12.

[20] E. R. Robbins, *Plane and Solid Geometry* (New York, [1906]), p. 16.

[21] *Report by the National Committee on Mathematical Requirements*, under the auspices of the Mathematical Association of America, Inc. (1923), p. 77.

[22] John Ward, *The Young Mathematicians' Guide* (9th ed.; London, 1752), p. 301, 369.

Diary[1] and in the writings of Reyer,[2] Bolyai,[3] and Ottoni.[4] This position is widely used in connection with one or three letters marking an angle. Thus, the angle ABC is marked by L. N. M. Carnot[5] $\widehat{ABC}$ in his *Géométrie de position* (1803); in the *Penny Cyclopedia* (1839), article "Sign," there is given $A\widehat{\ \ }B;$ Binet,[6] Möbius,[7] and Favaro[8] wrote $\widehat{ab}$ as the angle formed by two straight lines a and $b;$ Favaro wrote also $\widehat{PDC}$. The notation $a\widehat{\ \ }b$ is used by Stolz and Gmeiner,[9] so that $a\widehat{\ \ }b = -b\widehat{\ \ }a;$ Nixon[10] adopted $\widehat{A}$, also $\widehat{ABC};$ the designation $A\widehat{P}M$ is found in Enriques,[11] Borel,[12] and Durrell.[13]

362. Some authors, especially German, adopted the sign $\measuredangle$ for angle. It is used by Spitz,[14] Fiedler,[15] Halsted,[16] Milinowski,[17] Meyer,[18]

[1] *Leybourne's Ladies Diary*, Vol. IV, p. 273.

[2] *Samuel Reyhers . . . . Euclides, dessen VI. erste Bücher auf sonderbare Art mit algebraischen Zeichen, also eingerichtet, sind, dass man derselben Beweise auch in anderen Sprachen gebrauchen kann* (Kiel, 1698).

[3] Wolfgangi Bolyai de Bolya, *Tentamen* (2d ed.), Tome II (Budapestini, 1904; 1st ed., 1832), p. 361.

[4] C. B. Ottoni, *Elementos de Geometria e Trigonometria* (4th ed.; Rio de Janeiro, 1874), p. 67.

[5] See Ch. Babbage, "On the Influence of Signs in Mathematical Reasoning," *Transactions Cambridge Philos. Society*, Vol. II (1827), p. 372.

[6] J. P. Binet in *Journal de l'école polyt.*, Vol. IX, Cahier 16 (Paris, 1813), p. 303.

[7] A. F. Möbius, *Gesammelte Werke*, Vol. I (Leipzig, 1885), "Barycyentrischer Calcul, 1827," p. 618.

[8] A. Favaro, *Leçons de Statique graphique*, trad. par Paul Terrier, 1. Partie (Paris, 1879), p. 51, 75.

[9] O. Stolz und J. A. Gmeiner, *Theoretische Arithmetik* (Leipzig), Vol. II (1902), p. 329, 330.

[10] R. C. J. Nixon, *Euclid Revised* (3d ed.; Oxford, 1899), p. 9.

[11] Federigo Enriques, *Questioni riguardanti la geometria elementare* (Bologna, 1900), p. 67.

[12] Emile Borel, *Algèbre* (2d cycle; Paris, 1913), p. 367.

[13] Clement V. Durell, *Modern Geometry; The Straight Line and Circle* (London, 1920), p. 7, 21, etc.

[14] Carl Spitz, *Lehrbuch der ebenen Geometrie* (Leipzig und Heidelberg, 1862), p. 11.

[15] W. Fiedler, *Darstellende Geometrie*, 1. Theil (Leipzig, 1883), p. 7.

[16] George Bruce Halsted, *Mensuration* (Boston, 1881), p. 28; *Elementary Synthetic Geometry* (New York, 1892), p. vii.

[17] A. Milinowski, *Elem.-Synth. Geom. der Kegelschnitte* (Leipzig, 1883), p. 3.

[18] Friedrich Meyer, *Dritter Cursus der Planimetrie* (Halle, a/S, 1885), p. 81.

Fialkowski,[1] Henrici and Treutlein,[2] Brückner,[3] Doehlemann,[4] Schur,[5] Bernhard,[6] Auerbach and Walsh,[7] Mangoldt.[8]

If our quotations are representative, then this notation for angle finds its adherents in Germany and the United States. A slight modification of this sign is found in Byrne[9] ◁.

Among sporadic representations of angles are the following: The capital letter[10] L, the capital letter[11] V, or that letter inverted,[12] Λ, the inverted capital letter[13] $\forall$; the perpendicular lines[14] $\lrcorner$ or $\llcorner$, $\overline{pq}$ the angle made by the lines[15] p and q, (ab) the angle between the rays,[16] a and b, $\widehat{ab}$ the angle between the lines[17] a and b, or (u, v) the angle[18] formed by u and v.

363. Passing now to the designation of special angles we find $\measuredangle$ used to designate an oblique angle.[19] The use of a pictograph for the designation of right angles was more frequent in former years than now and occurred mainly in English texts. The two perpendicular lines $\llcorner$ to designate "right angle" are found in Reyher;[20] he lets

---

[1] N. Fialkowski, *Praktische Geometrie* (Wien, 1892), p. 15.

[2] J. Henrici und P. Treutlein, *Lehrbuch der Elementar-Geometrie*, 1. Teil, 3. Aufl. (Leipzig, 1897), p. 11.

[3] Max Brückner, *Vielecke und Vielfache-Theorie und Geschichte* (Leipzig, 1900), p. 125.

[4] Karl Doehlemann, *Projektive Geometrie*, 3. Aufl. (Leipzig, 1905), p. 133.

[5] F. Schur, *Grundlagen der Geometrie* (Leipzig und Berlin, 1909), p. 79.

[6] Max Bernhard, *Darstellende Geometrie* (Stuttgart, 1909), p. 267.

[7] Matilda Auerbach and Charles B. Walsh, *Plane Geometry* (Philadelphia, [1920]), p. vii.

[8] Hans V. Mangoldt, *Einführung in die höhere Mathematik*, Vol. I (Leipzig, 1923), p. 190.

[9] Oliver Byrne, *Elements of Euclid* (London, 1847), p. xxviii.

[10] John Wilson, *Trigonometry* (Edinburgh, 1714), "Characters Explained."

[11] A. Saverien, *Dictionnaire de math. et phys.* (Paris, 1753), "Caractere."

[12] W. Bolyai, *Tentamen* (2d ed.), Vol. I (1897), p. xi.

[13] Joseph Fenn, *Euclid* (Dublin, 1769), p. 12; J. D. Blassière, *Principes de géométrie élémentaire* (The Hague, 1782), p. 16.

[14] H. N. Robinson, *Geometry* (New York, 1860), p. 18; *ibid.* (15th ed., New York), p. 14.

[15] Charlotte Angas Scott, *Modern Analytical Geometry* (London, 1894), p. 253.

[16] Heinrich Schröter, *Theorie der Kegelschnitte* (2d ed; Leipzig, 1876), p. 5.

[17] J. L. S. Hatton, *Principles of Projective Geometry* (Cambridge, 1913), p. 9.

[18] G. Peano, *Formulaire mathématique* (Turin, 1903), p. 266.

[19] W. N. Bush and John B. Clarke, *Elements of Geometry* (New York, [1905]).

[20] *Samuel Reyhers Euclides* (Kiel, 1698).

$\wedge\!\!\wedge$I L stand for "angle A is a right angle," a symbolism which could be employed in any language. The vertical bar stands for equality (§ 263). The same idea is involved in the signs $a \uparrow b$, i.e., "angle a is equal to angle b." The sign $\llcorner$ for right angle is found in Jones,[1] Hatton,[2] Saverien,[3] Fenn,[4] and Steenstra.[5] Kersey[6] uses the sign $\perp$, Byrne[7] $\square$. Mach[8] marks right angles $\perp$. The Frenchman Hérigone[9] used the sign $\lrcorner$, the Englishman Dupius[10] $\urcorner$ for right angle.

James Mills Peirce,[11] in an article on the notation of angles, uses *"Greek letters* to denote the *directions* of lines, without reference to their length. Thus if ρ denotes the axis in a system of polar co-ordinates, the polar angle will be $\breve{\rho}$." Accordingly, $\dfrac{\beta}{\alpha} = -\dfrac{\alpha}{\beta}.$

More common among more recent American and some English writers is the designation "rt. $\angle$" for right angle. It is found in G. A. Wentworth,[12] Byerly's *Chauvenet*,[13] Hall and Stevens,[14] Beman and Smith,[15] Hopkins,[16] Robbins,[17] and others.

Some writers use instead of pictographs of angles abbreviations of the word. Thus Legendre[18] sometimes writes "Angl. ACB";

[1] William Jones, *Synopsis palmariorum matheseos* (London, 1706), p. 221.

[2] Edward Hatton, *An Intire System of Arithmetik* (London, 1721), p. 287.

[3] A. Saverien, *Dictionnaire*, "Caractere."

[4] Joseph Fenn, *Euclid* (Dublin, 1769), p. 12.

[5] Pibo Steenstra, *Grondbeginsels der Meetkunst* (Leyden, 1779), p. 101.

[6] John Kersey, *Algebra* (London, 1673), Book IV, p. 177.

[7] Oliver Byrne, *The Elements of Euclid* (London, 1847), p. xxviii.

[8] E. Mach, *Space and Geometry* (trans. T. J. McCormack, 1906), p. 122.

[9] P. Herigone, *Cursus mathematicus* (Paris, 1634), Vol. I, "Explicatio notarum."

[10] N. F. Dupius, *Elementary Synthetic Geometry* (London, 1889), p. 19.

[11] J. D. Runkle's *Mathematical Monthly*, Vol. I, No. 5 (February, 1859), p. 168, 169.

[12] G. A. Wentworth, *Elements of Plane and Solid Geometry* (3d ed.; Boston, 1882), p. 14.

[13] W. E. Byerly's edition of *Chauvenet's Geometry* (1887).

[14] H. S. Hall and F. H. Stevens, *Euclid's Elements*, Parts I and II (London, 1889), p. 10.

[15] W. W. Beman and D. E. Smith, *Plane and Solid Geometry* (Boston, 1896), p. 10.

[16] G. I. Hopkins, *Inductive Plane Geometry* (Boston, 1902), p. 12.

[17] E. R. Robbins, *Plane and Solid Geometry* (New York, [1906]), p. 16.

[18] A. M. Legendre, *Éléments de Géométrie* (Paris, 1794), p. 42.

A. von Frank,[1] *"Wkl,"* the abbreviation for *Winkel,* as in *"Wkl DOQ."*

The advent of non-Euclidean geometry brought Lobachevski's notation Π (ρ) for angle of parallelism.[2]

The sign $\overset{\vee}{=}$ to signify equality of the angles, and $\underset{=}{\perp}$ to signify the equality of the sides of a figure, are mentioned in the article "Caractere" by D'Alembert in Diderot' *Encyclopédie* of 1754 and of 1781[3] and in the Italian translation of the mathematical part (1800); also in Rees's *Cyclopaedia* (London, 1819), article "Characters," and in E. Stone's *New Mathematical Dictionary* (London, 1726), article "Characters," but Stone defines $\overset{\vee}{=}$ as signifying "equiangular or similar." The symbol is given also by a Spanish writer as signifying *angulos iguales.*[4] The sign $=°$ to signify "equal number of degrees" is found in Palmer and Taylor's *Geometry,*[5] but failed to be recommended as a desirable symbol in elementary geometry by the National Committee on Mathematical Requirements (1923), in their *Report,* page 79.

Halsted suggested the sign $\measuredangle$ for spherical angle and also the letter Ω to represent a "steregon," the unit of solid angle.[6]

364. *Signs for "perpendicular."*—The ordinary sign to indicate that one line is perpendicular to another, $\perp$, is given by Hérigone[7] in 1634 and 1644. Another Frenchman, Dulaurens,[8] used it in 1667. In 1673 Kersey[9] in England employed it. The inverted capital letter L was used for this purpose by Caswell,[10] Jones,[11] Wilson,[12] Saverien,[13]

[1] A. von Frank in *Archiv der Mathematik und Physik* von J. A. Grunert (2d ser.), Vol. XI (Leipzig, 1892), p. 198.

[2] George Bruce Halsted, *N. Lobatschewsky, Theory of Parallels* (Austin, 1891), p. 13.

[3] *Encyclopédie au Dictionnaire raisonné des sciences,* ... by Diderot, Vol. VI (Lausanne et Berne, 1781), art. "Caractere."

[4] Antonio Serra y Oliveres, *Manuel de la Tipografia Española* (Madrid, 1852), p. 70.

[5] C. I. Palmer and D. P. Taylor, *Plane Geometry* (1915), p. 16.

[6] G. B. Halsted, *Mensuration* (Boston, 1881), p. 28.

[7] Pierre Herigone, *Cursus mathematicus,* Vol. I (Paris, 1634), "Explicatio notarum."

[8] F. Dulaurens, *Specimina mathematica duobus libris comprehensa* (Paris, 1667), "Symbols."

[9] John Kersey, *Algebra* (London, 1673), Book IV, p. 177.

[10] J. Caswell's *Trigonometry* in J. Wallis' *Algebra* (1685).

[11] W. Jones, *op. cit.,* p. 253.

[12] J. Wilson, *Trigonometry* (Edinburgh, 1714), "Characters Explained."

[13] A. Saverien, *Dictionnaire,* "Caractere."

and Mauduit.[1] Emerson[2] has the vertical bar extremely short, ⊥. In the nineteenth century the symbol was adopted by all writers using pictographs in geometry. Sometimes ⌐ was used for "perpendiculars." Thomas Baker[3] adopted the symbol ⌐ for perpendicular.

365. *Signs for triangle, square, rectangle, parallelogram.*—The signs △, □, □ or ☐, ▱ are among the most widely used pictographs. We have already referred to their occurrence down to the time of Hérigone and Oughtred (§ 184). The ▱ for parallelogram is of rare occurrence in geometries preceding the last quarter of the nineteenth century, while the △, □, and □ occur in van Schooten,[4] Dulaurens,[5] Kersey,[6] Jones,[7] and Saverien.[8] Some authors use only two of the three. A rather curious occurrence is the Hebrew letter "mem," ☐, to represent a rectangle; it is found in van Schooten,[9] Jones,[10] John Alexander,[11] John I Bernoulli,[12] Ronayne,[13] Klügel's *Wörterbuch*,[14] and De Graaf.[15] Newton,[16] in an early manuscript tract on fluxions (October, 1666), indicates the area or fluent of a curve by prefixing a rectangle to the ordinate (§ 622), thus □ $\frac{axx-x^3}{ab+xx}$, where x is the abscissa, and the fraction is the ordinate.

After about 1880 American and English school geometries came to employ less frequently the sign □ for rectangle and to introduce more often the sign ▱ for parallelogram. Among such authors are

[1] A. R. Mauduit, *Inleiding tot de Kleegel-Sneeden* (The Hague, 1763), "Symbols."

[2] [W. Emerson], *Elements of Geometry* (London, 1763).

[3] Thomas Baker, *Geometrical Key* (London, 1684), list of symbols.

[4] Fr. van Schooten, *Exercitationvm mathematicorvm liber primus* (Leyden, 1657).

[5] F. Dulaurens, *loc. cit.*, "Symbols."

[6] J. Kersey, *Algebra* (1673).

[7] W. Jones, *op. cit.*, p. 225, 238.

[8] Saverien, *loc. cit.*

[9] Franciscus van Schooten, *op. cit.* (Leyden, 1657), p. 67.

[10] W. Jones, *op. cit.*, p. 253.

[11] *Synopsis Algebraica, opus posthumum Iohannis Alexandri* (London, 1693), p. 67.

[12] John Bernoulli in *Acta eruditorum* (1689), p. 586; *ibid.* (1692), p. 31.

[13] Philip Ronayne, *Treatise of Algebra* (London, 1727), p. 3.

[14] J. G. Klügel, *Math. Wörterbuch*, 5. Theil (Leipzig, 1831), "Zeichen."

[15] Abraham de Graaf, *Algebra of Stelkonst* (Amsterdam, 1672), p. 81.

[16] S. P. Rigaud, *Historical Essay on . . . . Newton's Principia* (Oxford, 1838), Appendix, p. 23.

Halsted,[1] Wentworth,[2] Byerly,[3] in his edition of Chauvenet, Beman and Smith,[4] Layng,[5] Nixon,[6] Hopkins,[7] Robbins,[8] and Lyman.[9] Only seldom do both $\square$ and $\square$ appear in the same text. Halsted[10] denotes a parallelogram by $\|g'm$.

Special symbols for right and oblique spherical triangles, as used by Jean Bernoulli in trigonometry, are given in Volume II, § 524.

366. *The square as an operator.*—The use of the sign $\square$ to mark the operation of squaring has a long history, but never became popular. Thus N. Tartaglia[11] in 1560 denotes the square on a line tc in the expression "il $\square$ de. tc." Cataldi[12] uses a black square to indicate the square of a number. Thus, he speaks of $8\frac{3}{4}\frac{1}{4}$, "il suo ■ è 75 $\frac{1489}{1936}$." Stampioen[13] in 1640 likewise marks the square on BC by the "$\square\ BC$." Caramvel[14] writes "$\square 25$. est Quadratum Numeri 25. hoc est, 625."

A. de Graaf[15] in 1672 indicates the square of a binomial thus: $\sqrt{a}\pm\sqrt{b}$, "zijn $\square$ is $a+b\pm2\sqrt{ab}$." Johann I Bernoulli[16] wrote $3\ \square\ \sqrt[3]{ax+xx}$ for $3\sqrt{(ax+x^2)^2}$. Jakob Bernoulli in 1690[17] designated

[1] G. B. Halsted, *Elem. Treatise on Mensuration* (Boston, 1881), p. 28.

[2] G. A. Wentworth, *Elements of Plane and Solid Geometry* (3d ed.; Boston, 1882), p. 14 (1st ed., 1878).

[3] W. E. Byerly's edition of *Chauvenet's Geometry* (1887), p. 44.

[4] W. W. Beman and D. E. Smith, *Plane and Solid Geometry* (Boston, 1896), p. 10.

[5] A. E. Layng, *Euclid's Elements of Geometry* (London, 1890), p. 4.

[6] R. C. J. Nixon, *Euclid Revised* (3d ed., Oxford, 1899), p. 6.

[7] G. J. Hopkins, *Inductive Plane Geometry* (Boston, 1902), p. 12.

[8] E. R. Robbins, *Plane and Solid Geometry* (New York, [1906]), p. 16.

[9] E. A. Lyman, *Plane and Solid Geometry* (New York, 1908), p. 18.

[10] G. B. Halsted, *Rational Geometry* (New York, 1904), p. viii.

[11] N. Tartaglia, *La Quinta parte del general trattato de nvmeri et misvre* (Venice, 1560), fols. 82*AB* and 83*A*.

[12] *Trattato del Modo Brevissimo de trouare la Radice quadra delli numeri*, ... Di Pietro Antonio Cataldi (Bologna, 1613), p. 111.

[13] J. Stampioen, *Wis-Konstich ende Reden-maetich Bewys* ('S Graven-Hage, 1640), p. 42.

[14] *Joannis Caramvelis mathesis biceps. vetus, et nova* (1670), p. 131.

[15] Abraham de Graaf, *Algebra of Stelkonst* (Amsterdam, 1672), p. 32.

[16] Johannis I Bernoulli, *Lectiones de calculo differentialium* . . . . von Paul Schafheitlin, Separatabdruck aus den *Verhandlungen der Naturforschenden Gesellschaft in Basel*, Vol. XXXIV (1922).

[17] Jakob Bernoulli in *Acta eruditorum* (1690), p. 223.

the square of $\frac{5}{6}$ by $\boxed{2}\frac{5}{6}$, but in his collected writings[1] it is given in the modern form $(\frac{5}{6})^2$. Sometimes a rectangle, or the Hebrew letter "mem," is used to signify the product of two polynomials.[2]

367. *Sign for circle.*—Although a small image of a circle to take the place of the word was used in Greek time by Heron and Pappus, the introduction of the symbol was slow. Hérigone used $\odot$, but Oughtred did not. One finds $\odot$ in John Kersey,[3] John Caswell,[4] John Ward,[5] P. Steenstra,[6] J. D. Blassière,[7] W. Bolyai,[8] and in the writers of the last half-century who introduced the sign $\square$ for parallelogram. Occasionally the central dot is omitted and the symbol $\bigcirc$ is used, as in the writings of Reyher[9] and Saverien. Others, Fenn for instance, give both $\bigcirc$ and $\odot$, the first to signify circumference, the second circle (area). Caswell[10] indicates the perimeter by $\dot{\bigcirc}$. Metius[11] in 1623 draws the circle and a horizontal diameter to signify *circulus.*

368. *Signs for parallel lines.*—Signs for parallel lines were used by Heron and Pappus (§ 701); Hérigone used horizontal lines $=$ (§ 189) as did also Dulaurens[12] and Reyher,[13] but when Recorde's sign of equality won its way upon the Continent, vertical lines came to be used for parallelism. We find $\parallel$ for "parallel" in Kersey,[14] Caswell, Jones,[15] Wilson,[16] Emerson,[17] Kambly,[18] and the writers of the last

---

[1] *Opera Jakob Bernoullis,* Vol. I, p. 430, 431; see G. Eneström, *Bibliotheca mathematica* (3d ser.), Vol. IX (1908–9), p. 207.

[2] See P. Herigone, *Cursus mathematici* (Paris, 1644), Vol. VI, p. 49.

[3] John Kersey, *Algebra* (London, 1673), Book IV, p. 177.

[4] John Caswell in Wallis' *Treatise of Algebra,* "Additions and Emendations," p. 166. For "circumference" Caswell used the small letter *c.*

[5] J. Ward, *The Young Mathematician's Guide* (9th ed.; London, 1752), p. 301, 369.

[6] P. Steenstra, *Grondbeginsels der Meetkunst* (Leyden, 1779), p. 281.

[7] J. D. Blassière, *Principes de géométrie élémentaire* (The Hague, 1723), p. 16.

[8] W. Bolyai, *Tentamen* (2d ed.), Vol. II (1904), p. 361 (1st ed., 1832).

[9] *Samuel Reyhers, Euclides* (Kiel, 1698), list of symbols.

[10] John Caswell in Wallis' *Treatise of Algebra* (1685), "Additions and Emendations," p. 166.

[11] Adriano Metio, *Praxis nova geometrica* (1623), p. 44.

[12] Fr. Dulaurens, *Specimina mathematica* (Paris, 1667), "Symbols."

[13] S. Reyher, *op. cit.* (1698), list of symbols.

[14] John Kersey, *Algebra* (London, 1673), Book IV, p. 177.

[15] W. Jones, *Synopsis palmariorum matheseos* (London, 1706).

[16] John Wilson, *Trigonometry* (Edinburgh, 1714), characters explained.

[17] [W. Emerson], *Elements of Geometry* (London, 1763), p. 4.

[18] L. Kambly, *Die Elementar-Mathematik,* 2. Theil, *Planimetrie,* 43. Aufl. (Breslau, 1876), p. 8.

fifty years who have been already quoted in connection with other pictographs. Before about 1875 it does not occur as often as do $\triangle$, $\square$, $\sqsubset$. Hall and Stevens[1] use "par[1] or $\parallel$" for parallel. Kambly[2] mentions also the symbols $\#$ and $\neq$ for parallel.

A few other symbols are found to designate parallel. Thus John Bolyai in his *Science Absolute of Space* used $\parallel\mid$. Karsten[3] used $\#$; he says: "Man pflege wohl das Zeichen $\#$ statt des Worts: *Parallel* der Kürze wegen zu gebrauchen." This use of that symbol occurs also in N. Fuss.[4] Thomas Baker[5] employed the sign $\backsim$.

With Kambly $\#$ signifies rectangle. Häseler[6] employs $\#$ as "the sign of parallelism of two lines or surfaces."

369. *Sign for equal and parallel.*—$\#$is employed to indicate that two lines are equal and parallel in Klügel's *Wörterbuch;*[7] it is used by H. G. Grassmann,[8] Lorey,[9] Fiedler,[10] Henrici and Treutlein.[11]

370. *Signs for arcs of circles.*—As early a writer as Plato of Tivoli (§ 359) used $\overset{\frown}{ab}$ to mark the arc ab of a circle. Ever since that time it has occurred in geometric books, without being generally adopted. It is found in Hérigone,[12] in Reyher,[13] in Kambly,[14] in Lieber and Lühmann.[15] W. R. Hamilton[16] designated by $\frown LF$ the arc "from F to L." These

[1] H. S. Hall and F. H. Stevens, *Euclid's Elements*, Parts I and II (London, 1889), p. 10.

[2] L. Kambly, *op. cit.*, 2. Theil, *Planimetrie*, 43. Aufl. (Breslau, 1876), p. 8.

[3] W. J. G. Karsten, *Lehrbegrif der gesamten Mathematik*, 1. Theil (Greifswald, 1767), p. 254.

[4] Nicolas Fuss, *Leçons de géométrie* (St. Petersbourg, 1798), p. 13.

[5] Thomas Baker, *Geometrical Key* (London, 1684), list of symbols.

[6] J. F. Häseler, *Anfangsgründe der Arith., Alg., Geom. und Trig.* (Lemgo), *Elementar-Geometrie* (1777), p. 72.

[7] G. S. Klügel, *Mathematisches Wörterbuch*, fortgesetzt von C. B. Mollweide, J. A. Grunert, 5. Theil (Leipzig, 1831), "Zeichen."

[8] H. G. Grassmann, *Ausdehnungslehre von 1844* (Leipzig, 1878), p. 37; *Werke* by F. Engel (Leipzig, 1894), p. 67.

[9] Adolf Lorey, *Lehrbuch der ebenen Geometrie* (Gera und Leipzig, 1868), p. 52.

[10] Wilhelm Fiedler, *Darstellende Geometrie*, 1. Theil (Leipzig, 1883), p. 11.

[11] J. Henrici und P. Treutlein, *Lehrbuch der Elementar-Geometrie*, 1. Teil, 3. Aufl. (Leipzig, 1897), p. 37.

[12] P. Herigone, *op. cit.* (Paris, 1644), Vol. I, "Explicatio notarum."

[13] *Samuel Reyhers, Euclides* (Kiel, 1698), Vorrede.

[14] L. Kambly, *op. cit.* (1876).

[15] H. Lieber und F. von Lühmann, *Geometrische Konstructions-Aufgaben*, 8. Aufl. (Berlin, 1887), p. 1.

[16] W. R. Hamilton in *Cambridge & Dublin Math'l. Journal*, Vol. I (1846), p. 262.

references indicate the use of ⌒ to designate arc in different countries. In more recent years it has enjoyed some popularity in the United States, as is shown by its use by the following authors: Halsted,[1] Wells,[2] Nichols,[3] Hart and Feldman,[4] and Smith.[5] The National Committee on Mathematical Requirements, in its *Report* (1923), page 78, is of the opinion that "the value of the symbol ⌒ in place of the short word *arc* is doubtful."

In 1755 John Landen[6] used the sign $(P\hat{Q}R)$ for the circular arc which measures the angle $P\hat{Q}R$, the radius being unity.

371. *Other pictographs.*—We have already referred to Hérigone's use (§ 189) of 5< and 6< to represent pentagons and hexagons. Reyher actually draws a pentagon. Occasionally one finds a half-circle and a diameter ⌓ to designate a segment, and a half-circle without marking its center or drawing its diameter to designate an arc. Reyher in his *Euclid* draws ⟁ for trapezoid.

Pictographs of solids are very rare. We have mentioned (§ 359) those of Metius. Saverein[7] draws ▓, ▲, ◪, ▓ to stand, respectively, for cube, pyramid, parallelopiped, rectangular parallelopiped, but these signs hardly belong to the category of pictographs. Dulaurens[8] wrote ③ for cube and ④ for *aequi quadrimensum.* Joseph Fenn[9] draws a small figure of a parallelopiped to represent that solid, as Metius had done. Halsted[10] denotes symmetry by ·⊦·.

Some authors of elementary geometries have used algebraic symbols and no pictographs (for instance, Isaac Barrow, Karsten, Tacquet, Leslie, Legendre, Playfair, Chauvenet, B. Peirce, Todhunter), but no author since the invention of symbolic algebra uses pictographs without at the same time availing himself of algebraic characters.

372. *Signs for similarity and congruence.*—The designation of "similar," "congruent," "equivalent," has brought great diversity of notation, and uniformity is not yet in sight.

Symbols for similarity and congruence were invented by Leibniz.

[1] G. B. Halsted, *Mensuration* (Boston, 1881).

[2] Webster Wells, *Elementary Geometry* (Boston, 1886), p. 4.

[3] E. H. Nichols, *Elements of Constructional Geometry* (New York, 1896).

[4] C. A. Hart and D. D. Feldman, *Plane Geometry* (New York, [1911]), p. viii.

[5] Eugene R. Smith, *Plane Geometry* (New York, 1909), p. 14.

[6] John Landen, *Mathematical Lucubrations* (London, 1755), Sec. III, p. 93.

[7] A. Saverein, *Dictionnaire*, "Caractere."

[8] F. Dulaurens, *op. cit.* (Paris, 1667), "Symbols."

[9] Joseph Fenn, *Euclid's Elements of Geometry* (Dublin, [ca. 1769]), p. 319.

[10] G. B. Halsted, *Rational Geometry* (New York, 1904), p. viii.

In Volume II, § 545, are cited symbols for "coincident" and "congruent" which occur in manuscripts of 1679 and were later abandoned by Leibniz. In the manuscript of his *Characteristica Geometrica* which was not published by him, he says: "similitudinem ita notabimus: $a \backsim b$."[1] The sign is the letter S (first letter in *similis*) placed horizontally. Having no facsimile of the manuscript, we are dependent upon the editor of Leibniz' manuscripts for the information that the sign in question was $\sim$ and not $\backsim$. As the editor, C. I. Gerhardt, interchanged the two forms (as pointed out below) on another occasion, we do not feel certain that the reproduction is accurate in the present case. According to Gerhardt, Leibniz wrote in another manuscript $\backsimeq$ for congruent. Leibniz' own words are reported as follows: "ABC $\backsimeq CDA$. Nam $\sim$ mihi est signum similitudinis, et = aequalitatis, unde congruentiae signum compono, quia quae simul et similia et aequalia sunt, ea congrua sunt."[2] In a third manuscript Leibniz wrote $|\backsimeq|$ for coincidence.

An anonymous article printed in the *Miscellanea Berolinensia* (Berlin, 1710), under the heading of "Monitum de characteribus algebraicis," page 159, attributed to Leibniz and reprinted in his collected mathematical works, describes the symbols of Leibniz; $\backsim$ for similar and $\backsimeq$ for congruent (§ 198). Note the change in form; in the manuscript of 1679 Leibniz is reported to have adopted the form $\sim$, in the printed article of 1710 the form given is $\backsim$. Both forms have persisted in mathematical writings down to the present day. As regards the editor Gerhardt, the disconcerting fact is that in 1863 he reproduces the $\backsim$ of 1710 in the form[3] $\sim$.

The Leibnizian symbol $\sim$ was early adopted by Christian von Wolf; in 1716 he gave $\sim$ for *Aehnlichkeit*,[4] and in 1717 he wrote "$= et$ $\sim$" for "equal and similar."[5] These publications of Wolf are the earliest in which the sign $\sim$ appears in print. In the eighteenth and early part of the nineteenth century, the Leibnizian symbols for "similar" and "congruent" were seldom used in Europe and not at all in England and America. In England $\sim$ or $\backsim$ usually expressed "difference," as defined by Oughtred. In the eighteenth century the signs for congruence occur much less frequently even than the signs

---

[1] Printed in *Leibnizens Math. Schriften* (ed. C. I. Gerhardt), Vol. V, p. 153.

[2] *Op. cit.*, p. 172.

[3] *Leibnizens Math. Schriften*, Vol. VII (1863), p. 222.

[4] Chr. Wolffen, *Math. Lexicon* (Leipzig, 1716), "Signa."

[5] Chr. V. Wolff, *Elementa Matheseos universalis* (Halle, 1717), Vol. I, § 236; see Tropfke, *op. cit.*, Vol. IV (2d ed., 1923), p. 20.

for similar. We have seen that Leibniz' signs for congruence did not use both lines occurring in the sign of equality =. Wolf was the first to use explicitly $\sim$ and = for congruence, but he did not combine the two into one symbolism. That combination appears in texts of the latter part of the eighteenth century. While the $\cong$ was more involved, since it contained one more line than the Leibnizian $\sim$, it had the advantage of conveying more specifically the idea of congruence as the superposition of the ideas expressed by $\sim$ and =. The sign $\sim$ for "similar" occurs in Camus' geometry,[1] $\backsim$ for "similar" in A. R. Mauduit's conic sections[2] and in Karsten,[3] $\sim$ in Blassière's geometry,[4] $\cong$ for congruence in Häseler's[5] and Reinhold's geometries,[6] $\backsim$ for similar in Diderot's *Encyclopédie*,[7] and in Lorenz' geometry.[8] In Klügel's *Wörterbuch*[9] one reads, "$\backsim$ with English and French authors means difference"; "with German authors $\backsim$ is the sign of similarity"; "Leibniz and Wolf have first used it." The signs $\sim$ and $\cong$ are used by Mollweide;[10] $\sim$ by Steiner[11] and Koppe;[12] $\backsim$ is used by Prestel,[13] $\cong$ by Spitz;[14] $\sim$ and $\cong$ are found in Lorey's geometry,[15] Kambly's

[1] C. E. L. Camus, *Élémens de géométrie* (nouvelle éd.; Paris, 1755).

[2] A. R. Mauduit, *op. cit.* (The Hague, 1763), "Symbols."

[3] W. J. G. Karsten, *Lehrbegrif der gesamten Mathematik*, 1. Theil (1767), p. 348.

[4] J. D. Blassière, *Principes de géometrié élémentaire* (The Hague, 1787), p. 16.

[5] J. F. Häseler, *op. cit.* (Lemgo, 1777), p. 37.

[6] C. L. Reinhold, *Arithmetica Forensis*, 1. Theil (Ossnabrück, 1785), p. 361.

[7] Diderot *Encyclopédie ou Dictionnaire raisoné des sciences* (1781; 1st ed., 1754), art. "Caractere" by D'Alembert. See also the Italian translation of the mathematical part of Diderot's *Encyclopédie*, the *Dizionario enciclopedico delle matematiche* (Padova, 1800), "Carattere."

[8] J. F. Lorenz, *Grundriss der Arithmetik und Geometrie* (Helmstädt, 1798), p. 9.

[9] G. S. Klügel, *Mathematisches Wörterbuch*, fortgesetzt von C. B. Mollweide, J. A. Grunert, 5. Theil (Leipzig, 1831), art. "Zeichen."

[10] Carl B. Mollweide, *Euklid's Elemente* (Halle, 1824).

[11] Jacob Steiner, *Geometrische Constructionen* (1833); *Ostwald's Klassiker*, No. 60, p. 6.

[12] Karl Koppe, *Planimetrie* (Essen, 1852), p. 27.

[13] M. A. F. Prestel, *Tabelarischer Grundriss der Experimental-physik* (Emden, 1856), No. 7.

[14] Carl Spitz, *Lehrbuch der ebenen Geometrie* (Leipzig und Heidelberg, 1862), p. 41.

[15] Adolf Lorey, *Lehrbuch der ebenen Geometrie* (Gera und Leipzig, 1868), p. 118.

Planimetrie,[1] and texts by Frischauf[2] and Max Simon.[3] Lorey's book contains also the sign ≦ a few times. Peano[4] uses ⌣ for "similar" also in an arithmetical sense for classes. Perhaps the earliest use of ∼ and ≅ for "similar" and "congruent" in the United States are by G. A. Hill[5] and Halsted.[6] The sign ⌣ for "similar" is adopted by Henrici and Treutlein,[7] ≅ by Fiedler,[8] ∼ by Fialkowski,[9] ⌣ by Beman and Smith.[10] In the twentieth century the signs entered geometries in the United States with a rush: ≅ for "congruent" were used by Busch and Clarke;[11] ≦ by Meyers,[12] ≅ by Slaught and Lennes,[13] ∼ by Hart and Feldman;[14] ≅ by Shutts,[15] E. R. Smith,[16] Wells and Hart,[17] Long and Brenke;[18] ≦ by Auerbach and Walsh.[19]

That symbols often experience difficulty in crossing geographic or national boundaries is strikingly illustrated in the signs ∼ and ≅. The signs never acquired a foothold in Great Britain. To be sure, the symbol ⌣ was adopted at one time by a member of the University

---

[1] L. Kambly, *Die Elementar-Mathematik*, 2. Theil, *Planimetrie*, 43. Aufl. (Breslau, 1876).

[2] J. Frischauf, *Absolute Geometrie* (Leipzig, 1876), p. 3.

[3] Max Simon, *Euclid* (1901), p. 45.

[4] G. Peano, *Formulaire de mathématiques* (Turin, 1894), p. 135.

[5] George A. Hill, *Geometry for Beginners* (Boston, 1880), p. 92, **177.**

[6] George Bruce Halsted, *Mensuration* (Boston, 1881), p. 28, 83.

[7] J. Henrici und P. Treutlein, *Elementar-Geometrie* (Leipzig, 1882), p. 13, 40.

[8] W. Fiedler, *Darstellende Geometrie*, 1. Theil (Leipzig, 1883), p. 60.

[9] N. Fialkowski, *Praktische Geometrie* (Wien, 1892), p. 15.

[10] W. W. Beman and D. E. Smith, *Plane and Solid Geometry* (Boston, 1896), p. 20.

[11] W. N. Busch and John B. Clarke, *Elements of Geometry* (New York, 1905]).

[12] G. W. Meyers, *Second-Year Mathematics for Secondary Schools* (Chicago, 1910), p. 10.

[13] H. E. Slaught and N. J. Lennes, *Plane Geometry* (Boston, 1910).

[14] C. A. Hart and D. D. Feldman, *Plane Geometry* (New York, 1911), p. viii.

[15] G. C. Shutts, *Plane and Solid Geometry* [1912], p. 13.

[16] Eugene R. Smith, *Solid Geometry* (New York, 1913).

[17] W. Wells and W. W. Hart, *Plane and Solid Geometry* (Boston, [1915]), p. x.

[18] Edith Long and W. C. Brenke, *Plane Geometry* (New York, 1916), p. viii.

[19] Matilda Auerbach **and Charles** Burton Walsh, *Plane Geometry* (Philadelphia, [1920]), p. xi.

of Cambridge,[1] to express "is similar to" in an edition of Euclid. The book was set up in type, but later the sign was eliminated from all parts, except one. In a footnote the student is told that "in writing out the propositions in the Senate House, Cambridge, it will be advisable not to make use of this symbol, but merely to write the word short, thus, *is simil.*" Moreover, in the Preface he is informed that "more competent judges than the editor" advised that the symbol be eliminated, and so it was, except in one or two instances where "it was too late to make the alteration," the sheets having already been printed. Of course, one reason for failure to adopt $\backsim$ for "similar" in England lies in the fact that $\backsim$ was used there for "difference."

373. When the sides of the triangle ABC and $A'B'C'$ are considered as being vectors, special symbols have been used by some authors to designate different kinds of similarity. Thus, Stolz and Gmeiner[2] employ $\stackrel{1}{\backsim}$ to mark that the similar triangles are uniformly similar (*einstimmig ähnlich*), that is, the equal angles of the two triangles are all measured clockwise, or all counter-clockwise; they employ $\overline{\backsim}$ to mark that the two triangles are symmetrically similar, that is, of two numerically equal angles, one is measured clockwise and the other counter-clockwise.

The sign $\sim$ has been used also for "is [or are] measured by," by Alan Sanders;[3] the sign $\cong$ is used for "equals approximately," by Hudson and Lipka.[4] A. Pringsheim[5] uses the symbolism $a_\nu \cong ab_\nu$ to express that $\lim\limits_{\nu = +\infty} \dfrac{a_\nu}{b_\nu} = a$.

374. The sign $\cong$ for congruence was not without rivals during the nineteenth century. Occasionally the sign $\equiv$, first introduced by Riemann[6] to express identity, or non-Gaussian arithmetical congruence of the type $(a+b)^2 = a^2 + 2ab + b^2$, is employed for the expression of geometrical congruence. One finds $\equiv$ for congruent in W. Bolyai,[7]

---

[1] *Elements of Euclid . . . . from the Text of Dr. Simson.* By a Member of the University of Cambridge (London, 1827), p. 104.

[2] O. Stolz und J. A. Gmeiner, *Theoretische Arithmetik* (Leipzig), Vol. II (1902), p. 332.

[3] Alan Sanders, *Plane and Solid Geometry* (New York, [1901]), p. 14.

[4] R. G. Hudson and J. Lipka, *Manual of Mathematics* (New York, 1917), p. 68.

[5] A. Pringsheim, *Mathematische Annalen*, Vol. XXXV (1890), p. 302; *Encyclopédie des scien. Math.*, Tom. I, Vol. I (1904), p. 201, 202.

[6] See L. Kronecker, *Vorlesungen über Zahlentheorie* (Leipzig, 1901), p. 86; G. F. B. Riemann, *Elliptische Funktionen* (Leipzig, 1899), p. 1, 6.

[7] W. Bolyai, *Tentamen* (2d ed.), Tom. I (Budapest, 1897), p. xi.

H. G. Grassmann,[1] Dupuis,[2] Budden,[3] Veronese,[4] Casey,[5] Halsted,[6] Baker,[7] Betz and Webb,[8] Young and Schwarz,[9] McDougall.[10] This sign $\equiv$ for congruence finds its widest adoption in Great Britain at the present time. Jordan[11] employs it in analysis to express equivalence.

The idea of expressing similarity by the letter S placed in a horizontal position is extended by Callet, who uses ∽, ᗡ, ᗡ̄, to express "similar," "dissimilar," "similar or dissimilar."[12] Callet's notation for "dissimilar" did not meet with general adoption even in his own country.

The sign $\equiv$ has also other uses in geometry. It is used in the Riemannian sense of "identical to," not "congruent," by Busch and Clarke,[13] Meyers,[14] E. R. Smith,[15] Wells and Hart.[16] The sign $\equiv$ or $\times$ is made to express "equivalent to" in the *Geometry* of Hopkins.[17]

The symbols $\sim$ and $\frown$ for "similar" have encountered some competition with certain other symbols. Thus "similar" is marked ||| in the geometries of Budden[18] and McDougall.

The relation "coincides with," which Leibniz had marked with $|\sim|$, is expressed by $\doteq$ in White's *Geometry*.[19] Cremona[20] denotes by

[1] H. G. Grassmann in *Crelle's Journal*, Vol. XLII (1851), p. 193–203.

[2] N. F. Dupuis, *Elementary Synthetic Geometry* (London, 1899), p. 29.

[3] E. Budden, *Elementary Pure Geometry* (London, 1904), p. 22.

[4] Guiseppe Veronese, *Elementi di Geometria*, Part I (3d ed.; Verona, 1904), p. 11.

[5] J. Casey, *First Six Books of Euclid's Elements* (7th ed.; Dublin, 1902).

[6] G. B. Halsted, *Rational Geometry* (New York, 1904), p. vii.

[7] Alfred Baker, *Transactions of the Royal Society of Canada* (2d ser., 1906–7), Vol. XII, Sec. III, p. 120.

[8] W. Betz and H. E. Webb, *Plane Geometry* (Boston, [1912]), p. 71.

[9] John W. Young and A. J. Schwartz, *Plane Geometry* (New York, [1905]).

[10] A. H. McDougall, *The Ontario High School Geometry* (Toronto, 1914), p. 158.

[11] Camille Jordan, *Cours d'analyse*, Vol. II (1894), p. 614.

[12] François Callet, *Tables portatives de logarithmes* (Paris, 1795), p. 79. Taken from Désiré André, *Notations mathématiques* (Paris, 1909), p. 150.

[13] W. N. Busch and John B. Clarke, *Elements of Geometry* (New York, [1905]).

[14] G. W. Meyers, *Second-Year Mathematics for Secondary Schools* (Chicago, 1910), p. 119.

[15] Eugene R. Smith, *Solid Geometry* [1913].

[16] W. Wells and W. W. Hart, *Plane and Solid Geometry* (Boston, [1905]), p. x.

[17] Irving Hopkins, *Manual of Plane Geometry* (Boston, 1891), p. 10.

[18] E. Budden, *Elementary Pure Geometry* (London, 1904), p. 22.

[19] Emerson E. White, *Elements of Geometry* (New York City, 1895).

[20] Luigi Cremona, *Projective Geometry* (trans. Ch. Leudesdorf; 2d ed.; Oxford, 1893), p. 1.

$a.BC \equiv A'$ that the point common to the plane a and the straight line BC coincides with the point A'. Similarly, a German writer[1] of 1851 indicates by $a \equiv b$, $A \equiv B$ that the two points a and b or the two straights A and B coincide (*zusammenfallen*).

375. *The sign* $\backsimeq$ *for equivalence.*—In many geometries congruent figures are marked by the ordinary sign of equality, $=$. To distinguish between congruence of figures, expressed by $=$, and mere equivalence of figures or equality of areas, a new symbol $\backsimeq$ came to be used for "equivalent to" in the United States. The earliest appearance of that sign known to us is in a geometry brought out by Charles Davies[2] in 1851. He says that the sign "denotes equivalency and is read *is equivalent to*." The curved parts in the symbol, as used by Davies, are not semicircles, but semiellipses. The sign is given by Davies and Peck,[3] Benson,[4] Wells,[5] Wentworth,[6] McDonald,[7] Macnie,[8] Phillips and Fisher,[9] Milne,[10] McMahon,[11] Durell,[12] Hart and Feldman.[13] It occurs also in the trigonometry of Anderegg and Roe.[14] The signs $\backsimeq$ and $=$ for equivalence and equality (i.e. congruence) are now giving way in the United States to $=$ and $\cong$ or $\backsimeq$.

We have not seen this symbol for equivalence in any European book. A symbol for equivalence, $\backsimeq$, was employed by John Bolyai[15] in cases like $AB \backsimeq CD$, which meant $\angle CAB = \angle ACD$. That the line BN is parallel and equal to CP he indicated by the sign "$BN \| \backsimeq CP$."

---

[1] *Crelle's Journal*, Vol. XLII (1851), p. 193–203.

[2] Charles Davies, *Elements of Geometry and Trigonometry from the Works of A. M. Legendre* (New York, 1851), p. 87.

[3] Charles Davies and W. G. Peck, *Mathematical Dictionary* (New York, 1856), art. "Equivalent."

[4] Lawrence S. Benson, *Geometry* (New York, 1867), p. 14.

[5] Webster Wells, *Elements of Geometry* (Boston, 1886), p. 4.

[6] G. A. Wentworth, *Text-Book of Geometry* (2d ed.; Boston, 1894; Preface, 1888), p. 16. The first edition did *not* use this symbol.

[7] J. W. Macdonald, *Principles of Plane Geometry* (Boston, 1894), p. 6.

[8] John Macnie, *Elements of Geometry* (ed. E. E. White; New York, 1895), p. 10.

[9] A. W. Phillips and Irving Fisher, *Elements of Geometry* (New York, 1896), p. 1.

[10] William J. Milne, *Plane and Solid Geometry* (New York, [1899]), p. 20.

[11] James McMahon, *Elementary Geometry (Plane)* (New York, [1903]), p. 139.

[12] Fletcher Durrell, *Plane and Solid Geometry* (New York, 1908), p. 8.

[13] C. A. Hart and D. D. Feldman, *Plane Geometry* (New York, [1911]), p. viii.

[14] F. Anderegg and E. D. Roe, *Trigonometry* (Boston, 1896), p. 3.

[15] W. Bolyai, *Tentamen* (2d ed.), Vol. II, Appendix by John Bolyai, list of symbols. See also G. B. Halsted's translation of that Appendix (1896).

376. *Lettering of geometric figures.*—Geometric figures are found in the old Egyptian mathematical treatise, the Ahmes papyrus (1550 B.C. or older), but they are not marked by signs other than numerals to indicate the dimensions of lines.

The designation of points, lines, and planes by a letter or by letters was in vogue among the Greeks and has been traced back[1] to Hippocrates of Chios (about 440 B.C.).

The Greek custom of lettering geometric figures did not find imitation in India, where numbers indicating size were written along the sides. However, the Greek practice was adopted by the Arabs, later still by Regiomontanus and other Europeans.[2] Gerbert[3] and his pupils sometimes lettered their figures and at other times attached Roman numerals to mark lengths and areas. The Greeks, as well as the Arabs, Leonardo of Pisa, and Regiomontanus usually observed the sequence of letters a, b, g, d, e, z, etc., omitting the letters c and f. We have here the Greek-Arabic succession of letters of the alphabet, instead of the Latin succession. Referring to Leonardo of Pisa's *Practica geometriae* (1220) in which Latin letters are used with geometric figures, Archibald says: "Further evidence that Leonardo's work was of Greek-Arabic extraction can be found in the fact that, in connection with the 113 figures, of the section *On Divisions*, of Leonardo's work, the lettering in only 58 contains the letters c or f; that is, the Greek-Arabic succession a b g d e z . . . . is used almost as frequently as the Latin a b c d e f g . . . . ; elimination of Latin letters added to a Greek succession in a figure, for the purpose of numerical examples (in which the work abounds), makes the balance equal."[4]

Occasionally one encounters books in which geometric figures are not lettered at all. Such a publication is Scheubel's edition of Euclid,[5] in which numerical values are sometimes written alongside of lines as in the Ahmes papyrus.

An oddity in the lettering of geometric figures is found in Ramus' use[6] of the vowels a, e, i, o, u, y and the employment of consonants only when more than six letters are needed in a drawing.

---

[1] M. Cantor, *op. cit.*, Vol. I (3d ed., 1907), p. 205.

[2] J. Tropfke, *op. cit.*, Vol. IV (2d ed., 1923), p. 14, 15.

[3] *Œuvres de Gerbert* (ed. A. Olleris; Paris, 1867), Figs. 1–100, following p. 475.

[4] R. C. Archibald, *Euclid's Book on Divisions of Figures* (Cambridge, 1915), p. 12.

[5] *Evclides Megarensis . . . . sex libri priores . . . .* authore Ioanne Schevbelio (Basel, [1550]).

[6] *P. Rami Scholarvm mathematicorvm libri vnus et triginta* (Basel, 1569).

In the designation of a group of points of equal rank or of the same property in a figure, resort was sometimes taken to the repetition of one and the same letter, as in the works of Gregory St. Vincent,[1] Blaise Pascal,[2] John Wallis,[3] and Johann Bernoulli.[4]

377. The next advancement was the introduction of indices attached to letters, which proved to be an important aid. An apparently unconscious use of indices is found in Simon Stevin,[5] who occasionally uses dotted letters $\dot{B}$, $\ddot{B}$ to indicate points of equal significance obtained in the construction of triangles. In a German translation[6] of Stevin made in 1628, the dots are placed beneath the letter B, B. Similarly, Fr. van Schooten[7] in 1649 uses designations for points:

$$C, 2C, 3C; \; S, 2S, 3S; \; T, 2T, 3T; \; V, 2V, 3V .$$

This procedure is followed by Leibniz in a letter to Oldenburg[8] of August 27, 1676, in which he marks points in a geometric figure by $_1B$, $_2B$, $_3B$, $_1D$, $_2D$, $_3D$. The numerals are here much smaller than the letters, but are placed on the same level with the letters (see also § 549). This same notation is used by Leibniz in other essays[9] and again in a treatise of 1677 where he lets a figure move so that in its new position the points are marked with double indices like $1\textcircled{D}$ and $1 \textit{D}$. In 1679 he introduced a slight innovation by marking the points of the principal curve $3b$, $6b$, $9b$ . . . . , generally yb, the curves of the entire curve $\overline{yb}$. The point $3b$ when moved yields the points 1 $3b$, 2 $3b$, 3 $3b$; the surface generated by $\overline{y}b$ is marked $\overline{zy}b$. Leibniz used indices also in his determinant notations (Vol. II, § 547).

[1] Gregory St. Vincent, *Opus geometricum* (Antwerp, 1647), p. 27, etc. See also Karl Bopp, "Die Kegelschnitte des Gregorius a St. Vincentio" in *Abhandlungen zur Gesch. d. math. Wissensch.*, Vol. XX (1907), p. 131, 132, etc.

[2] Blaise Pascal, "Lettre de Dettonville a Carcavi," *Œuvres complètes*, Vol. III (Paris, 1866), p. 364–85; *Œuvres* (ed. Faugere; Paris, 1882), Vol. III, p. 270–446.

[3] John Wallis, *Operum mathematicorum pars altera* (Oxford, 1656), p. 16–160.

[4] Johann Bernoulli, *Acta eruditorum* (1697), Table IV; *Opera omnia* (1742), Vol. I, p. 192.

[5] S. Stevin, *Œuvres* (éd. A. Girard; Leyden, 1634), Part II, "Cosmographie," p. 15.

[6] See J. Tropfke, *op. cit.*, Vol. II (2d ed., 1921), p. 46.

[7] F. van Schooten, *Geometria à Renato des Cartes* (1649), p. 112.

[8] J. Collins, *Commercium epistolicum* (ed. J. B. Biot and F. Lefort, 1856), p. 113.

[9] *Leibniz Mathematische Schriften*, Vol. V (1858), p. 99–113. See D. Mahnke in *Bibliotheca mathematica* (3d ser.), Vol. XIII (1912–13), p. 250.

I. Newton used dots and strokes for marking fluxions and fluents (§§ 567, 622). As will be seen, indices of various types occur repeatedly in specialized notations of later date. For example, L. Euler[1] used in 1748

$$x'\quad x''\quad x'''$$
$$y'\quad y''\quad y'''$$

as co-ordinates of points of equal significance. Cotes[2] used such strokes in marking successive arithmetical differences. Monge[3] employed strokes, $K`$, $K``$, $K```$, and also ‘K', “K'.

378. The introduction of different kinds of type received increased attention in the nineteenth century. Wolfgang Bolyai[4] used Latin and Greek letters to signify quantities, and German letters to signify points and lines. Thus, $\widetilde{\mathfrak{ab}}$ signifies a line $\mathfrak{ab}$ infinite on both sides; $\mathfrak{a}\widetilde{\mathfrak{b}}$ a line starting at the point $\mathfrak{a}$ and infinite on the side $\mathfrak{b}$; $\widetilde{\mathfrak{a}}\mathfrak{b}$ a line starting at $\mathfrak{b}$ and infinite on the side $\mathfrak{a}$; $\overset{\bullet}{\mathrm{P}}$ a plane P extending to infinity in all directions.

379. A remarkable symbolism, made up of capital letters, lines, and dots, was devised by L. N. M. Carnot.[5] With him,

$A, B, C, \ldots$ marked points

$\overline{AB}, \widehat{AB}$ marked the segment AB and the circular arc AB

$\overline{BCD}$ marked that the points B, C, D are collinear, C being placed between B and D

$\overline{AB} \cdot \overline{CD}$ is the point of intersection of the indefinite lines AB, CD

$\widehat{ABCD}$ marked four points on a circular arc, in the order indicated

$\widehat{AB} \cdot \widehat{CD}$ is the point of intersection of the two arcs AB and CD

$F \overline{AB} \cdot \overline{CD}$ is the straight line which passes through the points F and $\overline{AB} \cdot \overline{CD}$

---

[1] L. Euler in *Histoire de l'Academie r. d. sciences et d. belles lettres*, année 1748 (Berlin, 1750), p. 175.

[2] Roger Cotes, *Harmonia mensurarum* (Cambridge, 1722), “Aestimatio errorum,” p. 25.

[3] G. Monge, *Miscellanea Taurinensia* (1770/73). See H. Wieleitner, *Geschichte der Mathematik*, II. Teil, II. Hälfte (1921), p. 51.

[4] Wolfgangi Bolyai de Bolya, *Tentamen* (2d ed.), Tom. I (Budapestini, 1897), p. xi.

[5] L. N. M. Carnot, *De la Corrélation des figures de géométrie* (Paris, an IX = 1801), p. 40–43.

$= | =$ signifies equipollence, or identity of two objects

$\overset{\wedge}{ABC}$ marks the angle formed by the straight lines, AB, BC, B being the vertex

$\overline{AB}\,\overset{\wedge}{\overline{CD}}$ is the angle formed by the two lines AB and CD

$\triangle ABC$ the triangle having the vertices A, B, C

$\varangle ABC$ is a right triangle

$\overline{\overline{ABC}}$ is the area of the triangle ABC

A criticism passed upon Carnot's notation is that it loses its clearness in complicated constructions.

Reye[1] in 1866 proposed the plan of using capital letters, A, B, C, for points; the small letters a, b, c, ... , for lines; a, β, γ, ... , for planes. This notation has been adopted by Favaro and others.[2] Besides, Favaro adopts the signs suggested by H. G. Grassmann,[3] AB for a straight line terminating in the points A and B, Aa the plane passing through A and a, aa the point common to a and a; ABC the plane passing through the points A, B, C; $a\beta\gamma$ the point common to the planes a, β, γ, and so on. This notation is adopted also by Cremona,[4] and some other writers.

The National Committee on Mathematical Requirements (1923) recommends (*Report*, p. 78) the following practice in the lettering of geometric figures: "Capitals represent the vertices, corresponding small letters represent opposite sides, corresponding small Greek letters represent angles, and the primed letters represent the corresponding parts of a congruent or similar triangle. This permits speaking of a (alpha) instead of 'angle A' and of 'small a' instead of BC."

380. *Sign for spherical excess.*—John Caswell writes the spherical excess $c = A + B + C - 180°$ thus: "$E = \angle \angle \angle - 2 \lrcorner$." Letting π stand for the periphery of a great circle, G for the surface of the sphere, R for the radius of the sphere, he writes the area $\triangle$ of a spherical triangle thus:[5]

$$2\pi \triangle = EG = 2R\pi E,$$
$$\triangle = RE.$$

[1] Reye, *Geometrie der Lage* (Hannover, 1866), p. 7.

[2] Antonio Favaro, *Leçons de Statique graphique* (trad. par Paul Terrier), 1. Partie (Paris, 1879), p. 2.

[3] H. Grassmann, *Ausdehnungslehre* (Leipzig, Berlin, 1862).

[4] Luigi Cremona, *Projective Geometry* (trans. Charles Leudesdorf; Oxford, 1885), chap. i.

[5] John Wallis, *Treatise of Algebra* (London, 1685), Appendix on "Trigonometry" by John Caswell, p. 15.

The letter E for spherical excess has retained its place in some books[1] to the present time. Legendre,[2] in his *Éléments de géométrie* (1794, and in later editions), represents the spherical excess by the letter S. In a German translation of this work, Crelle[3] used for this excess the sign ε. Chauvenet[4] used the letter K in his *Trigonometry*.

381. *Symbols in the statement of theorems.*—The use of symbols in the statement of geometric theorems is seldom found in print, but is sometimes resorted to in handwriting and in school exercises. It occurs, however, in William Jones's *Synopsis palmariorum*, a book which compresses much in very small space. There one finds, for instance, "An $\angle$ in a Segment $>$, $=$, $<$ Semicircle is Acute, Right, Obtuse."[5]

To Julius Worpitzky (1835–95), professor at the Friedrich Werder Gymnasium in Berlin, is due the symbolism $S.S.S.$ to recall that two triangles are congruent if their three sides are equal, respectively; and the abbreviations $S.W.S.$, $W.S.W.$ for the other congruence theorems.[6] Occasionally such abbreviations have been used in America, the letter a ("angle") taking the place of the letter W (*Winkel*), so that asa and sas are the abbreviations sometimes used. The National Committee on Mathematical Requirements, in its *Report* of 1923, page 79, discourages the use of these abbreviations.

382. *Signs for incommensurables.*—We have seen (§§ 183, 184) that Oughtred had a full set of ideographs for the symbolic representation of Euclid's tenth book on incommensurables. A different set of signs was employed by J. F. Lorenz[7] in his edition of Euclid's *Elements;* he used the Latin letter C turned over, as in $A \supset B$, to indicate that A and B are commensurable; while $A \cup B$ signified that A and B are incommensurable; $A \cap B$ signified that the lines A and B are commensurable only in power, i.e., A^2 and B^2 are commensurable, while A and B were not; $A \cup B$, that the lines are incommensurable even in power, i.e., A and B are incommensurable, so are A^2 and B^2.

[1] W. Chauvenet, *Elementary Geometry* (Philadelphia, 1872), p. 264; A. W. Phillips and I. Fisher, *Elements of Geometry* (New York, [1896]), p. 404.

[2] A. M. Legendre, *Éléments de géométrie* (Paris, 1794), p. 319, n. xi.

[3] A. L. Crelle's translation of Legendre's *Géométrie* (Berlin, 1822; 2d ed., 1833). Taken from J. Tropfke, *op. cit.*, Vol. V (1923), p. 160.

[4] William Chauvenet, *Treatise on Plane and Spherical Trigonometry* (Philadelphia, 1884), p. 229.

[5] William Jones, *Synopsis palmariorum matheseos* (London, 1706), p. 231.

[6] J. Tropfke, *op. cit.*, Vol. IV (2d ed., 1923), p. 18.

[7] Johann Friederich Lorenz, *Euklid's Elemente* (ed. C. B. Mollweide; Halle, 1824), p. xxxii, 194.

383. *Unusual ideographs in elementary geometry.*—For "is measured by" there is found in Hart and Feldman's *Geometry*[1] and in that of Auerbach and Walsh[2] the sign ∞, in Shutt's *Geometry*[3] the sign $\sqsubset$. Veronese[4] employs $\equiv | \equiv$ to mark "not equal" line segments.

A horizontal line drawn underneath an equation is used by Kambly[5] to indicate *folglich* or "therefore"; thus:

$$\angle r + q = 2R$$
$$\angle s + q = 2R$$
$$\overline{\angle r + q = s + q}$$
$$\angle r = s$$

384. *Algebraic symbols in elementary geometry.*—The use of algebraic symbols in the solution of geometric problems began at the very time when the symbols themselves were introduced. In fact, it was very largely geometrical problems which for their solution created a need of algebraic symbols. The use of algebraic symbolism in applied geometry is seen in the writings of Pacioli, Tartaglia, Cardan, Bombelli, Widman, Rudolff, Stifel, Stevin, Vieta, and writers since the sixteenth century.

It is noteworthy that printed works which contained pictographs had also algebraic symbols, but the converse was not always true. Thus, Barrow's *Euclid* contained algebraic symbols in superabundance, but no pictographs.

The case was different in works containing a systematic development of geometric theory. The geometric works of Euclid, Archimedes, and Apollonius of Perga did not employ algebraic symbolism; they were purely rhetorical in the form of exposition. Not until the seventeenth century, in the writings of Hérigone in France, and Oughtred, Wallis, and Barrow in England, was there a formal translation of the geometric classics of antiquity into the language of syncopated or symbolic algebra. There were those who deplored this procedure; we proceed to outline the struggle between symbolists and rhetoricians.

[1] C. A. Hart and Daniel D. Feldman, *Plane Geometry* (New York, [1911]), p. viii.

[2] M. Auerbach and C. B. Walsh, *Plane Geometry* (Philadelphia, [1920]), p. xi.

[3] George C. Shutt, *Plane and Solid Geometry* [1912], p. 13.

[4] Giuseppe Veronese, *Elementi di geometria*, Part I (3d ed., Verona), p. 12.

[5] Ludwig Kambly, *Die Elementar-Mathematik*, 2. Theil: *Planimetrie* (Breslau, 1876), p. 8, 1. Theil: *Arithmetik und Algebra*, 38. Aufl. (Breslau, 1906), p. 7.

PAST STRUGGLES BETWEEN SYMBOLISTS AND RHETORICIANS IN ELEMENTARY GEOMETRY

385. For many centuries there has been a conflict between individual judgments, on the use of mathematical symbols. On the one side are those who, in geometry for instance, would employ hardly any mathematical symbols; on the other side are those who insist on the use of ideographs and pictographs almost to the exclusion of ordinary writing. The real merits or defects of the two extreme views cannot be ascertained by a priori argument; they rest upon experience and must therefore be sought in the study of the history of our science.

The first printed edition of Euclid's *Elements* and the earliest translations of Arabic algebras into Latin contained little or no mathematical symbolism.[1] During the Renaissance the need of symbolism disclosed itself more strongly in algebra than in geometry. During the sixteenth century European algebra developed symbolisms for the writing of equations, but the arguments and explanations of the various steps in a solution were written in the ordinary form of verbal expression.

The seventeenth century witnessed new departures; the symbolic language of mathematics displaced verbal writing to a much greater extent than formerly. The movement is exhibited in the writings of three men: Pierre Hérigone[2] in France, William Oughtred[3] in England, and J. H. Rahn[4] in Switzerland. Hérigone used in his *Cursus mathematicus* of 1634 a large array of new symbols of his own design. He says in his Preface: "I have invented a new method of making demonstrations, brief and intelligible, without the use of any lan-

---

[1] Erhard Ratdolt's print of *Campanus' Euclid* (Venice, 1482). Al-Khowârizmî's algebra was translated into Latin by Gerard of Cremona in the twelfth century. It was probably this translation that was printed in Libri's *Histoire des sciences mathématique en Italie*, Vol. I (Paris, 1838), p. 253–97. Another translation into Latin, made by Robert of Chester, was edited by L. C. Karpinski (New York, 1915). Regarding Latin translations of Al-Khowârizmî, see also G. Eneström, *Bibliotheca mathematica* (3d ser.), Vol. V (1904), p. 404; A. A. Björnbo, *ibid.* (3d ser.), Vol. VII (1905), p. 239–48; Karpinski, *Bibliotheca mathematica* (3d ser.), Vol. XI, p. 125.

[2] Pierre Herigone, *op. cit.*, Vol. I–VI (Paris, 1634; 2d ed., 1644).

[3] William Oughtred, *Clavis mathematicae* (London, 1631, and later editions); also Oughtred's *Circles of Proportion* (1632), *Trigonometrie* (1657), and minor works.

[4] J. H. Rahn, *Teutsche Algebra* (Zürich, 1659), Thomas Brancker, *An Introduction to Algebra* (trans. out of the High-Dutch; London, 1668).

guage." In England, William Oughtred used over one hundred and fifty mathematical symbols, many of his own invention. In geometry Oughtred showed an even greater tendency to introduce extensive symbolisms than did Hérigone. Oughtred translated the tenth book of Euclid's *Elements* into language largely ideographic, using for the purpose about forty new symbols.[1] Some of his readers complained of the excessive brevity and compactness of the exposition, but Oughtred never relented. He found in John Wallis an enthusiastic disciple. At the time of Wallis, representatives of the two schools of mathematical exposition came into open conflict. In treating the "Conic Sections"[2] no one before Wallis had employed such an amount of symbolism. The philosopher Thomas Hobbes protests emphatically: "And for . . . . your Conic Sections, it is so covered over with the scab of symbols, that I had not the patience to examine whether it be well or ill demonstrated."[3] Again Hobbes says: "Symbols are poor unhandsome, though necessary scaffolds of demonstration";[4] he explains further: "Symbols, though they shorten the writing, yet they do not make the reader understand it sooner than if it were written in words. For the conception of the lines and figures . . . . must proceed from words either spoken or thought upon. So that there is a double labour of the mind, one to reduce your symbols to words, which are also symbols, another to attend to the ideas which they signify. Besides, if you but consider how none of the ancients ever used any of them in their published demonstrations of geometry, nor in their books of arithmetic . . . . you will not, I think, for the future be so much in love with them. . . . ."[5] Whether there is really a double translation, such as Hobbes claims, and also a double labor of interpretation, is a matter to be determined by experience.

386. Meanwhile the *Algebra* of Rahn appeared in 1659 in Zurich and was translated by Brancker into English and published with additions by John Pell, at London, in 1668. The work contained some new symbols and also Pell's division of the page into three columns. He marked the successive steps in the solution so that all steps in the process are made evident through the aid of symbols, hardly a word

[1] Printed in Oughtred's *Clavis mathematicae* (3d ed., 1648, and in the editions of 1652, 1667, 1693). See our §§ 183, 184, 185.

[2] John Wallis, *Operum mathematicorum*, Pars altera (Oxford), *De sectionibus conicis* (1655).

[3] Sir William Molesworth, *The English Works of Thomas Hobbes*, Vol. VII (London, 1845), p. 316.

[4] *Ibid.*, p. 248.   [5] *Ibid.*, p. 329.

of verbal explanation being necessary. In Switzerland the three-column arrangement of the page did not receive enthusiastic reception. In Great Britain it was adopted in a few texts: John Ward's *Young Mathematician's Guide*, parts of John Wallis' *Treatise of Algebra*, and John Kirkby's *Arithmetical Institutions*. But this almost complete repression of verbal explanation did not become widely and permanently popular. In the great mathematical works of the seventeenth century—the *Géométrie* of Descartes; the writings of Pascal, Fermat, Leibniz; the *Principia* of Sir Isaac Newton—symbolism was used in moderation. The struggles in elementary geometry were more intense. The notations of Oughtred also met with a most friendly reception from Isaac Barrow, the great teacher of Sir Isaac Newton, who followed Oughtred even more closely than did Wallis. In 1655, Barrow brought out an edition of Euclid in Latin and in 1660 an English edition. He had in mind two main objects: first, to reduce the whole of the *Elements* into a portable volume and, second, to gratify those readers who prefer "symbolical" to "verbal reasoning." During the next half-century Barrow's texts were tried out. In 1713, John Keill of Oxford edited the *Elements* of Euclid, in the Preface of which he criticized Barrow, saying: "Barrow's Demonstrations are so very short, and are involved in so many notes and symbols, that they are rendered obscure and difficult to one not versed in Geometry. There, many propositions, which appear conspicuous in reading Euclid himself, are made knotty, and scarcely intelligible to learners, by his Algebraical way of demonstration. . . . . The *Elements* of all Sciences ought to be handled after the most simple Method, and not to be involved in Symbols, Notes, or obscure Principles, taken elsewhere." Keill abstains altogether from the use of symbols. His exposition is quite rhetorical.

William Whiston, who was Newton's successor in the Lucasian professorship at Cambridge, brought out a school *Euclid*, an edition of Tacquet's *Euclid* which contains only a limited amount of symbolism. A more liberal amount of sign language is found in the geometry of William Emerson.

Robert Simson's edition of Euclid appeared in 1756. It was a carefully edited book and attained a wide reputation. Ambitious to present Euclid unmodified, he was careful to avoid all mathematical signs. The sight of this book would have delighted Hobbes. No scab of symbols here!

That a reaction to Simson's *Euclid* would follow was easy to see. In 1795 John Playfair, of Edinburgh, brought out a school edition of Euclid which contains a limited number of symbols. It passed through many editions in Great Britain and America. D. Cresswell, of Cambridge, England, expressed himself as follows: "In the demonstrations of the propositions recourse has been made to symbols. But these symbols are merely the representatives of certain words and phrases, which may be substituted for them at pleasure, so as to render the language employed strictly comformable to that of ancient Geometry. The consequent diminution of the bulk of the whole book is the least advantage which results from this use of symbols. For the demonstrations themselves are sooner read and more easily comprehended by means of these useful abbreviations; which will, in a short time, become familiar to the reader, if he is not beforehand perfectly well acquainted with them."[1] About the same time, Wright[2] made free use of symbols and declared: "Those who object to the introduction of Symbols in Geometry are requested to inspect Barrow's *Euclid*, Emerson's *Geometry*, etc., where they will discover many more than are here made use of." "The difficulty," says Babbage,[3] "which many students experience in understanding the propositions relating to ratios as delivered in the fifth book of Euclid, arises entirely from this cause [tedious description] and the facility of comprehending their algebraic demonstrations forms a striking contrast with the prolixity of the geometrical proofs."

In 1831 R. Blakelock, of Cambridge, edited Simson's text in the symbolical form. Oliver Byrne's *Euclid* in symbols and colored diagrams was not taken seriously, but was regarded a curiosity.[4] The Senate House examinations discouraged the use of symbols. Later De Morgan wrote: "Those who introduce algebraical symbols into elementary geometry, destroy the peculiar character of the latter to

---

[1] *A Supplement to the Elements of Euclid*, Second Edition . . . . by D. Cresswell, formerly Fellow of Trinity College (Cambridge, 1825), Preface. Cresswell uses algebraic symbols and pictographs.

[2] J. M. F. Wright, *Self-Examination in Euclid* (Cambridge, 1829), p. x.

[3] Charles Babbage, "On the Influence of Signs in Mathematical Reasoning," *Transactions Cambridge Philos. Society*, Vol. II (1827), p. 330.

[4] Oliver Byrne, *The Elements of Euclid in which coloured diagrams and symbols are used* (London, 1847). J. Tropfke, *op. cit.*, Vol. IV (1923), p. 29, refers to a German edition of Euclid by Heinrich Hoffmann, *Teutscher Euclides* (Jena, 1653), as using color. The device of using color in geometry goes back to Heron (*Opera*, Vol. IV [ed. J. L. Heiberg; Leipzig, 1912], p. 20) who says: "And as a surface one can imagine every shadow and every color, for which reason the Pythagoreans called surfaces 'colors.'" Martianus Capella (*De nuptiis* [ed. Kopp, 1836], No. 708) speaks of surfaces as being "ut est color in corpore."

every student who has any mechanical associations connected with those symbols; that is, to every student who has previously used them in ordinary algebra. Geometrical reasons, and arithmetical process, have each its own office; to mix the two in elementary instruction, is injurious to the proper acquisition of both."[1]

The same idea is embodied in Todhunter's edition of Euclid which does not contain even a plus or minus sign, nor a symbolism for proportion.

The viewpoint of the opposition is expressed by a writer in the *London Quarterly Journal* of 1864: "The amount of relief which has been obtained by the simple expedient of applying to the elements of geometry algebraic notation can be told only by those who remember to have painfully pored over the old editions of Simson's *Euclid*. The practical effect of this is to make a complicated train of reasoning at once intelligible to the eye, though the mind could not take it in without effort."

English geometries of the latter part of the nineteenth century and of the present time contain a moderate amount of symbolism. The extremes as represented by Oughtred and Barrow, on the one hand, and by Robert Simson, on the other, are avoided. Thus a conflict in England lasting two hundred and fifty years has ended as a draw. It is a stupendous object-lesson to mathematicians on mathematical symbolism. It is the victory of the golden mean.

387. The movements on the Continent were along the same lines, but were less spectacular than in England. In France, about a century after Hérigone, Clairaut[2] used in his geometry no algebraic signs and no pictographs. Bézout[3] and Legendre[4] employed only a moderate amount of algebraic signs. In Germany, Karsten[5] and Segner[6] made only moderate use of symbols in geometry, but Reyher[7] and Lorenz[8]

---

[1] A. de Morgan, *Trigonometry and Double Algebra* (1849), p. 92 n.

[2] A. C. Clairaut, *Élémens de géométrie* (Paris, 1753; 1st ed., 1741).

[3] E. Bézout, *Cours de Mathématiques*, Tom. I (Paris: n. éd., 1797), *Élémens de géométrie*.

[4] A. M. Legendre, *Éléments de Géométrie* (Paris, 1794).

[5] W. J. G. Karsten, *Lehrbegrif der gesamten Mathematik*, I. Theil (Greifswald, 1767), p. 205–484.

[6] I. A. de Segner, *Cursus mathematici Pars I: Elementa arithmeticae, geometriae et calculi geometrici* (editio nova; Halle, 1767).

[7] *Samuel Reyhers . . . . Euclides* (Kiel, 1698).

[8] J. F. Lorenz, *Euklid's Elemente*, auf's neue herausgegeben von C. B. Mollweide (5th ed., Halle, 1824; 1st ed., 1781; 2d ed., 1798).

used extensive notations; Lorenz brought out a very compact edition of all books of Euclid's *Elements*.

Our data for the eighteenth and nineteenth centuries have been drawn mainly from the field of elementary mathematics. A glance at the higher mathematics indicates that the great mathematicians of the eighteenth century, Euler, Lagrange, Laplace, used symbolism freely, but expressed much of their reasoning in ordinary language. In the nineteenth century, one finds in the field of logic all gradations from no symbolism to nothing but symbolism. The well-known opposition of Steiner to Plücker touches the question of sign language.

The experience of the past certainly points to conservatism in the use of symbols in elementary instruction. In our second volume we indicate more fully that the same conclusion applies to higher fields. Individual workers who in elementary fields proposed to express practically everything in ideographic form have been overruled. It is a question to be settled not by any one individual, but by large groups or by representatives of large groups. The problem requires a consensus of opinion, the wisdom of many minds. That widsom discloses itself in the history of the science. The judgment of the past calls for moderation.

The conclusion reached here may be stated in terms of two schoolboy definitions for salt. One definition is, "Salt is what, if you spill a cupful into the soup, spoils the soup." The other definition is, "Salt is what spoils your soup when you don't have any in it."

ALPHABETICAL INDEX

(Numbers refer to paragraphs)

Abacus, 39, 75, 119

Abu Kamil, 273; unknown quantity, 339

Acta eruditorum, extracts from, 197

Adam, Charles, 217, 254, 300, 344

Adams, D., 219, 286, 287

Addition, signs for: general survey of, 200–216; Ahmes papyrus, 200; Al-Qalasâdî, 124; Bakhshālī MS, 109; Diophantus, 102; Greek papyri, 200; Hindus, 106; Leibniz, 198; *et in* Regiomontanus, 126

Additive principle in notation for powers, 116, 124, 295; in Pacioli, 135; in Gloriosus, 196

Additive principles: in Babylonia, 1; in Crete, 32; in Egypt, 19, 49; in Rome, 46, 49; in Mexico, 49; among Aztecs, 66

Adrain, R., 287

Aepinus, F. V. T., parentheses, 352

Aggregation of terms: general survey of, 342–56; by use of dots, 348; Oughtred, 181, 183, 186, 251; Romanus, 320; Rudolff, 148; Stifel, 148, 153; Wallis, 196. By use of comma, 189, 238; *communis* radix, 325; *Ra. col.* in Scheubel, 159; aggregation of terms, in radical expressions, 199, 319, 332, 334; redundancy of symbols, 335; signs used by Bombelli, 144, 145; Clavius, 161; Leibniz, 198, 354; Macfarlane, 275; Oughtred, 181, 183, 251, 334; Pacioli. *See* Parentheses, Vinculum

Agnesi, M. G., 253, 257

Agrippa von Nettesheim, 97

Ahmes papyrus, 23, 260; addition and subtraction, 200; equality, 260; general drawings, 357, 376; unknown quantity, 339; fractions, 22, 23, 271, 274

Akhmim papyrus, 42

Aladern, J., 92

Alahdab, 118

Al-Battani, 82

Albert, Johann, 207

Alexander, Andreas, 325, 326; aggregation, 343, 344

Alexander, John, 245, 253, 254; equality, 264; use of star, 356

Algebraic symbols in geometry, 384

Algebras, Initius, 325

Al-Ḥaṣṣâr, 118, 235, 272; continued fractions, 118

Ali Aben Ragel, 96

Al-Kalsadi. *See* Al-Qalasâdî

Al-Karkhî, survey of his signs, 116, 339

Al-Khowârizmî, survey of his signs, 115; 271, 290, 385

Allaize, 249

Alligation, symbols for solving problems in, 133

Al-Madjrītī, 81

Alnasavi, 271

Alphabetic numerals, 28, 29, 30, 36, 38, 45, 46, 87; for fractions, 58, 59; in India, 76; in Rome, 60, 61

Al-Qalasâdî: survey of his signs, 124; 118, 200, 250; equality, 124, 260; unknown, 339

Alsted, J. H., 221, 225, 229, 305

Amicable numbers, 218, 230

Anatolius, 117

Anderegg, F., and E. D. Rowe: equivalence, 375

André, D., 95, 243, 285

Andrea, J. V., 263

Angle: general survey of, 360–63; sign for, in Hérigone, 189, 359; oblique angle, 363; right angle, 363; spherical angle, 363; solid angle, 363; equal angles, 363

Anianus, 127

Apian, P., 148, 222, 223, 224, 278

Apollonius of Perga, 384

Arabic numerals. *See* Hindu-Arabic numerals

Arabs, early, 45; Al-Khowârizmî, 81, 115, 271, 290, 385; Al-Qalasâdî, 118, 124, 200, 250, 260, 339; Al-Madjrītī, 81; Alnasavi, 271; Al-Karkhî, 116, 339; Ali Aben Ragel, 96

433

Birks, John, omicron-sigma, 307
Björnbo, A. A., 385
Blakelock, R., 386
Blassière, J. J., 248, 249; circle, 367; parentheses, 351; similar, 372
Blundeville, Th., 91, 223
Bobynin, V. V., 22
Boethius, 59; apices, 81; proportions, 249, 250
Boëza, L., 273
Boissiere, Claude de, 229
Bolognetti, Pompeo, 145
Bolyai, John, 368
Bolyai, Wolfgang, 212, 268; angle, 361, 362; circle, 367; congruent in geometry, 374; different kinds of type, 378; equivalence, 376
Bombelli, Rafaele: survey of his signs, 144, 145; 162, 164. 190, 384; aggregation, 344; use of R, 319, 199
Bomie, 258
Boncompagni, B., 91, 129, 131, 132, 219, 271, 273, 359
Boon, C. F., 270
Borel, E., angle, 361
Borgi (or Borghi) Pietro, survey of his signs, 133; 223, 278
Bortolotti, E., 47, 138, 145, 344, 351
Bosch, Klaas, 52, 208
Bosmans, H., 160, 162, 172, 176, 297
Boudrot, 249
Bouguer, P., 258
Bourke, J. G., 65
Braces, 353
Brackets, 347; in Bombelli, 351, 352
Brahmagupta: survey of his signs, 106–8; 76, 80, 112, 114; unknown quantities, 339
Brancker, Thomas, 194, 237, 252, 307, 386; radical sign, 328, 333, unknown quantity, 341
Brandis, 88
Brasch, F. E., 125
Brasser, J. R., 343
Briggs, H., 261; decimal fractions, 283; use of l for root, 322
Brito Rebello, J. I., 56
Bronkhorst, J. (Noviomagus), 97
Brouncker, W., 264
Brown, Richard, 35
Brückner, Mac, angle, 362
Brugsch, H., 16, 18, 200

Bryan, G. H., 334, 275
Bubnov, N., 75
Budden, E., similar, 374
Bühler, G., 80
Bürgi, Joost, 278, 283; powers, 296
Burja, Abel, radical sign, 331
Bush, W. N., and John B. Clarke: angle, 363; congruent in geometry, 372; ≡, 374
Buteon, Jean: survey of his signs, 173; 132, 204, 263; equality, 263
Byerly, W. E.: angle, 360; parallelogram, 365; right angle, 363
Byrne, O.: angle, 362; edition of Euclid, 386; right angle, 363

Cajori, F., 75, 92
Calculus, differential and integral, 365, 377
Calderón (Span. sign), 92
Callet, Fr., 95; similar, 374
Cambuston, H., 275
Campanus' *Euclid*, 385
Camus, C. E. L., 372
Cantor, Moritz, 27, 28, 31–34, 36, 38, 46, 47, 69, 71, 74, 76, 81, 91, 96, 97, 100, 116, 118, 136, 144, 201, 238, 263, 264, 271, 304, 324, 339; per mille, 274; lettering of figures, 376
Capella, Martianus, 322, 386
Cappelli, A., 48, 51, 93, 94, 208, 274
Caramuel, J., 91, 92; decimal separatrix, 262, 283; equality, 265; powers, 303, 305, 366; radical signs, 328; unknowns, 341
Cardano (Cardan), Hieronimo: survey of his signs, 140, 141; 152, 161, 176, 166, 177, 384; aggregation, 343, 351; equality, 140, 260; use of R, 199, 319; use of round parentheses once, 351
Carlos le-Maur, 96
Carmichael, Robert, calculus, 348
Carnot, L. N. M.: angle, 705; geometric notation, 379
Carra de Vaux, 75
Carré, L., 255, 266; parentheses, 351
Cartan, E., 247
Casati, P., 273
Casey, John: angle, 360; congruent in geometry, 374
Cassany, F., 254
Castillon, G. F., 286; radical signs, 330

Curtze, M., 81, 85, 91, 123, 126, 138
219, 250, 290, 318, 325. 340, 359
Cushing, F. H., 65

Dacia, Petrus de, 91
Da Cunha, J. A., 210, 236, 307
Dagomari, P., 91
D'Alembert, J., 258; angle, 360, 363;
imaginary $\sqrt{-1}$, 346; parentheses,
352; similar, 372
Dash. *See* Line
Dasypodius, C., 53
Datta, B., 75
Davies, Charles, 287; equivalence, 375
Davies, Charles, and W. G. Peck:
equivalence, 375, repeating deci-
mals, 289
Davila, M., 92
Debeaune, F., 264, 301
De Bessy, Frenicle, 266
Dechales, G. F. M., 206, 225; decimals,
283; equality, 266; powers, 201
Decimal fractions: survey of, 276–89;
186, 351; in Leibniz, 537; in Stevin,
162; in Wallis, 196; repeating deci-
mals, 289
Decimal scale: Babylonian, 3; Egyp-
tian, 16; in general, 58; North Ameri-
can Indians, 67
Decimal separatrix: colon, 245; com-
ma, 282, 284, 286; point, 287, 288;
point in Austria, 288
Dee, John: survey of his signs, 169;
205, 251, 254; radical sign, 327
De Graaf, A. *See* Graaf, Abraham de
Degrees, minutes, and seconds, 55; in
Regiomontanus, 126, 127
De Gua. *See* Gua, De
Deidier, L'Abbé, 249, 257, 269, 285,
300, 351
De Lagny. *See* Lagny, T. F. de
Delahire, 254, 258, 264
De la Loubere, 331
Delambre, 87
De la Roche, E.: survey of his signs,
132; 319, radical notation, 199; use
of square, 132, 357
Del Sodo, Giovanni, 139
De Moivre, A., 206, 207, 257; aggrega-
tion, 354
De Montigny. *See* Montigny, De
De Morgan, Augustus, 202, 276, 278,
283; algebraic symbols in geometry,

386; complicated exponents, 313;
decimals, 286, 287; equality, 268;
radical signs, 331; solidus, 275
Demotic numerals, 16, 18
Descartes, René: survey of his signs,
191; 177, 192, 196, 205, 207, 209, 210,
217, 256, 386; aggregation, 344, 353;
equality, 264, 265, 300; exponential
notation, 294, 298–300, 302–4, 315;
geometrical proportion, 254; plus
or minus, 262; radical sign, 329, 332,
333; unknown quantities, 339, 340;
use of a star, 356
Despiau, L., 248
Determinants, suffix notation in Leibniz, 198
De Witt, James, 210, 264
Dibuadius, Christophorus, 273, 327,
332; aggregation, 348
Dickson, W., 286
Diderot, Denys, 255; *Encyclopédie*, 254
Didier. *See* Bar-le-Duc
Diez de la Calle, Juan, 92
Diez freyle, Juan, 290
Difference (arithmetical): $=$ symbol
for, 164, 177, 262; in Leibniz, 198,
344; in Oughtred, 184, 372
Digges, Leonard and Thomas: survey
of their signs, 170; 205, 221, 339;
aggregation, 343; equality, 263;
powers, 296; radical signs, 199, 334
Dilworth, Th., 91, 246, 287
Diophantus: survey of his signs, 101–5;
41, 87, 111, 117, 121, 124, 135, 200,
201, 217, 235; equality, 260, 104, 263;
fractions, 274; powers of unknown,
295, 308, 339
Distributive, ideogram of Babylonians,
15
Division, signs for: survey of, 235–47;
Babylonians, 15; Egyptians, 26;
Bakhshālī, 109; Diophantus, 104;
Leonardo of Pisa, 235. 122; Leibniz,
197, 198; Oughtred, 186; Wallis, 196;
complex numbers, 247; critical esti-
mate, 243; order of operations involving $\div$ and $\times$, 242; relative
position of dividend and divisor,
241; scratch method, 196; $\div$, 237,
240; :, 238, 240; $\mathfrak{D}$ 154, 162, 236
Dixon, R. B., 65
Dodson, James, 354
Doehlmann, Karl, angle, 362
Dot: aggregation, 181, 183, 251, 348;
as radical sign, 324–26; as separatrix
in decimal fractions, 279, 283–85;

Oliver, Wait, and Jones (joint authors), 210, 213

Olleris, A., 61,

Olney, E., 287

Omicron-sigma, for involution, 307

Oppert, J., 5

Oresme, N., survey of his signs, 123; 129, 308, 333

Ottoni, C. B., 258; angle, 361

Oughtred, William: survey of his signs, 180–87; 91, 148, 169, 192, 196, 205, 210, 218, 231, 236, 244, 248, 382, 385; aggregation, 343, 345, 347–49; arithmetical proportion, 249; cross for multiplication, 285; decimals, 283; equality, 261, 266; geometrical proportion, 251–53, 255, 256; greater or less, 183; pictographs, 359; powers, 291; radical signs, 329, 332, 334; unknown quantity, 339

Ozanam, J., 257, 264; equality, 264, 265, 266, 277; powers, 301, 304; radical sign, 328

Pacioli, Luca: survey of his signs, 134–38; 91, 117, 126, 132, 145, 166, 177, 200, 219, 220, 221, 222, 223, 225, 226, 294, 297, 359, 384; aggregation, 343; equality, 138, 260; powers, 297, 322; radix, 292, 297, 318, 199; unknown, 339

Pade, H., 213

Palmer, C. I., and D. P. Taylor, equal number of degrees, 363

Panchaud, B., 249, 259

Paolo of Pisa, 91

Pappus, 55; circle, 367; pictographs, 357

Parallel lines, 359, 368

Parallelogram, pictograph for, 357, 359, 365

Pardies, G., 206, 253, 255

Parent, Antoine, 254, 255, 258; equality, 263; unknowns, 341

Parentheses: survey of, 342–52; braces, 188, 351; brackets, 347, 351; round, in Clavius, 161; Girard, 164; Hérigone, 189; Leibniz, 197, 238; marking index of root, 329; Oughtred, 181, 186. *See* Aggregation

Paricius, G. H., 208, 262

Parker, 331

Pascal, B., 261, 304, 307; lettering figures, 376

Pasquier, L. Gustave du, 269

Pastor, Julio Rey, 165, 204

Paz, P., 274

Peano, G., 214, 275, 288; aggregation, 348; angle, 362; principal values of roots, 337; "sgn," 211; use of ω, 372

Peet, T. E., 23, 200, 217

Peirce, B., 247, 259, 287; algebraic symbols, 371

Peise, 14

Peletier, Jacques: survey of his signs, 172; 174, 204, 227, 292; aggregation in radicals, 332

Pell, John, 194, 237, 307, 386

Pellizzati. *See* Pellos, Fr.

Pellos, Fr., 278

Penny, sign for, 275

Per cent, 274

Pereira, J. F., 258

Perini, L., 245

Perkins, G. R., 287

Perny, Paul, 69

Perpendicular, sign for, 359, 364

Peruvian knots, 62–64, 69; Peru MSS, 92

Peruzzi, house of, 54

Peurbach, G., 91, 125

Phillips, A. W., and Irving Fisher: equivalent, 375; spherical excess, 380

Phoenicians, 27, 36

Pi (π): for "proportional," 245; $\frac{4}{\pi}$ and $\square$, 196

Picard, J., 254

Piccard, 96

Pictographs, 357–71, 384, 385

Pihan, A. P., 25, 30, 73

Pike, Nicolas, 91, 289

Pires, F. M., 258

Pitiscus, B., 279–81

Pitot, H., 255, 341

Planudes, Maximus, survey of his signs, 121; 87

Plato, 7

Plato of Tivoli, 290, 322; arcs of circles, 359, 370

Playfair, John: angle, 360; algebraic symbols, 371; edition of Euclid, 386

Pliny, 50

Plücker, J., 387

"Plus or minus," 210, 196; Leibniz, 198; Descartes, 262, 210